Principles of Modern Radar Missile Seekers

For a listing of recent titles in the *Artech House Radar Library*,
turn to the back of this book.

Principles of Modern Radar Missile Seekers

Evgeny Markin

ARTECH
HOUSE

BOSTON | LONDON
artechhouse.com

Library of Congress Cataloging-in-Publication Data
A catalog record of this book is available from the U.S. Library of Congress.

British Library Cataloguing in Publication Data
A catalog record for this book is available from the British British Library.

ISBN 978-1-63081-777-0

Cover design by Charlene Stevens

685 Canton Street
Norwood, MA 02062

10 9 8 7 6 5 4 3 2 1

Contents

CHAPTER 12

CHAPTER 13

Preface

This book is devoted to the problems of protection against interference that affect the homing heads of anti-aircraft guided missiles.

The book analyzes the current military-political situation in the world and identifies three threats that can lead to a military conflict on a global scale. An analysis of the world arms market and the prospects for the development of aerospace attack weapons is carried out, and the state and prospects for development of technology for air defense forces are considered.

The book examines the basic principles of building a modern anti-aircraft missile system, guidance methods, direction finding systems; problems when firing over the horizon; and the issues of accuracy and resolution are analyzed. The book contains mathematical calculations and results of calculations of the ability of modern anti-aircraft missile systems to intercept hypersonic targets and space targets, for example, satellites of a potential enemy.

The book proposes new methods of protection against interference that are critical for the airborne radar of an anti-aircraft missile; namely: polarization interference, interference along the sidelobes of the onboard antenna pattern, and interference from two points in space, including interference generated by reflections from the Earth's surface (water) when firing at low-flying targets.

The theoretical aspects of the synthesis of solutions to improve the interference immunity of monopulse radars are discussed, and an analysis of existing anti-jamming measures for monopulse radars and the solutions that have been synthesized in this book are presented. A synthesis method and a technical solution for protection against interference for low-flying targets, which illuminate the surface of the Earth or water located in front of them, are proposed. And finally, a synthesis of a technical solution that protects the onboard radar from the effects of polarization interference is presented.

This book is intended for radar developers, engineers, and students, and is recommended as a textbook for students specializing in the field of radar technology.

Acknowledgments

I dedicate this book to my wife Irina Brandausova.

I would like to express my deep gratitude to my wife, who saved my life in July 2019, when I went into cardiac arrest on the way from Orlando to St. Augustine while I was driving a car.

By some miracle, while in the passenger seat, she was able to stop the car, after which she stopped one of the passing cars and asked for help. Several men from this car helped her get me out of the driver's seat, put me on the ground, called emergency services, and gave me artificial respiration before help arrived. In a very short time, an ambulance, police, and firemen arrived almost at the same time and I was quickly taken to the hospital.

I thank the Lord God for keeping me alive and giving me the opportunity to live and work!

I am grateful to these strangers who took part in my rescue. I want to express my deep gratitude and appreciation to the doctors, nurses, and all the hospital staff at AdventHealth Daytona Beach (formerly known as Florida Hospital Memorial Medical), who saved my life and then looked after me in the ward for seven days in a coma and then when I regained consciousness.

I again want to thank my wife, and also my relatives, my friends, and loved ones who helped me and my wife during this difficult period of our life!

Without the help of all these wonderful people, this book simply would not have been!

Many thanks to all of them.

I wish them health and good luck in everything!

In addition, I would like to express my gratitude to the former and current employees of Artech House: Mark Walsh, Aileen Story, Soraya Nair, Natalie McGregor, and Kathryn Klotz, who were very supportive of me in writing the book, as well to all employees of Artech House who took part in the preparation and release of the book for printing.

I would especially like to thank Dr. Merlin Fox for his all-around help, understanding, and support, thanks to which I was able to finish writing the book on time, despite the fact that I could not work on the book for a long time due to health reasons.

I would like to say a huge thank you to my reviewer for comments, advice, and invaluable help in writing this book!

Dear friends, thank you so much for your help, support, and understanding!

CHAPTER 1

Introduction

One of the main reasons for the successful actions of German aviation against the Soviet Army in 1941 was the small number of anti-aircraft artillery and the weakness of the entire air defense of the USSR as a whole.

In 1967, Israel's small Air Force launched a sudden and powerful blow to the Egyptian-Syrian troops. First, they destroyed the Egypt-Syria Air Force, which provided the Israeli Air Force an air supremacy and deprived their opponents of aerial reconnaissance. Subsequent massive air strikes by the Israeli Air Force on the Arab ground forces ensured the victory of the Israeli army in several days. This war became known as the Six-Day War.

The main striking force of the U.S. Air Force in Vietnam was also aviation.

Currently, during local conflicts with use of conventional defeat means, aviation is given increasing importance as the most mobile and long-range method. The problems solved by aviation in modern warfare are far from being exhausted by traditional bomb-assault strikes against troops and targets or fighting with an air enemy. In conditions of high maneuverability of troops on the battlefield, deep location methods of attack and completeness and reliability of intelligence data will guarantee correct decisions in the conduct of the battle and, ultimately, its success. An important role in solving this problem belongs to aerial reconnaissance; that is, aviation. An increasing place in combat operations in modern warfare is assigned to airborne assault forces, where the main method of transporting personnel and the composition and equipment of the airborne troops is via aviation. Aviation is widely used to ensure continuous control of troops on the battlefield. This function is performed by air command posts and communications helicopters. This suggests that current aviation tasks in modern warfare cannot be done even partially by other types of armed forces.

It is reasonable to assume that any army in the world will be considered modern and combat-ready if it is prepared to successfully repel the first strike of an enemy air force. This thesis is confirmed by the experience of all modern wars and local conflicts. However, even with a full understanding of the importance and necessity of a strong air defense, by the beginning of a war, some armies may find themselves with weak air defense due to the complexity and high cost of these weapons.

A weak air defense will inevitably entail significant losses and defeats of troops. The forced creation of a strong air defense after the start of hostilities seems impossible due to the practically unlimited range of potential destruction combined with limited resources (including financial) that occur during the war.

Until recently, aviation as a method of attack systematically outstripped the means of air defense.

However, with the development of surface-to-air guided missiles, the effectiveness of air defense equipment has increased dramatically, which led to complications in applying the principles and methods of aviation. As the experience of using air defense systems in real combat operations shows, modern anti-aircraft missile units in terms of their technical equipment can conduct successful combat operations both against individual aircraft and against groups of aircraft, including in conditions of enemy interference.

Fighting aviation methods is an integral part of the armed struggle in general and anti-aircraft missile combat in particular.

This struggle is waged by all types of armed forces and includes:

1. Fighting aviation on the ground, where it is a potential enemy, by:
    (a) Defeating aircraft at airfields;
    (b) Destroying airfield control and guidance points;
    (c) Destroying aviation fuel, aerial bombs, air-to-ground missiles, air-to-air missiles, etc.
2. Fighting in the air when aviation becomes a real enemy during a combat mission.

Fighting aviation on the ground can be carried out by any means that have the possibility to reach home airfields. This could include aviation, class missiles ground-to-ground, artillery, airborne troops, and naval forces.

A successful fight against aviation in the air requires special fire and intelligence assets, which is primarily due to the air combat transience. These means are at the disposal of the air defense forces. The basis of their firepower are anti-aircraft missile units and regiment fighter aircraft, operating on the basis of information about air targets that is extracted by radio technical troops. Fighter aircraft have a longer range than most anti-aircraft missile units, as well as high maneuverability and effective weapons, including air-to-air missiles.

When conducting joint combat operations with fighter aircraft, it is necessary to take into account that fighters cannot be used at any time and in any situation, as is required when fighting against an air enemy. These fighter aviation restrictions are dictated by weather conditions and time of day. In addition, it is not always possible to detect an air enemy in low altitudes and such ranges at which timely interception of air target is possible by fighters, and therefore some of the fighters must be in the air all the time during combat operations. Therefore, the command post of an anti-aircraft missile battalion should have continuous information about the air environment, including about fighter aircraft, as well as their areas of duty (loitering zones). It should be noted that currently air combat can take place only with radar guidance of the fighter to the target.

This makes it possible to implement information interaction between the fighter aviation and anti-aircraft missile forces, which is made possible by radio technical troops. These units are constantly monitoring the airspace entrusted to them, as well as radio intelligence. In the Soviet Union, the air defense system consisted of air defense of objects and Armed Forces air defense. Air defense of objects protects large industrial and cultural centers, silos for intercontinental ballistic missiles, and any separate enterprise with a strategic purpose. Air defense of objects also includes the forces and means of missile defense.

The task of the units of the air defense of objects is to prevent breakthrough of the forces and means of air attack to the protected object. All targets that approach the protected object must be destroyed with a probability not lower than the specified one.

A similar task is performed by armed forces air defense in relation to the troops located on a march or deployed in battle formations. Anti-aircraft missile units today form the basis of anti-aircraft defense, as they can fight enemy aircraft in any conditions weather and time of day, having missiles at launch (firing) positions and ensuring the continuity of air defense; that is, constant readiness of forces and means of anti-aircraft missile forces to repel a strike of aerospace attack.

The means of destruction, or active means of air defense, are anti-aircraft missile units that fight an air enemy (using airplanes, helicopters, and cruise missiles) with the aim of destroying them. Some are even capable of destroying intercontinental ballistic missiles and hypersonic unmanned aerial vehicles [1].

The target for the anti-aircraft missile unit is the enemy's aerospace attack equipment and personnel.

The destruction of aerospace attack weapons is the main task of the combat activities of the anti-aircraft missile unit. At the same time, there is a derivative from the fact of the destruction of an air target, which often leads to results that are many times greater than the combat effect and cost of the downed vehicle. This is the danger (possibility) of a destruction target in a given combat area. The danger (probability) of the aircraft destruction, as a rule, forces the remaining crews to abandon the combat mission, or significantly reduce the effectiveness of the blow, due to premature or inaccurate dropping of bombs, launching missiles, premature exit from a dive, or forced performance of an anti-missile maneuver.

The main point is that to overcome anti-aircraft fire in modern conditions, most of the aviation forces are used to support the actions of shock groups. So in a grouping of 100 aircraft, 60 to 70 of them do not participate in the direct strike and instead perform the tasks of distracting (demonstration) groups, jammers, cover groups, and so on [2].

Air defense firing pushes the aircraft forces to abandon comfortable and safe average heights and go to extremely uncomfortable and very dangerous heights for piloting low and ultralow altitudes.

During the tests of the S-200D system at the Sary-Shagan test site test flights were carried out on a Su-24M aircraft, which flew to the airfield (site No. 7) from another region of the USSR. These aircraft had a variable sweep wing, as well as equipment that ensured flights at low and extremely low altitudes, with rounding a terrain relief, and are very similar to the American FB-111 aircraft. These vehicles and their crews participated in tests to determine the capabilities of the S-200D to detect and track targets at different speeds and flight altitudes, including at low and extremely low altitudes. In a conversation with the crews, I asked about whether it is difficult to fly at extremely low altitudes and at such high speeds. In response, the squadron commander asked me to imagine that I was driving along the road in a car at a speed of 100 km per hour and suddenly the car door opened, my face

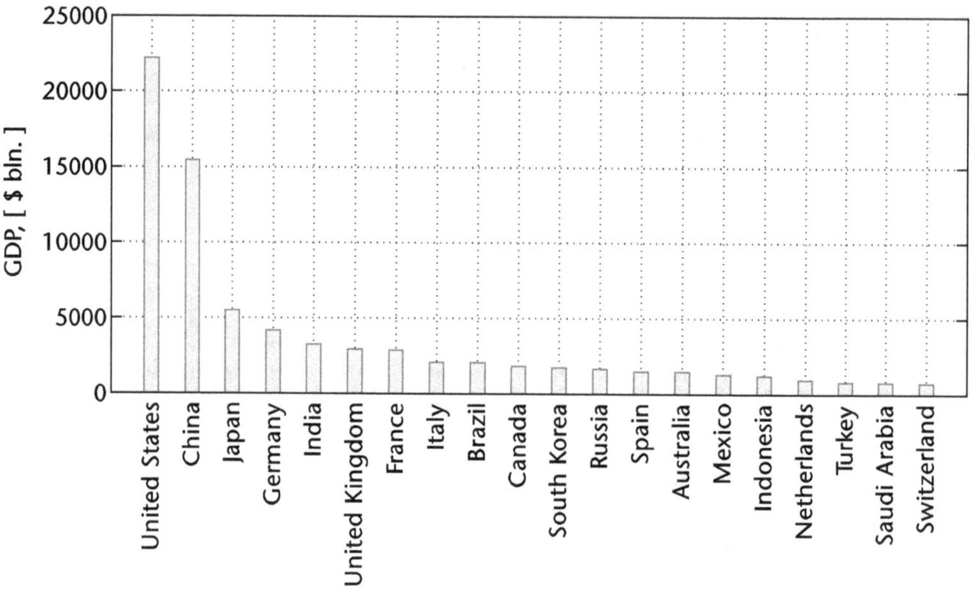

Figure 1.1 Countries of the world sorted by their GDP.

is 20 cm from the road surface, and in these conditions I must drive the car while maintaining speed mode.

This example very clearly showed the complexity and danger of piloting an aircraft at high speeds and extremely low altitudes. This requires incredible effort on the part of the crew; after such a flight, the pilot and navigator are soaked through with sweat. Therefore, the duration of the flight at extremely low altitudes is limited by the capabilities of the crew and can only occur within a few minutes to several tens of minutes.

1.1 Threats

According to the International Monetary Fund (IMF), in 2019, the U.S. gross domestic product (GDP) was $22,198.12 billion, while the total GDP for all countries of the world was $91,976.44 billion [3]. Therefore, the U.S. GDP in 2019 amounted to 24.13% of all goods and services in the world, while the U.S. population is 4.248% of the world [4]. Countries of the world sorted by their GDP are shown in Figure 1.1.

According to the BP *Statistical Review of World Energy 2020* [5], the United States ranks first in the world for natural gas production. This is 920.9 billion cubic meters with a specific weight of 23.1% compared to 679.0 billion cubic meters from Russia and its share in the total volume of 17.0%, respectively.

The United States produces 17,045 thousand barrels daily, which corresponds to a share of 17.9% of the total world daily oil production. In 2019, oil production in the United States was 746.7 million tons, which amounted to 16.7% of the global total. At the same time, for Russia it was 568.1 million tons or 12.7%, and for Saudi Arabia it was 556.6 million tons or 12.4%. Thus, the United States is the largest consumer of gas and oil in the world.

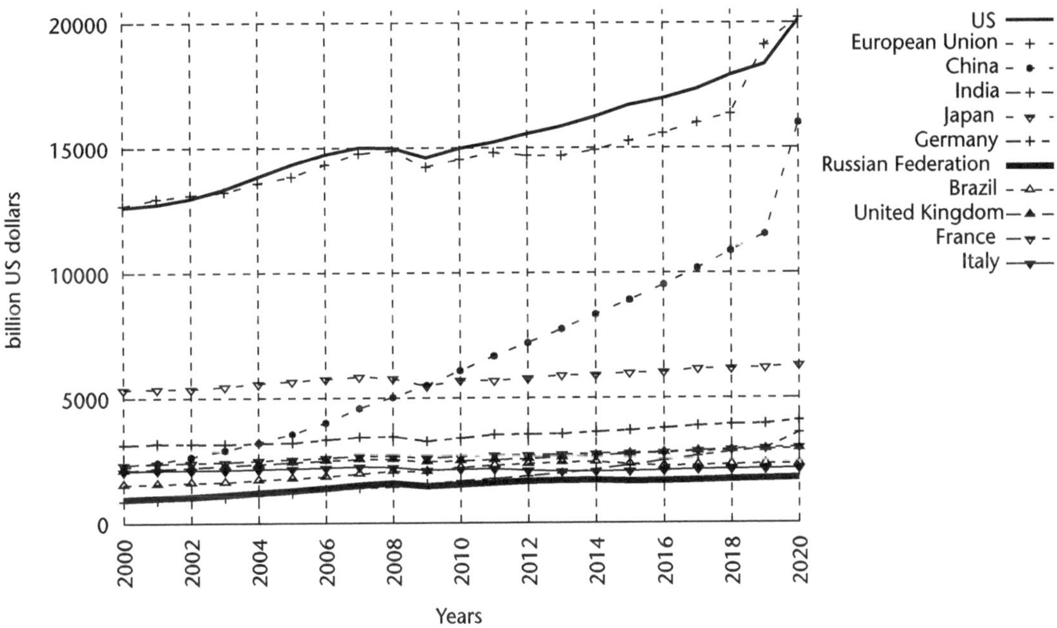

Figure 1.2 International Monetary Fund data.

But more importantly, the United States is the main center for patents and inventions in the world. Today it is the knowledge economy.

After World War II, the United States became the backbone of the world economy and the global financial system, where the measure of value is the U.S. dollar.

At present, the entire world is the strategic sphere of U.S. national interests, but not all countries agree with this, so there are several geopolitical threats.

1.1.1 The First Threat

Currently, many experts talk about the growing threat from China, which they see as a potential enemy and urge to prepare for confrontation. Their position is largely based on the growth of the Chinese GDP, and their forecasts are based on the assumption that this trend will continue.

Figure 1.2 shows the countries of the world sorted by their GDP, which is the value of all final goods and services produced in the country in a given year. The dollar GDP estimates here are based on data calculated by the IMF [3]. The longer-term outlook was prepared by PricewaterhouseCoopers (PWC) LLP [6]. This forecast is shown in Figure 1.3. It should be noted that the data shown in Figures 1.2 and 1.3 characterize a certain spread in assessments of the Chinese economy and the pace of its development. Figure 1.2 shows that the functions “US,” “European Union,” and “China” have an inflection point in 2019, and from 2020 onward these functions become smooth again. This is due to the fact that at the time of preparation of this material, 2020 was not yet completed, and therefore there was no verified GDP data yet, so the data presented for 2020 represents forecasted data. In addition, it should be noted that China’s economy is increasingly burdened with outstanding loans. Currently there are more than $900 billion in accounts

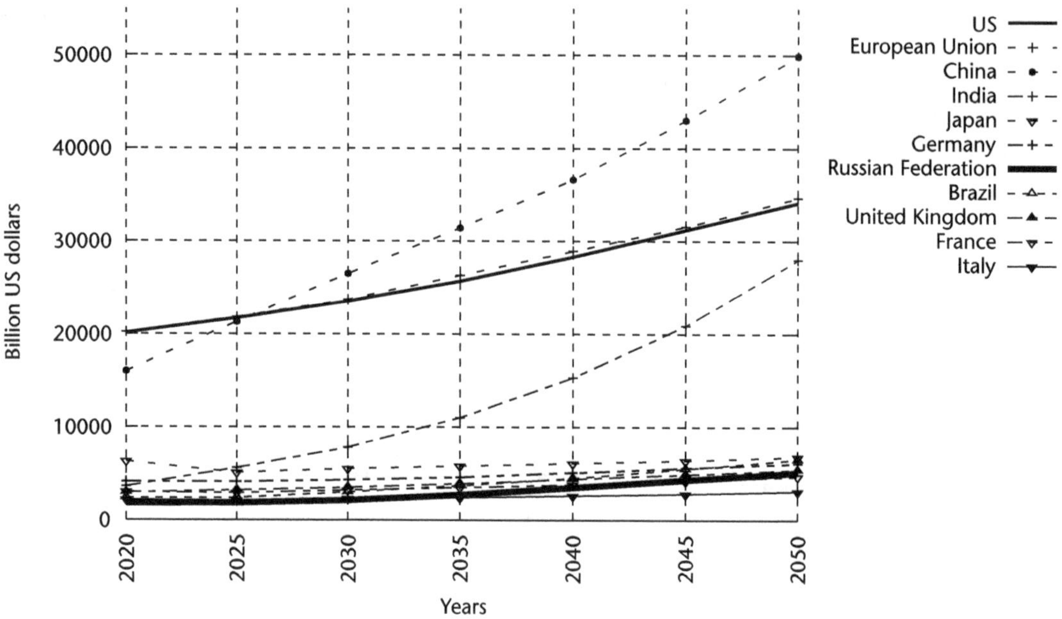

Figure 1.3 The forecast from PWC.

receivable, and only the export of goods at a low price can maintain high rates of production growth. In addition, China is not moving to free yuan convertibility, which in turn does not allow an objective assessment of China's GDP in U.S. dollars. In addition, the tight pegging of the yuan exchange rate to the U.S. dollar promotes high rates of production of cheap goods for subsequent export. Today, the People's Republic of China is a world center for the production of various goods. The main market for Chinese exports is the United States.

Thus, the United States can determine the parameters of the Chinese economy and the conditions under which China will be forced to refinance U.S. debt. A reduction in Chinese exports will lead to a drop in production, and as a result, to a drop in the incomes of the Chinese population. In addition, the persisting inequality in living standards in the central and coastal regions of China can lead, in the event of a drop in incomes of the population, to social instability, which will in turn lead to an even greater decline in production.

This state of affairs, of course, is not to the liking of the leadership of the People's Republic of China, and therefore, the differences between China and the United States could grow for the foreseeable future. It looks very dangerous, because on the one hand, China, with the help of the United States, has modernized its economy, which today is the second biggest in the world, and on the other hand, China received almost free of charge some modern military technologies of the USSR after its collapse.

1.1.2 The Second Threat

The second threat is the contradiction between the U.S. dollar and the common European currency, the euro. According to the quantitative equation of exchange,

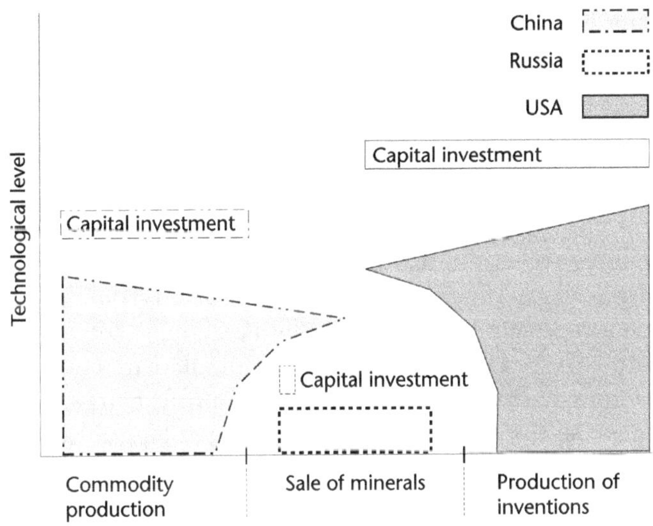

Figure 1.4 China–Russia–United States.

known as the Fisher equation:

$$\hat{p} \times Q = M \times \upsilon \tag{1.1}$$

for consumer price index of goods $\hat{p}$ (the price level) multiplied by its volume Q (which is an index of real expenditures on newly produced goods and services) must be identically equal to the money supply M (which is the total nominal amount of money in circulation on average in an economy) multiplied by money velocity υ (which is the average frequency with which a unit of money is spent). Due to the fact that Europe's share in the GDP of the world is about 20%, after the introduction of the euro in 1999, this volume of goods no longer used the U.S. dollar. This led to an excess supply of U.S. dollars and then to an exchange rate of 1.17 USD per euro.

Since the purpose of any system is striving for balance, the velocity of U.S. dollar circulation was reduced, thus reducing the U.S. GDP and decreasing the standard of living for U.S. citizens.

During the conflict in the Balkans, consumer prices in Europe rose by an average of 1.21. This made it possible to compensate for the harm caused by the introduction of the euro, but did not solve the problem. Currently, Europe's total GDP is $19.1 trillion, which is higher than that of the United States. Today the countries of the European Union are the centers of invention and production of many goods.

1.1.3 The Third Threat

As shown above, there are countries that produce goods and there are countries that produce inventions. But there is a country that produces nothing and does not produce inventions, but only sells gas and oil and other minerals. This is Russia, which has the remnants of the scientific and technical potential of the USSR, and has nuclear weapons and their delivery vehicles. This situation is shown in Figure 1.4, with examples of the United States, Russia, and China. In the past 15 years, the United States invented the mining technology for shale gas. In this research field considerable money was invested, and as was stated above, since 2010, the United States ranks first in the world in natural gas mining. Figure 1.4 shows this by a change to the rectangle on the left side. China is the world leader in factory production but in the last 15 years they have been making investments to produce inventions and purchase technologies. In the Figure 1.4, this trend is characterized by a change to the shape of the figure on the right. Russia has outdated technology and it currently is not investing in new technologies. In oil mining, for example, high-speed pumps are used instead of an oil pumping unit, and as a result, oil fields are being depleted rapidly. The share of oil extracted from the oil reservoir is low. This situation is represented in Figure 1.4 by a rectangle that has not changed shape since the collapse of the USSR. Russia must accumulate money and invest in the development of commodity production with high-profit margins (high-tech), in research, and in new development. Therefore, looking at Figure 1.4, USSR placement should ideally move up and left in the direction of increasing commodity production with a high level of technology. This goal is related to the development of science and education. However, the current Russian government is doing the opposite. Russia strongly implemented the commercialization of education and science, drastically reducing the number of research, research-technical, and educational institutions. The population is forced to work for older private companies for very little money. Russia also opened the border for migrants from Central Asia. All of this creates a competitive environment for the organization of slave labor. The high world prices for oil and gas, which existed until recently, could give Russia the opportunity to modernize technology and equipment. However, in reality this has not happened, since this money was taken out of the economy by various methods.

Many conflicts between countries are repeated over time because they do not have a solution to the contradictions that gave rise to them in the first place. Usually it is a struggle for resources (natural, human, industrial, etc.).

Scientific and technological progress is moving toward the creation of new energy supply technologies, including those based on new physical principles. It is quite clear that after these new technologies are introduced, the price of oil will fall dramatically and economies based on oil and gas will collapse.

This creates conditions for a new conflict. The danger is that Russia has nuclear weapons. Russian leadership's unwillingness to invest money into the economy and a penchant for simple solutions could lead to nuclear blackmail or to a war.

Using data from the International Monetary Fund and data from Goldman Sachs, you can build GDP forecasts for blocks of countries. On the one hand, we have the United States and the European Union, and on the other hand, we have Brazil, Russia, India, and China (the BRIC countries). The data is presented in Figure 1.5, where we see that the GDP growth rates of the BRIC countries are higher than the growth rates of the United States and the European Union, as well as higher than the growth rates of the total GDP of the United States and the European Union countries. Also from Figure 1.5 it can be seen that 2020 is the transition point when the GDP of the BRIC countries should have exceeded the GDP of the United States and the GDP of the European Union separately, and 2038 is the assumed transition point when the GDP of the BRIC countries will exceed the total GDP of United States and European Union. Such high growth rates of the BRIC countries are mainly driven by the high growth rates of the Chinese

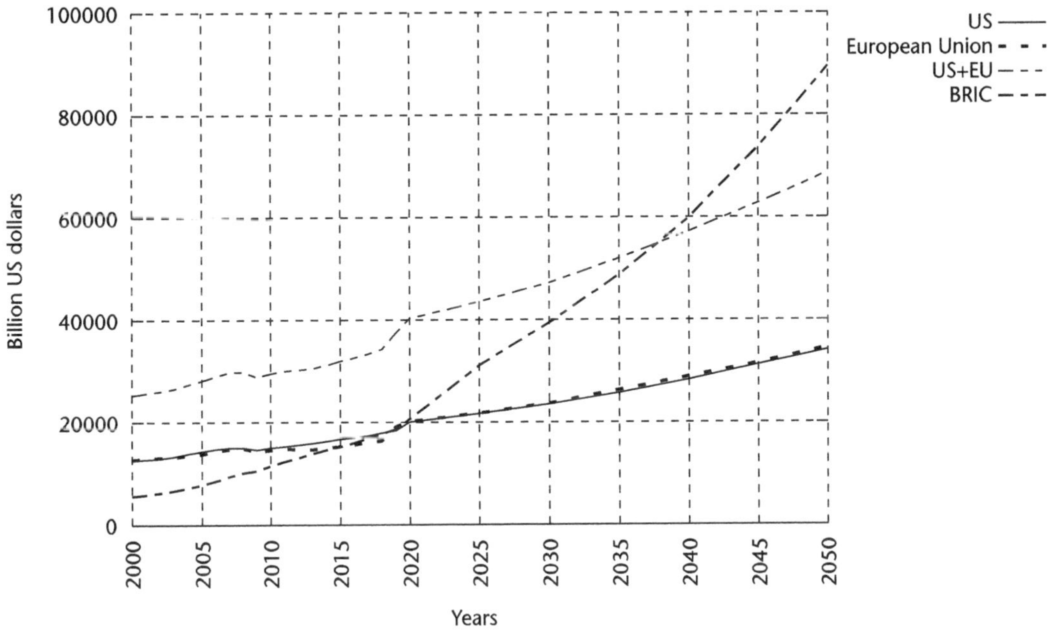

Figure 1.5 US + EU–BRIC.

economy. In this regard, note that over the past 20 years, Russia has sold a license to China to manufacture the Su-27. As a result, China copied the plane, and now China produces its own J-11 and J-11B (Jianjiji-11) fighters in any quantities [7]. In 2011, China confirmed that there are flight tests of the Shenyang J-15 [8] aircraft. The prototype of this aircraft is the Russian Su-33 (USSR T-10K project) [9]. The unfinished prototype of the Su-33, T-10K-3, was purchased from Ukraine some time in 2001 and, was allegedly studied extensively, and development of the J-15 began immediately thereafter [9–11]. China has also built an upgraded version of the HQ-10 anti-aircraft missile system [12], which is a copy of the Soviet S-300PM anti-aircraft missile system, designated HQ-15, with a maximum missile range increased from 150 km (93 mi) to 200 km (120 mi). There are unconfirmed reports claiming that this version is a Chinese-made S-300PMU-2 [13,14]. The total number of S-300PMU/1/2 and HQ-15/18 batteries in the People's Liberation Army of China (PLA) is approximately 40 and 60, respectively. The total number of missiles is well over 1,600, with about 300 launcher platforms [14]. Five of these air defense battalions are deployed and are on active duty in the Beijing area, six battalions in the Taiwan Strait area, and the remaining battalions are in other major cities such as Shanghai, Chengdu, and Dalian. Two RIF systems (SA-N-6) were purchased in 2002 for the Chinese Navy for Type 051C destroyers [15].

Currently, we are seeing a rapid increase in military cooperation between Russia and India. Russia is helping to build and launch Indian missiles into space. Russia has built the *Vikramaditya* aircraft carrier for India. This is the new name of the former Soviet aircraft carrier *Admiral Gorshkov,* which was purchased by India and entered service with the Indian Navy on November 16, 2013 [16,17]. *Vikramaditya* is a modified Kiev class aircraft carrier, built in 1978–1982 at the

Black Sea Shipyard, Nikolaev, Ukraine. Currently, the ship has undergone a large-scale re-equipment at the Sevmash shipyard in Severodvinsk, Arkhangelsk region, Russia, and since November 16, 2013 has been included in the Indian Navy.

Both India and Russia plan to build, purchase, and export large quantities of Sukhoi Su-57 stealth fighters [18]. An example of this is the Fifth Generation Fighter Aircraft (FGFA) program. This is the Fifth Generation Fighter Joint Construction Program (FGFA) [19], the contract for which involves the transfer of technology to India for the production of fifth generation fighters. However, in October 2017, the Indian Air Force command demanded the closure of the FGFA. The Indian Air Force believes that the Russian Su-57 does not meet the stated requirements for stealth. In addition, the Indian side believes that the military avionics, radars, and sensors of the Russian development do not meet fifth generation [19] standards.

The Indian company Hindustan Aeronautical Ltd. (HAL) is developing a $600 million multipurpose transport aircraft together with Russian partners Rosoboronexport and the United Aircraft Corporation (UAC). This transport aircraft is another example of two countries planning to build, purchase, and export weapons and military equipment on a 50:50 [20] cost-sharing basis. In general, Russian–Indian arms programs are estimated at about $3 billion per year [21].

As follows from the above, Russia is the source of Soviet military technology, while India and China are absorbers of these technologies. Consequently, there are prerequisites for the creation of military–political blocs. The likelihood that a missile with weapons of mass destruction will be used today is higher than during most of the Cold War. Some countries are technically capable of developing missile launching mechanisms for use from forward-based ships or other platforms to launch short-range ballistic missiles (SRBM) and medium-range ballistic missiles (MRBM) or air-to-ground cruise missiles against their opponents [22]. Some countries may develop and deploy the system ahead of time. A short- or medium-range ballistic missile can be launched from a forward-based sea platform located within a few hundred kilometers of a potential enemy's territory.

1.2 Goals and Objectives

The targets for air defense systems are just the main strike forces of the aerospace attack:

- Nuclear submarines with submarine-launched ballistic missiles;
- Ground-based intercontinental ballistic missiles;
- Air- and sea-based missiles;
- Unmanned aerial vehicles (UAVs).

In addition, new means and methods of air attack are currently emerging.

Thus, on September 14, 2019, an attack of aircraft drones and cruise missiles was carried out on the world's largest oil refinery in Abkaik, Saudi Arabia, and the Khurais oil field located there. As a result of the attack, oil production in Saudi Arabia dropped significantly for some time, which immediately affected the global hydrocarbon market. This successfully implemented drone attack raised many questions about the capabilities of modern air defense encountering new types of threats and the level of preparation of modern states for such challenges. The attacks on the Saudi Aramco facilities were carried out using two types of air vehicles:

1. Unmanned aerial vehicles with explosives (the so-called kamikaze drones);
2. Long-range land-based cruise missiles.

Apparently, all these weapons were Iranian-made, but such weapons systems are also used by the Yemeni Houthi rebels, who, with the support of Iran, are opposing the Arab coalition led by Saudi Arabia.

What happened and how did the Saudi air defense systems work? The first impression was that the defense systems were not working at all. Perhaps, so to repel such an unexpected and massive attack, it would be necessary to support the means of defense in a state of heightened combat readiness. In addition, the attacks on small-sized vehicles may have turned out to be undetectable for some of the radars of the air defense system. It is also possible that the peculiarities of the main anti-aircraft system of the Saudis—the American complex Patriot—also played a role. This air defense system is designed to repel an air or missile attack in a specific sector and, accordingly, is usually deployed to repel an attack from specific directions. If the means of air attack made a long-range bypass maneuver or were used on a side from which they were not expected, then it was not at all necessary to count on repelling the strike by the Patriot complex.

Something similar had already happened in Syria before: on April 7, 2017, U.S. naval forces launched 60 cruise missiles at the Al-Shayrat airbase in Syria. Estimates of the effectiveness of this strike vary, but according to reports from independent sources, at least 44 targets were hit, some of them twice. The objectives of the strike were clearly for demonstration—to show the ability to make powerful high-precision strikes against ground targets using a limited naval contingent and despite the presence of more or less modern air defense systems in the region.

It is possible that the air defense systems during the attack of Ash-Shayrat acted separately and were not united under a single command, which actually led to the achievement of all the objectives of the attack.

The purpose of any air defense system is to ensure the safety of specific, varied objects that are the responsibility of the corresponding air defense units.

This is achieved because, in the area of responsibility of each air-defense fire weapon, all targets that pose a threat to the protected object must be destroyed with a probability not lower than the specified one [23].

So in the Military Doctrine of the Russian Federation, approved by the Decree of the President of the Russian Federation on February 5, 2010, among the main tasks of the development of the military organization is the task of improving air defense along with creating aerospace defense of the Russian Federation.

In accordance with the military doctrine, it is necessary to ensure the integration of the existing systems of air and missile defense, missile attack warning, and space control. They must operate under a single strategic command. In accordance with the instructions of the President of the Russian Federation, a new branch of the armed forces was formed in the Armed Forces—the Aerospace Defense Forces, which began combat duty on December 1, 2011. This was a fundamentally new branch of the military, designed to ensure the security of Russia in the aerospace sphere [24].

The modern aerospace defense system is designed to ensure that the following tasks are carried out:

- Warning of air, missile, and space attacks, reconnaissance of the air and space situation, and notification of troops about it;
- Protection of the state border in the airspace and control over the use of airspace;
- Repulsion of aggression in the aerospace sphere, anti-aircraft, and anti-missile defense of the most important objects of state and military command, key facilities of the Armed Forces, economy, and infrastructure;
- Implementation of spacecraft launches, control of the orbital group in flight and their use in the interests of the troops and forces of the state;
- Support of national and international space programs;
- Maintaining the established composition and state of preparation of space systems and their means of launch and control;
- Maintaining the forces and means of anti-aircraft, anti-missile, and anti-space defense in the mode of constant combat readiness.

1.3 Basic Concepts

Using the detailed market data collected by ICD Research in its report *Global Missile and Missile Defense Systems Market 2011–2021* [25] and my own research based on open sources and media data, it is possible to report and list several interesting trends in this area.

The global missile and missile defense market, estimated at US $15.2 billion in 2011, is projected to grow 2.71% over the forecast period to reach US $19.8 billion in 2021. The demand for missiles is expected to be driven by internal and external security threats, territorial disputes, and modernization initiatives undertaken by armed forces all over the world. Taken together, the market for missiles and missile defense systems is expected to reach US $187.8 billion over the forecast period. Major sponsors include North America and Europe, and the global market for missile and air defense systems is likely to be dominated by the United States throughout the forecast period. Missile defense systems are likely to account for the largest share of the global missile market; namely, surface-to-air missiles (SAM) and air-to-surface missiles, in that order. Demand for missile defense systems is expected to be higher than for other categories, mainly due to the increasing number of programs being developed by countries around the world in response to the threats posed by the proliferation of weapons of mass destruction.

My research is based on publicly available information only. The period from 1960 to the present is considered.

Figure 1.6 shows the distribution of missile types and the weight of each missile type on the market. As seen in Figure 1.6, the largest market share is occupied by anti-ship missiles (32.22%) and anti-aircraft missiles (35.56%).

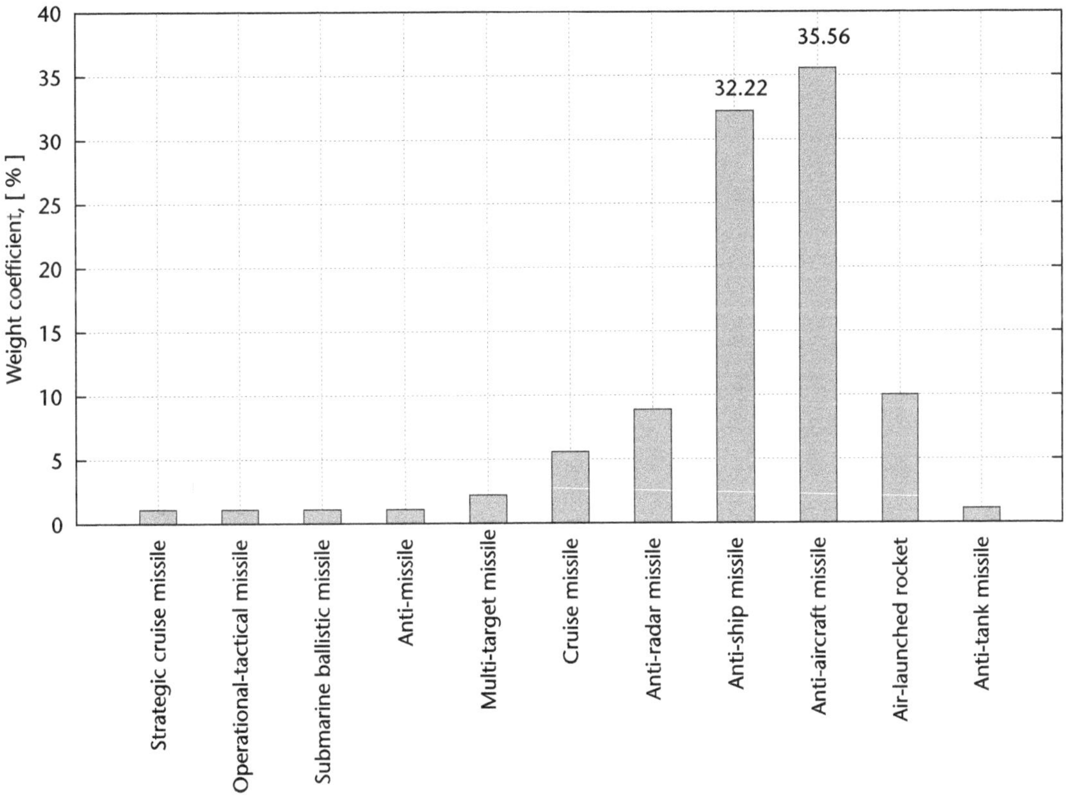

Figure 1.6 Missile types.

The next classification is the classification of missiles by basing type. This data is shown in Figure 1.7. It can be seen that the greatest weight is carried by missiles placed on chassis (22.22%), on airplanes (27.78%), and missiles placed on surface ships (13.33%). The sum of these three groups was 63.33% of the total number of missiles. The next classification is the classification for different environments. This classification covers air-launched missiles, land-based missiles, and sea-launched missiles. The results are shown in Figure 1.8, which shows that the largest market share (36.67%) belongs to the Navy. Figure 1.9 shows the distribution of missiles by country of origin. As follows from Figure 1.9, Russia has the largest weight (46%). The next histogram in Figure 1.10 shows the distribution of missiles according to their flight speed. It can be seen that missiles with a flight speed from 1M to 4M have the greatest weight, where M is the Mach number.

It is interesting to analyze the trend in the development of missile technology in terms of flight speed. The groups of missiles shown in Figure 1.10 are broken down by year. Then table functions were created for these groups. The solutions of the matrix equations for the coefficients of the approximating functions for each missile group were obtained. Polynomials of the second degree are used as approximating functions.

For missiles with a flight speed of less than 1M, the approximating function is represented by the following expression:

$$y_1(t) = -25259.03 + 25.39t - 0.0063t^2 \tag{1.2}$$

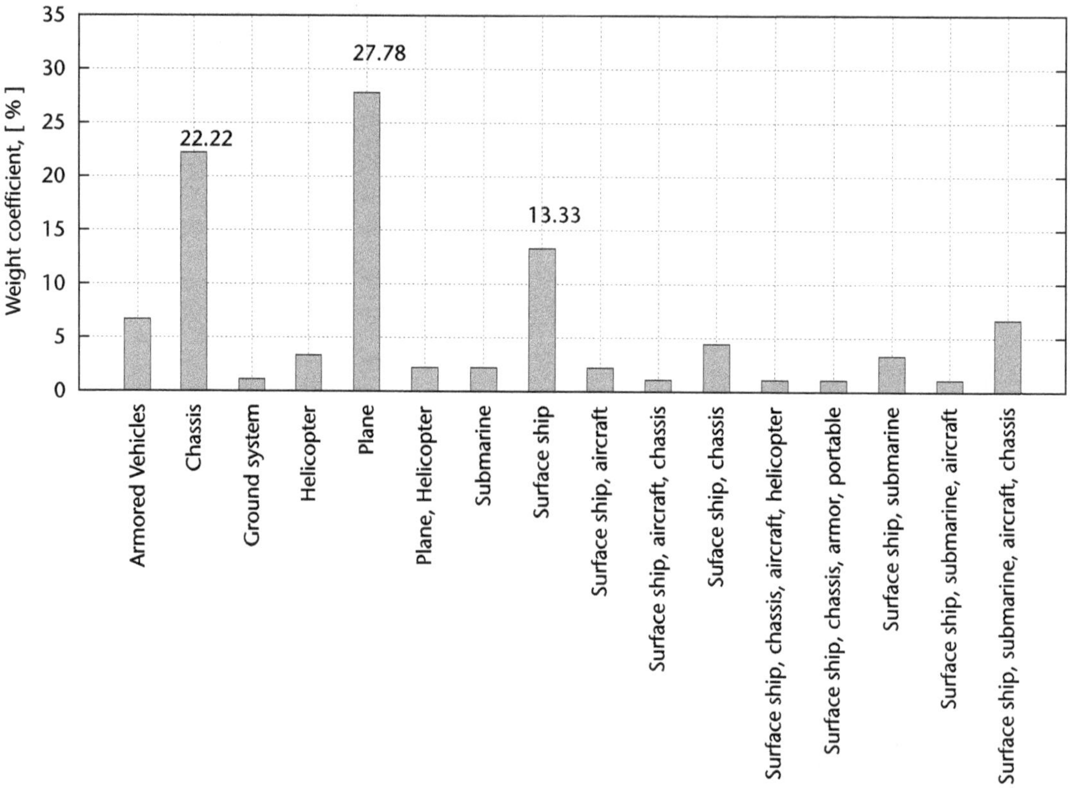

Figure 1.7 Basing missiles.

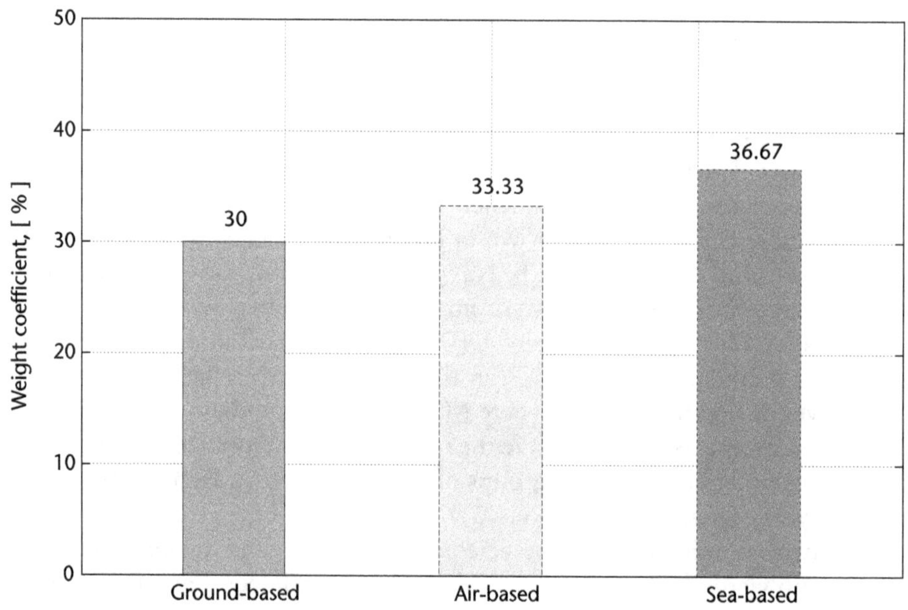

Figure 1.8 Different environments.

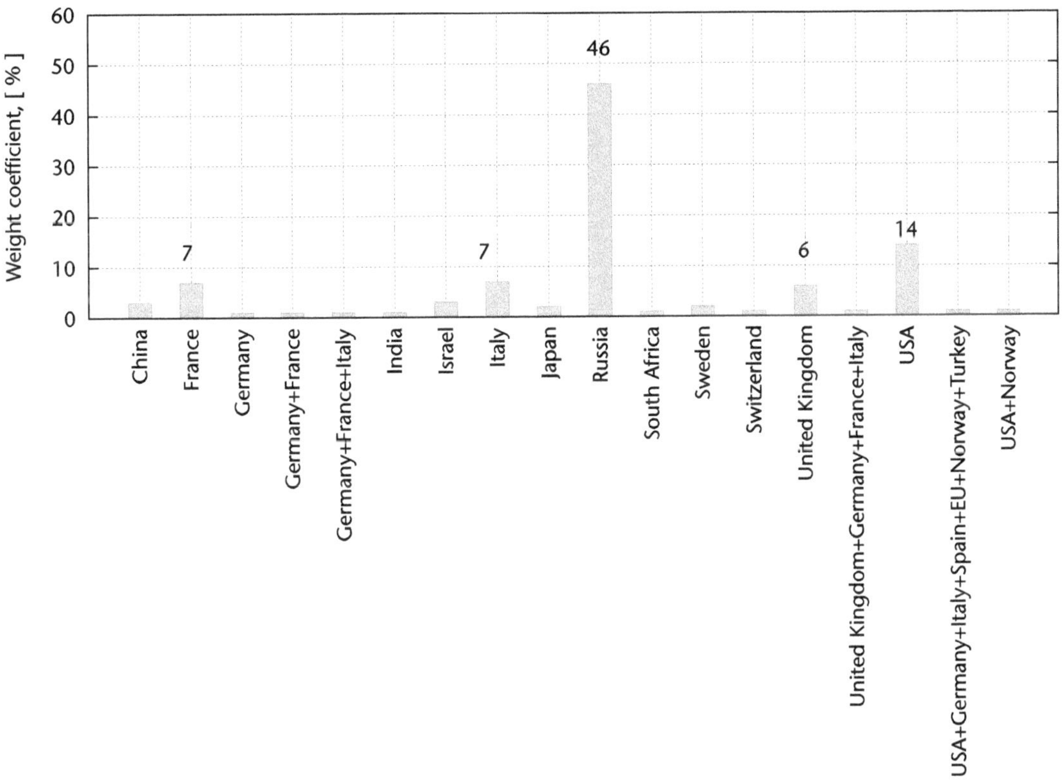

Figure 1.9 Countries that are developing missiles.

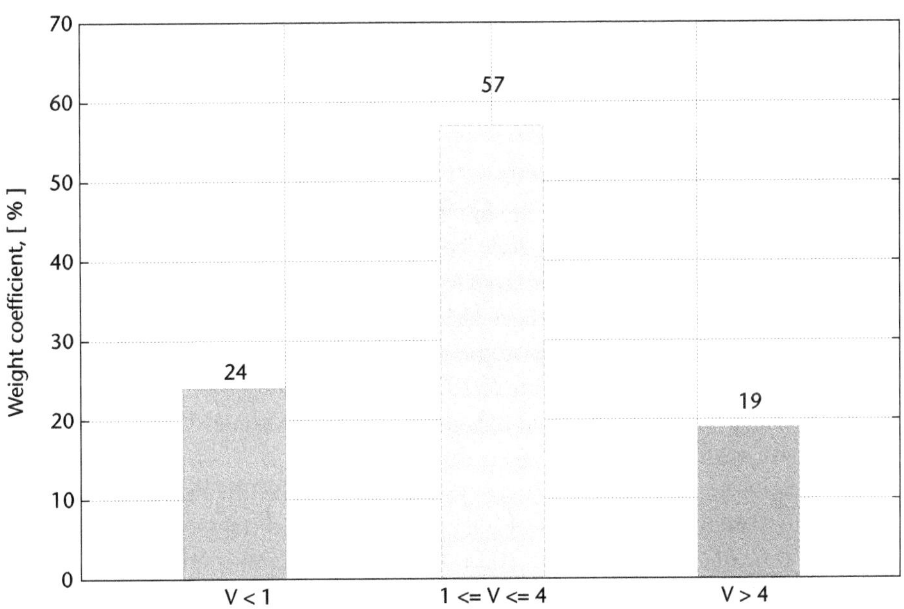

Figure 1.10 Missile speed (M).

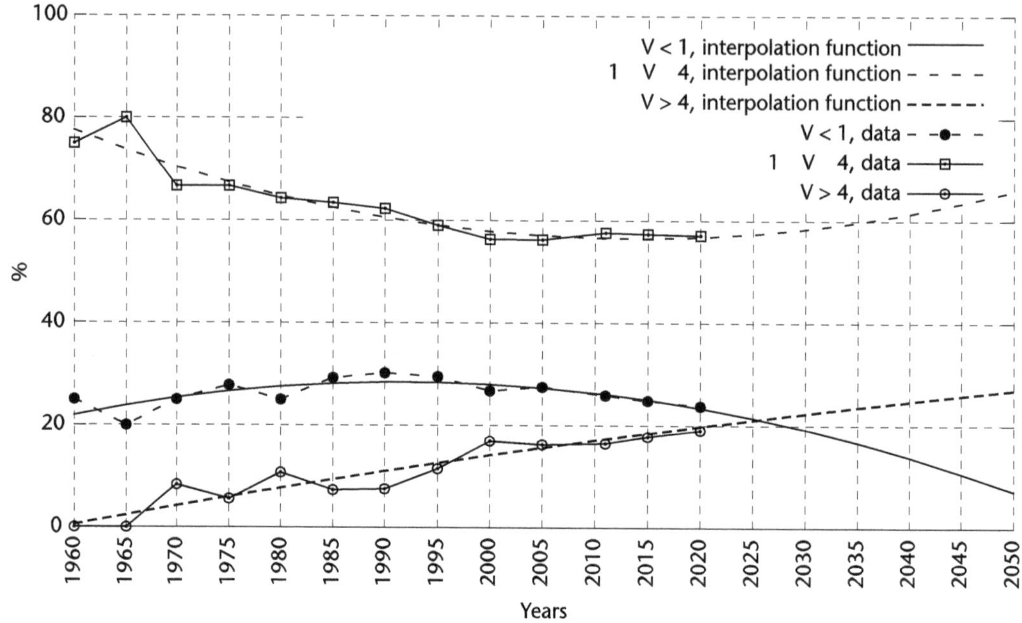

Figure 1.11 Forecast of the development of missile weapons.

For missiles with a flight speed of 1M to 4M, the approximating function is:

$$y_2(\mathrm{t}) = 29741.89 - 29.48\mathrm{t} + 0.0073\mathrm{t}^2 \tag{1.3}$$

For missiles with a flight speed of more than 4M, the approximating function is presented in the following form:

$$y_3(\mathrm{t}) = -4404.9 + 4.11\mathrm{t} - 0.00095\mathrm{t}^2 \tag{1.4}$$

The functions $y_1(\mathrm{t})$, $y_2(\mathrm{t})$, and $y_3(\mathrm{t})$ are shown in Figure 1.11, where t is the time, expressed in calendar years.

It is easy to show that the extremum of the function $y_1(x)$ is 1991.901; that is, November 25, 1991. This date corresponds to the maximum number of missiles with a flight speed of less than 1M. Therefore, after 1992, the number of missiles with a flight speed of less than 1M will only decrease over time.

In the same way, the extremum of the function $y_2(x)$ is 2013.56, which corresponds to the date of July 23, 2013. This date corresponds to the minimum number of missiles with a flight speed in the range from 1M to 4M. In the future, the number of these missiles will only increase.

Solving the equation $y_1(x) - y_2(x) = 0$ will give the date after which the number of missiles flying at a speed of over 4M will exceed the number of subsonic missiles. The root of this equation is 2025.43, and, accordingly, the date is June 5, 2025.

2025 is an important year because after this year the number of missiles with a flight speed of more than 4M will increase rapidly.

Thus, the development of missile technology will go in the direction of increasing hypersonic flight speeds.

Figure 1.12 The U.S. air-launched hypersonic cruise missile AGM-183A.

This hypothesis is evidenced by the fact that officials of the Air Force Life Cycle Management Center at Eglin Air Force Base, Florida, issued a formal notice (FA8682-20-R-0008) on Wednesday, August 12, 2020, concerning upcoming contractor selection for the Future Hypersonics Program. The contractors are the Boeing Co. Phantom Works in Washington, the Lockheed Martin Corp. Skunk Works segment in Palmdale, California, and the Raytheon Technologies Corp. Missiles & Defense segment in Tucson, Arizona. These three companies will develop an air-launched hypersonic cruise missile, and according to Air Force officials, these companies are considered to be the only responsible sources of hypersonic cruise missiles. Hypersonic weapons typically travel at Mach 5 to Mach 10 (or 3,840 to 7,680 miles per hour, 6,179.8656 km/h to 12,359.7312 km/h, or 1,716.6293 m/s to 3,433.2586 m/s). A missile traveling at Mach 5 could hit a target 100 miles (160.934 km) away in about 1 minute and 34 seconds. A missile moving at Mach 10, meanwhile, could hit a target 100 miles (160.934 km) away in about 47 seconds. Attempts to defend against incoming missiles traveling at these speeds leave very little time to deploy countermeasures (Figures 1.12 and 1.13).

One of the most difficult challenges of developing hypersonic weapons involve crafting rugged electronics for guidance and navigation that can withstand the extreme shock, vibration, and heat of hypersonic flight. Moving an object through the air at hypersonic speeds has been likened to a spacccraft reentering the Earth's atmosphere. These kinds of superfast munitions typically will be for high-value targets such as enemy aircraft carriers, relocatable strategic missiles, and forward-deployed military bases.

The U.S. Department of Defense (DOD) has identified developing hypersonic munitions and the ability to defend against them as one of the department's highest priorities. Potential U.S. adversaries China and Russia also are developing hypersonic weapons [26].

Figure 1.13 The U.S. hypersonic unmanned aerial vehicle X-51 [39].

Figure 1.14 The U.S. LTAMDS missile-defense radar.

Raytheon won a U.S. Army contract in October 2019 to provide the next-generation Lower Tier Air and Missile Defense Sensor (LTAMDS) advanced air and missile defense radar to help defeat advanced missile threats such as hypersonic weapons (Figure 1.14).

LTAMDS is the next-generation, 360-degree missile-defense radar that ultimately will replace the current U.S. Army's Patriot missile radars. The radar has gallium nitride components and is scheduled to reach initial operational capability with the Army in 2022 [27].

Prices for some missiles are shown in Figure 1.15, according to open sources [28–37]. The missiles shown in Figure 1.15 are equipped with radar guidance. Radar homing is semi-active radar homing (SARH), with NATO short code "Fox One" or active radar targeting with NATO short code "Fox Three." Figure 1.16 shows a histogram of average missile prices based on Figure 1.15 by their types. It is not difficult to see that the cost of radar homing in the total cost of the missile

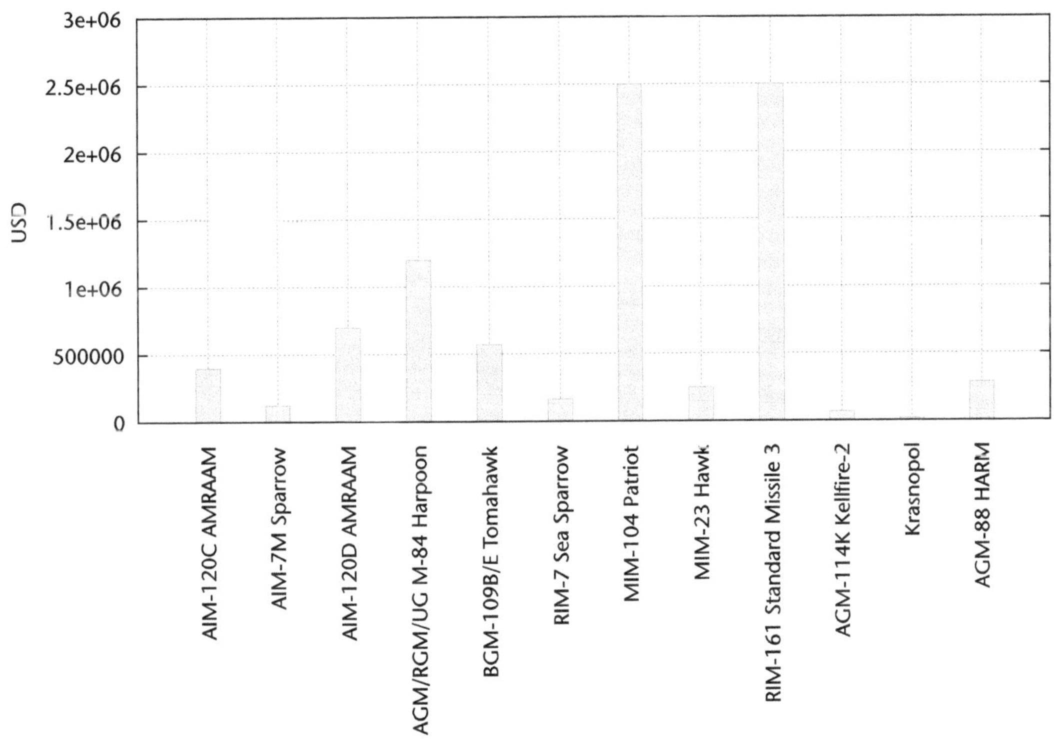

Figure 1.15 Prices for missiles with a radar homing.

can be estimated as 0.4037 [38], and thus the estimate of the average cost of radar homing is 233,888 U.S. dollars per unit.

ICD Research estimated the global missile market at U.S. \$15.2 billion in 2011 and forecast U.S. \$19.8 billion in 2021 [25], so it is easy to write a linear function of the market dynamics:

$$P(\mathtt{t}) = 0.46 \times \mathtt{t} - 909.86 \tag{1.5}$$

where $P(\mathtt{t})$ is the world missile market in billion U.S. dollars, and $\mathtt{t}$ is the time in calendar years. The world missile market in 2011–2021 is shown in Figure 1.17.

As seen from Figure 1.17, the global missile market in 2020 is estimated at \$19.34 billion U.S. dollars [25]; therefore, the market for radar homing in 2020 can be estimated at 7.79 billion U.S. dollars. Using the distribution data shown in Figure 1.6, we can evaluate the market missile-type radar homing. This estimate is shown in Figure 1.18.

For further analysis, we will introduce an efficiency indicator:

$$\hat{p}_i = \frac{\hat{w}_i \hat{M}}{\hat{n}_i \hat{W}} - 1, \qquad \forall i \in [1, ...N], \tag{1.6}$$

where N is the missile type number, $\hat{M}$ is an estimate of the total number of missiles produced this year, $\hat{W}$ is an estimate of the total cost of the homing radars produced this year, $\hat{w}_i$ is an estimate of the cost of a homing radar by the type i, and $\hat{n}_i$ is an

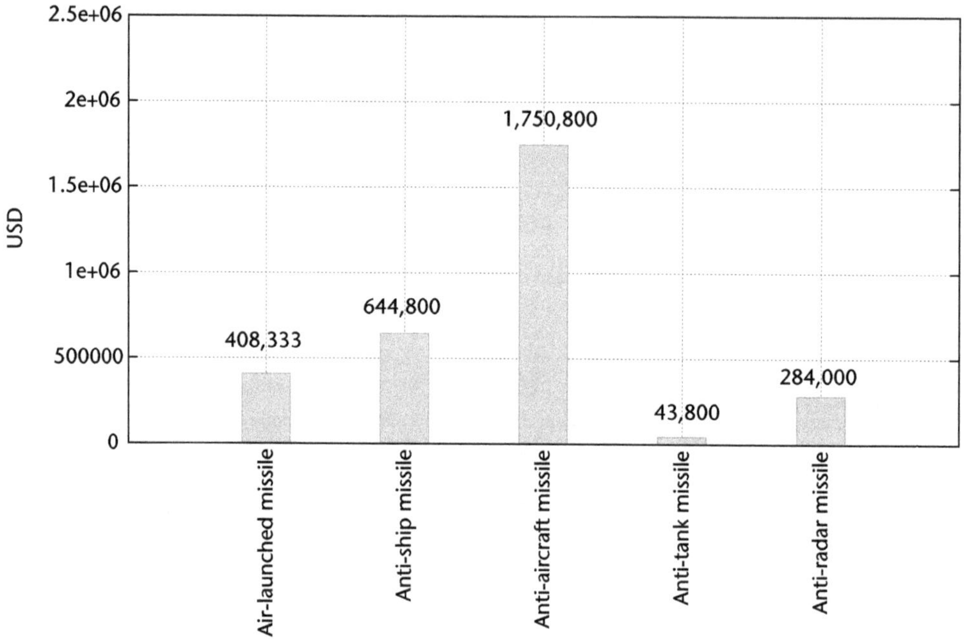

Figure 1.16 The average price by type for missiles with radar homing.

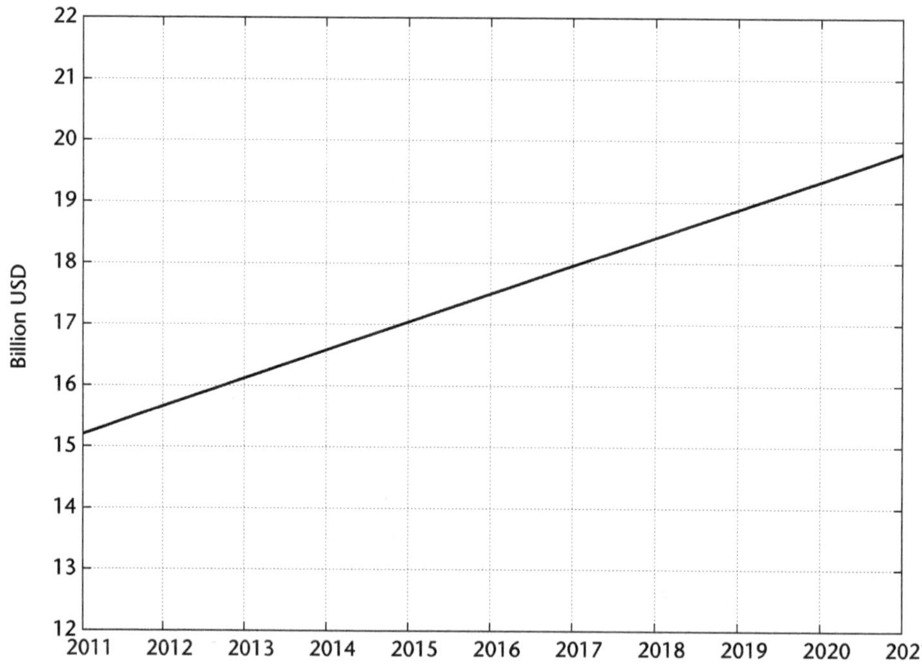

Figure 1.17 The global missile market from 2011–2021.

estimate of the sum of the homing radar by the type i. It is obvious that the criterion of efficiency is the condition

$$\hat{p}_i > 0, \qquad \forall i \in [1, ...N], \tag{1.7}$$

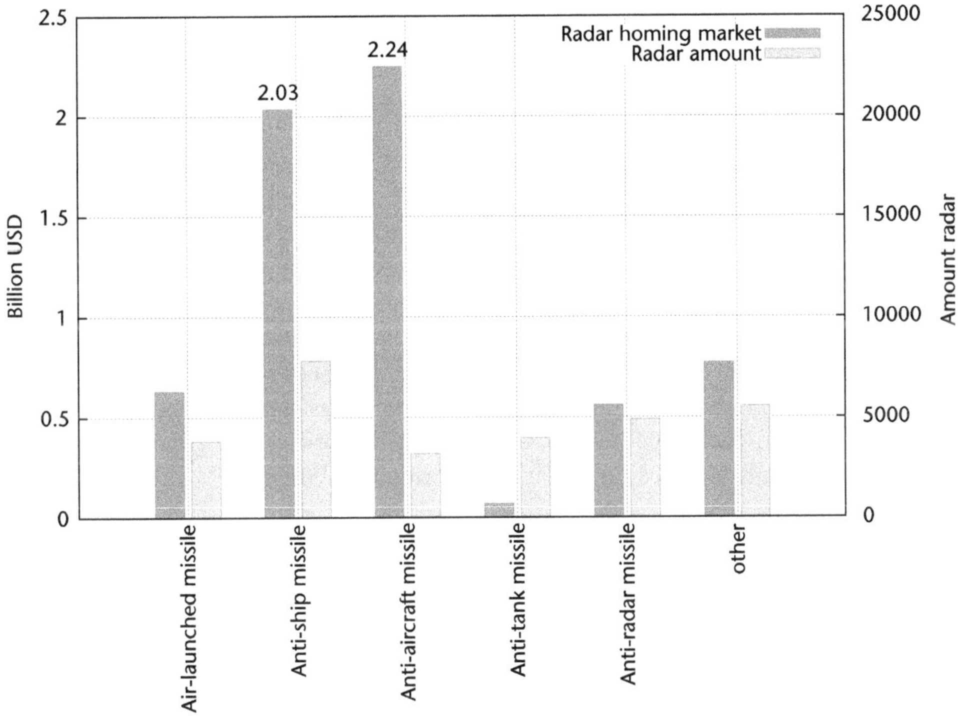

Figure 1.18 The global market of radar homing by type of missile.

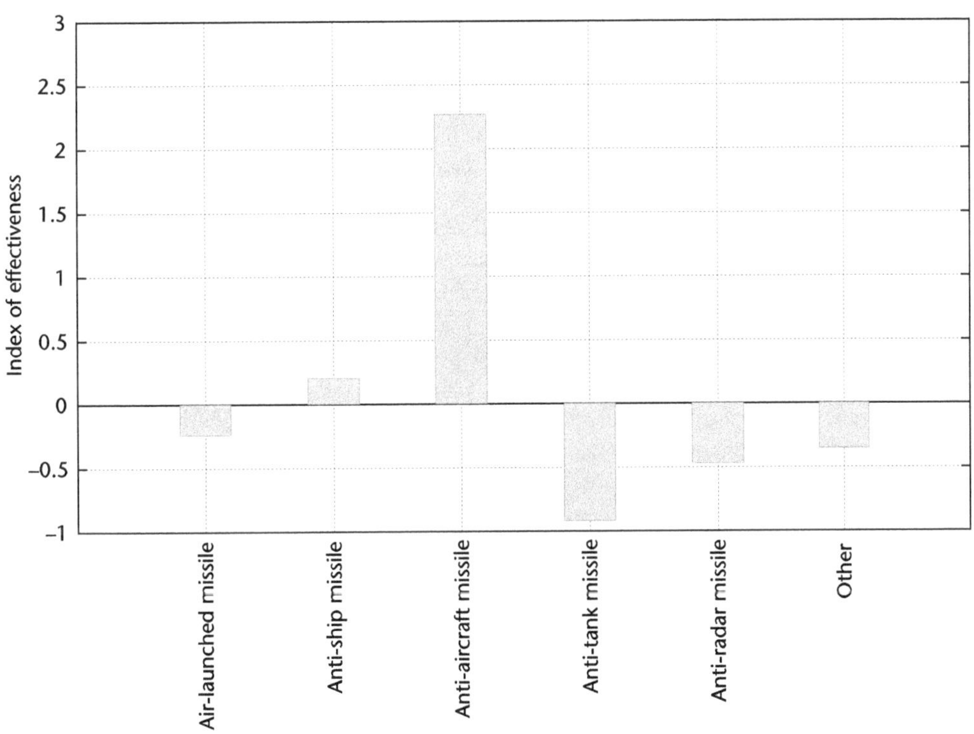

Figure 1.19 Indicators of effectiveness by type of missile with radar homing.

A histogram with performance indicators is shown in Figure 1.19 and we see that the most efficient business is the radar design and production business for anti-aircraft missiles, where the indicator of effectiveness has a maximum positive value of $\hat{p}_i$.

Thus, anti-aircraft missile systems armed with missiles with on-board radars are the main means of combating enemy aerospace attack forces. Under these conditions, a potential enemy will try with all their might to try to reduce the effectiveness of anti-aircraft missile fire against its means of aerospace attack by jamming airborne anti-aircraft missile radars.

In this regard, the task of ensuring noise immunity of air defense systems becomes the main one in the development and design of air defense radars in general and homing heads of anti-aircraft guided missiles in particular.

The main focus of this book is solving the problem of increasing the noise immunity of airborne radars of anti-aircraft guided missiles.

References

[1] DefenceTalk, Global Defence, Aerospace & Military Portal, http://www.defencetalk.com/us-air-force-tests-hypersonic-uav-26609.

[2] Gatsolaev, V. A., *Anti-Aircraft Units in Battle,* Moscow: Military Publishing, 1974.

[3] International Monetary Fund, *World Economic Outlook Update, June 2020: A Crisis Like No Other, An Uncertain Recovery*, June 24, 2020, https://www.imf.org/en/Publications/WEO/Issues/2020/06/24/WEOUpdateJune2020.

[4] World Population Review, https://worldpopulationreview.com/countries/countries-by-gdp.

[5] *Statistical Review of World Energy*, 2020, 69th Edition, https://www.bp.com/content/dam/bp/business-sites/en/global/corporate/pdfs/energy-economics/statistical-review/bp-stats-review-2020-full-report.pdf, http://www.bp.com/statisticalreview.

[6] PricewaterhouseCoopers LLP, The World in 2050, *The Long View: How Will the Global Economic Order Change by 2050*, https://www.pwc.com/gx/en/world-2050/assets/pwc-the-world-in-2050-full-report-feb-2017.pdf, https://www.pwc.com/gx/en/issues/economy/the-world-in-2050.html.

[7] Chinese Aircraft-J-11 (Su-27 FLANKER), http://www.globalsecurity.org/military/world/china/j-11.htm.

[8] Chinese Aircraft J-15, http://en.wikipedia.org/wiki/Shenyang_J-15.

[9] Chapligina, M., "Russia Downplays Chinese J-15 fighter Capabilities," RIA Novosti, June 4, 2010, http://en.rian.ru/mlitary_news/20100604/159306694.html.

[10] "Revealing Shenyang J-XX Stealth Fighter of China—What's on Xiamen," http://www.whatsonxiamen.com/tech111.html.

[11] "China or the J-15 Aircraft Under Development Is Similar to the Su-33," http://news.ifeng.com/mil/2/200909/0917_340_1353365.shtml.

[12] "Missile Threat 'Hongqi-10 (HQ-10)'," http://www.missilethreat.com/systems/hq-10_china.html.

[13] "Missile Threat 'Hongqi-15 (HQ-15)'," http://www.missilethreat.com/systems/hq-15_china.html.

[14] "S-300 (SA-10) Surface-to-Air Missile," *Sino Defence Today,* http://www.sinodefence.com/army/surfacetoairmissile/s300.asp.

[15] S-300, http://en.wikipedia.org/wiki/S-300_(missile).

[16] "Aircraft Carrier 'Vikramaditya'—The Flagship of the Indian Fleet," http://en.wikipedia.org/wiki/INS_Vikramaditya, http://www.india-defence.com/reports-4077.

[17] NDTV, October 10, 2012, http://www.ndtv.com/article/india/delivery-of-ins-vikram-adityafurther-delayed-by-a-year-admits-russia-277861.

[18] *Russia & India Report*, October 2, 2012, http://indrus.in/articles/2012/10/02/ building_indias_defence_base_offsets_vs_joint_development_18039.html.

[19] Karpenko, A. V., "VTS 'BASTION'," August 8, 2020, http://bastion-karpenko.ru/fgfa/.

[20] "HAL Signs Design Contract for Multirole Transport Aircraft," *The Hindu*, October 13, 2012, http://www.thehindu.com/news/national/\\hal-signs-design-contract-for-multirole-transport-aircraft/ article3992279.ece.

[21] RIA Novosti, June 29, 2012, http://en.rian.ru/mlitary_news/20120629/174308214.html.

[22] "Foreign Missile Developments and the Ballistic Missile Threat," Statement for the Record to the Senate Foreign Relations Committee on Foreign Missile Developments and the Ballistic Missile Threat to the United States Through 2015 by Robert D. Walpole, September 16, 1999, https://www.cia.gov/news-information/speeches-testimony/1999/walpole.htm.

[23] "Reflection of Hybrid Threats–SAM in the Air Defense and Missile Defense System, New Defence Order," *Strategy*, Vol. 6, No. 59, 2019, https://dfnc.ru/ekspertnoe-mnenie/otrazhenie-gibridnyh-ugroz-zrk-v-sisteme-pvo-i-pro/.

[24] Lyahov, N., "Aerospace Defense of Russia: History of Creation and Main Tasks," http://mil.ru/files/Anti-ballistic/07_Lyahov.pdf.

[25] "The Global Missiles and Missile Defense Systems Market 2011–2021," ICD Research, August 2012, https://www.asdreports.com/shopexd.asp?id=28275&desc=The+Global+Missiles+and+Missile+Defense+Systems+Market+2011-2021+-+SWOT+Analysis \%3A+Market+Profile.

[26] "Boeing, Lockheed Martin, and Raytheon Chosen to Develop Hypersonic Cruise Missile and Rugged Electronics," *Military & Aerospace Electronics*, August 7, 2020, https://www.militaryaerospace.com/sensors/article/14181178/hypersonic-cruise-missile-rugged-electronics?utm_source=MAE+Weekly&utm_medium=email&utm_campaign=CPS200811029&o_eid=0652E6064734B4U&rdx.ident%5Bpull%5D=omeda%7C0652E6064734B4U&oly_enc_id=0652E6064734B4U

[27] "Orolia Chosen to Provide Positioning, Navigation, and Timing (PNT) for Hypersonic Missile-Defense Radar," *Military & Aerospace Electronics*, July 9, 2020, https://www.militaryaerospace.com/communications/article/14179149/ positioning-navigation-and-timing-pnt-missiledefense-radar-hypersonic-weapons.

[28] AIM-120, http://en.wikipedia.org/wiki/AIM-120.

[29] AIM-7M Sparrow, http://en.wikipedia.org/wiki/AIM-7_Sparrow.

[30] M-84 Harpoon, http://en.wikipedia.org/wiki/Harpoon_(missile).

[31] BGM-109 Tomahawk, http://en.wikipedia.org/wiki/BGM-109_Tomahawk.

[32] RIM-7-Sea Sparrow, http://en.wikipedia.org/wiki/RIM-7_Sea_Sparrow.

[33] MIM-104-Patriot, http://en.wikipedia.org/wiki/MIM-104_Patriot.

[34] MIM-23 Hawk, http://en.wikipedia.org/wiki/MIM-23.

[35] RIM-161 Standard Missile 3, http://en.wikipedia.org/wiki/RIM-161_Standard_Missile_3.

[36] AGM-114, http://en.wikipedia.org/wiki/AGM-114.

[37] AGM-88 HARM, http://en.wikipedia.org/wiki/AGM-88.

[38] "The Day," Udmurtian Information Agency, Izhevsk, Russia. http://www.dayudm.ru/paper/248/index.php?ELEMENT_ID=16949.

[39] "US Air Force's Hypersonic Test Aircraft Fails During Attempt to Fly at Mach 6," *The Guardian*, August 15, 2012, http://www.guardian.co.uk/world/2012/aug/15/us-air-force-hypersonic-aircraft.

CHAPTER 2

Tactical and Technical Characteristics of an Anti-Aircraft Missile System

An anti-aircraft missile is the main component of anti-aircraft guided missile weapons, which is a set of the minimum required number of components (including the anti-aircraft guided missile) that are functionally interconnected and provide:

- Detection;
- Identification;
- The choice of target for shelling;
- Launch and guidance of a missile at a selected target;
- Destruction of the selected target [1].

The target for the anti-aircraft missile system is the enemy's aerospace means of attack. Therefore, destruction of the aerospace attack assets of a potential enemy is the main task of the combat activity of any anti-aircraft missile system.

For the development of anti-aircraft guided weapons, the creation of radar stations (radars), which are part of any anti-aircraft missile system, is of great importance. These radar stations (radars) provide the ability to hit a target with an anti-aircraft guided missile.

In modern anti-aircraft missile systems, radars perform a number of important tasks:

1. Detection of the aerospace means of attack of a potential enemy; that is, targets in flight toward the protected object;
2. Identification of targets according to the principle of friend or foe or according to the principle of true target–false target;
3. Determination of the spatial coordinates of targets and missiles, which are guided toward these targets;
4. Calculation of missile control commands that aim at a target or at several targets simultaneously;
5. Transmission on board each missile of the control commands calculated for it.

Target detection and identification is carried out, as a rule, by radar detection and target designation, which are part of the anti-aircraft missile system, as well as detection and target designation radars, which are part of the higher control levels: the anti-aircraft missile regiment, anti-aircraft missile brigade, air defense corps, air defense army, and central command post air defense troops of the country. At the

Figure 2.1 Radar detection P-18.

same time, the detection (reconnaissance) and target designation radars that are part of the anti-aircraft missile battalion are their own means of reconnaissance and target designation. The detection and target designation radars of higher control levels other than the anti-aircraft missile battalion are external means of reconnaissance and target designation. Bear in mind that the Air Defense Corps and the Air Defense Army also include air defense fighter aircraft aviation.

Thus, the command post of an anti-aircraft missile battalion receives information about the air situation in automatic mode or from its own means of reconnaissance and target designation and/or from external ones.

So, for example, the elements of reconnaissance and target designation in the S-75 anti-aircraft missile system were:

- Meter range radar P-18 ($\lambda \approx 1.2\ m$) [2];
- Radio altimeter of decimeter range PRV-13 ($\lambda \approx 1.1\ dm$) [3].

The P-18 radar (Figure 2.1), when coupled with the PRV-13 (Figure 2.2), forms a target designation complex that provides the determination of three target coordinates with high precision [2]. For the MiG-21 fighter, the P-18 detection range of a clean target (i.e., a target without interference) by means of a P-18 radar was the following [4]:

- 260 km for a target flying at an altitude of 20–27 km;
- 180 km for a target flying at an altitude of 10 km;
- 60 km for a low-altitude target flying at an altitude of 500 m;
- The number of simultaneously tracked targets from 7 to 12 targets;
- Information update rate from 10 sec to 30 sec;
- Accuracy of determining the target coordinates:
    - At a distance of 1,800 m;
    - In azimuth 6°.

Figure 2.2 Radio altimeter PRV-13.

The main tactical and technical characteristics of an anti-aircraft missile system are:

1. The defeat zone, which is determined by the minimum and maximum destruction ranges of targets and the minimum and maximum heights of their destruction. So for the S-75M3 anti-aircraft missile system (SA-2 GUIDELINE mod.4), the characteristics of the target defeat zone are as follows [5,6]:
   - Maximum target destruction range: 29–34 km;
   - Minimum target destruction range: 7 km;
   - Maximum target destruction height: 25–27 km;
   - Minimum target destruction height: 100 m.

   Other tactical and technical characteristics of an anti-aircraft missile system S-75M3 (SA-2 GUIDELINE mod.4):

2. The number of channels:
   - By target, 1;
   - By missile, 3.
3. The target's maximum speed on a head-on course is 3,700 km/h or 1,000 m/sec.
4. Missile speed is 3.5 M or 1,200 m/sec.
5. The deployment time of the anti-aircraft missile system is 6 hours, but this is a standard; in practice it takes a little more time, since only about 150 signal cables must be reconnected between semitrailers A and U, and this can be done only while lying on the ground, because hatches with connectors are located under the semitrailers.
6. The time interval between firing of targets is 1 min.

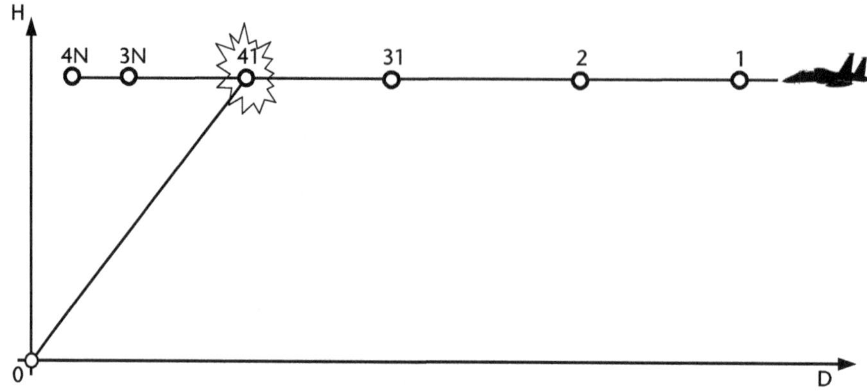

Figure 2.3 SAM firing scheme.

7. The time interval between missile launches is 6 sec.
8. Interference immunity (S-75M3):
    - From active interference: up to 100 W/MHz;
    - From passive interference: 2–3 packs per 100 meters.
9. Combat efficiency is 6.5% (based on results of the Vietnam War [5, 6]).
10. Probability of the target defeat:
    - One missile: 60%–80% (no interference);
    - By three missiles: 50%–60% (in conditions of strong interference).
11. The ammunition load of the anti-aircraft missile battalion is 12 anti-aircraft guided missiles (including 6 missiles on launchers).

Despite the variety of types of modern anti-aircraft missile systems (SAM), their firing pattern is fundamentally equal. Shooting an anti-aircraft missile system in an air target can be divided into the following interrelated sequential stages:

- Target detection;
- Identification;
- The choice of target for shelling;
- Determination of the moment of launch and the actual launch of an anti-aircraft guided missile;
- Guidance of anti-aircraft guided missiles at the target;
- Defeat of the target.

Figure 2.3 shows these successive stages of firing an anti-aircraft missile system at a target and the simplest case of firing at a single target flying horizontally at a constant speed in coordinates H-D (height-range).

The following notations are used in Figure 2.3:

0: Location point of the air defense missile system;

1: The point at which the detection and target designation system begins to work;

2: The point at which the anti-aircraft guided missile guidance systems begin to work (i.e. when the air defense system radar takes a target for tracking);

31: The point at which the first missile is launched;

41: The point where the first missile meets the target;

3N: The point where the *N*th missile is launched;

4N: The point where the *N*th missile meets the target.

Each stage of shooting is deliberately timed. The detection and targeting system begins to work when the target (targets) enters its coverage area (point 1). The detected targets are identified by the own–alien principle or by the true target–false target principle.

It should be emphasized that target identification according to the true target–false target principle is carried out, as a rule, in the anti-aircraft missile systems of anti-missile defense, since in this case, the aerospace attack weapons move along a ballistic trajectory and to increase the probability of breaking through the anti-missile defense use false targets. False targets are divided into two groups: light false targets and heavy false targets. Light false targets are used in the extra-atmospheric flight segment and represent, for example, inflatable balloons with a metallized shell simulating the effective dispersion surface of a real warhead. In addition, a real warhead can be placed inside an inflatable false target, which makes it difficult to select a real warhead for high-resolution ground-based missile defense radars. Heavy false targets are used in the extra-atmospheric and atmospheric sections of the intercontinental ballistic missile flight. They are used because light false targets significantly slow down their speed when entering the atmosphere and therefore can be speed-selected by the high-resolution ground-based radar of the anti-missile defense complex. To make it difficult, and ideally to prevent selection of the warhead a number of false targets, a heavy false target can either be equipped with its own solid-propellant rocket engine to accelerate the false target to the speed of movement of a real warhead and therefore compensate for the deceleration of the false target in the atmosphere (quasi-heavy false target), or it can be equipped with a special plasma-forming charge, the combustion of which produces an amount of energy no less than the amount of energy released by a real warhead when it is decelerated in the atmosphere (plasma false target). It is quite understandable that when complexing a quasi-heavy false target and a plasma false target for a high-resolution ground-based anti-missile defense radar, it becomes very difficult to select a real warhead, provided that the probability of correct detection of the warhead is not lower than the specified one.

Target identification based on the own–alien principle is carried out by anti-aircraft defense systems, since in their defeat zone, not only can the aircraft of a potential enemy be located, but also the own aircraft (e.g., fighter aircraft). Active radiolocation with automatic transponder-repeaters (answering machines) installed on board own aircraft are used to identify other own aircraft. The planes of the potential enemy are also equipped with answering machines; however, because the signals of the radar interrogators are coded and classified, the responder of the enemy aircraft will either not respond at all to the air defense system's request, or its response will be incorrectly coded. An own–alien radar identification system consists of an interrogator and a responder. Aircraft radar transponders are of two types:

1. Transponders providing air traffic control, which are designed to automatically transmit information about the location of an air target to the air traffic controller;
2. Transponders of state identification, which are designed to determine the state nationality of an air target.

The state identification system consists of a radar transponder, which sends a request to the aircraft according to the principle of "Who are you?". The aircraft radar responder should answer with, "I am own." The answer must contain one of the codes that it has in the set, which change over time according to a predetermined schedule. Only two countries in the world have developed and are using own–alien systems of identification:

- The United States, with the MK12 system, which is used by all NATO members;
- The USSR, with the Password system, which is used today in Russia and the CIS countries.

These are completely different systems that are incompatible with each other. The algorithms and the principle of operation of the own–alien recognition systems are the most important state secrets. Until September 6, 1976 in the USSR, the own–alien state identification system was used under the name Silicon2. This system operated at a carrier frequency of 668 MHz, the radio pulses of which were modulated by video pulses in the form of a comb, and the frequency of which changed depending on the number of codes in units of MHz [7]. However, everything changed after September 6, 1976. On this day, a senior lieutenant of the fighter regiment of the 11th Separate Air Defense Army of the USSR Viktor Belenko hijacked the newest high-altitude fighter-interceptor MiG-25P to Japan, which was named *Flying Fox* according to NATO classification. Great damage was caused by the disclosure of information on the own–alien identification system, which had to be urgently changed throughout the country. The official inflicted material damage was estimated at two billion Soviet rubles [8]. The S-75 anti-aircraft missile system in the village Mir, Belarus, where I had served since August 1983 as the head of the second part of the radio technical battery, received a new autonomous NRZ-4P radio interrogator with separate equipment, code 1L22B [9], only at the beginning of 1984. This radio interrogator was combined with the P-18 radar and the S-75M3 anti-aircraft missile guidance station. The scale of the damage and the task of rearming the Armed Forces of the USSR and then Russia can be evidenced by the fact that the complete rearmament to the new radar interrogator NRZ-4P took about 20 years and was completed only at the beginning of 1995 [7]. Thus:

1. There is only an own request signal and a response signal. No one can fake this signal in the time required for the target to complete its combat mission.
2. On anti-aircraft missile systems and combat aircraft, when the missile control button Start is pressed, an automatic request is made, and if the Own signal arrives in combat mode, the missile launch will be blocked.

3. An airplane equipped with own–alien equipment has the right to reply "Own" only to the request of its radar [10].

Thus, after detecting a target by the radar surveillance (detection) or missile guidance station (point 1, Figure 2.3), the operator of this radar presses the Request button. In this case, the radar interrogator sends an encrypted interrogation signal in the direction of the target, the target identification equipment receives it, decrypts it, and if the request is correct (i.e., came from Own radar), then the responder generates an encrypted response signal and sends it to the radar interrogator. The interrogator receives the response signal, decrypts it and if it is correct, then the mark "Own" appears on the radar display near the target mark, and if the target is someone else's, the mark will not appear [10].

Further, the degree of threat to the protected object from identified targets and the possibility of their firing by an anti-aircraft missile system is established. For this, the direction, altitude, range, and flight speed of each target are determined. Depending on the technical capabilities of the anti-aircraft missile system, targets are selected by the operator (in manual mode) or automatically (in automatic mode). The number of targets assigned for firing is determined by the number of target channels of the complex. If the number of targets exceeds the target channels capacity of the complex; that is, when the number of targets exceeds the capabilities of the complex for simultaneous shelling of several targets, then the targets that most threaten the protected object are selected for shelling.

The detection and target designation system takes time to carry out the operations listed above where $t_d = t_1 - t_2$ is the detection time, during which the target flies the distance $D_d = |1, 2|$ from the moment the target is detected until the moment it is selected for automatic tracking by the missiles guidance station of the anti-aircraft missile system (Figure 2.3). In modern air defense systems, this time ranges from several seconds to several tens of seconds.

Target designation and the necessary data on the target(s) selected for firing are transmitted to the anti-aircraft missile system and the air defense missile system's radar selects the target(s) for tracking, while the guidance means of the SAM begin to work (point 2, Figure 2.3).

Determining when to launch the missiles and then actually launching them is carried out by the commander of the anti-aircraft missile system (battalion), who is located at the battalion command post during combat operation (manual mode), or the decision is made automatically by an electronic computer based on data that is received from the target designation and reconnaissance equipment (automatic mode). In addition, the so-called semi-automatic mode is also possible, when an electronic computer offers the commander of an anti-aircraft missile battalion possible options for firing at a target, and the commander themself then makes a decision.

The time from the moment the target is transferred for tracking the radar of the anti-aircraft missile system to the point of launching the first missile is called the decision time $t_{decision} = t_2 - t_{31}$, and when the target flies the distance is $D_{decision} = |2, 31|$ (Figure 2.3).

The time interval from the moment in which the target is detected by the detection radar of the air defense system (point 1, Figure 2.3) to the moment in which the first anti-aircraft guided missile is launched (point 31, Figure 2.3) is called the working time or reaction time of the air defense missile system $t_r = t_1 - t_{31}$.

Modern anti-aircraft missile systems can have a reaction time of up to several seconds. During the working time (the reaction time of the air defense missile system), the target flies the distance $D_r = |1, 31|$ (Figure 2.3). And in the case of uniform rectilinear movement of the target with constant velocity V_t can be written as:

$$D_r = V_t \times t_r \tag{2.1}$$

When the target is at point 31, the air defense missile system is ready for launch. In order to fire the anti-aircraft guided missiles at the maximum range, it is necessary that during the flight of the air defense missile at a distance $D_{max} = |0, 41|$, the target from point 31 also flies to point 41. It is in this case that the anti-aircraft guided missile meeting with the target will occur at the maximum combat range of the anti-aircraft guided missile flight, which is determined by the tactical and technical characteristics of the anti-aircraft guided missile and the features of its design.

The operating time of the anti-aircraft guided missile guidance system t_o is the time interval $t_o = t_2 - t_{41}$, during which the target will fly the distance $D_o = |2, 41|$.

If the target will fly at a speed exceeding the maximum speed of the anti-aircraft guided missile specified during design, then when the anti-aircraft guided missile is launched when the target is at point 31, it will meet with the target at a distance less than its maximum combat range.

If, under the same conditions, the target will fly at a speed lower than the speed set in advance, then when the anti-aircraft guided missile is launched at point 31, it will not meet with the target, since by the time the target arrives from point 31 to point 41, the anti-aircraft guided missile will have exhausted its maximum range resource.

Thus, the characteristics of the detection and target designation system, the characteristics of the missile control, as well as the characteristics of the anti-aircraft guided missile itself must be strictly linked in time and in range.

It should be especially noted that the tactical and technical characteristics of any anti-aircraft missile system depend primarily on the flight-technical characteristics of the targets for which this anti-aircraft missile system is designed to fight.

2.1 The Defeat Zone

The defeat zone of an anti-aircraft missile system is the main generalized characteristic of its combat capabilities.

The defeat zone connects the combat reach of an anti-aircraft guided missile in height, range and heading parameter with the probability of defeat of a target by an anti-aircraft missile system.

This defeat zone is a part of the space within which it is ensured that a typical target by anti-aircraft guided missile defeat with a probability not lower than a given one.

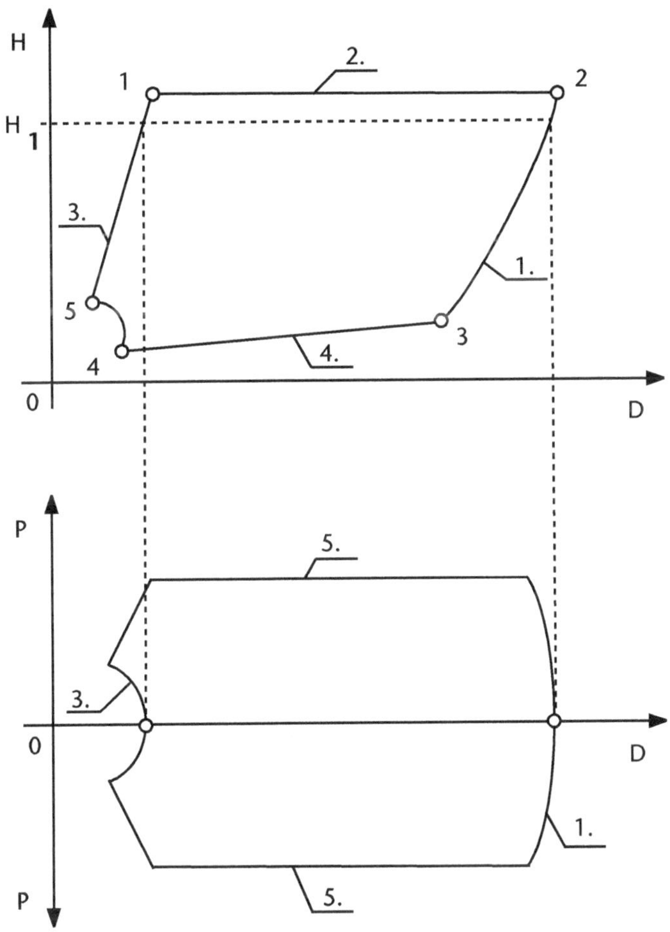

Figure 2.4 The defeat zone of an anti-aircraft missile system.

In the terms of reference for the design of an anti-aircraft missile system, typical air (aerospace) attributes of a potential enemy are set on the basis of their known tactical and technical characteristics.

The dimensions of the approximate defeat zone of the anti-aircraft missile system are shown in Figure 2.4 and are characterized by its boundaries: far = 1, upper = 2, near = 3, lower = 4, and maximum parameter value = 5.

2.1.1 Choosing the Anti-Aircraft Missile System Defeat Zone Boundaries

The defeat zone of any anti-aircraft missile system can be considered based on the following criteria:

1. The choice of the anti-aircraft missile system defeat zone boundaries;
2. Providing the selected boundaries of the defeat zone with the technical characteristics of combat assets of an anti-aircraft missile system.

The choice of the boundaries of the defeat zone precedes the design and/or modernization stage of each anti-aircraft missile system. The initial data for this stage, as shown above, are the characteristics of typical air and aerospace targets of a potential enemy, taking into account the prospects for their development, the impact of which will be on the objects protected by this anti-aircraft missile system.

This requires typical targets to have the following characteristics:

- Maximum and minimum target flight speed;
- Maximum and minimum possible target flight altitude;
- Maneuverability of the target in flight (available overload);
- Types of airborne weapons installed or planned to be installed on the target;
- Heights and maximum ranges for the use of airborne weapons;
- Methods of combat use of airborne weapons;
- Target vulnerability (i.e., it is necessary to know the probability of defeat of a target by a warhead), which is used or planned for use in an anti-aircraft guided missile;
- Effective scattering surface of a typical target (in square meters);
- The expected jamming situation during the combat use of this target;
- The minimum admissible conditional or total probability of target defeat when firing one or several anti-aircraft guided missiles within a given defeat zone. This characteristic is set based on the condition of the required degree of coverage for defended objects, taking into account the tactics of using this anti-aircraft missile system, as well as the combat capabilities of other anti-aircraft missile systems operating simultaneously and are considered as part of the general air defense system.

In modern conditions, for the same anti-aircraft missile system, several typical targets are given, and therefore the defeat zone of these targets will differ in size, shape, and probability of destruction.

1. The far boundary of the defeat zone of the anti-aircraft missile system $D_{2,3}$ (Figure 2.4) can be selected, for example, from the condition that a typical target will be defeated with a probability not lower than a given at ranges exceeding the line of using enemy aviation onboard weapons against the protected (defended) object. Due to the fact that the improvement of aerospace attack means of a potential enemy is carried out in the direction of increasing ranges use of on-board weapons, namely "air-to-ground" missiles, therefore a universal anti-aircraft missile system should have a far border of the defeat zone exceeding launching lines of modern and advanced aviation onboard weapons. This, in turn, necessitates a constant increase in the far border of the defeat zone of the anti-aircraft missile system.

2. The upper boundary of the defeat zone of the anti-aircraft missile system $H_{1,2}$ (Figure 2.4) is responsible for the range of heights which, covered by the anti-aircraft missile system and is chosen to ensure that the probability of destruction of targets not lower than that specified in the entire range of heights of the defeat zone of the system. At present, the upper boundary of the defeat zone is located at a distance of units and tens of kilometers from the Earth's surface for anti-aircraft defense systems and up to several hundred kilometers from the Earth's surface for anti-missile defense systems.

3. The near border of the defeat zone of an anti-aircraft missile system $D_{4,5,1}$ (Figure 2.4) is chosen so that it is as close as possible to a flat area that passes through the position of the anti-aircraft missile system (point 0, Figure 2.4) and that is perpendicular to the direction of the target flight. In this case, the depth of the defeat zone increases, which allows for a longer time to fire at the target and provides better opportunities for sequential firing of targets flying after the initial firing, and also improves the firing capabilities of late-detected targets.

4. The lower boundary of the defeat zone of the anti-aircraft missile system $H_{3,4}$ (Figure 2.4) is responsible for a target's defeat at low and extremely low altitudes. As shown in Chapter 1, one of the ways to overcome anti-aircraft defense systems is the flight of the potential enemy's aviation at low and extremely low altitudes, and therefore, it is reasonable to require the target defeat on these altitudes with a probability not lower than the specified one. This explains the desire to bring the lower boundary of the defeat zone of modern anti-aircraft missile systems as close as possible to the Earth's surface. At present, this zone ranges from units to several tens of meters above the Earth's surface. For anti-missile defense systems, the lower boundary of their defeat zone is currently in the range from several kilometers to several tens of kilometers from the Earth's surface. This ensures the safety of their protected objects from the impact of the nuclear warheads of attacking enemy missiles, undermined by the anti-missile system of the anti-missile defense, which, in turn, can also be equipped with a nuclear charge.

5. Due to the fact that the parameter P is values of the P axis, determined on this axis by the position of the segments $+P_5$ and $-P_5$, there is an increase in the parameter $+/-P_5$. This allows us to reduce the number of SAM systems in the air defense grouping consisting of several SAM systems, located, for example, along the P axis, while maintaining effectiveness of all air defense grouping.

It should be noted that the size of the defeat zone of an anti-aircraft missile system and the value of the probability target defeat in it can be set not only for a target flying at a constant altitude with a constant speed, but also for a target that is making a maneuver in course and speed. In addition, the defeat zone of the anti-aircraft missile system can be set for targets that are on catch-up courses; that is, for shooting in the direction when $D_t < 0$ is relative to the position of the air defense missile system (point 0 in Figure 2.4).

Thus, the selected location of the boundaries of the defeat zone of the anti-aircraft missile system determines the characteristics of all its combat assets.

2.1.2 Providing a Given Defeat Zone with the Technical Characteristics of Anti-Aircraft Missile System Combat Requirements

The main characteristics of the combat requirements of an anti-aircraft missile system in terms of ensuring a given defeat zone include:

- The reach of an anti-aircraft guided missile to the boundaries of the defeat zone and to all of points located within the boundaries of the defeat zone.
- The average speed of an anti-aircraft guided missile, ensuring its flight to all points of the defeat zone, as well as its speed along its trajectories.
- The range of the target's means of detection.

- The range of the anti-aircraft missile system's guidance equipment, which ensures the stable guidance of the missiles to the target.
- The working time of detection and target designation equipment $t_{detection} = t_d = t_1 - t_2$.
- The working time of anti-aircraft guided missile guidance equipment $t_{control} = t_c = t_2 - t_{4N}$.
- The working time of the entire complex as a whole $t_{total} = t_1 - t_{4N}$, during which the target flies the distance $D_{total} = D_1 - D_{4N}$ (see Figure 2.3).
- The probability of failure-free operation of the combat equipment of the complex. The probability of no-failure operation $P(t)$ is the probability that the equipment will not fail in a given time interval [11]. This function is decreasing: $P(0) = 1$, $P(\infty) = 0$; therefore $0 \leq P(t) \leq 1$. Since an air defense missile system is a complex system, breaking its electronic parts into N_0 separate blocks can be written as:

$$P(t) = 1 - \frac{n(t)}{N_0}, \qquad \forall t \in [0, ..., \infty], \tag{2.2}$$

where $n(t)$ is the number of inoperative blocks by the time t and N_0 is the number of monitored blocks. It should be emphasized that the probability of failure-free operation $P(t)$ is a very important criterion for the combat readiness of anti-aircraft missile forces to repel an air attack. By integrating the probability of uptime $P(t)$ over time, we can get the mean time between failures T_0, which is mathematical expectation of the system operation time before failure [11]:

$$T_0 = \int_0^{\infty} P(t)dt \tag{2.3}$$

Because the probability of no-failure operation of a group of elements is equal to the product of the probabilities of no-failure operation of each element in this group:

$$P(t) = \prod_{k=1}^{N} P_k(t) \tag{2.4}$$

Consequently, the more elements in a group, the lower the probability of failure-free operation, and hence the reliability of the entire group.

For example, the S-75 air defense system was based on analog technology using electronic tubes. In addition, the installation of radio electronic elements was carried out by a hinged method without the use of printed circuit board technology. With this installation, the resistances, capacitors, chokes, and other elements were separate radio-electronic elements interconnected by wires (Figure 2.5) [12]; therefore, by virtue of the expression (2.4) the probability of failure-free operation of the S-75 air defense system was extremely low. From personal experience, I can say that sometimes it took days to eliminate malfunctions. This was due to the fact that the air defense forces are troops of constant combat readiness, and so in the anti-aircraft missile forces of the USSR there was an unwritten rule that an officer could not leave to rest if their equipment was not combat-ready.

Figure 2.5 One of the blocks of the coordinate system of the S-75 air defense missile system.

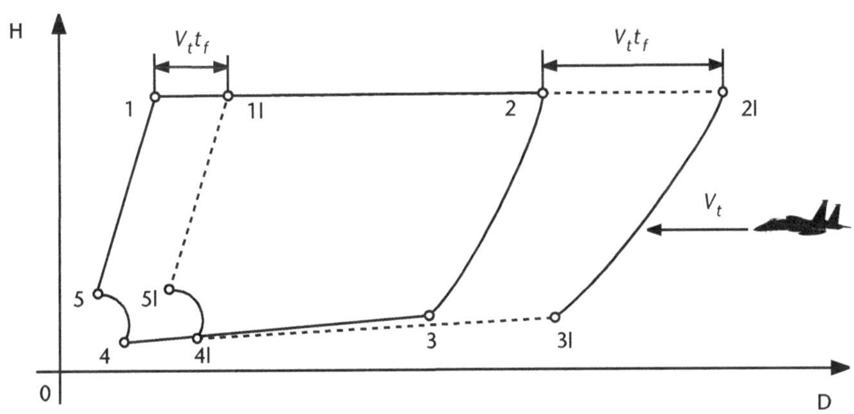

Figure 2.6 An air defense missile system launch zone.

2.2 Missile Launch Zone

To successfully target defeat in the defeat area of a given air defense system, it is first necessary that the anti-aircraft missile be able to reach all of its points lying inside the polygon 1-2-3-4-5 (see Figure 2.6), including those lying on its borders 1-2-3-4-5. However, this is not enough. To defeat a target with a given level of probability, it is necessary to ensure the required speed and controllability of the anti-aircraft missile, which is determined by the adopted guidance method.

The average flight speed of an anti-aircraft guided missile to the boundaries of the defeat zone determines the launch zone of the missile, which is a part of the space where a target is located in which, at the time of launching the missile, it meets the target within the specified defeat zone.

In Figure 2.6, the case of firing toward a nonmaneuvering target flying horizontally at a constant speed V_t is considered.

The launch zone $1l, 2l, 3l, 4l, 5l$, which corresponds to given defeat zone 1, 2, 3, 4, 5, can be built for each vertical section by put-off at different heights H from the near, far and lower boundary of horizontal segments which equal, in the corresponding scale, the product of the target's flight speed V_t and the time of an anti-aircraft guided missile flight t_f to the corresponding points boundaries of the SAM defeat zone.

The distance along the horizon from the position of the anti-aircraft missile system 0 to the point of the launch zone D_l, corresponding to certain points of encounter between the missile and the target in the defeat zone, can be calculated when a target is flying with a zero course parameter as follows:

$$D_l = D_m \frac{k+1}{k}, \tag{2.5}$$

where D_m is the distance along the horizon from the position of the anti-aircraft missile system 0 to the point where the missile meets the target in the defeat zone of the anti-aircraft missile system, $k = \frac{V_m}{V_t}$ is the ratio of the average flight velocity of the anti-aircraft guided missile to the meeting, and V_t is the point of the missile with the target V_m to the average flight velocity of the target V_t.

Obviously, with an increase in the average flight speed of an anti-aircraft guided missile, the depth of the launch zone decreases and the time of the air defense missile system occupation for firing at the target decreases, and therefore, the fire performance of the air defense system α increases:

$$\alpha = \frac{N_f}{T}, \tag{2.6}$$

where N_f is the number of fired targets in a given time interval $T = [0; T]$. Here, N_f is the total number of targets at which missiles were fired during time T. Moreover, it is not necessary that all targets are defeated by these missiles.

We can see that an obstacle to a significant increase in the average flight speed of anti-aircraft guided missiles is the limitation of its launch weight, as well as the fact that solid-propellant rocket engines currently have reached their peak of development, reaching a speed of 7.33 M for the 48N6E3 missile of the S-400 [13] anti-aircraft missile system. Higher speeds can be achieved by using a hypersonic ramjet engine (scramjet engine) in missiles, which is currently installed on experimental aircraft that burn fuel in a supersonic air stream. Therefore, a prerequisite for launching a hypersonic ramjet engine is the preliminary acceleration of an anti-aircraft guided missile to a supersonic speed greater than $5M$. It follows that a hypersonic anti-aircraft guided missile should be two-stage. As the first stage, it is reasonable to use a solid-propellant rocket booster, which separates after reaching a speed sufficient to start the main engine.

The first actual flying missile with a ramjet engine in world history was the La-350 intercontinental cruise missile *Tempest* conceived and created under the leadership of general designer S. A. Lavochkin and chief designer N. S. Chernyakov in the USSR.

This missile was two-stage:

- The first stage: Two side boosters operating on two four-chamber liquid-propellant rocket engines R-11 (S2.1100) and providing traction of about 65 tons each. The chief designer for OKB-2 was A. M. Isaev (the Design Bureau of Chemical Engineering is named after him).
- The second stage (marching): This was a cruise missile with a mid-located thin delta wing of low aspect ratio and symmetrical profile. The wing had a 70° sweep along the leading edge and a straight trailing edge. The cruciform tail was placed

in the tail section. The missile body was cylindrical, slightly tapered in front and behind. Inside it, along its entire length, there was an air intake channel of the RD-012U sustainer supersonic ramjet engine, designed by OKB-670 (Chief Designer M. M. Bondaryuk). The engine provided 7.75 tons of thrust. The front of the body was a supersonic diffuser with a three-stage cone.

On the acceleration stage, the missile was controlled by gas rudders, which were dropped and the control was switched to air rudders. After acceleration, when the flight speed reached the desired value, the air-jet engine entered full thrust mode, and at an altitude of 17.5 kilometers the accelerators from the main stage were separated. After that, the flight was corrected with the help of an Earth-type automatic astronavigation control system. The entire system was designed for a range of 8,000 kilometers.

The *Tempest* flew at a constant speed from Mach 3.1 to 3.2 [14]. The absolute record for flight range was previously achieved in 1959, and on December 16, 1960, having covered the distance of 6,500 km between the Vladimirovka test site and the Kamchatka Peninsula, it deviated from the target by a distance of just over 4 km [15]. This was a world record for accuracy of aiming at a target located at such a great distance from the launch point of the missile. However, despite the successful tests of the *Tempest,* the project was closed on the personal instructions of the then head of the USSR, N. S. Khrushchev, in favor of the R-7 project, which was directed by S. P. Korolev.

Later in the USSR, the Central Institute of Aviation Motors (TsIAM), which was named after P. I. Baranov, was engaged in the development of ramjet engines. In the mid-1980s, a hypersonic flying laboratory called Kholod was created in the USSR [16] with a hypersonic ramjet engine based on the missile in the S-200 air defense missile system. On November 28, 1991, at the Sary-Shagan test site, Kazakhstan, the world's first unique successful flight test of a hypersonic ramjet engine was carried out. During these flight tests, a speed of $5.7M$ was achieved. In 1998, a hypersonic ramjet engine successfully passed flight tests as part of the Igla installation. The engine was turned on at a flight speed of $3M$, and at the end of the flight the speed reached $6.47M$ [17].

Note that the missile launch zone when firing at a nonmaneuvering target flying horizontally at a constant speed is often not the only zone that allows you to determine the optimal characteristics of air defense missile systems.

Due to the high efficiency of modern air defense, the air attack means of a potential enemy widely use various course and speed maneuvers when flying through the air defense zone and attacking ground targets, as well as when evading missiles launched at them.

In these cases, a guaranteed missile launch zone is considered, which is a missile launch zone for a given SAM defeat zone. This is determined from the condition that for any possible or specified maneuver, the target will be destroyed in the SAM defeat zone with a probability not lower than the specified one.

All other things being equal, the guaranteed launch zone is much smaller in volume than the launch zone when firing at a nonaneuvering target. It is obvious that the increase in the guaranteed launch zone depends on the increase of the average flight speed of the anti-aircraft guided missile [1].

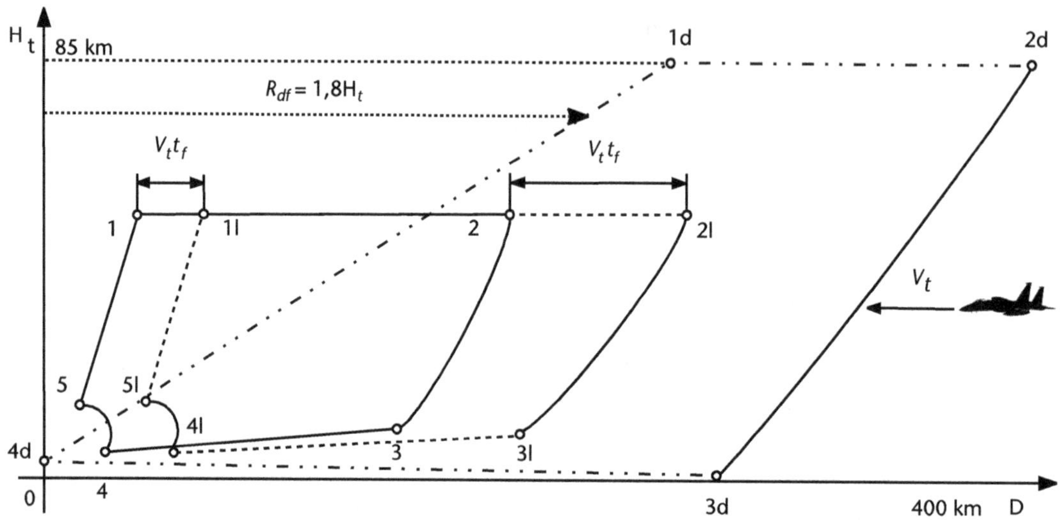

Figure 2.7 A target detection zone.

2.3 A Target Detection Zone

The detection zone of an air defense missile system is an area of space within which the specified means of an air-space attack of a potential enemy are detected at a maximum distance from the standing point of the air defense missile system, with a probability not lower than a given one, while determining their coordinates and characteristics.

To determine the required maximum horizontal range of the target detection means D_d^{max}, the length of the segments which when deferred from the lower and far boundaries of the missile launch zone should be equal to the product of the target flight speed V_t by the working time of the anti-aircraft missile system $t_r = t_1 - t_{31}$, which is also the SAM system reaction time. Using Figure 2.3 for uniform rectilinear movement of the target, we can write:

$$D_d^{max} = D_{31} + V_t \times t_r \tag{2.7}$$

So, for a complex of own reconnaissance and target designation compenents, consisting of a P-18 radar and a PRV-13 radio altimeter, the detection zone $1_d, 2_d, 3_d, 4_d$ will have an approximate view that is shown in Figure 2.7. This detection zone is primarily characterized by the tactical and technical characteristics of the reconnaissance and target designation assets being integrated.

Therefore the P-18 radar operates in the meter wavelength range and has a fan-shaped antenna pattern: wide in the elevation plane and narrow in the azimuth plane (6° wide), which determines the resolution of the radar detection in the azimuthal plane. Since the P-18 radar does not have the ability to resolve targets in elevation, this radar gives out only two coordinates of the detected target:

D_t: Target distance;

β_t: Target azimuth.

Compared to radars of centimeter and decimeter wavelengths, radar detecting meter range has a number of advantages that contribute to the successful solution of problems in detecting the air attack weapons of a potential enemy:

1. In the radar of the meter wavelength range, a large distance and height of air target detection can be provided for the following reasons.
    - A radar in the meter wavelength range has a target effective scattering surface of 1–2 orders greater than that of a radar in the centimeter and decimeter wavelength ranges for the same target. This is because the structural elements of a real target have dimensions commensurate with the wavelength of a meter-range radar, and when a target is irradiated with such a radar, resonance phenomena arise that increase the intensity of the reflected wave. In addition, these radars have shown themselves to be very effective in detecting targets made using stealth technology. This is because the geometry of these aircraft is designed in such a way that a minimum of reflected energy is provided in the forward hemisphere along the target's velocity vector due to the addition of radiation from different parts of the target in antiphase. The geometry of the aircraft is optimized according to the criterion of the minimum signal reflected from the target into the forward hemisphere for tracking and guidance radars of air defense missile systems operating in the centimeter wavelength range. When using the meter range, the detection radar wavelength is much greater than the linear dimensions of the target's geometric shapes, which reduces the target's effective scattering surface in the centimeter wavelength range. The geometry of the aircraft is optimized according to the criterion of the minimum signal reflected from the target into the forward hemisphere for tracking and guidance radars of air defense missile systems operating in the centimeter wavelength range. The effectiveness of the use of the meter wavelength range in detection radar can be evidenced by the fact that on March 27, 1999, the American stealth bomber F-117 was shot down by a 5V27D missile launched by the Serbian air defense system S-125M Neva in the sky over Yugoslavia [18]. The F-117 was detected by P-12 Enisey and/or P-18 Terek meter range radars. After detection, the target was accompanied by the P-18 radar, and when less than 15 km remained to the launch point of the anti-aircraft guided missile and the target was in the guaranteed defeat zone of the anti-aircraft missile system, the S-125M Neva missile guidance station was turned on, to which target designation was transmitted from the P-18 detection radar at that time. After the target was locked at auto-tracking in angles and range, two missiles were launched. After the point of meeting between the missile and the target, the F-117 bomber was shot down, and the radiation from the missile guidance station S-125M was turned off. At the same time, a false transmitter was turned on, simulating the signal of a real S-125M missile guidance station. This false transmitter was installed a few hundred meters from the position of the anti-aircraft missile system and enemy anti-radar missiles hit to the false transmitter without damaging it, since it was behind concrete fortifications.
    - The formation of the radar radiation pattern in the elevation plane is carried out using the Earth's surface. In this case, the interference phenomenon

plays an important role in the radar antenna pattern formation. Due to this phenomenon, in some elevation directions the signal power increases and, accordingly, the target detection range increases, and in some elevation directions, in contrast, minimums in the radar antenna pattern are observed and the detection range at these elevation directions will be reduced.

- Since the working (main) polarization of the detection radar is usually horizontal (i.e., the vector of the electric field strength is parallel to the Earth's surface), therefore, in such detection radars of the meter range, there are less losses for absorption and dispersion of electromagnetic energy.
- The noise coefficient of meter range radar receivers is usually slightly less than the noise coefficient of centimeter and decimeter radar receivers. Consequently, the detection range of these radars will also be somewhat greater, with other technical characteristics being equal.

2. Meter wavelength range detection radars generally have more stable tracking of detected targets, since the probability of target detection is less dependent on the range than in the centimeter range due to less irregularity of the secondary radiation pattern of the target, which has a complex geometric shape.
3. In the meter wavelength range practically no reflections from hydrometeorological formations are observed, since their dimensions are much smaller than the wavelength of the detection radar.
4. Meter wavelength range radars are usually more resistant to interference from active jammers, since it is often very difficult to place a meter range antenna with a good directional coefficient on an aircraft.
5. Meter wavelength range detection radars generally have higher absolute stability transmitting device.

Thus, the indicated advantages of meter wavelength range radars make it possible to use these radars as radars for detection, reconnaissance and target designation, and the disadvantage of the meter wavelength range associated with low resolution in angular coordinates can be compensated by working together with other detection radars of decimeter and centimeter ranges.

An example of such a combination of radar detection of the meter and decimeter wavelength ranges is the joint work on target detection of the P-18 radar and the PRV-13 radio altimeter. The PRV-13 operates in the decimeter wavelength range and has an antenna pattern that is narrow in the elevation plane, $55'$ wide, while in the azimuthal plane the antenna pattern width is 2°.

In the mode No2: SECTOR, SWAYING 30°, the antenna of the PRV-13 radio altimeter in azimuth is automatically controlled in a given apriori azimuth sector within 10°–170° with the speed of its azimuthal movement 54°/min. The swaying of the radio altimeter antenna in elevation occurs in the 30° sector. The position of the bisector of the swaying sector in the elevation plane is selected manually within 13°–17°, while zero degrees on the scale corresponds to the elevation angle of the bisector equal to 15°. Therefore, by moving the position of the bisector of the scanning sector in space, you can view the space within the sector of elevation angles from $-2°$ to $+32°$. This mode is used when searching for targets and when working with several targets located at different azimuths within a given sector.

The parameters of the PRV-13 elevation scanning sector determine for the entire complex of reconnaissance and target designation the value of the radius of the dead funnel, which is $R_{df} = 1.8H_t$, which, in turn, determines the shape of the detection zone of the integrated system as a whole (see Figure 2.7).

So when placing the PRV-13 radio altimeter on the combat position of the S-75M3 air defense system in the forest, near the village Mir, Belarus, negative closing angles were ensured by installing the B-1 PRV-13 trailer on a horizontal platform with a diameter of 10 meters, on an embankment with a height of 24 meters from the Earth's surface, which has a ramp 8 meters wide, with an angle of inclination of the ramp to the horizon of 25°. In this case, the phase center of the PRV-13 antenna was located at a height of $\approx 27m$. This turned out to be enough to ensure negative closing angles with an average pine height of 20 meters. However, over time, it was necessary to cut the tops of the pine trees around the position of the air defense missile system as they grew.

The PRV-13 can issue three coordinates of the target located in the sector viewed by the radio altimeter:

- β_t: Target azimuth;
- H_t: Target height;
- D_t: Target range.

This happens in the mode No6: SECTOR SWAYING 2°. In this mode, the antenna is automatically controlled in azimuth from the SNR-75 missile guidance station or from the P-18 radar. In elevation, the antenna is swaying with an amplitude of 3° and a frequency of 72 antenna oscillations in elevation per minute. The sector bisector can be smoothly set manually by means of the Elevation Angle knob in the range from +1° to +27°.

At the same time, the PRV-13 has the following resolution [3]:

- In elevation (ε): 1° at ranges over 200 km;
- In height (H):
  - 1 km at a distance of up to 100 km;
  - 2 km at a distance of up to 200 km.
- In azimuth (β): 2°;
- By range (D): 2 km.

Thus, the integration of the P-18 radar and the PRV-13 radio altimeter makes it possible to select targets in the battle formations of a potential enemy's aviation, provided that the azimuths of the targets differ by 2°, and their ranges differ by 2 km.

References

[1] Peresada S. A., *Anti-Aircraft Missile Systems*, Military Publishing, 1973.

[2] Garin, E. N., D. D. Dmitriev, V. N. Tyapkin, et al., in *Military Technical Training: The Device of the Radar Station of the Radio-Technical Forces of the Air Force, Radar Station P-18R*, E. N. Garin (ed.), Krasnoyarsk: Siberian Federal University 2012 http://vii.sfu-kras.ru/images/pdf/ustrojstvo-p-18.pdf.

[3] Kashcheev, M. M., D. D. Dmitriev, B. K. Sargin, et al., in *Combat Use of Subunits of Radio Engineering Troops: Mobile Radio Altimeter PRV-13*, A. D. Sosnovsky (ed.), Krasnoyarsk: Siberian Federal University, 2014, http://vii.sfu-kras.ru/images/libs/sosnovski_BPPRV13.PDF.

[4] *Tactical and Technical Characteristics of the P-18 Radar,* http://library.voenmeh.ru/jirbis2/files/materials/ifour/book3/book_on_main_page/5.3.11.htm.

[5] *S-75 -SA-2 Guideline,* http://militaryrussia.ru/blog/topic-172.html.

[6] "SAM S-75: The First Among Equals, Part III," *Aerospace Defense Magazine* http://www.vko.ru/oruzhie/zrk-s-75-pervyy-sredi-ravnyh-chast-iii.

[7] Zakrevskyi O., "Own-Stranger," https://dou.ua/forums/topic/10097/.

[8] Sidorchik A., "JSC Arguments and Facts Escape on the Flying Fox. How Pilot Belenko Betrayed His Homeland and His Family" September 6, 2016, https://aif.ru/society/history/pobeg_na_letuchey_lisice_kak_letchik_belenko_predal_rodinu_i _svoyu_semyu.

[9] Nurulin, R. N., *Basic Construction of Equipment Navigation, Topographic Reference, Orientation and Radar Identification,* PERO, 2019.

[10] Galaburda I., "Own? Stranger? How It Works," https://zen.yandex.ru/media/id/5daad4a0-c7e50c00b1294ff0/svoi-chujoi-kak-eto-rabotaetchast-1-5dabefd1e6cb9b00b112afbd.

[11] Lelikov O. P. "Basics of Calculation and Design of Parts and Units of Machines," Lecture Notes for the Course Machine Parts Mechanical Engineering (ISBN 5-217-03077-1), 2002.

[12] "SAM S-75: The First Among Equals, Part II *Aerospace Defense Magazine,* http://www.vko.ru/oruzhie/zrk-s-75-pervyy-sredi-ravnyh-chast-ii.

[13] S-400 Missile System, https://https://en.wikipedia.org/w/index.php?title=S-400_missile_system&stable=1.

[14] Pervushin A I., *Battle for the Stars:2. Cosmic Confrontation* Part I https://tech.wikireading.ru/598.

[15] "Burya" and "Dal" by Constructor Lavochkin, https://www.sovsekretno.ru/articles/burya-i-dal-konstruktora-lavochkina.

[16] "Holod" Hypersonic Flying Laboratory, http://www.testpilot.ru/russia/tsiam/holod/.

[17] NPO "Energomash" im. Academician V. P. Glushko Hypersonic Ramjet Engine (Scramjet Engine), http://engine.space/science/develop/giperzvukovoy-pryamotochnyy-vozdushno-reaktivnyydvigatel-gpvrd8700/.

[18] Sazonova, O., "'We Did Not Know that He Was Invisible': How Serbian Missilemen Made Stealth Technology Meaningless with the Help of the Soviet Air Defense System," *Balkanist,* April 16, 2020 https://balkanist.ru/my-ne-znali-chto-on-byl-nevidimyj-kak-serbskie-raketchiki-obessmysliliamerikanskuyu-tehnologiyu-stels-s-pomoshhyu-sovetskogo-zrk/.

CHAPTER 3

Firing over the Horizon

Modern radar is developing in the following directions:

1. Increasing the target detection range by means of radar;
2. Increasing the accuracy of determining the spatial coordinates of the detected air, ground, surface, and aerospace targets, as well as in improving other essential characteristics of modern radars;
3. Resolution in angular coordinates, range, and speed;
4. Increasing the noise immunity (i.e., increasing the resistance of the radar to interference organized by the enemy);
5. Increasing reliability (i.e., radar uptime probability (see Chapter 2)).

Modern air defense radars operate, in the meter and centimeter wavelength ranges, which, in accordance with the electromagnetic wave scale, refers to the ultrashort wavelength range (VHF) [1].

The currently achieved operating ranges of radars operating in the ultrashort wave range ($\lambda \leq 10\ m$) make it possible to observe targets located many hundreds, thousands, and millions of kilometers from the Earth, which made it possible to receive signals reflected from the Moon, the Sun, and the planets of the solar system [2].

However, VHF radars are usually not capable of detecting targets located in the area below the radio horizon, which is hidden from these radars by the bulge of the Earth.

3.1 Basic Principles

Since the early 1970s, active work has been underway to create over-the-horizon radars capable of detecting the launch of an intercontinental ballistic missile from a firing position located deep in the territory of a potential enemy. These radars typically use the short wavelength (HF) range.

Possible detection areas for VHF and HF radars are shown in Figure 3.1, where:

1 is the location point of VHF and HF radar;

2 is the upper boundary of the target detection area of the VHF radar;

R_d is targets maximum detection range of VHF radar;

$R_E = 6371$ km, Earth radius;

$H_i = 300$ km, the effective height of the ionosphere.

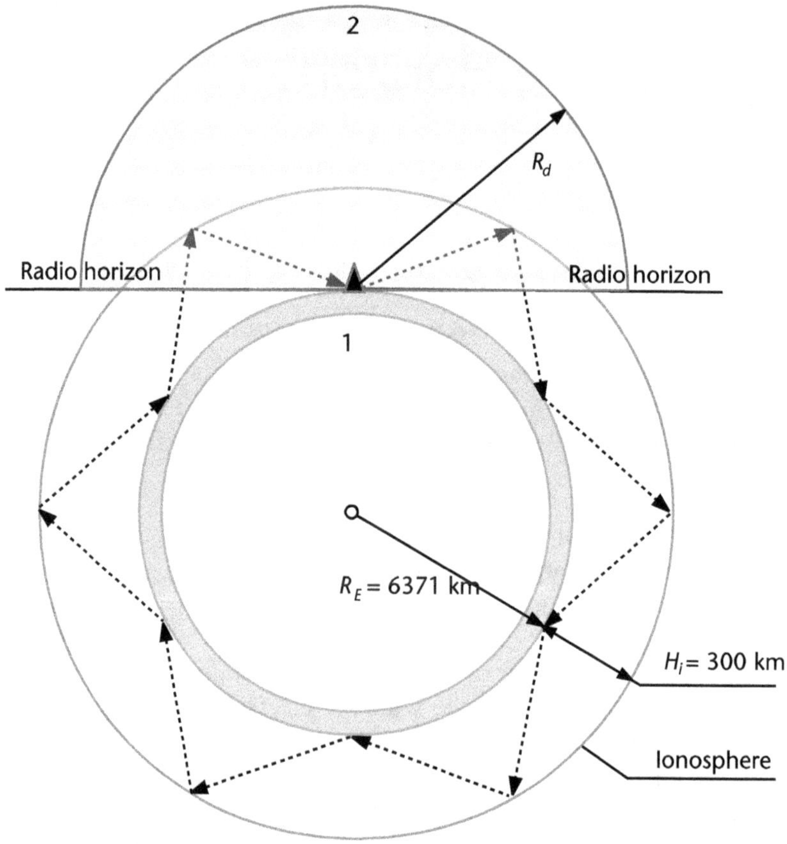

Figure 3.1 Radar detection areas of VHF and HF diapasones.

It is quite clear that the detection area of VHF radars is ideally a hemisphere with a radius R_d, equal to the maximum detection range, which is limited from below by the radio horizon. The propagation of radio waves in this range is carried out in a straight line and, therefore, its lower limit is determined by zero closing angles; that is, when the target elevation angle ϵ_t is equal to zero or slightly less than zero, if the phase center of the antenna is at a height of several tens of meters above the Earth's surface, thus the radio horizon condition can be written as:

$$\varepsilon_t \approx 0^\circ \tag{3.1}$$

It is known that the Earth does not have the correct geometric shape compared to a sphere. It is in the shape of a geoid [3]. It is generally accepted that the shape of the Earth is close to an ellipsoid, somewhat flattened along the axis of rotation (i.e., the shape of the Earth is obtained by rotating the ellipse around the minor axis). Terrestrial arc measurements are capable of yielding a value of the equatorial radius of the Earth, but satellite measurements are greatly superior for determining the flattening. After 10 years of satellite observations, the International Union of Geodesy and Geophysics adopted the Geodetic Reference System 1967, defining equatorial radius $R_{aequatorial}$ and flattening value $J_{2,0}$. Minor revisions to the numerical values were made in 1983. The revised value is $R_{aequatorial} = (6{,}378{,}136 \pm 1)$ m. The adopted value of $J_{2,0}$, corresponds to a flattening of 1/298.257223563 [3,4]. When solving most radar problems, the value of the flattening of the Earth is neglected and the Earth is taken as a ball whose volume is equal to the volume of the Earth's ellipsoid. Substituting the values of the semi-major axis a and semi-minor axis b into the formula $R_E = \sqrt[3]{a^2 b}$, we get the radius of such a ball $R_E = 6{,}370.999786$ km, and then we will take the integer value of the Earth's radius further in the calculations to $R_E = 6371$ km [5].

Unlike VHF radar, which directs its beam directly to the target, the HF radar antenna directs its beam to a specific area of the reflecting layer of the ionosphere (Figure 3.1). The beam tilt angle in the elevation plane (ϵ_t) and the operating wavelength of the radar are chosen to ensure the reflection of electromagnetic waves from the ionosphere toward the target located beyond the radio horizon. The target, in turn, reflects the energy of radio waves, which partially returns to the radar also passing through the ionosphere.

The possible target detection area of the over-the-horizon radar is limited from below by the surface of the Earth, and from above by the layer of the ionosphere. The bulge of the Earth is not an obstacle to the propagation of the shortwave radar signal. The propagation of short waves during the operation of the over-the-horizon radar can be represented as a series of reflections occurring successively from the ionized space layers—the ionosphere—located on average at an altitude of about 300 km from the Earth's surface.

The maximum length of one jump; that is, the distance on the ground between two adjacent points of radio wave reflections from the Earth's surface is approximately 3,000 km. This data determines the size of the area of near-earth space in which it is possible to detect targets by means of the over-the-horizon radar (Figure 3.1).

From a comparison of detection zones the VHF radar and HF radar built in the vertical plane, it can be concluded that in order to timely detect the means of aerospace attack of a potential enemy, it is reasonable to share these radars, combining information coming from them into a unified automated system. This system has the following tasks:

- Detection and calculation of targets tracks;
- Distribution of targets between the weapons of the air defense system;
- Issuance of target designation to a specific fire subunit to carry out combat work to destroy the selected target.

It should be noted that the solution to the problems of detecting and measuring the coordinates of a target located beyond the horizon differs from the solution of the same problems in a VHF radar. In reality, there are two types of over-the-horizon radars:

1. Colocated radar is when the transmitter and receiver of the radar signal are colocated (i.e., they are in the same geographic location).
2. Radar diversity is when the receiver is separated from the transmitter by a considerable distance.

Each type of over-the-horizon radar has its own radar signal propagation scheme. In a colocated radar, the radar signal (1) from the T transmitter is reflected

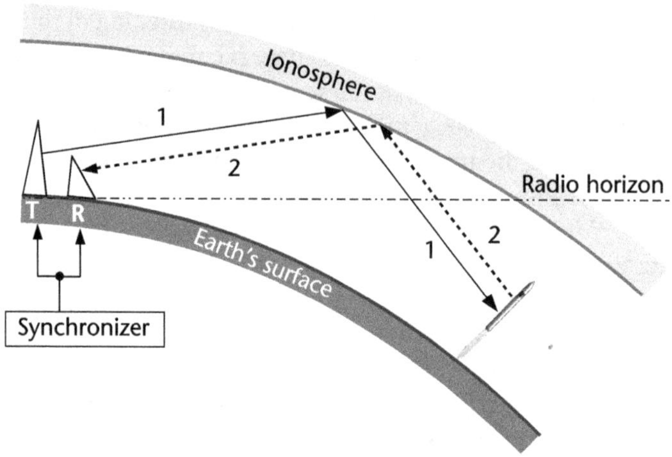

Figure 3.2 Radar with backward ionospheric scattering.

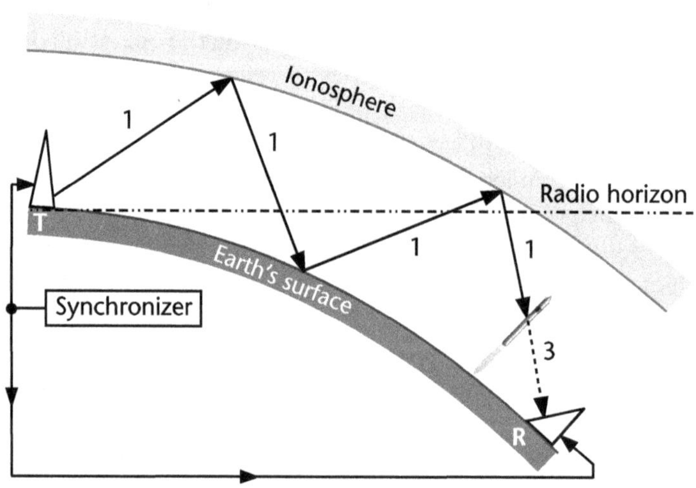

Figure 3.3 Radar with direct ionospheric scattering.

from the ionosphere, then reflected from the target (2), and then follows in the opposite direction through the ionosphere to the receiver R, which is located near the transmitter T. This type of radar is called an over-the-horizon ionospheric backscatter radar (Figure 3.2).

In a spaced system, the radar signal only travels in the forward direction, coinciding with the location of the transmitter T, the target, and the receiver R. The signal reaches the receiver sequentially, reflecting from the ionosphere and parts of the Earth's surface, while the paths of the probing signal (1) and the signal reflected from the target (3) are generally not the same. Due to the one-way signal propagation, the spaced system is called an over-the-horizon radar with forward ionospheric scatter (Figure 3.3).

In Figures 3.2 and 3.3, the following notation is used:

T: Radar transmitter;

R: Radar receiver;

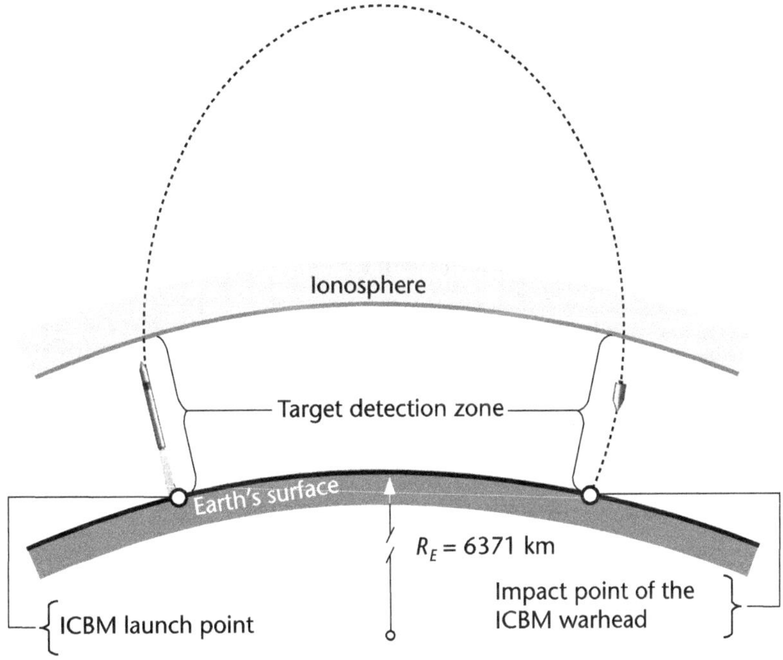

Figure 3.4 Detection zone of ICBM.

1: Probing signal;

2: A signal reflected from the target;

3: A signal containing information about the target.

The most typical application of over-the-horizon radars is the detection of ICBM launches with their help, the detection of their warheads, as well as the detection of artificial earth satellites and atomic explosions.

The detection zone of ICBMs consists of two sections of the ballistic missile's flight path: ascending and descending (Figure 3.4). It is quite clear that in order for the ballistic trajectory to be realized, it is necessary that the velocity of the missile V_t would be less than the first cosmic velocity V_{1s}, also known as the orbital velocity, which would put the missile into orbit around the Earth [6]:

$$V_t \leq V_{1s}, \tag{3.2}$$

The orbital velocity is found from the formula:

$$V_{1s} = \sqrt{G \times \frac{M_E}{R_E}} = 7910.5665597796 \text{ [m/s]} \approx 7.9 \text{ [km/s]}, \tag{3.3}$$

Here $G = 6.67430 \times 10^{-11}$ is Newton's gravitational constant $[\frac{\text{m}^3}{\text{kg}\cdot\text{s}^2}]$ [7], $M_E = 5.9724 \times 10^{24}$ [kg] is the mass of the Earth [5], and $R_E = 6.370 \times 10^6$ [m] is the radius of the Earth.

The velocity that a missile must reach in order to escape from the Earth's gravitational field and fly to other planets is the so-called escape velocity from Earth or the second cosmic velocity V_{2s}. Taking into account that in the initial state the missile speed is equal to V_{2s}, and in its final state the missile will be at an infinite distance from the Earth, and its speed will be negligible and taken equal to zero, therefore the potential energy at the end of the path will be zero due to infinite distance, and the kinetic energy at the end of the path will be zero due to zero velocity. According to the law of conservation of energy, the sum of the kinetic and potential energy at the beginning of the path must be equal to the sum of the kinetic and potential energy at the end of the path. Therefore, we can write:

$$\frac{1}{2}mV_{2s}^2 + \frac{-mGM_E}{R_E} = 0 + 0, \qquad (3.4)$$

after excluding the missile mass m from the equation, the expression for the escape velocity V_{2s} takes the form:

$$V_{2s} = \sqrt{2G \times \frac{M_E}{R_E}} = 11187.2305148954 \text{ [m/s]} \approx 11.2 \text{ [km/s]}, \qquad (3.5)$$

Thus, the velocities of movement of artificial earth satellites will be in the interval:

$$V_{1s} \leq V_t < V_{2s} \qquad (3.6)$$

The detection zone of satellites by means of over-the-horizon radar is an area of space through which the satellite trajectory passes, the height of which is located between the Earth's surface and the reflecting layer of the ionosphere (Figure 3.5).

Target detection by means of over-the-horizon radar on the descending segment of the trajectory of an intercontinental ballistic missile, when its velocity $V_t \approx$ 6.5 [km/s] corresponds to the reaction time of the anti-aircraft missile system $t_r =$ 41.53 s (Chapter 2), subject to the detonation of a target at an altitude of at least 30 km from the Earth's surface, which should correspond to the minimum damage when the enemy uses a nuclear warhead, makes intercepting such a target a very difficult task. Therefore, over-the-horizon radars cannot be used as a means of reconnaissance and target designation for anti-aircraft and missile defense systems. It is advisable to use these radars as missile attack warning devices included in the general missile attack warning system because their main task is to detect the moment of ICBM launch and the approximate coordinates of this launch.

For example, the North American Air Defense Command (NORAD) is obliged to ensure completion of the following tasks [8]:

- Tactical warning and attack assessment of bomber or ballistic missile attack on North America;
- Space surveillance, tracking, and cataloging of all human-made objects in space;
- Peacetime surveillance, detection, and identification of aircraft;
- Operational control of U.S. and Canadian air defence forces.

NORAD also has the responsibility to the United States Joint Chiefs of Staff (JCS) to provide worldwide detection of missile launches and nuclear events. This

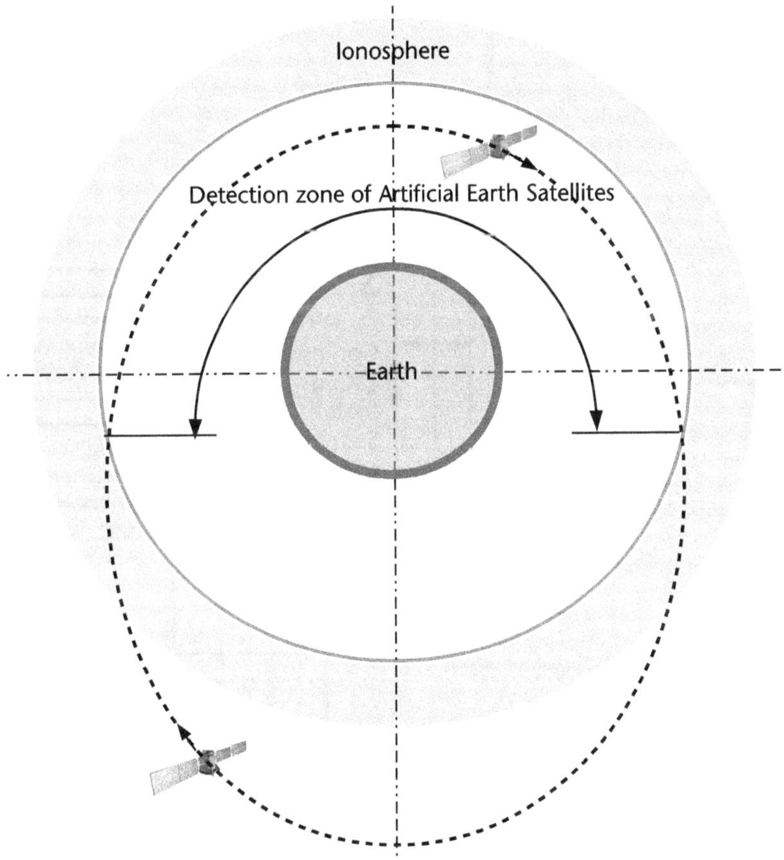

Figure 3.5 Detection zone of artificial Earth satellites.

includes the Pacific area and Europe, as well as the area adjacent to the North American continent.

To accomplish this, NORAD exercises operational control of the detection and communications systems and operates and maintains the analytical systems in the NORAD Cheyenne Mountain Complex that provide the National Command Authorities and Commander in Chief of the Strategic Air Command with real-time missile warning messages and NORAD confidence assessments. New NORAD requirements are to provide satellite attack warning and verification for all U.S. satellites and support the Space Shuttle missions. The NORAD command post is in the underground Cheyenne Mountain Complex in Colorado. Missile warning information gathered from a worldwide network of sensors is processed on the 427M system computers and warning messages are distributed to U.S. and Canadian command posts as depicted in Figure3.6.

The core of the NORAD Missile Warning System is the 427M computer system. The 427M system is divided into three main segments:

1. The Communication System Segment (CSS);
2. The NORAD Computer System (NCS);
3. The Space Computational Center (SCC).

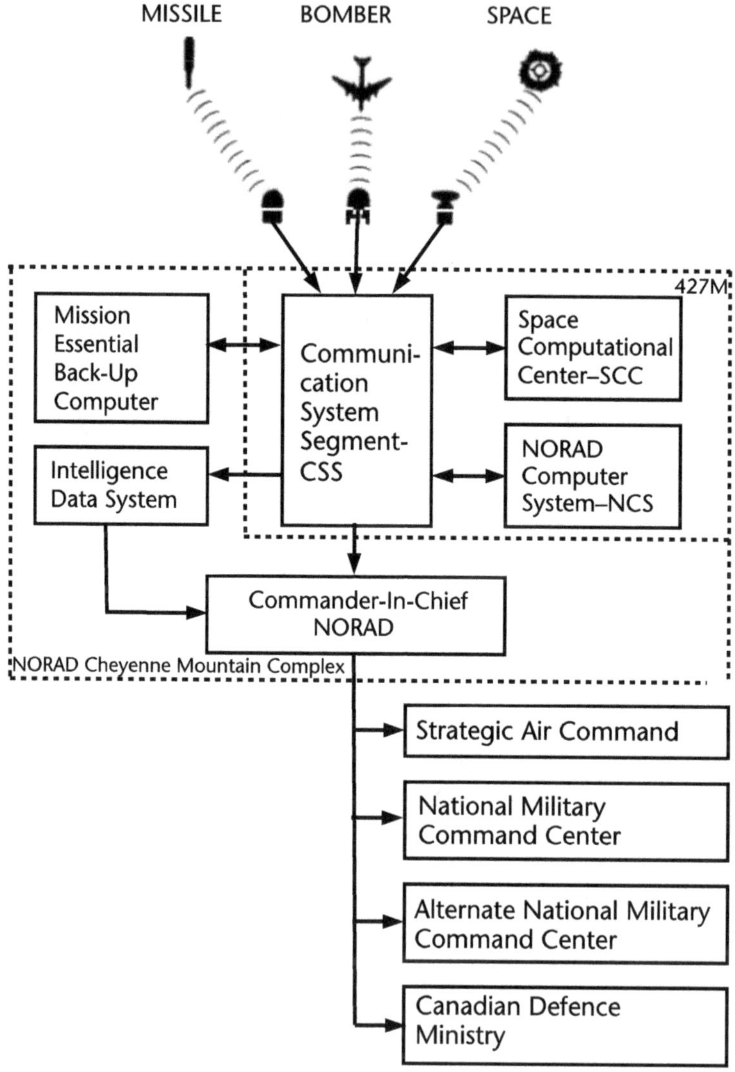

Figure 3.6 NORAD Missile Warning System.

The CSS provides essential communications support for NORAD Headquarters. It interfaces with all external facilities and sensors serving or served by NORAD as well as internal circuits. CSS provides complete message processing, formatting, line code conversions, and routing of messages. Due to critical availability requirements, one of the two computers processes live data while the other runs in a backup mode. Since it controls all communications, CSS is the single most critical element of the 427M system.

The NCS provides central command and control information for NORAD for the aerospace defense of North America. Its highly time-sensitive mission essential functions include the receipt, processing, display, and output of missile warning data; nuclear detonation reports; aircraft surveillance and warning; and weapons and sensor systems status. Its less time-sensitive functions include logistics, personnel, airfield data, and environmental support.

The SCC is the focal point for integrated U.S. and Canadian activities that detect and track all human-made satellites in space. This includes the entire satellite life cycle: launch, orbit injection, normal tracking, maneuver, decay, orbital decay, and impact prediction. The SCC also maintains the official United Nations inventory of all human-made orbiting satellites. The development of satellite attack warning requirements and support for the employment of space defense weapons and the Space Shuttle have also been added to the SCC mission.

The task of determining the spatial coordinates of the target is the main task of the Intelligence Data System, which makes a decision on the detection of a target and provides information about the measured coordinates of the detected target through the NORAD Commander-in-Chief to the appropriate command post, for example, to the command post of the Strategic Air Command.

This task is reduced to measuring the values of geometric coordinates that uniquely characterize the spatial position of the target relative to the location point A of the over-the-horizon radar (Figure 3.7(a)). An over-the-horizon backscatter radar typically uses a polar coordinate system $D_t, \varphi\}$ (Figure 3.7(b)). The distance from the transceiver center of the over-the-horizon radar located at point A to the target C along the broken line ABC is the distance to the target along the beam D_b. This distance is measured by the time delay of the signal reflected from the target.

The line of the shortest distance between point A and the projection of point C onto the surface of the globe (point D), equal to $|AD|$, is part of the great circle with the radius of the globe R_E, which is the distance to the target on the ground D_t (Figure 3.7(b)).

The target azimuth line φ, which lies in the horizontal plane, is aligned with the meridian passing through the location of the over-the-horizon radar (point A). The north direction is taken as the initial direction when measuring the target azimuth φ by any over-the-horizon radar (Figure 3.7(b)). The target distance on the ground, D_t, is used for mapping the flight path of the object. D_t values are obtained, as a rule, by numerically solving the corresponding analytical expressions. It should be noted that the difference between the calculated ranges along the beam D_b and along the ground D_t decreases with increasing range along the beam D_b. Consequently, for ground-based, and in some cases for small-altitude objects, the range along the beam D_b can be replaced by the range along the ground D_t with an accuracy acceptable for practical calculations [9]. Target flight altitude h (Figure 3.7(a)), which is required for unambiguous determination of its location, is not measured in over-the-horizon backscatter radars.

One of the many reasons that it is difficult to determine the height of the target is the large number of highly complex paths for propagation of radio waves through the ionosphere. This makes it impossible to accurately identify the areas of reflection of the electromagnetic wave along the entire path of its movement from the radar to the target and back. An over-the-horizon forward scatter radar (Figure 3.3) defines only one coordinate of the object -φ- azimuth. Determining the location of the target in this case is carried out by the triangulation method; that is, by simultaneous direction finding of the target by three receiving points, which are located relative to the radar transmitter in a certain way. The measured bearings are sent to the computing center, which determines the polar coordinates of the target. It is quite clear that the operation of over-the-horizon radars should be integrated

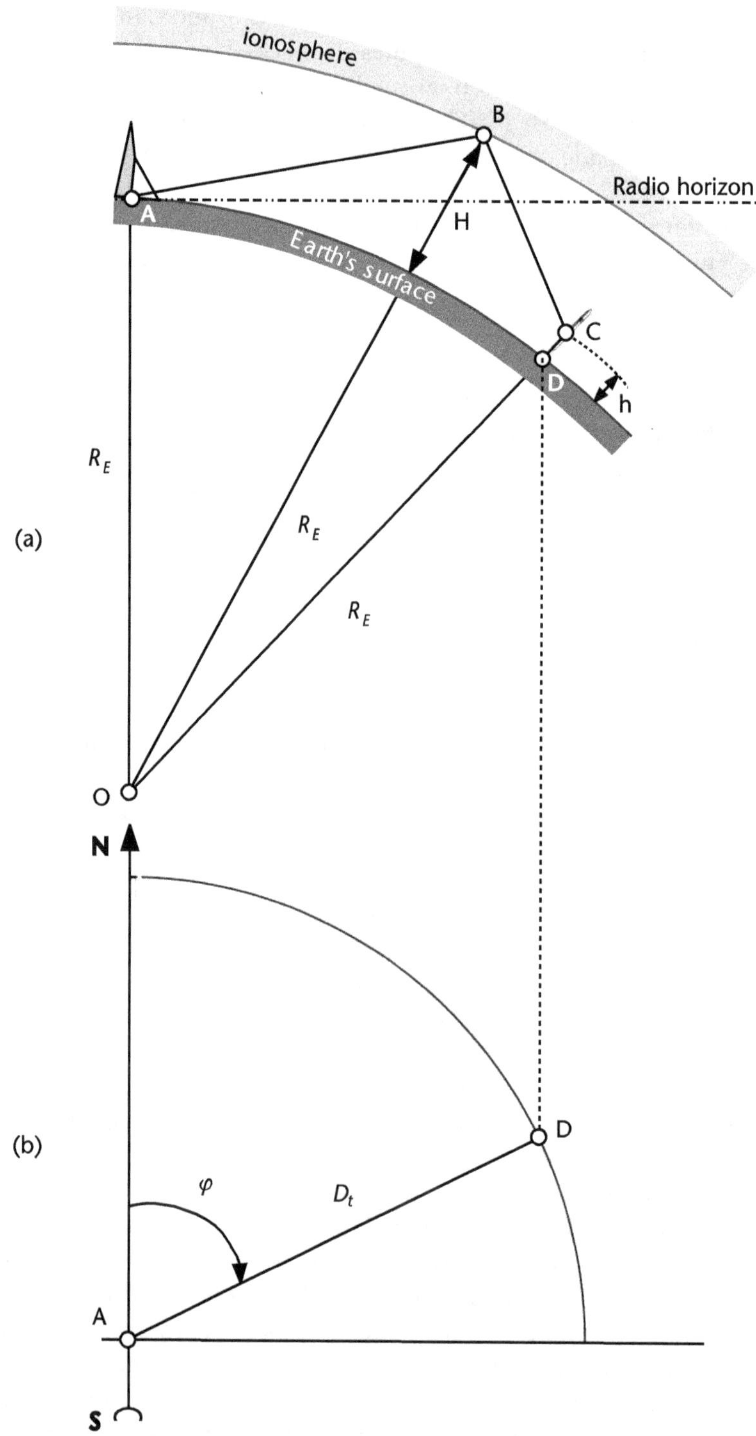

Figure 3.7 Target coordinates determined by over-the-horizon radar from point A.

into the general early-warning system missile attack. The concept of an integrated early-warning system that would include radars as well as satellites and that would be capable of providing the command and control of the strategic forces with the capability necessary for implementing a launch-on-warning option did not appear until about 1972. The draft project, prepared in 1972, called for development of an integrated early-warning system that would include above-the-horizon and over-the-horizon (OTH) radars, as well as early-warning satellites. The early-warning satellites and over-the-horizon radars were supposed to detect launches of ballistic missiles during boost phase of their flight (i.e., almost immediately after their launch). The satellites rely on infrared sensors that can directly detect radiation emitted by the missile plume. An over-the-horizon radar, like all radars, detects reflections of electromagnetic signal that it sends in the direction of a target. OTH radars deployed on the Soviet territory were able to detect missile launches on the territory of the United States, due to the fact that their signal, undergoing reflections from the Earth's ionosphere, could propagate over long distances [10]. In 1972, within the framework of the concept of an integrated missile attack warning system, it was decided to start building several radars of various types at once, including two stations of the 5N32 Duga project—one near Chernobyl, and near Komsomolsk-on-Amur. In the same year, the construction of the Chernobyl station began, and in 1976 it went into operation.

The NATO Reporting Name for the over the horizon radar Duga is often referred to as STEEL YARD, and also Russian Woodpecker or Woodpecker. This radar was built not far from the Chernobyl nuclear power plant: due to the large losses when the signal passed more than 3,000 km to the target and back, it was required to provide a huge pulse power, so the station was consuming voracious amount of energy. The receiving complex of antennas—Chernobyl-2—consumed about 10 MW, and the estimates of the consumption of the transmitter Lubech-1 in the Chernihiv region vary greatly, from 8 to 160 MW per pulse. Both Duga stations—located in Ukraine and in the Far East—blocked the entire U.S. territory through the North Pole (Figure 3.8). An experimental Duga-N near Nikolaev shone in the direction of China, which confirmed the effectiveness of the project by tracking the launches at Baikonur.

The viewing angle of each station was 50°–75° depending on the current state of the ionosphere along the signal path.

The antennas of the Duga over-the-horizon radar were built on the principle of a phased array. They were an openwork wall made of steel structures 85m high. The radar operated in the frequency range of $3.26 - 17.54$ MHz ($\lambda \approx 92.02 - 17.103$m). There were two receiving antennas in the Duga over-the-horizon radar: one antenna 90m high and about 250m long received a higher-frequency signal, and the second antenna 140m high and about 450m long received a lower frequency signal.

It should be noted that the ionosphere at high latitudes is very unstable, and the signal losses and distortions on a two-hop route turn out to be quite large. In 1983, the modernization of the Chernobyl-2 radar began in order to increase the efficiency of the radar when operating through the North Pole of the planet.

This modernization was planned to be completed at the end of 1986. At the beginning of the 1986, Duga near Chernobyl began to confidently detect launches

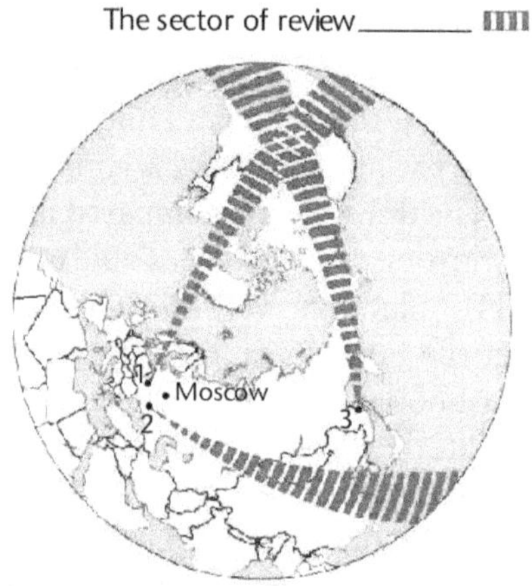

Figure 3.8 Sectors of review for OTH radars.

Figure 3.9 OTH radar antennas.

of shuttles and ICBMs at a range of 7,000–9,000 km. It even registered the explosion of the *Challenger* in January 1986. However, the modernization was never completed—on April 26, the Chernobyl nuclear power plant exploded, which is located only 9 km north of the radar [11].

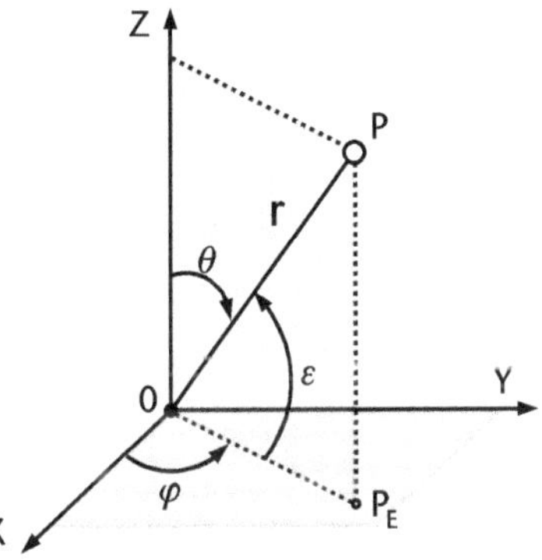

Figure 3.10 Target coordinates, which are determined by VHF radar from point 0.

In 1988, the U.S. Federal Communications Commission (FCC) conducted a study on the Woodpecker radar signal. Data analysis showed an inter-pulse period of about 90 ms, a frequency range of 7 to 19 MHz, a bandwidth of 0.02 to 0.8 MHz, and typical transmission time of 7 minutes. The signal was observed using three repetition rates: 10, 16, and 20 Hz. The most common rate was 10 Hz, while the 16- and 20-Hz modes were rather rare. The pulses transmitted by the Woodpecker had a wide bandwidth, typically 40 kHz [12]. This radar signal contained a clearly recognizable structure in each pulse, which was eventually identified as a 31-bit pseudorandom binary sequence, with a bit-width of 100 μs resulting in a 3.1-ms pulse [13]. This sequence is usable for a 100-μs chirped pulse amplification system, giving a resolution of 15 km (10 mi) (i.e., the distance light travels in 50 μs). It is therefore quite obvious that the operation of over-the-horizon radars should be integrated into the general missile attack warning system, which includes space detection equipment and VHF radars.

For any VHF radar, the coordinates of the target are determined in the local coordinate system in which a spherical coordinate system is usually used. The position of any point P in a spherical coordinate system is determined by three quantities: r, θ, φ, where: $r \geq 0$ is the distance from coordinates origin to the given point P (i.e., $r = |0P|$), and $0° \leq \theta \leq 180°$ is the angle between the normal to the surface XOY $\vec{n} = 0\vec{Z}$ (axis $0\vec{Z}$) and the direction $|0\vec{P}|$ to the given point P. The angle θ lies in the vertical plane $Z0P$ passing through the radar station (point 0).

In radars that are part of the anti-aircraft missile battalion, as the coordinate characterizing the position of the target (point P) in the vertical plane, ϵ is used as the target elevation angle, which is related to the angle θ by the following relationship:

$$\theta = 90° - \varepsilon \tag{3.7}$$

In this case, the target elevation angle is in the range $-90° \leq \varepsilon \leq +90°$.

$0° \leq \varphi < 360°$ is the target azimuth, that is, this is the angle between the direction to North (axis $0\vec{X}$) and the projection of the vector $|\vec{r}| = |0\vec{P}|$ onto the plane $X0Y$, representing is a segment connecting the origin 0 with point P_E. The angle φ lies in the horizontal plane $X0Y$.

The origin of this coordinate system is located at the radar station location point (point 0, Figure 3.10), which is characterized by geographical coordinates: latitude ϕ and longitude α, which uniquely determine the position of the radar on the Earth's

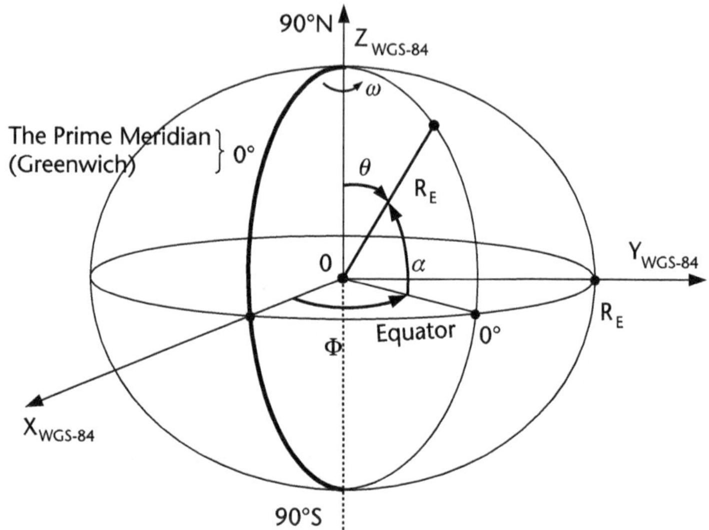

Figure 3.11 WGS-84 reference frame.

surface (Figure 3.11). At present, the satellite navigation system Global Positioning System ((GPS), United States) and/or GLONASS (Russia) makes it possible to determine with great accuracy the latitude ϕ and longitude α of the GPS signal receiver installed on the radar station. GPS uses the 1984 World Geodetic System (WGS-84). The world system WGS-84 is an astronomical-geodetic-gravimetric reference system inscribed in the figure of the Earth [14]. Latitude ϕ and longitude α are angular values that uniquely determine the position of terrain points on the surface of the Earth's sphere relative to the plane of the equator $X_{\mathtt{WGS-84}}0Y_{\mathtt{WGS-84}}$ and the prime meridian, which is considered to be the Greenwich meridian, passing through the astronomical observatory in Greenwich near London (Figure 3.11) [14,15]. Determining the location of the receiver of signals from satellites using GPS or GLONASS is carried out by the spatial intersection method.

Satellites act as control points, the coordinates of which are always known, and they are predicted and calculated for each moment in time. The distance to the satellite is determined by the time delay of the radio signal emitted by the satellite. The GPS receiver (GLONASS), installed in this case on the radar, uses an accurate clock that is corrected by the atomic clock of the satellites.

The accuracy of determining the coordinates in an absolute way using GPS in the Standard Positioning Service (SPS) mode, available to all consumers, is several meters [16]. The Precise Positioning Service (PPS) mode is intended exclusively for use by armed U.S. forces, U.S. federal agencies, and some Allied forces. The positioning accuracy in PPS mode is one or two orders of magnitude higher than in SPS mode. The accuracy depends on the code and the length of the pseudorandom sequence of the navigation signal, which is emitted by each spacecraft in the system, as well as from its processing algorithm.

Let the target move from North to South in the direction of an anti-aircraft missile battalion at an altitude of 35,000 ft (10.668 km) starting from a distance of 1,000 km from its radar located at point **A** (Figure 3.12). In this case, the target is under the radio horizon line at an altitude of −68.168 km. It is quite clear that the

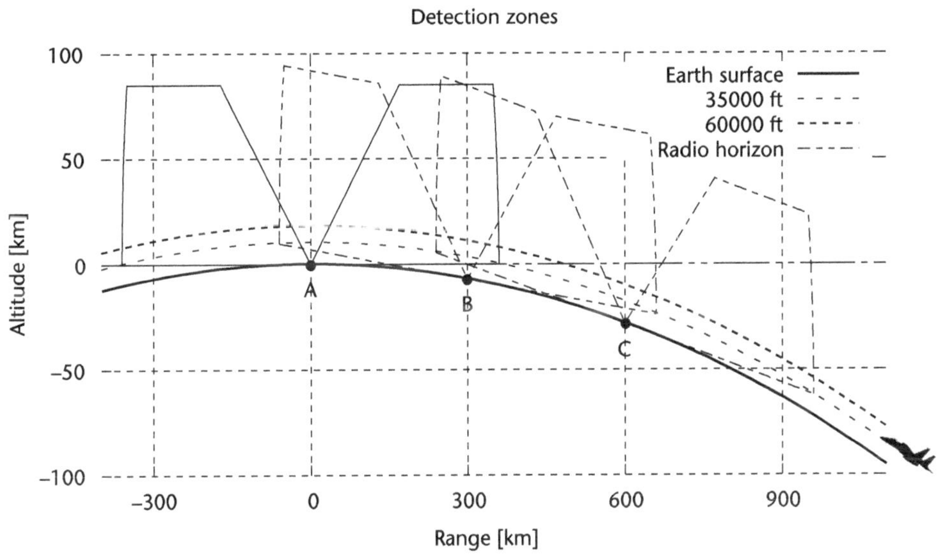

Figure 3.12 Detection zones.

detection radar located at point **A**, does not see this target due to the curvature of the Earth.

Taking into account the growing danger from hypersonic targets, at target speed $V_t \approx 5.1\text{M} = 1668.1$ m/s, its flying time from the 1,000 km line will be $t_f = 592.38$s or 9.87 min. In this case, the flight time of the target to point **A** after its timely detection at a distance of 360 km will be $t_d = 213.26$s, and the target flight time outside the radar detection zone will be $t_{oth} = 379.12$s. In order to defeat such a target at a range of $D_{max} = 380$ km with a 48N6E missile in the S-400 anti-aircraft missile system, which has a missile speed of $V_m = 1190$ m/s, it is necessary to detect this target at a range $D_{detection}$ no closer than 919.057 km:

$$D_{detection} = D_{max} + \frac{V_t \times D_{max}}{V_m} \tag{3.8}$$

We can see that a radar detection network is needed to shoot beyond the radio horizon. In order to avoid missing targets, it is necessary to overlap the detection zones of these radars (Figures 3.12 and 3.13). All elements of this network must be united by a single communication system and a single computer system. In addition, it is desirable that the detection radars lay routes in the mode of determining the three coordinates of targets and transmit information about the location of targets to the command post of the anti-aircraft missile battalion in automatic mode. Each of the radars, for example the radar at point **A**, has corresponding GPS coordinates: latitude ϕ_A and longitude α_A.

Let the location line of detection radars A-B-C be directed to the North, therefore $\phi = Const$ and coordinate transformations can be performed only in the vertical plane $X0Z$ of the receiving base of each radar (Figure 3.10):

$$\Delta_{AB} = \alpha_B - \alpha_A = X_{AB}/R_E;$$
$$\Delta_{AC} = \alpha_C - \alpha_A = X_{AC}/R_E.$$

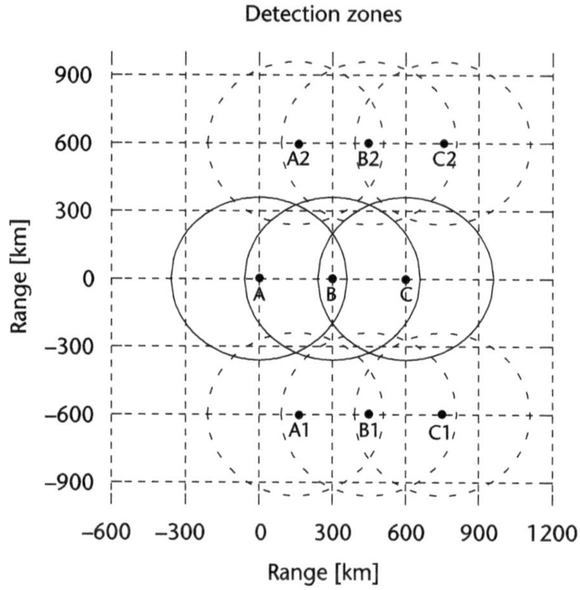

Figure 3.13 Detection zones.

The coordinates of the target in the radar basis located at point **A** are obtained from the coordinates of the target in the radar basis located at the point **B** as follows:

$$X_A = X_B\cos(\Delta) - H_B\sin(\Delta) - X_{0B}; \quad H_A = X_B\sin(\Delta) - H_B\cos(\Delta) - Z_{0B},$$

where: $\Delta = \Delta_{AB}$; $X_{0B} = 300$ km; $Z_{0B} = \sqrt{R_E^2 - X_{0B}^2} - R_E = -7.067175$ km, and $R_E = 6371$ km.

X_B is the horizontal target range along the $0X$ axis, and H_B is the target height along the OZ axis in the radar receiving basis located at the point **B** (Figure 3.10). Similar calculations are carried out on the coordinates coming from the radar located at the point **C**: $X_{0C} = 600$ km, $\Delta = \Delta_{AC}$ $Z_{0C} = \sqrt{R_E^2 - X_{0C}^2} - R_E = -28.3159467$ km. When the detection radar is located with a distance between the radar of 300 km, the continuity of target detection is ensured practically over the entire possible range of heights in the $Y0Z$ plane. When radar stations are located with a distance of 600 km between them, a zone of invisibility of targets is observed when they are behind the radio horizon. The equations of the lower boundary of the radar detection zones located at the points **A**, **B**, and **C** are written as:

$$\begin{aligned} H_A &= Z_A = 0 \\ H_B &= -\Delta_{AB} \cdot D_{\text{detection}} + Z_B \\ H_C &= -\Delta_{AC} \cdot D_{\text{detection}} + Z_C \end{aligned} \tag{3.9}$$

Accordingly, the range of the beginning and end of the invisibility zone in the direction of the target movement for $|X_C - X_B| = |X_B - X_A| = 600$ km will have the following values:

$$D_{begin} = D_{BC} = \frac{Z_C - Z_B}{\Delta_{BC}} = 448.422 \text{ km}$$
$$D_{middle} = D_B = \frac{-Z_C}{\Delta_{AC}} = 298.900 \text{ km} \tag{3.10}$$
$$D_{end} = D_{AB} = \frac{Z_B}{\Delta_{AB}} = 150.027 \text{ km}$$

In this case, the total length of the invisibility zone is $D_{Stealth} = 298.395$ km, and the maximum height of the invisibility zone is at the point B and is $H_{Stealth} = 7.067175$ km. It is quite clear that such a location of the detection radar is unacceptable to ensure firing over the radio horizon.

Thus, the first condition for ensuring over the horizon firing is availability of reliable radar information on targets located beyond the radio horizon of the radar for tracking and guiding missiles of the anti-aircraft missile battalion, located at ranges not lower than the ranges that ensure the defeat of hypersonic targets on the far border of the affected area of this anti-aircraft missile battalion, according to the expression (3.8).

3.2 Estimate of Ballistic and Flight Characteristics of the Missile According to Data From Open Sources

As mentioned in Section 2.2 in Chapter 2, in order to secure the affected area, which is indicated in the performance characteristics of the air defense missile system, it is necessary that its anti-aircraft guided missile can reach all points in SAM defeat area, but still more must be done.

Suppose that the requirement (3.8) is currently implemented by creating a single radar field with a probability level of correct target detection not lower than the specified one, by means of information interfacing of a long-range anti-aircraft missile system with a large number of ground-, air-, and space-based information sources. One of the additional sources of information for a long-range anti-aircraft missile system with the ability to fire over the radio horizon is the airborne early warning and control (AEW&C) aircraft complex, also called the airborne warning and control system ((AWACS) United States) or a similar system A-50 (Russia). Because the antenna of the circular-looking radar station installed on board rises to the flight altitude of the carrier aircraft, the line-of-sight range of this radar increases. It is logical to assume that such an aviation complex carries out reconnaissance of the enemy air in order to support the combat operations of fighter aircraft and long-range anti-aircraft missile systems. In addition, having increased survivability in comparison with ground-based radars, this complex can be used to build up and quickly restore the radar field. It is quite clear that with the location of the patrol zone of the airborne complex of the radar patrol and guidance at an altitude of $H_{AWACS} = 35000$ ft or $H_{AWACS} = 10.668$m directly above the position of the anti-aircraft missile system (point A, Figures 3.12 and 3.13), detection range of D_{det_AWACS} targets flying at the same altitude $H_t = H_{AWACS}$ for an early warning and guidance aircraft will be:

$$D_{det\_AWACS} = 2 \times \sqrt{(R_E + H_{AWACS})^2 - R_E^2} = 737.687 \text{ km} \tag{3.11}$$

Obviously, in order to ensure the timely detection of targets by means of an aircraft complex for long-range radar detection and guidance, the following condition must be met:

$$D_{det\_AWACS} \geq D_{det\_min} = 960.706 \text{ km}, \tag{3.12}$$

therefore, to ensure the fulfillment of the condition (3.12), it is necessary to move the near border of the loitering zone forward by at least $D_{AWACS} = 223.019$ km at $H_{AWACS} = 10.668$m, or raise a carrier aircraft of a high-altitude all-round radar. Quadratic equation:

$$-(H_{AWACS})^2 - 2R_E H_{AWACS} + (D_{det\_min} - D_{det\_t})^2 \tag{3.13}$$

where $D_{det\_t} = 377.587$ km is the detection range of a target flying at an altitude of $H_t = 10.668$m, when the phase center of the multifunctional radar antenna is located at point **A** at an altitude of $H_A = 6$m has the following solution:

$$H_{AWACS} = 26.630 \text{ km}, \qquad for \quad D_{AWACS} = 0 \text{ km} \tag{3.14}$$

Obviously, removing the loitering zone of a long-range radar detection and guidance aircraft capable of controlling and directing anti-aircraft guided missiles to air targets at a distance of $D_{AWACS} = 223.019$ km at $H_{AWACS} = 10.668$ km toward the target is impractical for safety reasons, and the rise of this aircraft to an altitude of $H_{AWACS} = 26.630$ km at $D_{AWACS} = 0$ km from the point where the multifunctional radar for the guidance of anti-aircraft missile systems (point A, Figure 3.12), is currently technically impossible.

Thus, the only solution capable of providing the mode of firing anti-aircraft guided missiles beyond the radio horizon is an algorithm that implements the following principles:

1. Availability of reliable and high-precision coordinate information about targets flying beyond the radio horizon relatively multifunctional radar anti-aircraft missile system;
2. Use of an active homing missile capable of autonomously performing additional search and homing at a target designated for destruction;
3. Implementation of the missile's flight and ballistic characteristics, providing reference kinematic trajectories realizing the height of the start of homing $H_0 = 26$ km and the maximum range of the point start homing $D_{0\_max} = 350$ km relative to the multifunctional radar;
4. Implementation of the given power-to-weight ratio of the missile, which provides homing at the start point missile flight speed not less than $V^m_{0\_min} = 1{,}000$ m/s;

5. Availability of a stable communication line multifunctional radar-missile for transfer to the missile board the spatial coordinates of the target and the subsequent orientation of the antenna of the active homing head missiles in the direction of the target;
6. Search for a target by angular coordinates in a given solid angle;
7. Target detection;
8. Acquiring a target for automatic tracking by angular coordinates;
9. Approach of the missile to the target with the minimum miss $h_{min} < 100$ m;
10. Coordination of the area of dispersion of warhead fragments with the position of the target in space and the subsequent detonation of the warhead, causing maximum damage to the target.

Taking into account the requirement to ensure the survivability of the firing complex achieved by its mobility, deployment to a combat position and transfer to a stowed position equal to 5 minutes, then for missiles of a long-range anti-aircraft missile system, it is advisable to choose a single-stage scheme, a solid-fuel power plant and placement of missiles in transport-launch containers (TLCs). It is clear that in the firing position, the transport and launch containers should be located vertically. When launched, the missile is catapulted from the transport and launch container to a height of 20–25 meters, after which the propulsion engine is turned on, and the solid fuel is fully developed and must form a specific impulse sufficient to provide the entire spectrum of reference trajectories and firing at the maximum range for the radio horizon.

The reference trajectories when shooting beyond the radio horizon are ballistic trajectories for firing to the designated point. The parameters of the ballistic trajectory are calculated on the basis of apriori radar information and the estimated target location and the calculated homing point are entered into the onboard computer complex of the missile immediately before its launch.

Difference equations of motion of the missile for firing into the calculated point of the start of homing can be written in the following form:

$$\begin{cases} x_i^m = x_{i-1}^m + \dfrac{x_i^m - x_{i-1}^m}{C} \times V_0^m dt - \dfrac{F_{air}(x_i^m - x_{i-1}^m)dt^2}{2M(t)C} \\ z_i^m = z_{i-1}^m + \dfrac{z_i^m - z_{i-1}^m}{C} \times V_0^m dt - \dfrac{g(dt)^2}{2} - \dfrac{F_{air}(z_i^m - z_{i-1}^m)dt^2}{2M(t)C} \\ x_0^m = 0 \\ z_0^m = 0 \\ V_0^m = 1000 \end{cases} \tag{3.15}$$

where x_i^m, z_i^m - missile coordinates, m; V_0^m - given missile speed, m/s; $g = 9.81$, free fall acceleration, m/s^2; dt, time differentiation step, s; $C = \sqrt{(x_i^m - x_{i-1}^m)^2 + (z_i^m - z_{i-1}^m)^2}$ is the modulus of the distance between two adjacent points of the missile trajectory, m; $F_{air} = \frac{1}{2} \times C_x \times S_{mid} \times \rho \times (V_i^m)^2$ is the resistance force of the air environment; C_x is the missile drag coefficient, Ns^2/m kg;

S_{mid} is the missile midsection area, m^2; $\rho = 1.29$ kg/m^3 is the air density, for temperature $t^\circ = 0°$C; $M(t) = M_1 - \frac{M1-M2}{t_{trd}} \times t$ is the assumption that the mass solid fuel decreases during engine operation according to the linear law; t_{trd} is the engine operation time until complete fuel burnout, s; $M_1 = 1900$ is the total mass of the missile before launch, kg; $M_2 = 636$ is the mass of the missile after full burnout fuel, kg.

The C_x coefficient generally consists of two components: $C_x = C_{x0} + C_{x\alpha\delta}$, where $C_{x\alpha\delta}$ is the drag coefficient at the angles of attack α and the angles of deflection of the rudders δ, while the angle of attack is understood in the sense of the angular distance between the velocity vector of the missile and the vector of its longitudinal axis.

In general form, the missile drag coefficient can be represented as the sum of three terms: $C_x = C_{xb} + C_{xw} + C_{xc}$, where: $C_{xb} = C_{xb0} + C_{xb\alpha\delta}$ is the missile body drag coefficient; $C_{xw} = C_{xw0} + C_{xw\alpha\delta}$ is the missile wing drag coefficient; $C_{xc} = C_{xc0} + C_{xc\alpha\delta}$ is the drag coefficient of the missile tail.

Taking into account the fact that the wings in their classical form, as, for example, on the S-200 missile or on the S-75 missile in this case are absent, the transverse size of the tail does not exceed 0.4m, oppositely located in the lower part of the missile body. The output to a given trajectory of movement is carried out by gas rudders that control the position of the thrust vector on the active section of the trajectory with duration $t_{trd} = 12$s, so therefore we can assume:

$$C_{xw} = 0, \qquad C_{xc} = 0, \qquad C_{xb\alpha\delta} = 0, \tag{3.16}$$

thus: $C_x = C_{xb0}$, where $C_{xb0} = C_{xfb} + C_{xwb} + C_{xdb}$.

Herewith: C_{xfb} is the frictional resistance, characterized by the behavior of the boundary layer of air; C_{xwb} is the wave resistance; C_{xdb} is the resistance of the bottom. In order to simplify the analysis, it is reasonable to accept:

$$C_{xwb} = 0, \qquad C_{xdb} = 0, \tag{3.17}$$

And in this case, the frictional resistance will be characterized by the behavior of the boundary layer of air, which in turn can be represented by the process of flow around a flat plate with the drag coefficient of the flat plate $\overline{C_{xfb}}$:

$$C_x = \overline{C_{xfb}} \times \frac{F_b}{S_M} \times \eta_M \times \eta_0, \tag{3.18}$$

where:

$$\overline{C_{xfb}} = \begin{cases} \dfrac{0.455}{(\lg Re)^{2.58}} &, \forall Re: \ 10^5 < Re \leq 10^9 \\ \dfrac{0.074}{(Re)^{0.2}} &, \forall Re: \ Re > 10^9 \end{cases} \tag{3.19}$$

$$Re = \frac{V_0^m L_m \rho}{\eta}, \tag{3.20}$$

where Re is thc Reynolds number; $V_0^m = 1{,}000$ is the flow velocity (given missile velocity), m/s; $L_m = 7.5$ is the missile length, m; $\rho = 1.2920$ is the air density,

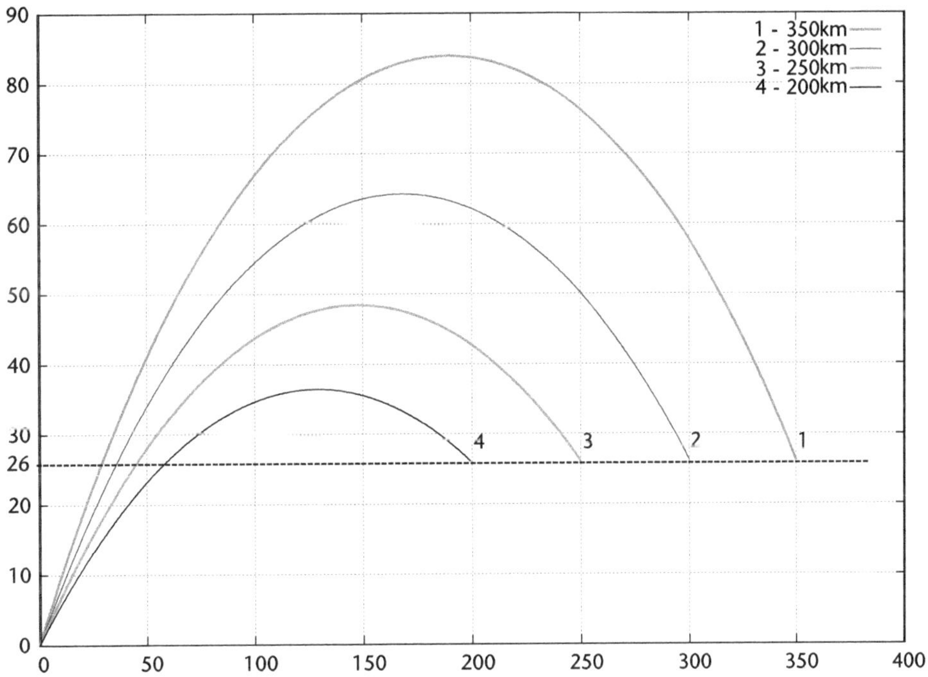

Figure 3.14 Missile reference trajectories for firing over the horizon.

kg/m^3; $\eta = 1.01–10^{-3}$ is the air viscosity coefficient; $F_b = (3–3.5)\lambda_M S_M$ is the wettable surface of the missile body; $\lambda_b = \frac{L_m}{D_m}$ is the elongation of the missile body, in relation to its diameter D_m; S_M is the area of the missile midsection, which is the area of its frontal projection; $\eta_M = \frac{1}{1+0.1615(M-0.25)^{1.325}}$, $\forall M = 0–4$ is the environment (air) compressibility coefficient; $\eta_0 = 1 + \frac{1}{\lambda_r}$ is the missile body shape influence coefficient; $\lambda_r = \frac{L_r}{D_r}$ is the head part of missile body lengthening; L_r is the missile fairing length; and D_r is the missile fairing diameter.

Ballistic trajectories for different declination angles calculated by the formulas (3.15)–(3.20) are shown in Figure 3.14.

From Figure 3.14 it can be seen that to ensure the destruction of a target beyond the radio horizon, when firing at the far border of the affected area using the missile speed $V_0^m = 1{,}000$ m/s, it is necessary to maximally raise the missile trajectory. In this case, the apogee of this trajectory is located at an altitude of 84.65 km (Figure 3.14).

In this case, the expression for the missile characteristic velocity, which can be attainable under the influence of the thrust of the missile engine, unchanged in direction and in the absence of all other forces (Tsiolkovsky's formula), has the form:

$$V_c = I_{trd} \ln\left(\frac{M_1}{M_2}\right) \tag{3.21}$$

and for given values of the specific impulse of a solid propellant missile engine $I_{trd} = 2650$ m/s, missile curb weight $M_1 = 1900$ kg and the mass of a missile with completely burnt out missile fuel $M_2 = 636.89$ kg:

$$V_c = 2896.50 \text{ m/s} \qquad \text{or} \qquad V_c = 8.75\text{M} \tag{3.22}$$

Writing down the Tsiolkovsky formula for the initial mass of fuel:

$$M_1 - M_2 = \frac{M_0 k_t (\exp^{V_m/I_{trd}} - 1)}{k_t + 1 - \exp^{V_m/I_{trd}}} \tag{3.23}$$

for the payload mass $M_0 = 230$ kg, which is a warhead with a mass of 200 kg and an active homing head with a mass of 30 kg, when using a specific impulse $I_{trd} = 2650$ m/s and a fuel mass $M_1 - M_2 = 1263.11$ kg, the value of the maximum permissible speed V_{max} for the 40N6 missile developed by the JSC Machine-Building Design Bureau Torch is:

$$V_{max} = 2,417.16 \text{ m/s} \qquad \text{or} \quad V_{max} = 7.3\text{M} \tag{3.24}$$

which, taking into account the losses for aerodynamic control and air resistance, corresponds to the data $V_m \approx 2{,}100$ m/s (6.34M) given by the Central Design Bureau Almaz for missiles of the 48N6 family used in the system S-300 PMU-2 Favorit.

Realizing that when firing at the maximum range, the missile at the start point of homing ($H_m = 26$ km, $D_m = 350$ km) is in the passive phase of the flight; that is, its propulsion system does not work. Therefore, in this case, the trajectory of the missile can be described using the equations of motion of a body thrown at an angle to the horizon. Taking into account the application of the homing method, when at any moment in time the missile velocity vector must be located along the line-of-sight of the missile-target and with uniform and rectilinear motion of the target, providing a given height of its flight above the Earth's surface, we can write:

$$\begin{cases} x_i^m = x_{i-1}^m + \dfrac{x_{i-1}^m - x_i^t}{R} \times V_0^m \times dt \times sign - \dfrac{F_{air}(x_i^m - x_{i-1}^m)dt^2}{2M(t)C} \\ z_i^m = z_{i-1}^m + \dfrac{z_{i-1}^m - z_i^t}{R} \times V_0^m \times dt - \dfrac{g(dt)^2}{2} - \dfrac{F_{air}(z_i^m - z_{i-1}^m)dt^2}{2M(t)C} \\ x_i^t = x_{i-1}^t - V_0^t dt \\ z_i^t = z_0^t \\ R = \sqrt{(x_{i-1}^m - x_i^t)^2 + (z_{i-1}^m - z_i^t)^2} \\ x_0^m = 0 \\ z_0^m = 26000 \\ V_0^m = 1000 \\ x_0^t = 60000 \\ z_0^t = 25 \cdots 26000 \\ V_0^t = 0 \cdots 2800 \end{cases} \tag{3.25}$$

The conditions for terminating the iterative computational process are:

- For reaching the value of the miss modulus h less than 40m:

$$h = \sqrt{(x_i^m - x_i^t)^2 + (z_i^m - z_i^t)^2} \leq 40\text{m}, \tag{3.26}$$

- For reaching the homing error modulus e_i greater than 6°:

$$e_i = \left[\left|\arctan\left(\frac{z_{i-1}^m - z_i^t}{x_{i-1}^m - x_i^t}\right)\right| - \left|\arctan\left(\frac{z_{i-2}^m - z_{i-1}^t}{x_{i-2}^m - x_{i-1}^t}\right)\right|\right]\frac{180}{\pi} \geq 6°, \quad \forall i \geq 2 \tag{3.27}$$

In this case, it is necessary to observe the condition of the sign transition:

$$sign = \begin{cases} -1, & \forall \arctan\left(\frac{z_{i-1}^m - z_i^t}{x_{i-1}^m - x_i^t}\right) > 0, \quad \forall i > 0 \\ +1, & \forall \arctan\left(\frac{z_{i-1}^m - z_i^t}{x_{i-1}^m - x_i^t}\right) < 0, \quad \forall i > 0 \end{cases} \tag{3.28}$$

where:

$$\begin{aligned} \arctan\left(\frac{z_{i-1}^m - z_i^t}{x_{i-1}^m - x_i^t}\right) &= \arctan\left(\frac{z_{i-1}^m - z_i^t}{x_{i-1}^m - x_i^t}\right) - \pi, \quad \forall\, sign = -1 \\ \arctan\left(\frac{z_{i-1}^m - z_i^t}{x_{i-1}^m - x_i^t}\right) &= -\frac{\pi}{2}, \quad \forall i : (x_{i-1}^m - x_i^t) = 0 \end{aligned} \tag{3.29}$$

When calculating, you can make a replacement $V_0^m \to V_i^m$:

$$V_i^m = \frac{\sqrt{(x_i^m - x_{i-1}^m)^2 + (z_i^m - z_{i-1}^m)^2}}{dt} \tag{3.30}$$

The upper and lower points of the affected area of the air defense system are critical from the point of view of the missile's flight and ballistic characteristics. The upper point of the affected zone is located on its near border, and the lower point of the affected zone is determined by the capabilities of the multifunctional radar station for stable auto-tracking of low-altitude targets.

The initial conditions of a missile flight to the near border of the affected zone (end of declination) are realized at the time of flight $t_m = 25$s at a missile flight speed $V_m = 550$ m/s and an inclination angle of the trajectory $\theta_{tr} = 45°$.

The best and worst trajectories of the missile, calculated for declination section and taking into account the disturbances acting on the missile, are shown in Figures 3.15 and 3.16, respectively.

Here, $Z_L O X_L$ is the coordinate system associated with the TLC, which is in a combat position ready to launch missiles.

The trajectories for the best flight conditions are in the area bounded by curves 1 and 2, and the trajectories for the worst flight conditions are in the area bounded

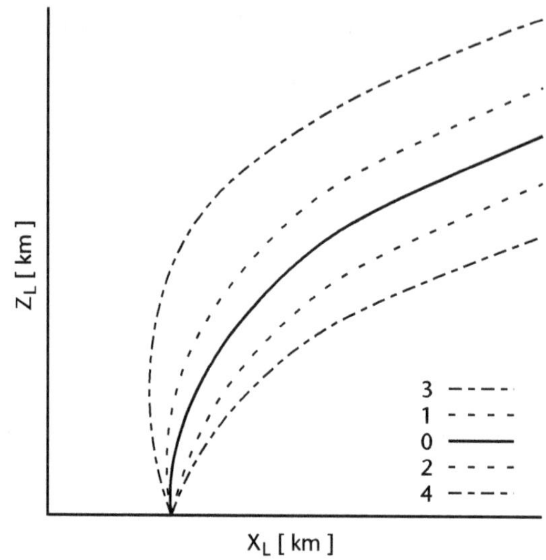

Figure 3.15 The best trajectories in the declination section.

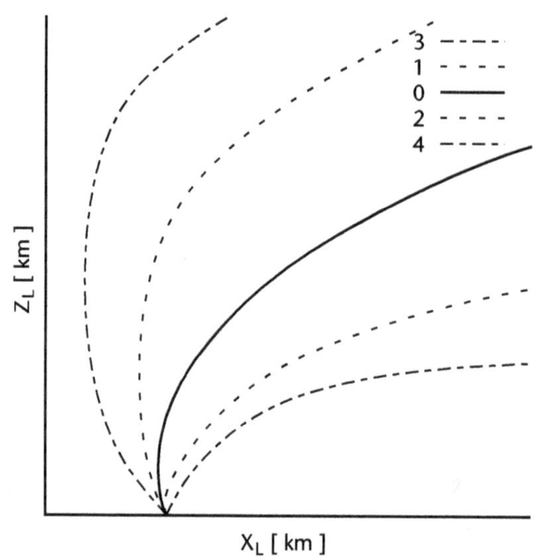

Figure 3.16 The worst trajectories in the declination section.

on one side by curves 1 and 3, and on the other side by curves 2 and 4. The optimal trajectories are represented by curves with the number 0.

The perturbing moments affecting the missile after it leaves the TLC are as follows:

- After ejection, the missile acquires the initial angular velocity of rotation ω_{z_0}, ω_{y_0} relative to the transverse axes;
- The longitudinal axis of the missile can be deflected relative to the building axis at an angle $\Delta\vartheta_0$;

- Firing from unprepared sites in engineering terms with an inclination angle of the TLC up to $\Delta\vartheta_L = 5°$;
- Wind load;
- Gas-dynamic asymmetry of the thrust line of action (summary gas dynamic eccentricity).

The estimate of the initial start angle can be presented in the form:

$\theta_0 = 90° + \Delta\vartheta_L + \Delta\vartheta_0 = 90° + 5° + 1° = 96°$

$\Delta\omega_{z0,y0} = 7°/s$ is the initial angular missile velocity relative to the transverse axes.

The accuracy of outputting h_{base} to the reference trajectory, when firing into any area of the affected zone, depending on the range D_{Lm} (Figures 3.15–3.18) between the start's point and the point of the required missile position in space, is:

$$\forall\, D_{Lm} = 8 \text{ km} : h_{base} \leq \pm 2.0 \text{ km}$$
$$\forall\, D_{Lm} = 12 \text{ km} : h_{base} \leq \pm 1.5 \text{ km}$$
$$\forall\, D_{Lm} = 16 \text{ km} : h_{base} \leq \pm 1.0 \text{ km}$$

The missile velocity vector error does not exceed $\pm 3°$.

When simulating firing over the horizon to the lowest point of the affected zone, a range of 400 km was obtained at a target velocity $V_t = 300$ m/s, a missile velocity $V_m = 1{,}000$ m/s and implementation the nominal ballistic characteristics of the missile.

Based on the reference trajectory of the missile and the speed of the target, it is possible to determine the meeting point of the missile with the target for the missile to fly without maneuvering.

Declination trajectories for best and worst in terms of ballistic characteristics of flight conditions for firing at the extreme points of the affected zone:

- To the top point: $D_m = 25$ km, $H_{T1} = 35$ km-trajectory T1
- To the bottom point: $D_m = 60$ km, $H_{T2} = 0.15$ km-trajectory T2

are presented in Figures 3.17 and 3.18, respectively.

The trajectories for the best flight conditions are in the area bounded by curves 1 and 2, and the trajectories for the worst flight conditions are in the area bounded on one side by curves 1 and 3, and on the other side by curves 2 and 4. The optimal trajectories are represented by curves with the number 0.

Estimation of the width of the region of possible reference trajectories relative to the optimal trajectory T1 is $\hat{h}_{base} = \pm 1.5$ km, and relative to the optimal trajectory T2 is estimated as $\hat{h}_{base} = \pm 1.3$ km.

3.3 Problems: Target Search and Auto-Tracking

As shown in Section 3.2 for firing over the horizon, it is necessary to bring the missile to the reference trajectory, which will provide the altitude of the start of homing $H_0 = 26$ km, which will in turn provide the start homing altitude $H_0 = 26$ km corresponding to the estimated range D_0 on which homing the missile at the target

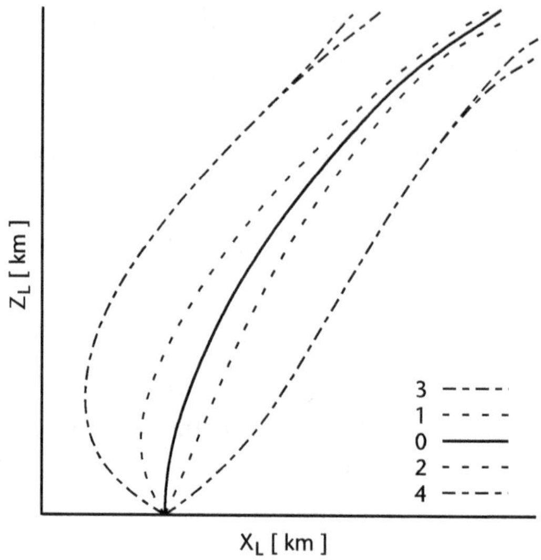

Figure 3.17 Missile trajectories to the top point of the affected zone.

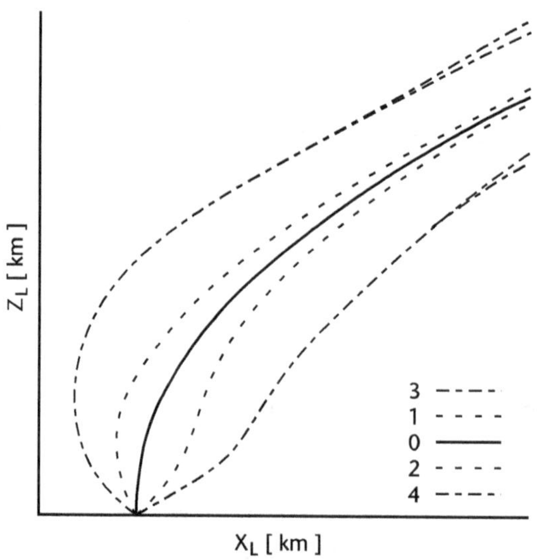

Figure 3.18 Missile trajectories to the lowest point of the affected zone.

begins. Due to the fact that the assumed coordinates of the target are entered into the on-board computer of the missile before its start, then during the flight of the missile to the start homing point, the target may not be at the assumed point. Therefore, in any case, it is necessary to carry out an additional search for the target by angular coordinates, by means of conical scanning of the active homing head antenna.

In a radar with conical beam scanning, the target position is determined by nutation by rotating the beam of the homing head antenna shifted relative to the axis of rotation by a certain angle of deflection of this beam β (Figure 3.19). If the target is located on the axis of rotation of the antenna beam, in all positions of this beam, signals of equal amplitudes arrive at the receiver input, and in this case the error signal $S_e(t) = 0$. If the target is displaced relative to the axis of rotation

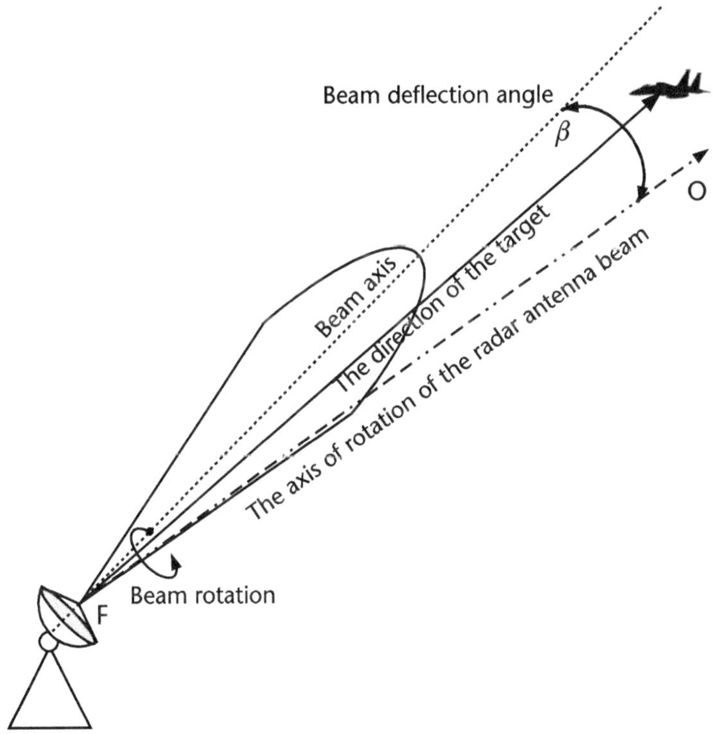

Figure 3.19 Target tracking by conical scanning.

of the antenna beam, then the error signal turns out to be amplitude modulated by the directional pattern of the main antenna lobe of the active homing head. Moreover, if the target is on the same side as the instantaneous position of the main lobe of the antenna directional pattern, the signal amplitude will be greater than the signal amplitude when the target is on the opposite side. To determine the sign of the error signal, this amplitude modulated signal is compared in phase with a quadrature reference signal from a generator coupled to the nutation drive. As a result, the magnitude of the error is proportional to the modulation amplitude, and the sign is characterized by the phase difference between the error signal and the reference signal. To simultaneously obtain error signals in two mutually orthogonal planes, it is necessary to have two phase detectors, a reference signal with a zero phase is fed to the second input of the first phase detector, and a reference signal shifted by $\pi/2$ relative to the first is fed to the second input of the second phase detector. Phase detectors are essentially devices that perform scalar multiplication of sinusoidal reference signals with a scanning frequency and envelopes of received signal bursts with appropriate phases to obtain the elevation error from one detector and the azimuthal error from the other.

The main parameter of a radar with conical beam scanning is the ratio of the dimensions of the scanning circle 2β and the width of the main beam of the radiation pattern Θ_B of the radar antenna (Figure 3.20). The radius of the circle β, which describes the position of the center of the main beam of the antenna pattern during scanning, corresponds to the deviation angle β of the main beam of the radar antenna from its rotation axis FO Figure 3.20). The choice of the value of

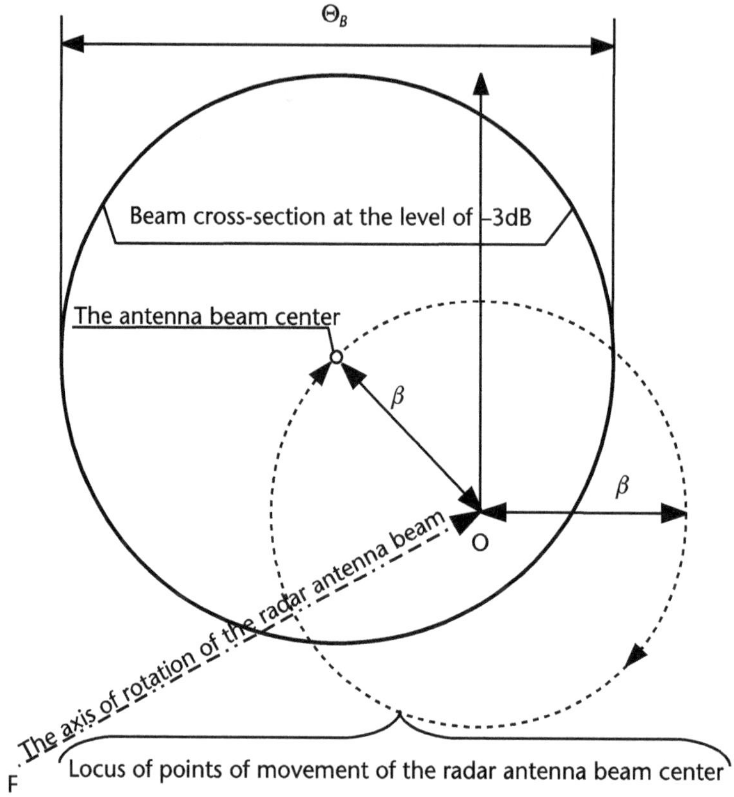

Figure 3.20 Cross section of the antenna beam at the level of -3dB and the path described by the center of the beam when it rotates.

β is determined by the trade-off between the decrease in the antenna gain L_k at the point of intersection of the main beam of the antenna pattern with the axis of its rotation FO and an increase in the angular sensitivity k_s (slope of the direction finding characteristic).

A large value of k_s is needed to obtain higher angular tracking system (phase detector) output voltages for a given real angular error compared to unwanted angular error voltages caused by receiver thermal noise. For a given signal/noise value, the effect of thermal noise is inversely proportional to the slope of the direction finding characteristic of the error signal k_s. However, as β increases in order to increase k_s, the losses L_k also increase, which entails a decrease in the signal-to-noise ratio [17].

The relative values of k_s and L_k depend on whether the target is provided with a transponder, which eliminates the modulation in the received signal that occurs during transmission when operating on a signal reflected from the target. Since the transponder is not used when firing over the horizon, the directional pattern of the radar antenna will be used twice—for transmission and for reception. In this case, for a given β, the modulation depth (angular sensitivity) increases, while the L_k loss (in dB) doubles. Figure 3.21 shows the dependence of the loss L_k and the steepness of the error signal k_s on the selected value β [18]. Root mean square error (RMS) error due to thermal noise is inversely proportional to k_s and proportional

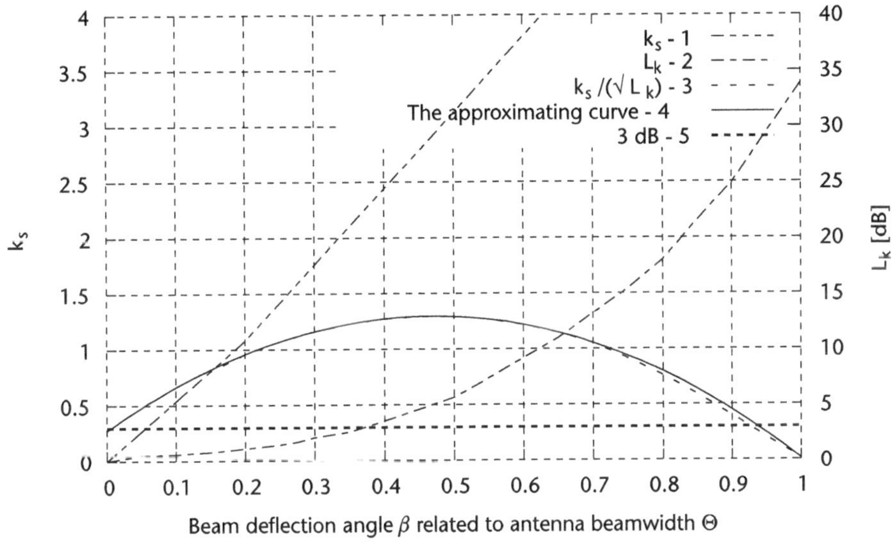

Figure 3.21 The error signal steepness k_s and the loss L_k at the intersection of the antenna beam patterns.

to $\sqrt{L_k}$, where L_k is the signal power loss. The approximating curve $k_s/\sqrt{L_k}$ (curve 4 Figure 3.21) has the form:

$$Y(\beta) = -4.57\beta^2 + 4.31\beta + 1.37 \tag{3.31}$$

The extremum of this function is found from the condition:

$$\frac{dY(\beta)}{d\beta} = 0, \qquad at \quad \beta_{max} = 0.47\Theta_B. \tag{3.32}$$

In this case, β_{max} is the value of the optimal deflection angle β, at which the influence of thermal noise for auto tracking along the angular coordinates is minimized. It should be noted that only L_k affects the range tracking system, so that the most accurate range tracking is obtained only for $\beta = 0$. Therefore, the optimal value of β should be chosen from a compromise between quality auto tracking by angular coordinates and by range. The approximating curve of the function $L_k^{app}(\beta)$ has the form:

$$L_k^{app}(\beta) = 43.99\beta^2 - 12.98\beta + 1.37 \tag{3.33}$$

Taking the level of 3 dB as the value of the admissible loss and solving the quadratic equation (3.33) under the condition $L_k^{app}(\beta) = 3$ dB, the optimal value of the tracking error angle when using conical scanning looks like:

$$\beta_{opt} = 0.39\Theta_B. \tag{3.34}$$

In radars that use beam scanning for additional target search, the tracking angular error information is the change in time of the signal amplitude reflected from the target. Other factors causing changes in the amplitude of the signal reflected from the target may be fluctuations in the target's effective scattering surface σ, which can cause false tracking error information. The unwanted fluctuation rate is usually of the same order of magnitude as the scanning frequency. Due to the fact that the frequency of fluctuations of the effective scattering surface targets σ are usually below 100 Hz, it is desirable to increase the scanning frequency. The maximum possible scanning frequency is 1/4 of the pulse repetition period at which there are four pulses per scanning period: one pulse each from the top, bottom, right, and left. In turn, the maximum pulse repetition rate, and therefore the maximum scanning frequency, are limited by the maximum range of targets that must be detected and tracked by the airborne radar of the anti-aircraft missile. By setting an unambiguous target detection range at a given range, the signal reflected from the target must have time to be received by the homing head receiver until the next pulse is emitted by the missile homing head transmitter.

When firing over the horizon to the far border of the affected zone (400 km), it is necessary to deliver the missile to the starting point of the search, which is at an altitude of $H_m = 26$ km and a distance of $D_m = 350$ km from the location point of the multifunctional radar. Using the expression (3.8), we can determine the unambiguous detection range of a hypersonic target flying at a speed of $V_t = 1668$ m/s, which will be $D_{det\_min} = 120$ km from the start point search or $D_{det\_OTH} = 470$ km from the location point of the anti-aircraft missile system multifunctional radar. Thus, the maximum scanning frequency when the active homing head operates in the conical scanning mode considering that the light speed $c = 3 \times 10^8$ m/s will be:

$$f_{max} = \frac{c}{4 \times 2 \times D_{det\_min}} = \frac{3 \times 10^8 \text{ m/s}}{4 \times 2 \times 120 \times 10^3 m} = 312.5 \text{ Hz} \tag{3.35}$$

Taking into account that to measure the speed of rotation of the antenna directional pattern beam, the circular frequency $\omega_{max} = 2\pi f_{max}$, having the dimension revolutions/minute [rpm]:

$$\omega_{max} = 60 \times f_{max} = 18.75 \times 10^3 \tag{3.36}$$

This is a sufficiently high angular velocity for mechanical rotation of the missile's active homing antenna. Due to the fact that the onboard power supply of the missile is carried out through a DC circuit with a voltage of -12V to -48V, and the antenna is controlled in two mutually orthogonal planes (azimuth - ϕ and elevation ϵ), two powerful DC motors are needed, which will rotate the antenna axis each in its own plane according to a certain law to ensure fast conical rotation of the antenna. Currently, these are, as a rule, brushless DC motors with a maximum shaft speed of $\omega_{DC} \leq 30 \times 10^3$ rpm.

From (3.35) and (3.36) it follows that with decreasing range to the target D_{det_min}, the maximum frequency of conical scanning ω_{max} will increase. Therefore, conical scanning can be used only for target detection by means of its additional search by angular coordinates. In the future, it is reasonable to carry out auto-tracking of a detected target by an active homing head in a pulsed mode using a sum-and-difference monopulse direction finder, having the following width of the main beam of the antenna directional pattern:

$$\Theta_B = \Delta(\beta) = \Delta(\epsilon) \approx 5^\circ \ldots 6^\circ \tag{3.37}$$

Therefore, the accuracy $\{\delta Z_t, \delta Y_t, \delta X_t\}$ of target designation for targets, which flying beyond the radio horizon on range $D_{det\_OTH} \geq 470$ km should be no worse than δ_{max}:

$$\begin{aligned} \delta_{max} &= D_{det\_min} \times sin(0.39\Theta_B) = 4.899 \text{ km}, \\ \Rightarrow \quad \delta H_t &= \delta Z_t = \delta Y_t = \delta X_t \leq |\ 4.899 \text{ km}\ | \end{aligned} \tag{3.38}$$

Thus, this chapter has shown the fundamental possibility of firing over the horizon, outlined some of the ballistic and flight characteristics of anti-aircraft guided missiles, and also defined the reference trajectories for firing over the horizon and the problems of finding targets over the radio horizon.

References

[1] Litvinov, O. S., K. B. Pavlov, and V. S. Gorelik, "Electromagnetic Waves and Optics," Physics at the Technical University, Vol. 4, 2002, Bauman Moscow State Technical University, http://fn.bmstu.ru/data-physics/library/physbook/tom4/ch1/texthtml/ch1_3.htm.

[2] Mishchenko, Y. A., *Over-the-Horizon Radar,* Military Publishing, 1972.

[3] "The Concept of the Geoid," *Encyclopaedia Britannica,* 2020, https://www.britannica.com/science/geoid/The-contribution-of-orbiting-satellites.

[4] Information and Analysis Center for Positioning, Navigation and Timing, JSC TsNIIMash, Korolyov, Russia, https://www.glonass-iac.ru/guide/gnss/gps.php.

[5] Williams, D. R., Earth Fact Sheet. NASA Goddard Space Flight Center, Greenbelt, MD, https://nssdc.gsfc.nasa.gov/planetary/factsheet/earthfact.html.

[6] Bakhshiyan, B. T., and K. S. Fedyaev, *Fundamentals of Space Ballistics and Navigation,* Federal State Budgetary Institution of Science, Space Research Institute of the Russian Academy of Sciences, Moscow, 2013, http://www.iki.rssi.ru/books/2013bakhshiyan.pdf.

[7] NIST National Institute of Standards and Technology, U.S. Department of Commerce, Gaithersburg, MD, https://physics.nist.gov/cgi-bin/cuu/Value?bg.

[8] *NORAD's Missile Warning System: What Went Wrong?,* Report to the Chairman Committee on Government Operations House of Representatives of the United States, https://www.gao.gov/assets/140/133240.pdf.

[9] Alabastrov, V. A., E. S. Goikhman, I. M. Zamorin, et al., "The Basics of Over-the-Horizon Radar," A. A. Kolosov (ed.), Radio and Communication, 1984.

[10] Podvig, P., "History and the Current Status of the Russian Early-Warning System," *Science & Global Security,* Vol. 10, No. 1, 2002, pp. 21–60, DOI: 10.1080/08929880290008395, http://scienceandglobalsecurity.org/archive/2002/01/history_and_the_current_status.html.

[11] "Barrayar Over-the-Horizon Woodpecker: A Brief History of the Chernobyl-2 Installation," https://habr.com/ru/company/mailru/blog/351212/.

[12] Duga Radar, https://en.wikipedia.org/wiki/Duga_radar.

[13] Martinez, J. P., "Letter from J. P. Martinez," *Wireless World,* April 1982, http://www.mysterysignals.signalshed.com/Wpecker6.html.

[14] Coordinate System WGS-84, Basic Parameters, https://geostart.ru/post/324.

[15] Eshtokin, A. N., "Basic Coordinate Systems Used in Geodesy," Luhansk National Taras Shevchenko University, http://topography.ltsu.org/geodezy/g3.html.

[16] GNSS EXPERT, "Information Resource on High-Precision Positioning Technologies Using Global Navigation Satellite Systems (GNSS)," http://gnss-expert.ru/?page_id=520.

[17] Skolnik, M. I., *Radar Handbook,* Vol. 4, Radar Stations and Systems, New York: McGraw-Hill, 1970.

[18] Armstrong, J. A., "Measurement of Picosecond Laser Pulse Widths," *Appl. Phys. Letters,* Vol. 10, January 1967, pp. 16–18.

CHAPTER 4

Radio Control of Atmospheric Anti-Aircraft Missile

The combat missile control system is a complex of means designed for the automatic guidance of the missile at the target to be hit.

The main feature of firing anti-aircraft guided missiles is that the missile is launched in such a way as to arrive not at the point in space where the target is at the moment of launch, but at another point, which is called the look-ahead point of space, to which the missile must fly simultaneously with the target.

Another feature of firing anti-aircraft guided missiles is the need to choose a hypothesis about the trajectory of the target during the lead time.

At the same time, it is reasonable to take the time from the moment the missile is launched until the moment as the missile meets the target for the lead time.

It is also reasonable to consider a certain assumption about the nature of the target's flight as a hypothesis of the trajectory of the target during the lead time: about the height of the target, speed, and direction of its movement.

For example, according to the three-term hypothesis, the target flies horizontally during the anticipatory time, at a constant altitude H_t, in a straight line, with a constant heading Q_t and with a constant speed V_t [1]:

$$H_t(t) = Q_t(t) = V_t(t) = Const \tag{4.1}$$

With an increase in the maneuverability of air targets, a two-term hypothesis arose, in which the target remains unchanged in terms of its speed and straightness of flight, and changes the altitude; for example, it dives:

$$Q_t(t) = V_t(t) = Const; \qquad H_t(t) = var \tag{4.2}$$

The following hypothesis keeps only one parameter unchanged: the target's course, while the speed and altitude of its flight can vary according to a certain law:

$$Q_t(t) = Const; \qquad V_t(t) = var, \qquad H_t(t) = var \tag{4.3}$$

With the constant development of air attack weapons, today it makes sense to consider the null hypothesis, when all three parameters change over time:

$$Q_t(t) = var, \qquad V_t(t) = var, \qquad H_t(t) = var \tag{4.4}$$

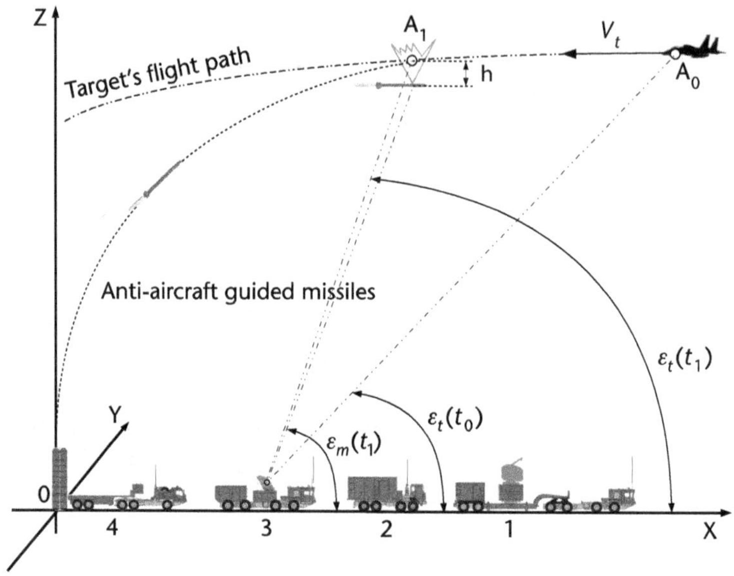

Figure 4.1 Anti-aircraft missiles shooting.

The choice of hypothesis is based on what coordinates the target had in space at the moment preceding the launch of the missile, as well as on what law and at what speed these coordinates changed.

At present, it makes sense to consider the hypotheses (4.3) and (4.4) as a target flight hypothesis, as well as the assumption that the missile flight time from the launch point O to the anticipatory point A_1 is equal to the time it takes for the target to move from point A_0, where it was at the moment of the missile launch, to the anticipatory point A_1 (see Figure 4.1).

This figure shows an anti-aircraft missile system with the following composition:

1. Radar detection;
2. The command post of the anti-aircraft missile system;
3. Radar illumination and guidance (multifunctional radar);
4. Launcher with missiles in a firing position, located in transport-launch containers.

When firing anti-aircraft guided missiles, the anticipatory time $t_a = t_{A_1} - t_{A_0}$ depends not only on the technical characteristics of the missile, but also on the nature of the target's flight. This is due to the fact that the missile's flight path is influenced by changes in the target's flight speed, altitude, and direction of flight. In addition, the lead time t_a depends on the way the missile is guided to the anticipatory point A_1, which, if the target is successfully hit, is called the missile meeting point with the target A_{mt}.

Thus, for the successful defeat of the target, it is necessary to ensure:

1. Early target detection;
2. Stable automatic target tracking by angular coordinates, range, and/or radial velocity;

3. Calculation of the characteristics of the target's movement in space (the laws of time variation of the coordinates of the target's movement): $x_t(t)$, $y_t(t)$, $z_t(t)$ and $V_t(t)$;
4. Determination of the optimal missile control method suitable for a specific type of anti-aircraft missile system (i.e., determination of the optimal method for guiding the missile to the target);
5. Stable control of the missile's flight to the target, ensuring the guidancc of the missile to the target with a given accuracy, at which the target will be hit with a probability not lower than the given one.

4.1 Generalized Block Diagram of the Radio System

A combat missile control system is a complex of means designed to automatically target a missile to the target [2]. The problem of keeping the missile close to the given flight path is solved by the control loop. It is quite clear that in order to successfully solve this problem, it is necessary to know the position of the target in space at any time in the entire time interval over which the missile is controlled, up to the moment t_w of its warhead detonation:

$$\exists\ \{\varepsilon_t(t), \beta_t(t), r_t(t)\}, \qquad \forall t \in [t_{at}, t_w], \tag{4.5}$$

or

$$\exists\ \{\varepsilon_t(t), \beta_t(t)\ \&\ V_t(t)\}, \qquad \forall t \in [t_{at}, t_w], \tag{4.6}$$

where t_{at} is the start time of target auto-tracking by angular coordinates and range $r_t(t)$ or by angular coordinates $\varepsilon_t(t)$, $\beta_t(t)$ and radial velocity $V_t(t)$:

$$\begin{cases} \varepsilon_t(t) = \dfrac{\pi}{2} - \theta_t(t) = \dfrac{\pi}{2} - \arctan\left(\dfrac{\sqrt{x_t^2(t) + y_t^2(t)}}{z_t(t)}\right); \\ \varphi_t(t) = \arctan\left(\dfrac{y_t(t)}{x_t(t)}\right); \\ r_t(t) = \sqrt{x_t^2(t) + y_t^2(t) + z_t^2(t)}. \end{cases} \tag{4.7}$$

Here, $\varepsilon_t(t)$, $\varphi_t(t)$, and $r_t(t)$ are functions of time variation of the target elevation angle ε_t, target azimuth φ_t and target range r_t, respectively. In addition, you need to know the method of guiding the missile to thc target that is used in this particular case. So, for homing, coordinates presented in (4.6) are quite enough, since the missile rudder control commands are generated on board the missile according to the data of the onboard radio direction finder, which captures the target for tracking at the radial velocity V_t and angular coordinates (ε_t and φ_t) on the basis of the receiving antenna of the airborne direction finder (missile homing head).

With command guidance, it becomes necessary to constantly monitor not only the position of each ith target in space, but also the spatial position of each kth

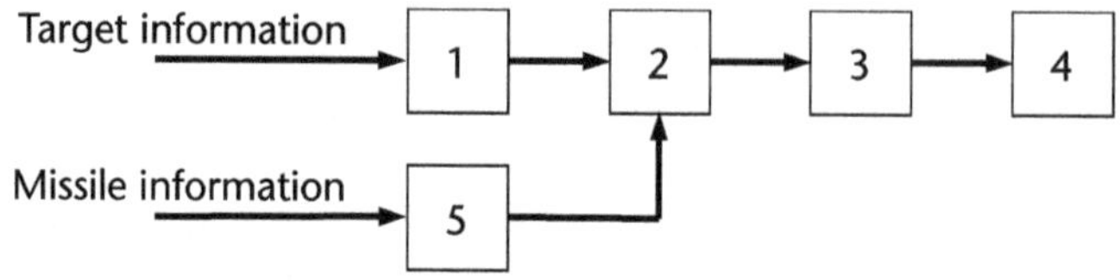

Figure 4.2 Generalized control system diagram.

missile launched at the ith target with the task of destroying it. Thus, for command guidance, the expression (4.5) must be supplemented with information about the missile, which, taking into account the multichannel nature of the target and the missile, will look like:

$$\begin{cases} \exists \{\varepsilon_t^i(t), \beta_t^i(t), r_t^i(t)\} & \forall t \in [t_{at}^i, t_w^{ki}], \quad \forall i \in [1, N]; \\ \exists \{\varepsilon_m^{ki}(t), \beta_m^{ki}(t), r_m^{ki}(t), \}, & \forall t \in [t_s^{ki}, t_w^{ki}], \quad \forall k \in [1, M], \end{cases} \tag{4.8}$$

where t_{at}^i is the start time of auto-tracking of the ith target in angular coordinates and range;
t_s^{ki} is the launch time of the kth missile toward the ith target;
t_w^{ki} is the detonation time of the kth missile near the ith target;
N is the maximum number of simultaneously fired targets (channels number by target);
M is the maximum number of missiles assigned to each ith target (channels number by missile).

Thus, in the most general form, the control system of one missile channel is a set of elements.

1. Scheme (method) of processing information about the target.
2. Guidance method.
3. Missile control loop.
4. Scheme for controlling the detonation of a warhead (operation principle of a radio fuse).
5. Scheme for processing information about a missile. A generalized control system diagram about a missile can be shown in Figure 4.2.

It is quite clear that the guidance system is part of the control system. Unlike the on-board control system, which is located only on board the missile, the guidance system devices can be located outside it: on the ground, on a ship, on an airplane, or on another missile.

The guidance system generally determines the position of the missile in space relative to the target, calculates the required flight path, and in accordance with this, generates control commands for the onboard control system, which changes the missile's flight path [3].

The accuracy of guiding a missile to a target is usually determined by the magnitude of the miss h (Figure 4.1), which largely depends on the functioning of the guidance system:

$$h = \sqrt{(x_t(t_{A_1}) - x_m(t_{A_1}))^2 + (y_t(t_{A_1}) - y_m(t_{A_1}))^2 + (z_t(t_{A_1}) - z_m(t_{A_1}))^2}, \quad (4.9)$$

where the rectangular coordinates are at the meeting point of the target $x_t(t_{A_1}), y_t(t_{A_1}), z_t(t_{A_1})$ and the missile $x_m(t_{A_1}), y_m(t_{A_1}), z_m(t_{A_1})$ at the moment of detonation t_w of its warhead are represented as:

$$\begin{cases} x_{t,m}(t_{A_1}) = r_{t,m}(t_w)\cos(\varepsilon_{t,m}(t_w))\cos\varphi_{t,m}(t_w) \\ y_{t,m}(t_{A_1}) = r_{t,m}(t_w)\cos(\varepsilon_{t,m}(t_w))\sin\varphi_{t,m}(t_w) \\ z_{t,m}(t_{A_1}) = r_{t,m}(t_w)\sin(\varphi_{t,m}(t_w)) \end{cases} \quad (4.10)$$

Depending on the tactical purpose of the missile, the nature of the target and the range of the target from the launch point, the missile can be controlled by various control systems, the name of which is determined by the principles of those guidance systems' operation.

The main types of missile automatic control systems can be classified by control methods:

- Autonomous;
- Telecontrol systems (command control systems);
- Homing systems;
- Combined guidance systems.

In autonomous systems, there is no element that processes current information about the target. The guidance method is replaced by the motion program, given apriori. The lack of current information about the coordinates of the target limits the use of autonomous systems only for firing at stationary objects or at targets whose movement can be predicted with great accuracy. The principle of operation is distinguished by magnetometric systems, inertial systems, and astronavigation systems.

Telecontrol systems are subdivided into guidance systems by radar beam, radio navigation systems, and command control systems. Command control systems are characterized by the fact that the law of motion of the missile is determined at the control center, which is outside the guided missile. For anti-aircraft missile systems, these are ground-based missile control centers. At the ground control station, the trajectory of the missile is generated in accordance with the selected guidance method according to the coordinates of the target and the missile.

In homing systems, the reception and processing of information about the target and the formation of control commands is carried out by the equipment installed on board the missile. Homing can be the only way to control a missile, as in the S-200D anti-aircraft missile system, or it can be used in the final guidance phase, as in the S-300 and S-400 systems. In the latter case, the control method is called combined.

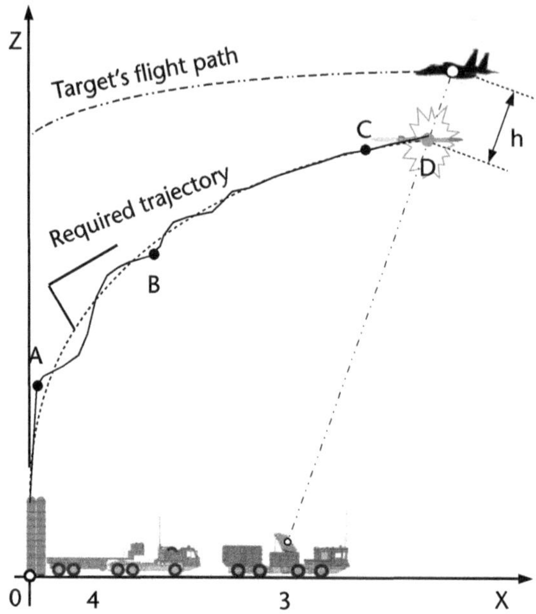

Figure 4.3 Anti-aircraft missile control stages.

Due to the fact that the coordinates of the target are determined on board the missile itself, the homing accuracy is practically continuous radar signals, for example, as in the S-200D system.

4.2 Radio Control System Classification

Controlling an anti-aircraft missile and ensuring target destruction, includes the following stages (Figure 4.3):

1. Launch control: $0A$ trajectory section;
2. Missile flight control: AC trajectory section, including:
   - A section of the missile exit to the required trajectory AB;
   - Missile guidance, trajectory section BC.
3. Warhead detonation control, trajectory section CD.

4.2.1 Launch Control

At the launch control site, the calculation of the anti-aircraft missile system is obliged to check the technical readiness of ground and on-board systems for launch, remove the protection stages, and control the launch line. Launch control includes preparing the missile for launch and launching it at a given moment in time.

At this stage, the system has to:

- Choose of the launcher from which the missile should be launched, while eliminating the possibility of the launched missile hitting the equipment and personnel of the anti-aircraft missile system.

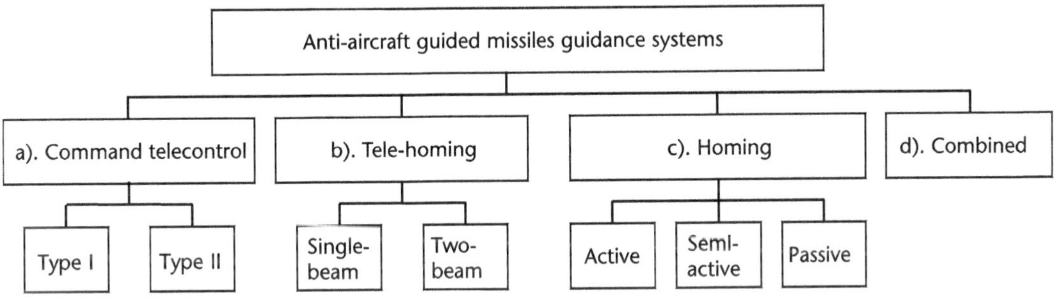

Figure 4.4 Anti-aircraft guided missiles guidance systems.

- Bring the on-board devices of the missile to a state that ensures its launch and flight on an uncontrolled section of the trajectory. It is envisaged to bring the onboard power supplies to the operating mode, remove the stages of protection against unintentional launch and ensure the operating modes of the missile stabilization system in flight. At this stage, the initial data can be entered into the missile, which determines the coordinates of the target.

The main part of bringing the on-board devices into operation is carried out according to a predetermined program. Some procedures may vary depending on the conditions of the combat situation.

4.2.2 Anti-Aircraft Missile Flight Control

At the stage of missile flight control, the missile is brought to the required trajectory and the missile movement along this trajectory is controlled until the missile meets the target.

The missile is guided to the target, as a rule, in automatic mode. This is due to missile and target high speeds, as well as the need to achieve high guidance accuracy.

A set of devices that provide a solution to problems arising during flight control is called a control system or guidance system. There are different types of systems: command telecontrol, telehoming, homing, and combined (Figure 4.4).

4.2.2.1 Command Telecontrol Systems

Command telecontrol systems are those in which the missile's flight is controlled by control commands that are calculated at the ground and transmitted to the missile board throughout the radio link ground-missile board.

Depending on the method of obtaining information about the target, command telecontrol systems are divided into two types: Type I and Type II. In the Type I system, the coordinates of the target are measured by a radar located on the ground (Figure 4.5), and in the Type II system the coordinates are measured by a radar located on board the missile (Figure 4.6).

Command guidance system, Type I. Detection radar 1 (Figure 4.5) detects the *i*target according to target designation data from a higher command post or independently. The coordinates of the detected target are then transmitted through the command post of the anti-aircraft missile battalion to the illumination and

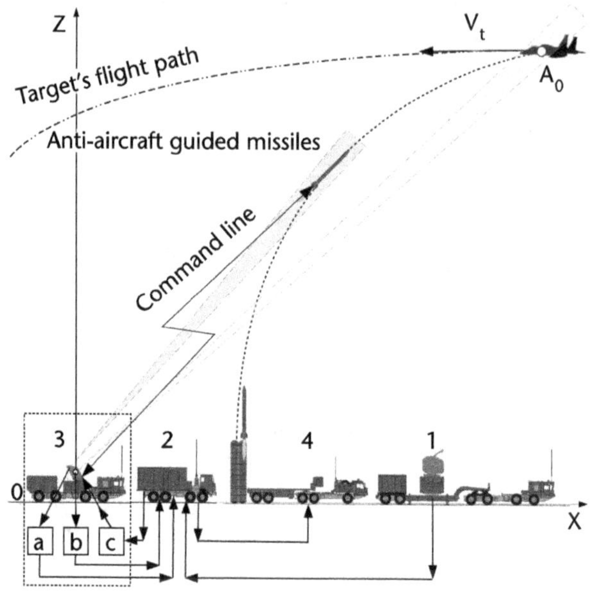

Figure 4.5 Command guidance system, Type I.

guidance radar. All coordinates are usually transmitted in a spherical coordinate system. Illumination and guidance radar 3 (block **b**, Figure 4.5) according to the coordinates, received from the detection radar also detects the target and takes it for automatic tracking along the angular coordinates $\varepsilon_t^i(t)$, $\varphi_t^i(t)$ and range $r_t^i(t)$; that is, in every sufficiently small time interval δt a vector of coordinates ith the target $\vec{C}_t^i$ is formed:

$$\vec{C}_t^i = \vec{C}_t^i(n \cdot \delta t) = \{\varepsilon_t^i(n \cdot \delta t),\ \varphi_t^i(n \cdot \delta t),\ r_t^i(n \cdot \delta t)\}, \qquad \forall n \in [0, L_t],$$
$$\text{where } L_t = |t_w^{ki} - t_{at}^i|/\delta t \tag{4.11}$$

The vector $\vec{C}_t^i$ enters the computing complex of the command post of the anti-aircraft missile battalion 2 (Figure 4.5) for the subsequent control commands formed by the missile. Based on this data, the moment of launching the kth missile assigned to destroy the ith target is determined. After launching this missile, illumination and guidance radar 3 (block **a**, Figures 4.5 and 4.6) captures the kth missile for auto-tracking and measured coordinates of the missile come into the computing complex of the anti-aircraft missile battalion command post 2. As a result of comparing the coordinates of the ith target and the kth missile, control commands are formed $\vec{u}_{i,k} = \vec{u}_{i,k}(n \cdot \delta t)$ for the kth missile:

$$\vec{P}_m^k = \vec{P}_m^k(n \cdot \delta t) = \{\varepsilon_m^k(n \cdot \delta t), \varphi_m^k(n \cdot \delta t), r_m^k(n \cdot \delta t)\}, \qquad \forall n \in [0, L_t] \tag{4.12}$$

$$\vec{u}_{i,k}(n \cdot \delta t) = \begin{cases} 0, & \forall n \in \left[0, \frac{t_s^{ki}}{\delta t}\right]; \\ \{\varepsilon_t^i(n \cdot \delta t) - \varepsilon_m^k(n \cdot \delta t), \varphi_t^i(n \cdot \delta t) - \varphi_m^k(n \cdot \delta t)\}, & \forall n \in \left[\frac{t_s^{ki}}{\delta t}, \frac{t_w^{ki} - t_{at}^i}{\delta t}\right] \end{cases} \tag{4.13}$$

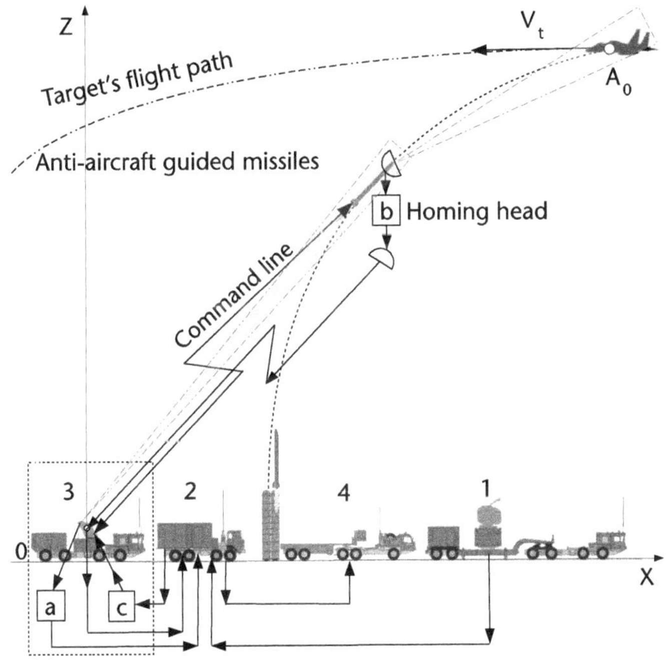

Figure 4.6 Command guidance system, Type II.

Calculation $\Delta r_{tm}^{ik} = r_t^i(n \cdot \delta t) - r_m^k(n \cdot \delta t)$ does not make sense, since speed control is not implemented in missiles, especially in missiles with solid missile engines. In these engines, fuel burns out in an uncontrolled manner. Having started once, this engine cannot stop burning and works until all the fuel has been consumed. The calculation of Δr_{tm}^{ik} makes sense only when the missile approaches the point of impact with the target. This is necessary to determine the moment of detonation of the warhead of the missile and the optimal coordination of the area of dispersion of the fragments to cause maximum damage to the target. But another device, located on the missile, is responsible for this function. This is a radio fuse. The command radio control line, block **c** (Figures 4.5 and 4.6), is a multichannel communication line designed to convert commands into radio signals and transmit them to missiles aimed at targets. In the case of a multichannel anti-aircraft missile system, a large number of commands have to be transmitted simultaneously to the target and the missile. In this regard, to avoid confusion, the commands are converted into the corresponding code by means of a certain encoding function $\mathcal{F}_k$. The code obtained in this way is unique for each missile. Moreover, each of the transmitted commands has the corresponding attribute of belonging, the protective attribute {ε,k} and {φ,k} (see (4.14) and (4.15)). The control and coding equipment (transmission path) is installed at the command post of an anti-aircraft missile system.

$$code[u_{i,k}^{\varepsilon}(n \cdot \delta t)] = \begin{cases} 0, & \forall n \in \left[0, \dfrac{t_s^{ki}}{\delta t}\right]; \\ \mathcal{F}_{\varepsilon,k}(\varepsilon_t^i(n \cdot \delta t) - \varepsilon_m^k(n \cdot \delta t), & \forall n \in \left[\dfrac{t_s^{ki}}{\delta t}, \dfrac{t_w^{ki} - t_{at}^i}{\delta t}\right] \end{cases} \quad (4.14)$$

$$code[u^{\varphi}_{i,k}(n \cdot \delta t)] = \begin{cases} 0, & \forall n \in \left[0, \dfrac{t^{ki}_{s}}{\delta t}\right]; \\ \mathcal{F}_{\varphi,k}(\varphi^{i}_{t}(n \cdot \delta t) - \varphi^{k}_{m}(n \cdot \delta t), & \forall n \in \left[\dfrac{t^{ki}_{s}}{\delta t}, \dfrac{t^{ki}_{w} - t^{i}_{at}}{\delta t}\right] \end{cases} \tag{4.15}$$

The decrypting and decoding equipment (receiving path) is installed on the missile board. After decrypting and decoding the commands (4.16):

$$\begin{aligned} u^{\varepsilon}_{i,k}(n \cdot \delta t) &= \mathcal{F}^{-1}_{\varepsilon,k}\{code[u^{\varepsilon}_{i,k}(n \cdot \delta t)]\} \\ u^{\varphi}_{i,k}(n \cdot \delta t) &= \mathcal{F}^{-1}_{\varphi,k}\{code[u^{\varphi}_{i,k}(n \cdot \delta t)]\} \end{aligned} \tag{4.16}$$

comes to the missile autopilot and then to rudders, which control the missile position in 3-D space.

The advantages of the command system, Type I, are:

- Sufficient guidance accuracy when firing at short and medium ranges (up to 120 km);
- The ability to target several missiles at the same time;
- Sufficient simplicity, and accordingly, low cost of onboard equipment.

The main disadvantage of the command system, Type I, is the deterioration of guidance accuracy when increasing the firing range. So when testing the S-300 PM system, it was found that the permissible horizontal range, at which the guidance accuracy is ensured, when the target is hit with a probability not lower than the specified one, is limited from above $\bar{D_w} = 120$ km.

Command guidance system, Type II. This system differs from the Type I system in that the device for obtaining information about the spatial position of the target is located on board the missile. This function is performed by the onboard direction finder **b** (Figure 4.6), the antenna of which receives signals reflected from the target emitted by the target illumination radar 3 (Figure 4.6). In this case, the missile should be equipped with a radio-transparent nose cone. Data on the ith target received by the kth missile in its own coordinate system $\{\varepsilon^{ik}_{t}(\delta t), \varphi^{ik}_{t}(\delta t)\}$, after appropriate coding, is transmitted via a secure radio link to the illumination and guidance radar 3 (Figure 4.6) and then they are transmitted to the command post of the anti-aircraft missile system 2 (Figure 4.6). This also comes to the coordinates of the missiles, the spatial position of which is determined using radio responders installed on board the missiles and the receiving antenna of the missile channel 3(a) (Figure 4.6) installed on illumination and guidance radar 3.

The rest of the guidance process is similar to that of the Type I system. The advantages of the Type II command system are:

- High guidance accuracy, independent of the firing range;
- The possibility to select and recognize targets;
- The ability to target several missiles at the same time.

The disadvantage of this system is more sophisticated onboard equipment.

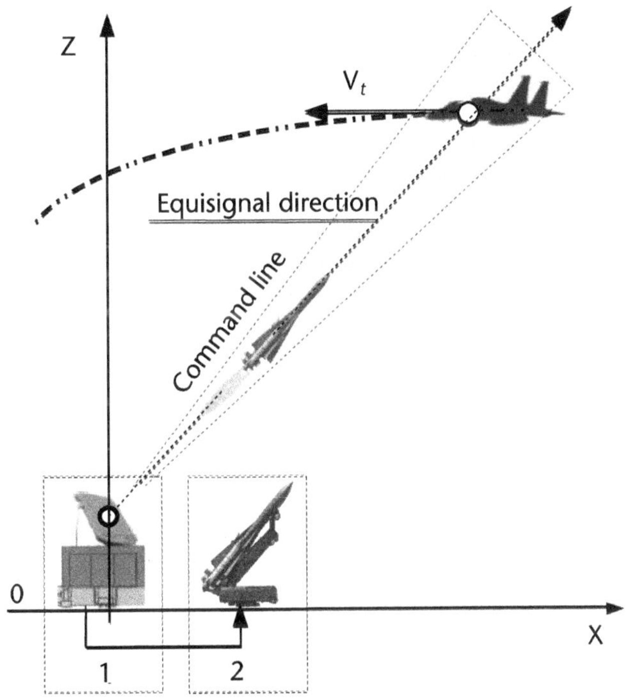

Figure 4.7 Single-beam guidance system.

4.2.2.2 Telehoming Systems

Telehoming systems are missile control systems, in which flight control commands are formed on board the missile. Their value is proportional to the deviation of the missile from the equal-signal direction created by control point radar sight 1. These systems are called guidance systems by radio beam. There are two types:

1. Single-beam guidance systems (Figure 4.7);
2. Two-beam guidance systems (Figure 4.8).

In a single-beam tracking system, the target must always be in the equal-signal direction of the beam formed by the tracking radar 1. The missile must also be in the same beam.

In a two-beam guidance system, there should be two radars: one radar 1 accompanies the target, and the second radar 3 controls the missile's flight and directs it to the target.

In any case, the missile starts from the launcher 2 only on the command "Start" coming from radar 1, which carries out target tracking. At the initial stage of development of air defense systems, the method of conical scanning was used with a single-beam guidance system for determining the equal-signal direction. Later, the conical scanning method began to be used when searching for a target in advance. The entered target coordinates into the onboard computer of the active homing head, which is used in long-range missiles.

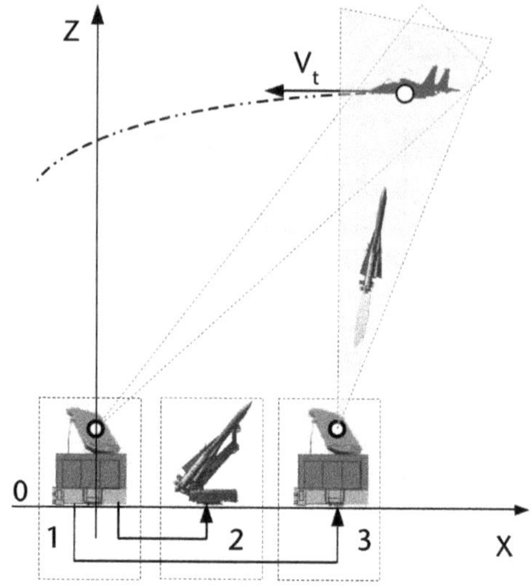

Figure 4.8 Two-beam guidance system.

4.2.2.3 Homing Systems

Homing systems are systems in which flight control is carried out by control commands formed on board the missile. These systems use homing missiles, in which the information necessary for the formation of control commands is issued by an onboard guidance device (i.e., a homing head). It should be especially noted that the ground control point does not take part in missile flight control!

Active homing systems (Figure 4.9). In these guidance systems, the source of the signal that irradiates the target is installed on board the missile. Signals reflected from the target are received by the onboard direction finder and are used to measure the mismatch between the missile velocity vector and the vector directed from the phase center of the seeker antenna to the target.

In Figure 4.9, the following numbering of the active homing system elements is adopted:

1. Active homing head, which carries out irradiation of the target designated for destruction with a radar signal, reception of the signal reflected from the target, and its processing and measurement of misalignment (pointing errors) in two perpendicular planes.
2. A digital computer complex, which participates in the processing of the received radar signal and calculates, according to the data received from the homing head, the missile rudder control commands.
3. Autopilot, which controls the missile rudders according to commands received from the digital computer complex. It also controls other parameters of the missile flight.

The same elements 1, 2, and 3 are used in the passive homing system (Figure 4.10), except that the homing head 1 does not emit any radar signals, but only works to receive signals emitted by the target.

Figure 4.9 Active homing system.

Figure 4.10 Passive homing system.

Semi-active homing systems (Figure 4.11). The most classic example of a semi-active homing system is the Soviet long-range system S-200 [4] (Figure 4.11). In semi-active homing systems, the target's radiation source is usually located on the ground. This is usually a target illumination radar 3 (Figure 4.11). The composition of the onboard equipment of the 5V28M missile of the system S-200 is the same as for missiles with a passive homing head (Figure 4.11). The radio signals reflected from the target are used by the seeker 1 placed on the 5V28M missile (Figure 4.11) to measure the parameters of the mismatch between the normal of the receiving

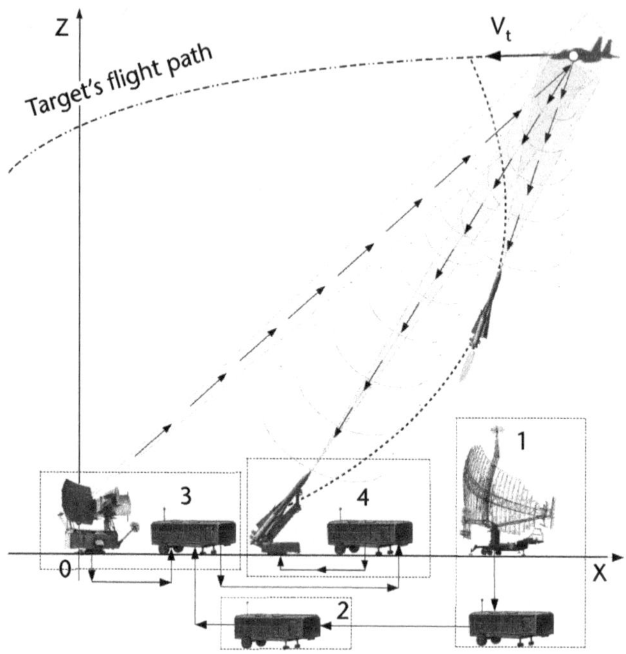

Figure 4.11 Homing SAM system.

antenna homing head and a straight line connecting the phase center of the homing head receiving antenna to the target. This figure shows an S-200 anti-aircraft missile division for one missile channel, with the following composition:

1. Early warning radar P-14 5N84A (TALL KING) as part of the antenna post and control cabin, with a detection range of 400 km for a target flying at an altitude of 30 km [5].
2. Command post, K9M cabin.
3. 5N62 target illumination radar (Square Pair), consisting of the K1 cabin (antenna post), and the cabin control K2.
4. 5P72 launcher with K3 missile preparation and launch control cabin. The launcher could be completed with two charging machines 5U24 moving along specially laid short rail tracks that are automated.

This system has shown its highest efficiency in working on real targets. It should be emphasized that in the period from 1985 to 1989, during the factory and state tests of the S-200D system at the Sary-Shagan test site, and then during the tests of the first production model from 1989 to 1991, all targets were successfully affected by the system C-200D. In addition, during the same period at the Sary-Shagan test site firing was successfully carried in a two-position mode, when the target was detected and tracking by a target illumination radar, which was located near the city of Balkhash, and the 5V28M missile was at the launch site located on the point 35 of the Sary-Shagan test site. The target was successfully taken for auto-tracking by the missile's homing head and was subsequently successfully destroyed.

It is especially necessary to emphasize the following:

1. In the S-200 system, the target illumination radar has a half-power main beam width of 0.7°, and therefore it cannot independently search for a target with such a narrow beam. The illumination and guidance radar can take only the target, the coordinates of which were obtained by automatic mode, either from its own detection radar or from a higher command post.
2. The homing head locks onto the target for auto-tracking when the missile is on the launcher and works on the reflected signal only from its target illumination radar.
3. If the homing head has ceased to see the signal reflected from the target, which was emitted by its target illumination radar, and within a specified amount of time (several seconds) this signal was not detected by the homing head, then the autopilot of the missile issues the command "maximum top" on the rudders, with the subsequent self-destruction of the missile in the upper atmosphere.

This algorithm has been tested at the Sary-Shagan test site and has proven to be highly effective. This is a so-called fool-proof system: It has been proven that the S-200 system cannot hit a target by mistake.

A flight accident with the destruction of an erroneous target by means of an S-200 air defense system is possible only if an erroneous external target designation is received for this particular target. In this case, "fool-proof protection" is useless!

Passive homing systems (Figure 4.10). The main distinguishing feature of passive homing systems is that they do not emit anything (i.e., only the receiving path exists in them). In passive homing systems, the energy emitted by the target itself is used to measure the parameters of the target's movement. Passive homing heads are classified according to that part of the electromagnetic wave spectrum to which the homing head receiver is optimally tuned. It can be secondary radar radiation of the target, and then the homing head is called a passive radar homing head. It can be ultraviolet radiation of a target or some object, and then it is called an ultraviolet homing head. This can be infrared radiation emitted by a running aircraft or missile engine. Satellites of the global missile attack warning system and many portable (wearable) anti-aircraft missile systems operate on this principle. In the missiles of these anti-aircraft missile systems, a thermal imaging head is installed homing, with or without cooling, which allows firing at a distance of several kilometers (no more than 6 km). Due to the fact that the mass of a portable air defense system is several kilograms, and as a rule does not exceed 20 kg, the firing range is also small. However, these portable air defense systems, such as the FIM-92 Stinger, became a formidable weapon in the fight against Soviet aviation during the war in Afghanistan. These complexes were especially effective in the mountains when a plane or helicopter came in for a landing.

4.2.2.4 Combined Guidance Systems

In combined guidance systems, guidance is carried out sequentially by several systems. They are used in long-range complexes. This can be a combination of a command telecontrol system at the initial stage of the missile's flight to the target and homing at the final stage of the missile's trajectory. Or, the option of radio beam guidance at the initial section of the missile flight path and homing at the final one is possible. The aggregation of various control systems with homing at the end of the trajectory allows missiles to be guided at long ranges with great accuracy.

4.2.3 Missile Warhead Detonation Control

At the stage of controlling the detonation of the warhead of the missile, the moment of detonation is selected, the direction of expansion of the striking elements of the warhead is controlled. The main task of this stage is to ensure that the target is hit with a probability not lower than the specified one. The control of the detonation of the warhead occurs in the area of the point where the missile meets the target. A special device, a radio fuse, controls the detonation of the warhead. The case of a direct missile hitting a high-speed, maneuverable, and in some cases low-flying target is unlikely. Therefore, it makes no sense to consider contact fuses. The warheads of anti-aircraft guided missiles are divided into fragmentation, high-explosive and cumulative. In missiles, fragmentation or high-explosive fragmentation warheads are used. The warhead is placed in a metal or plastic shell on top of which there are striking elements in the form of small cubes approximately 5 × 5 mm or small cylinders or balls, or striking elements of other geometric shapes. Each air defense system has its own unique handwriting characterized by the shape of the striking elements and the place of their maximum density on the body of the struck target. Based on these signs, it is most likely possible to determine the type and name of the air defense system that destroyed the target. There is a distinction between nondirectional and directional warheads [6]. For the reasons already mentioned above, it is reasonable to consider only directed combat units.

Warheads of directed action have a zone of dispersal of striking elements limited to a certain area of space. The scattering zone of the fragments can be symmetrical and asymmetrical relative to the longitudinal axis of the missile (relative to its velocity vector). The direction of the scattering of the fragments is achieved by choosing the points of initiation of the warhead detonation. At the same time, an autonomous missile fuse must solve not only the problem of determining the moment of detonation of the warhead, but also the problem of orienting the area of dispersion of fragments. Controlling the position of the scattering area relative to the target can be achieved by rotating the missile around its longitudinal axis [7]. The main characteristics of a fragmentation warhead are:

- The number of damaging elements n_{el};
- The mass of one striking element m_{el};
- The shape and size of striking elements (fragments);
- The density of striking elements distribution;
- The static area of fragments dispersion;
- The initial velocity of fragments, V_{el};
- The nature of the change in the speed of dispersion of fragments, depending on the conditions of undermining the warhead.

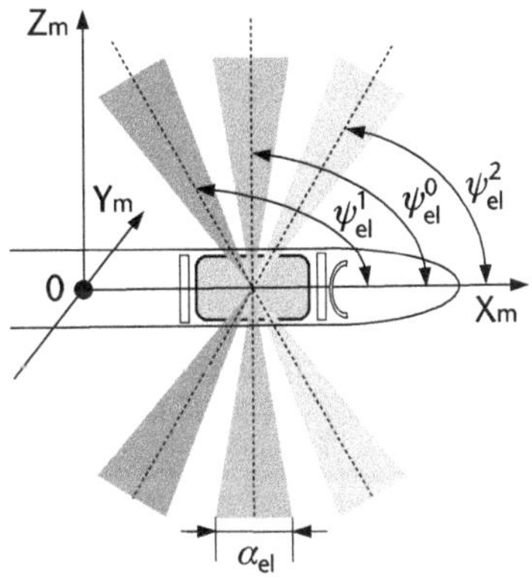

Figure 4.12 Fragments' scattering areas.

The number of striking elements (fragments) depends on the volume of the warhead and the optimal mass of each striking element. The mass of the striking element at a given expansion velocity determines its kinetic energy E^k_{el} and is selected from the condition:

$$E^k_{el} = \frac{m_{el}V^2_{el}}{2} \geq WSh, \tag{4.17}$$

where W is the specific crowding out energy per unit volume of the target's structural element, depending on the material, S is the area of the hole, and h is the thickness of the barrier. The shape and size of striking elements are selected, as a rule, empirically. The distribution density of striking elements depends on their number and the size of the area (cloud) of scatter. The density of the fragments' scattering cloud is maximum in the area of the warhead detonation point and decreases from it according to a certain law. The static region of fragments scattering in the coordinate system associated with the missile $X_mY_mZ_m$, in the center 0 of which is the center of the mass of the missile (Figure 4.12), is a volumetric figure, which is characterized by the static angle of the fragments scattering α_{el} and the angle of inclination of the bisector of the fragments scattering region Ψ_{el}. Here $\Psi^1_{el} > 90°$, $\Psi^0_{el} = 90°$, and $\Psi^2_{el} < 90°$. The angle of inclination of the bisector of fragments scattering area Ψ^i_{el} depends on the location of the initiation points: if the initiation point is located in the middle of the charge, then $\Psi^0_{el} = 90°$ and the area of fragment scattering is located perpendicular to the missile body. If it is in the rear, then $\Psi^2_{el} < 90°$ and the area of fragment scattering is directed forward in the direction of the missile movement. In the event that the detonation of the warhead is carried out in front of the warhead, then $\Psi^1_{el} > 90°$ and the area of fragment scattering is turned back in relation to the missile velocity vector. For control the angle of inclination of the bisector Ψ_{el} of the scattering angle of the fragments α_{el}, several initiation points are installed in the warhead and the warhead is detonated at the point that will provide maximum target coverage with a cloud of fragments. The choice of the point of initiation and the moment of detonation of the warhead is determined automatically by the command of the radio fuse. In real conditions, the warhead is detonated when the missile is flying at a certain speed $\vec{V}_m$. Since in the case of $\Psi^0_{el} = 90°$ the vector of the scattering velocity of the fragments $\vec{V}_{el}$ is directed perpendicular to the missile body, and then after detonation, the striking elements receive an additional velocity $\vec{V}_m$. As a result, the dynamic velocity of the scattering of fragments $\vec{V}_{el}^d$ can be written as:

$$\vec{V}_{el}^d = \vec{V}_m + \vec{V}_{el} \tag{4.18}$$

The dynamic region of fragment scattering is characterized in this case by the dynamic scattering angle of fragments α_{el}^d and the dynamic angle of inclination of the bisector Ψ_{el}^d of the fragment scattering region. It is quite clear that due to (4.18), $\alpha_{el}^d < \alpha_{el}$ and this inequality will increase as $|\vec{V}_m|$. In addition, the dynamic tilt angle of the region Ψ_{el}^d will also decrease with increasing $|\vec{V}_m|$. Thus, the dynamic region of the fragment scattering can be represented as two cones with tops at the point of detonation of the warhead, which are directed along the missile velocity vector $\vec{V}_m$. It should be understood that the angle between the generatrices of the outer and inner cones is equal to the dynamic scattering angle of striking elements α_{el}^d, and the bisector between these generators is equal to the dynamic angle of bisector inclination Ψ_{el}^d.

For successful target destruction, it is necessary to coordinate the spatial position of the target with the area of fragments dispersion in which the density of striking elements per unit volume is maximum. Therefore, the main tasks of a radio fuse are:

1. Warhead detonation moment control;
2. Spatial position control of the dynamic area of expansion of the warhead's striking elements.

For example, in 5Ya23 missile of the S-75 air defense missile system, control of the spatial position of the dynamic area of expansion of the warhead's striking elements was carried out by changing the spatial position of the radio fuse main beams' directivity patterns of four waveguide slot antennas, located on the missile body oppositely relative to its axis $0\vec{X}_m$ in two mutually orthogonal planes Y_m0X_m and Z_m0X_m. Figure 4.13 shows the process of good (standard) matching of the area of striking elements dispersion (1) with the area of operation of the radio fuse (2), determined by the spatial position of the main beams of the directional diagram of the radio fuse. Figure 4.13 shows an unacceptable matching when target destruction is not ensured due to late detonation, in the case when at the moment of detonation the directional diagram of the radio fuse is in position (3), or due to early detonation, when its directional pattern at the moment of the explosion is in position (4).

4.3 Flight Trajectories and Guidance Methods

The method of guiding an anti-aircraft guided missile is a certain law according to which the missile approaches the target (i.e., establishes a connection between the law of target movement and the law of missile movement). It is clear that aiming a missile at a target in time will inevitably occur with some error $\Delta(t)$:

$$\Delta(t) = X(t) - Y(t), \tag{4.19}$$

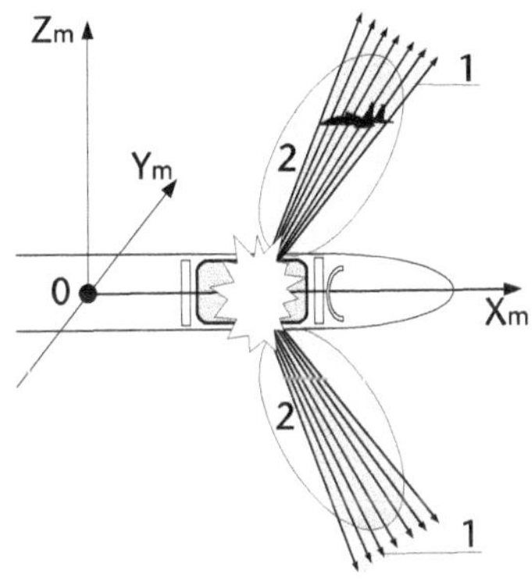

Figure 4.13 Good coordination of striking elements dispersion area and the radiation pattern of the radio fuse.

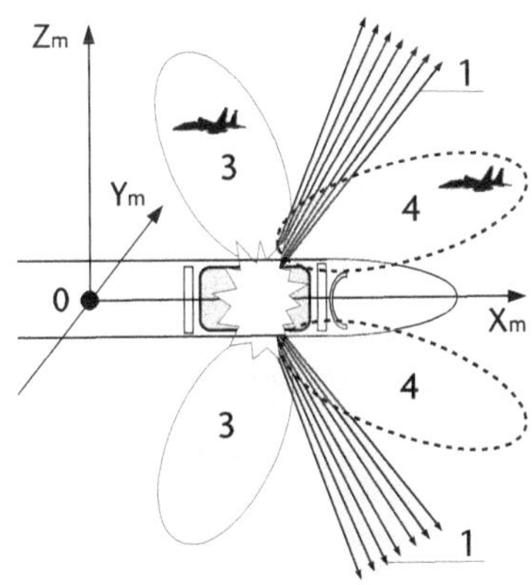

Figure 4.14 Unacceptable coordination of striking elements dispersion area and the radiation pattern of the radio fuse.

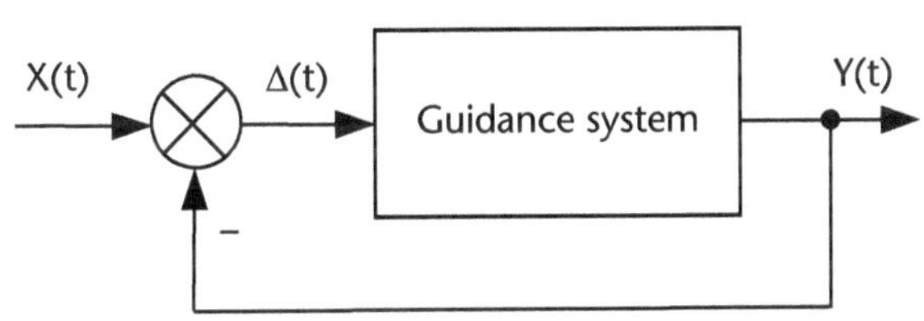

Figure 4.15 Guidance system.

where $X(t)$ is the given law missile motion, $Y(t)$ is the realized law missile motion, and $\Delta(t)$ is the mismatch parameter between given and actual laws of missile motion. The guidance system is a closed-loop control system (see Figure 4.15). The guidance method determines the form of representation of the mismatch parameter. With some guidance methods, the mismatch parameter is represented in angular measure, while others are in the form of the rotation angular velocity of the line of the missile-target, (i.e., the guidance method determines the law of the missile's approach to the target).

If, when aiming the missile at the target, the law of approach of the missile-target is fulfilled, then in this case the mismatch parameter is zero and the missile moves along a given trajectory. If the law is violated, guidance errors occur, which

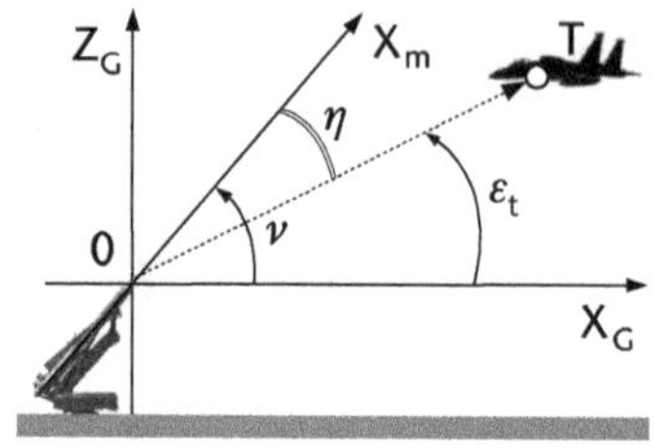

Figure 4.16 Direct guidance method.

ultimately reduce the likelihood of hitting the target. The mismatch parameter is an input for the guidance system, according to which control commands are subsequently formed, the execution of which returns the missile to the calculated trajectory. The mismatch parameter is measured by ground or airborne devices. As a result of measurements, $\Delta(t)$ is represented as voltage (current) in analog or digital form:

$$U_{\Delta(t)} = K\Delta(t), \tag{4.20}$$

where K is the gain of the control loop.
Basic requirements for guidance methods include:

- The smallest curvature of the missile trajectory, including firing at maneuvering targets;
- The simple instrumental implementation;
- A given firing range.

Guidance methods are divided into two large groups:

1. Two-point, used in homing systems;
2. Three-point, used in telecontrol systems [8].

4.3.1 Two-Point Guidance Methods: Homing Methods

The essence of the two-point guidance methods follows from the name itself, since it implies the constant determination of the relative position of two points: the target and the missile. The most widely used methods are direct guidance method, method of direct guidance with a constant lead angle, chase curve guidance method, parallel rapprochement, and proportional rapprochement methods.

4.3.1.1 Direct Guidance Method

The direct guidance method imposes a strict requirement on the spatial position of the longitudinal axis of the missile. It is necessary that for the entire time of the missile's flight to the target, its longitudinal axis must be matched with line of the missile-target. Let time t be when the missile is at the starting point, at point 0 (Figure 4.16), and the target is at point T, while the line of sight missile-target is located at an angle to the horizon ε_t. The angle between the longitudinal axis of the missile $[0, X_m)$ and the horizontal plane (straight line $[0, X_G)$ is equal to υ-the pitch angle. The angle between the longitudinal axis of the missile $[0, X_m)$ and the line of sight missile-target $[0, T)$ is η. From Figure 4.16 it follows that the mismatch parameter:

$$\Delta = \eta \tag{4.21}$$

$$\Delta = \upsilon - \varepsilon_t, \tag{4.22}$$

Based on the equation $\eta = 0$, you can write the guidance equation:

$$\upsilon = \varepsilon_t \tag{4.23}$$

From expressions (4.21) and (4.22) it can be seen that the mismatch parameter can be determined by measuring the angle η or by measuring the angles ε_t and υ with their subsequent subtraction. Both options for determining the mismatch parameter are implemented in the S-200 air defense system. The first option is used in the case when the homing head has taken the target for automatic tracking in angular coordinates. In this case, the missile can be on the launcher. The second option is implemented in the target designation mode to the missile, when the homing head has not yet captured the target and the missile is on the ground, on the launcher. In this case, the target elevation angle ε_t is determined with great accuracy by the target illumination radar, and the υ angle is formed by the angular position sensor of the launcher guide.

For the direct guidance method, the required normal acceleration of the missile increases as the missile approaches the target. Therefore, the required normal acceleration to ensure the movement of the missile along the kinematic trajectory is directly proportional to the target's speed and depends on the nature of its maneuver. The considered method can be implemented mainly for shooting at stationary, slightly mobile and nonmaneuvering targets.

4.3.1.2 Method of Direct Guidance with a Constant Lead Angle

With the direct guidance with a constant lead angle method, the angle η_0 between the longitudinal axis of the missile $[P_{i-1}, P_i],\ for\ i \in [1, N]$ and the line of sight missile-target $[P_i, T_i]$ remains constant during the entire time of the missile flight to the point of meeting with the target. This method is a development of the direct guidance method.

As can be seen from Figure 4.17, the equation of the mismatch parameter Δ can be represented as:

$$\Delta = \eta - \eta_0 \tag{4.24}$$

Taking into account the equality $\eta = \upsilon - \varepsilon_t$, obtained from (4.21) and (4.22), the equation for Δ will be rewritten as follows:

$$\Delta = \upsilon - \varepsilon_t - \eta_0, \tag{4.25}$$

where η is the angle between the current position of the longitudinal missile axis $[P_0, X_1]$ and line of sight missile-target $[P_0, T_0]$, and η_0 is the angle between the

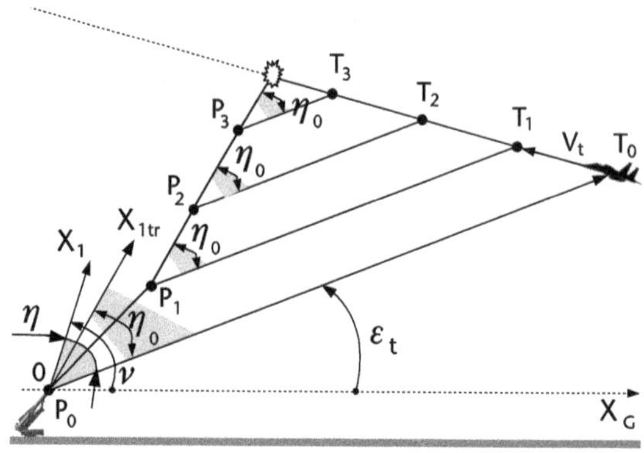

Figure 4.17 Direct guidance method in constant lead.

required position of the longitudinal missile axis $[P_0, X_{1tr}]$ and line of sight missile-target $[P_0, T_0]$. Hence the equation, which when fulfilled, the missile will move along a kinematic trajectory and will have the form:

$$\eta = \eta_0 \tag{4.26}$$

$$\varepsilon_t = \upsilon - \eta_0 \tag{4.27}$$

The technical implementation of this guidance method is similar to the direct guidance method (4.23). In anti-aircraft missile systems, this guidance method can be used only for firing at targets moving in a straight line and at low speed. In other cases, it can only be used as an auxiliary guidance method used in conjunction with other guidance methods.

4.3.1.3 Chase Curve Guidance Method

The chase curve guidance method requires that the vector of the true velocity of the missile $\vec{V}_m$ is continuously aligned with the line of sight missile-target $\vec{V}_{mt}$. Some variation of this method is the weathervane method, when the velocity vector of the missile moving in the air is continuously aligned with the line of sight $\vec{V}_{mt}$ $\vec{V}_{m,atm}$. Taking into account that the condition of the missile movement in an unperturbed atmosphere is fulfilled:

$$\vec{V}_m = \vec{V}_{m\_atm} \tag{4.28}$$

we will consider both methods to be identical (Figure 4.18). In the elevation plane ε: $[0, X_{adf}]$ is the equal-signal direction of the homing head antenna, $[0, X_1]$ is the direction of the longitudinal axis of the missile, $\vec{V}_m$ is the vector of the true missile velocity, and $\vec{V}_t$ - target velocity vector.

In this case, the equations of the mismatch parameter take the following form:

$$\Delta = \xi, \tag{4.29}$$

$$\Delta = \varepsilon_m - \varepsilon_t, \tag{4.30}$$

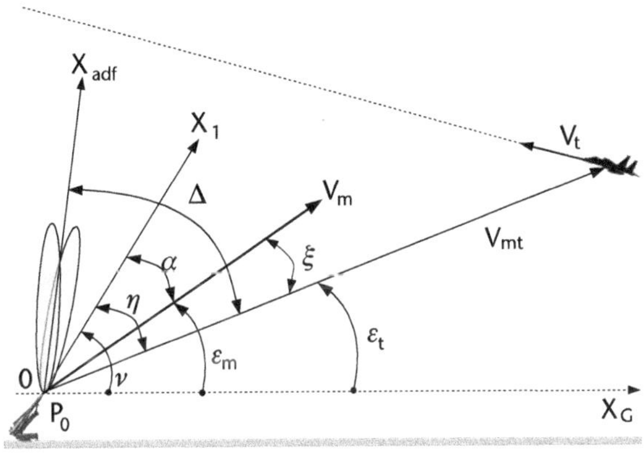

Figure 4.18 Chase curve guidance method.

$$\Delta = \eta - \alpha, \tag{4.31}$$

The equations of correspondence between the missile flight path and the kinematic trajectory will take the form:

$$\xi = 0 \tag{4.32}$$

$$\varepsilon_m = \varepsilon_t \tag{4.33}$$

$$\eta = \alpha \tag{4.34}$$

4.3.1.4 Parallel Rapprochement Method

When using the parallel rapprochement method, the line of sight of the missile-target remains parallel to the given direction during the entire flight time of the missile from the moment it is brought to the kinematic trajectory until it meets the target.

In the elevation plane ε, the line of sight of the missile-target will move parallel to itself if one of the three conditions is met (Figure 4.19):

$$\dot{\varepsilon} = 0 \tag{4.35}$$

$$\varepsilon = \varepsilon_{R0} \tag{4.36}$$

$$V_t \cdot \sin(q_t) = V_m \cdot \sin(q_m) \tag{4.37}$$

or

$$q_m = \arcsin\left(\frac{V_t}{V_m} \cdot \sin(q_t)\right), \tag{4.38}$$

where $\dot{\varepsilon}$ is the angular velocity of line of sight rotation of the missile-target, ε_{R_0} is the angle of inclination of the line of sight of the missile-target at the moment of starting guidance, and q_t, q_m is the angle between the line of sight of the missile-target and the velocity vector of the target and the missile, respectively. Equations of ideal

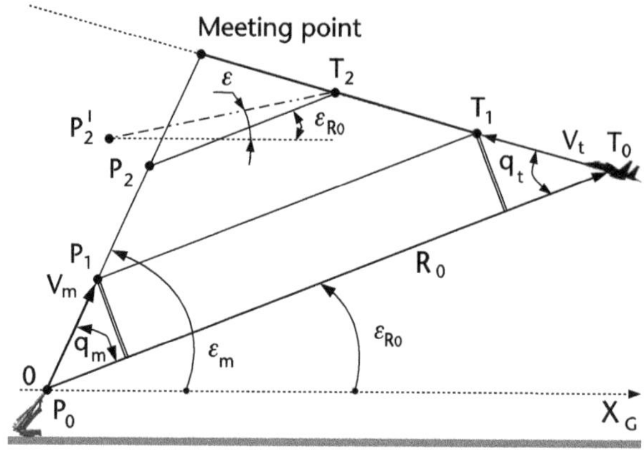

Figure 4.19 Parallel rapprochement method.

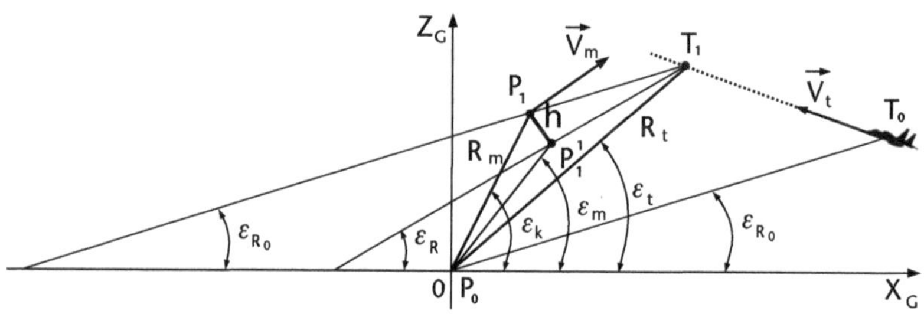

Figure 4.20 Parallel rapprochement method in teleguidance systems.

coupling (4.35), (4.36), and (4.38) will correspond to the following equations of the mismatch parameter Δ:

$$\Delta = \dot{\varepsilon} \tag{4.39}$$

$$\Delta = \varepsilon - \varepsilon_{R0} \tag{4.40}$$

$$\Delta = q_m - \arcsin\left(\frac{V_t}{V_m} \cdot \sin(q_t)\right) \tag{4.41}$$

The parallel rapprochement method can also be used in the guidance of anti-aircraft guided missiles in teleguidance systems, which use the generation of guidance commands at the ground control point and their subsequent transmission to the missile through a secure radio link ground control point-missile.

Parallel Approach Method, When Guiding Anti-Aircraft Guided Missiles in Command Guidance Systems

Suppose that the origin of the Earth coordinate system $X_G 0 Z_G$ coincides with the position of the ground control station (Figure 4.20).

At the start of control, the target is at the point T_0, and the missile is at the point P_0, the initial angle of inclination of the line of sight of the missile-target is equal to ε_{R_0}. After some time, the target moved to the T_1 point. It is clear that to ensure a plane-parallel movement, the missile must move to the point P_1 and in this case the equation $\varepsilon_R = \varepsilon_{R_0}$ will be fulfilled. Using the sine theorem and the triangle

$0P_1T_1$ we can write:

$$\frac{R_m}{\sin(\varepsilon_t - \varepsilon_{R_0})} = \frac{R_t}{\sin(\varepsilon_k - \varepsilon_t)} \tag{4.42}$$

Therefore, the required kinematic angle ε_k can be written as:

$$\varepsilon_k = \varepsilon_t + \arcsin\left[\frac{R_t}{R_m}\sin(\varepsilon_t - \varepsilon_{R_0})\right] \tag{4.43}$$

If the missile is at the point P_1^1 instead of the supposed point P_1, therefore, in this case, the mismatch equation will be written as:

$$\Delta = \varepsilon_k - \varepsilon_m = \Delta_\varepsilon \tag{4.44}$$

and for linear mismatch it will be true:

$$\Delta h = \Delta_\varepsilon R_m \tag{4.45}$$

Taking into account the expressions (4.44) and (4.45), the equation of the mismatch parameter can be written in the form:

$$\Delta\varepsilon = \left\{\varepsilon_t - \varepsilon_m + \arcsin\left[\frac{R_t}{R_m}\sin(\varepsilon_t - \varepsilon_{R_0})\right]\right\} \cdot R_m \tag{4.46}$$

For small values of the angle difference, the following can be written:

$$\begin{aligned} \sin(\varepsilon_t - \varepsilon_{R_0}) &\approx \varepsilon_t - \varepsilon_{R_0} \\ \sin(\varepsilon_k - \varepsilon_t) &\approx \varepsilon_k - \varepsilon_t \\ R &\approx R_t - R_m \end{aligned} \tag{4.47}$$

from where the equation for the mismatch parameter in the linear measure for the elevation angle control plane ε follows:

$$\Delta h_\varepsilon = R_m \cdot \left[(\varepsilon_t - \varepsilon_m) + \frac{(R_t - R_m)}{R_m} \cdot (\varepsilon_t - \varepsilon_{R_0})\right] \tag{4.48}$$

A similar expression can be written for the orthogonal control plane β:

$$\Delta h_\beta = R_m \cdot \left[(\beta_t - \beta_m) + \frac{(R_t - R_m)}{R_m} \cdot (\beta_t - \beta_{R_0})\right] \tag{4.49}$$

From an equation analysis of (4.48) and (4.49), it follows that at the ground control station, it is necessary to measure the angular coordinates and ranges of the target and several missiles that are guided at this target. Along with this, knowledge and memorization of the angle of inclination of the line of sight of the missile target at the moment of beginning control is required.

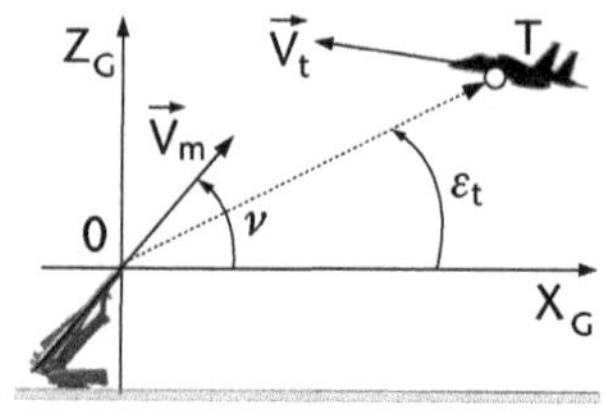

Figure 4.21 Proportional rapprochement method.

4.3.1.5 Proportional Rapprochement Method

With the Proportional rapprochement method, the required missile movement is determined by the following condition: the rotation speed of the missile velocity vector $\dot{\upsilon}$ remains proportional to the angular rotation velocity of the line of sight of the missile-target $\dot{\varepsilon_t}$ (Figure 4.21):

$$\Delta = k\dot{\varepsilon_t} - \dot{\upsilon_\varepsilon} \tag{4.50}$$

where k is the proportionality coefficient. In this case, the equation of ideal coupling, when the tracking error Δ is equal to zero, will have the form:

$$\dot{\upsilon_\varepsilon} = k\dot{\varepsilon_t} \tag{4.51}$$

The equations in this method for two mutually orthogonal coordinates β and ε, taking into account that $\dot{\upsilon_\varepsilon} = \dot{\varepsilon}_m$ and $\dot{\upsilon_\beta} = \dot{\beta}_m$, will be written as:

$$\begin{aligned} \dot{\varepsilon}_m &= k_\varepsilon \dot{\varepsilon_t} \\ \dot{\beta}_m &= k_\beta \dot{\beta}_t, \end{aligned} \tag{4.52}$$

This method belongs to the group of methods with variable lead angle. With ideal guidance, the law of variation of the required lead angle can be found from (4.51), provided that $q_\varepsilon = \upsilon_\varepsilon - \varepsilon_t$ and $q_\beta = \upsilon_\beta - \beta_t$ and oblique ranges r_t and r_m, respectively. Unlike the parallel approach method, when instant elimination of the angular rotation of the line of sight of the missile-target is required, with proportional approach it is allowed to decrease the angular rotation velocity of the line of sight of the missile-target.

Another equally important advantage of this method is that for $V_m/V_t > 1$ and $k > 4$, normal required overloads of the missile in the vicinity of the rendezvous point tend to be zero regardless of the initial conditions of its start [6].

It should be noted that parallel rapprochement and chase methods can be derived from the proportional rapprochement method equation (4.51). For $k = \text{inf}$ this equation turns into the equation of the parallel approach method ($\varepsilon = 0$), and for $k = 1$ this will be the equation of method chase ($\varepsilon = \upsilon$). It has been established from practice that for $k = 4...6$ the trajectory of the method differs slightly from the trajectory of the parallel approach method, and therefore it can be considered as an approximate implementation of the parallel approach method.

4.3.2 Three-Point Guidance Methods: Teleguidance Methods

When guiding a missile at a target, the three-point methods determine the relative position of three points: the target, the missile, and the point of control (guidance). At present, three-point methods and preemption methods are most widely used to solve the problems of anti-missile and anti-aircraft defense. The position of the target and the missile relative to the guidance point is uniquely characterized by the angular coordinates of the target (ε_t, β_t) and the missile (ε_m, β_m) and inclined ranges r_t and r_m, respectively. The relative position of the target and the missile is estimated by the difference between these coordinates: $\Delta\varepsilon = \varepsilon_t - \varepsilon_m$; $\Delta\beta = \beta_t - \beta_m$; $\Delta r = r_t - r_m$.

For the first time, the three-point missile guidance method was applied in the Soviet missile defense system A (General Designer, G. V. Kisunko) and it was then called the method of three ranges [9]. The missile defense system A was deployed at the Sary-Shagan test cite in 1957 and included

1. Radar Dunay-2 for early detection of ballistic missiles, located on the shore of Lake Balkhash, near the town of Priozersk.
2. Launching position with V-1000 anti-missile launchers (general designer S. A. Lavochkin), located on site 6, which was in the center of the test cite, west of Priozersk.
3. Three radars for accurate guidance of anti-missile missiles to the target, which were located on test sites 1, 2, and 3, were placed in the corners of an almost isosceles triangle. The farthest from Priozersk was test site 2, which was located in the western part of the Sary-Shagan test cite. In the 1980s, target-missiles were launched from site 2.

   Each of these three radars consisted of two radars:

   3.1: Radar for determining target coordinates;

   3.2: Radar for determining missile coordinates.
4. An anti-missile launching and sighting radar with an anti-missile flight control station combined with this radar.
5. The main command and computer center with a central computer M-40, located at site 40, in Priozersk.
6. Radio relay communication lines between all means of the missile defense system.

According to the algorithm of the A system, early warning radar Dunay-2 determines the current coordinates of the target and transmits them to the central computer M-40, which issues target designation for each of three precision guidance radars and determines the lead coordinates of the target. According to the coordinates of the target received from the central computer, each of the three precision guidance radars (3.1, radar for determining target coordinates) searches for and captures for automatic tracking of ballistic target. The specified coordinates of the target are given by these radars to the central computer, which calculates the parameters of guidance of the anti-missile to the target, and calculates and gives out to the launch position the angles of rotation for radar antennas of launching and sighting. The central computer determines the moment of launch and automatically issues the command "Start" on the anti-missile. After missile launch, an anti-missile launching and sighting radar captures it for auto-tracking and transmits the data to the central computer M-40. Coordinate information about the target is constantly supplied to the central computer from three precision-guided radars and about the anti-missiles from anti-missile launching and sighting radars, and the computer calculates their trajectories, and then, through the command transmission stations, sends commands to board for anti-missiles to output to the starting point of the precision guidance, on a trajectory close to the target flight path. On a command from the central computer, the precision guidance radar (3.2, radar for determining the coordinates of the missile) captures the anti-missile for automatic tracking, after which the stage of precise guidance begins in which the spatial positions of the target and the missile are determined by the three-range method. It is essentially a triangulation method using known ranges, elevation, and azimuths angles of the target and missile in their own coordinate systems corresponding to each of three separate radars, as well as the exact distances between the vertices of the measuring triangle.

This method of measuring the coordinates of the target and the missile made it possible to achieve a miss of 21 meters in launch No. 68 on November 24, 1960 [9], carried out on a real ballistic target, which was an R-5 medium-range ballistic missile known as SS -3 Shyster. This is an excellent result, almost a direct hit on a high-speed target, whose entry speed into the dense layers of the atmosphere was 2.992 km/sec or 9.067M [10].

4.3.2.1 Tree-Points Method

In the three-point method, the missile must be constantly on the line of sight of the ground radar-target and during the guidance process, since only one straight line can be drawn through two points, in this guidance method the missile (third point) behaves as if crawling along the line of sight of the ground radar-target from the start point to the target.

The condition for the implementation of this method is that the angular coordinates of the missile $(\varepsilon_{mk}, \beta_{mk})$ at any moment must be equal to the angular coordinates of the target (φ_t, β_t) [11]:

$$\begin{cases} \varepsilon_{mk} = \varepsilon_t \\ \beta_{mk} = \beta_t, \end{cases} \tag{4.53}$$

where $(\varepsilon_{mk}, \beta_{mk})$ are the angular coordinates of the points of the missile kinematic trajectory. If the condition (4.53) is exactly met, the missile will move along a kinematic trajectory $0 - 1^1 - 2^1 - 3^1 - 4^1 - 5$ (Figure 4.22). When using optical sighting devices, the operator must control the missile in such a way that its image is constantly aligned with the target image. The mismatch parameter in this case characterizes the angular mismatch of the true position of the missile from its kinematic trajectory (from the target sight line).

$$\begin{cases} \Delta_\varepsilon = \varepsilon_{mk} - \varepsilon_m \\ \Delta_\beta = \beta_{mk} - \beta_m, \end{cases} \tag{4.54}$$

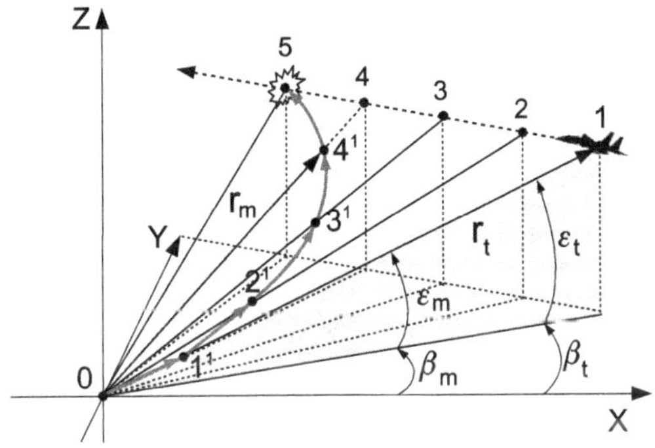

Figure 4.22 The three-points guidance method.

Because the condition of the three-point method is condition (4.53), therefore it is possible to write:

$$\begin{cases} \Delta_\varepsilon = \varepsilon_t - \varepsilon_m \\ \Delta_\beta = \beta_t - \beta_m, \end{cases} \tag{4.55}$$

The advantages of this method are:

- There is no need to measure the range to the target r_t and to the missile r_m in the process of guiding the missile to the target. This allows you to use only one radar and fire at a target when it is impossible to measure the range to the target, for example, in the case when the enemy uses interference against the air defense system. This targeting method is used in air defense systems, including as a means of increasing its noise immunity.
- Simple instrumental implementation of devices for measuring the coordinates of a target, a missile and a calculating device that calculates the error signals $\Delta_{\varepsilon,\beta}$.
- The ability to use optical sighting devices without measuring range to target r_t.

The disadvantages of this method are:

- The trajectory of the missile has a significant curvature. This requires large available overloads from the missile. Since this requirement is not always met, aiming accuracy and firing range decrease with an increase in target speed.
- The possibility of firing at low-flying targets is impeded, since being in the line of sight of local objects does not allow the missile to be aimed at the target.

In the Soviet S-75M3 air defense system, a three-point firing method was implemented called 3T in the troops. The shelling of targets was provided at a flight speed of up to 3,600 km/h (1,000 m/s) and the lower border of the affected zone of 100m. This air defense system also implemented the firing mode "in pursuit of" and the mode of "firing at a ground target." In addition, after it was equipped with a 9Sh33A television optical sight, this system received the code S-75M4 and made it possible to fire targets in optical mode. My experience of service in the Belarusian forests on this air defense system indicates that in order to eliminate the influence of local objects on the line of sight of the missile-target when firing at low elevation angles (at low-altitude targets) in our anti-aircraft missile battalion, the antenna post (cabin P) was raised on an embankment with a height of about 20m, plus it

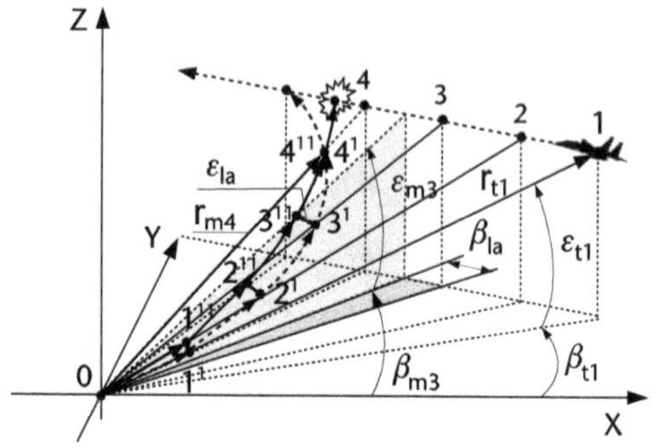

Figure 4.23 The preemption method.

was constantly necessary to saw pine treetops around the position. An antenna post equipped with a 9Sh33A television-optical sighting device and introducing an optical target tracking channel made it possible, under conditions of visual observation of an air target, to conduct its tracking and firing without the use of air defense radar systems in radiation mode. It was reported that it was precisely the presence of the thermal imaging sight that allowed the Yugoslav air defense crew to destroy the American F-117 aircraft in March 1999 [12].

4.3.2.2 Preemption Methods

In preemption methods, the missile is guided to the target with a certain lead angle along the target's velocity vector. Due to this, the curvature of the trajectory is reduced, which makes it possible to increase the guiding accuracy and firing range. The condition for the implementation of this method is the equality: (4.53). If condition (4.53) is exactly met, then the missile will move along a kinematic trajectory $0-1^{11}-2^{11}-3^{11}-4^{11}-$ *meeting point* (Figure 4.23). If at some moment in time t the target is at point 1, and the line of sight is at position [0,1], then with accurate guidance the missile should be at point 1^{11}, anticipating the line of sight target for some values of the elevation angle ε_{la1} and azimuth β_{la1}. When the target is at point 3, the missile must be at point 3^{11} with the lead ε_{la3}, β_{la3}. From Figure 4.23, it can be seen that as the missile approaches the target, the lead value decreases and at the meeting point is zero. Curve $0-1^{1}-2^{1}-3^{1}-4^{1}-$ *meeting point* is the kinematic trajectory of the missile movement according to the three-point guidance method.

The equation of the preemption method can be written as

$$\begin{cases} \varepsilon_{mk} = \varepsilon_t + \varepsilon_{la} \\ \beta_{mk} = \beta_t + \beta_{la} \end{cases} \tag{4.56}$$

The amount of lead depends on the adopted law of input of the components ε_{la} and β_{la}, in this regard, the methods of full straightening and partial straightening of the trajectory are distinguished. The lead is proportional to the rate of change in the angular coordinates of the target and rapprochement time missile to target t_r

$$\begin{cases} \varepsilon_{la} = \dot{\varepsilon}_t \cdot t_r \\ \beta_{la} = (\dot{\beta}_t \cdot \cos(\varepsilon_t)) \cdot t_r \end{cases} \tag{4.57}$$

where $t_r = (r_t - r_m)/\Delta\dot{r} = \Delta r/\Delta\dot{r}$, and $\Delta r = r_t - r_m$. Equation (4.57) can be rewritten as

$$\begin{cases} \varepsilon_{la} = \dfrac{\dot{\varepsilon}_t}{\Delta\dot{r}} \cdot \Delta r \\ \beta_{la} = \dfrac{(\dot{\beta}_t \cdot \cos(\varepsilon_t))}{\Delta\dot{r}} \cdot \Delta r \end{cases} \tag{4.58}$$

In (4.57) and (4.58), the factor $\cos(\varepsilon_t)$ allows for recalculating the rate of change of the target azimuth, measured in the horizontal plane, into the inclined plane of guiding. With the preemption method (lead method), the mismatch parameter in the angular measure will be equal to

$$\begin{cases} \Delta\varepsilon_{la} = \varepsilon_{mk} - \varepsilon_m = (\varepsilon_t + \varepsilon_{la}) - \varepsilon_m = \varepsilon_t - \varepsilon_m + \varepsilon_{la} \\ \Delta\beta_{la} = \beta_{mk} - \beta_m = (\beta_t + \beta_{la}) - \beta_m = \beta_t - \beta_m + \beta_{la} \end{cases} \tag{4.59}$$

Whereas $\varepsilon_t - \varepsilon_m = \Delta\varepsilon$ and $\beta_t - \beta_m = \Delta\beta$, it can be written as

$$\begin{cases} \Delta\varepsilon_{la} = \Delta\varepsilon + \varepsilon_{la} \\ \Delta\beta_{la} = \Delta\beta + \beta_{la} \end{cases} \tag{4.60}$$

From (4.60) it follows that the use of the mismatch parameter in the angular measure which will lead to an increase in guidance errors with increasing missile range (as in the three-point method), and therefore, the mismatch parameter is taken as a linear value equal to

$$\begin{cases} h_{\Delta\varepsilon_{la}} \approx (\Delta\varepsilon + \varepsilon_{la})r_m \\ h_{\Delta\beta_{la}} \approx (\Delta\beta + \beta_{la})r_m \end{cases} \tag{4.61}$$

or

$$\begin{cases} h_{\Delta\varepsilon_{la}} = h_{\Delta\varepsilon} + h_{la\ \varepsilon} \\ h_{\Delta\beta_{la}} = h_{\Delta\beta} + h_{la\ \beta} \end{cases} \tag{4.62}$$

The mismatch parameter in (4.62) for each of the control planes ε and β contains the values and $h_{\Delta\varepsilon}$ and $h_{\Delta\beta}$, respectively, which characterize the deviation of the missile from the line of sight of the target, and the components of the lead $h_{la\ \varepsilon}$ and $h_{la\ \beta}$

$$\begin{cases} h_{la\ \varepsilon} = h_{r\ \varepsilon} = \varepsilon_{la} r_m = (\dot{\varepsilon}_t/\Delta\dot{r}) \times r_m \Delta r \\ h_{la\ \beta} = h_{r\ \beta} = \beta_{la} r_m = (\dot{\beta}_t \cos(\varepsilon_t)/\Delta\dot{r}) \times r_m \Delta r \end{cases} \tag{4.63}$$

The guidance method in which the lead constituents $h_{la\ \varepsilon}$ and $h_{la\ \beta}$ are introduced according to (4.63) is called the trajectory full straightening method. With this guidance method, the missile's flight path remains curvilinear, and its curvature is minimal. In case the guidance method uses only half of the full straightening values $h_{la\ \varepsilon}$ and $h_{la\ \beta}$, this guidance method is called the method of half straightening missile trajectory:

$$\begin{cases} h_{la\ \varepsilon} = h_{r\ \varepsilon} = \dfrac{1}{2} \times (\dot{\varepsilon}_t / \Delta\dot{r}) \times r_m \Delta r \\ h_{la\ \beta} = h_{r\ \beta} = \dfrac{1}{2} \times (\dot{\beta}_t \cos(\varepsilon_t) / \Delta\dot{r}) \times r_m \Delta r \end{cases} \quad (4.64)$$

The advantages of the preemption methods are

- Less curvature of the missile flight path and, as a result, lower overloads are realized during the missile flight along this trajectory;
- Due to a more straight-line trajectory, the capabilities of the air defense system for firing high-speed targets are increasing;
- Long firing ranges are provided in comparison with the three-point method.

The disadvantages of the preemption methods are

- More complex hardware implementation;
- An increase in guidance errors due to the input into the missile flight control command derivatives of the measured values of the angular coordinates of the target, as well as the derivative of the measured difference between the distances to the target and to the missile;
- The need for continuous measurement of ranges to the target and to flying anti-aircraft guided missiles, which reduces the anti-aircraft missile system's noise immunity.

It should be noted that on December 16, 2020, the U.S. Space Command said Russia has tested a direct-ascent anti-satellite (DA-ASAT) missile [13]. The DA-ASAT system is capable of destroying satellites in low-earth orbit by launching missiles from ground-based missile launchers. In this case, the launch was apparently made from a standard mobile launcher. In addition, Russia already demonstrated kinetic weapons in orbit in 2017 and demonstrated them again in 2020. Anti-satellite missiles destroy targeted satellites by kinetic kill warheads [14]. According to analysts, the new layered territorial missile defense system A-235 (Nudol) was tested on December 16, 2020, at the Plesetsk cosmodrome, 800 km north of Moscow [15]. The A-235 anti-missile and anti-space defense system will use two-stage anti-missiles, providing it with the ability to shoot down hypersonic attack weapons, ballistic missiles and their warheads, as well as satellites in near space [16,17]. Based on this information, it can be concluded that the A-235 missile defense system uses, most likely, three-point guidance methods that implement sufficient guidance accuracy of the interceptor missile, which provides the ability to push the satellite off the trajectory it occupies. Thus, tests of mobile missile defense systems indicate that space threats are growing.

The possibility of intercepting satellites by means of the S-400 mobile air defense system can be considered. From Figure 4.24 it can be seen that the starting position of the system is at the point P_0, which is located in the northern hemisphere of the Earth with the coordinate $\Theta_P = 90° - \alpha$, where α is the latitude of this point

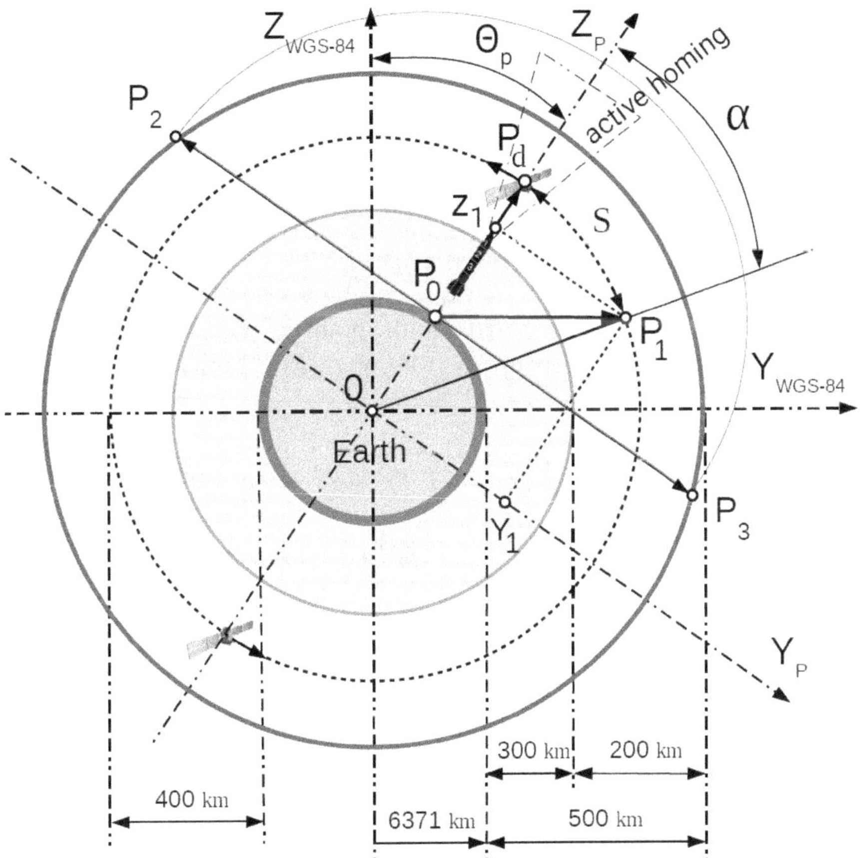

Figure 4.24 Satellites interception capabilities.

in the World Geodetic System WGS-84. To intercept a satellite in near-earth orbit with an altitude of H_s, for example at 400 km, it must first be detected. For this, it is necessary that the mobile anti-aircraft missile system has in its composition a powerful radar capable of detecting space objects with a given effective scattering surface σ_s with a probability no lower than a given one at a range no worse than $|P_0, P_3| = 2573.13$ km for $H_s = 500$ km. In order to understand to what maximum height the missile can rise at a speed of V_m when it starts vertically upwards $\varepsilon_m = 90°$, it is possible to use the formula:

$$h_m^{max} = \frac{V_m^2}{2g}, \tag{4.65}$$

where $g = 9.81$ m/s^2 - acceleration of gravity. For a $40N6E$ missile with a velocity of $V_m = 3.97\text{M} = 1350.57$ m/s, the maximum lift height will be: $h_m^{max} = 92.968$ km $\ll 400$ km, which is clearly not enough for completing this task. For a $48N6E3$ missile with a velocity of $V_m = 8.38\text{M} = 2850.83$ m/s, the maximum lift height will be:

$$h_m^{max} = 414.233 \text{ km} \geq 400 \text{ km}, \tag{4.66}$$

Consequently, this missile can be considered as a means of destruction of space objects located in low-earth orbits. According to the formula

$$t_{max} = \frac{V_m}{g}, \tag{4.67}$$

the maximum flight time of this missile will be $t_{max} = 290.6$s, in this case, the distance that the satellite flies at a speed $V_s = 7910.56$ m/s during the same time t_{max}, moving in the orbit $|P_1, P_d|$, will be $S = 2298.849$ km. It is clear that the missile guidance must be started in the point P_1 (Figure 4.24). The detection range of a satellite located at point P_1 from point P_0 is possible to determine in the local coordinate system $\{Z_p, 0, Y_p\}$:

$$|P_0, P_1| = \sqrt{(Z_1 - 6371)^2 + Y_1^2} = 2254.978 \text{ km} = R_{start}, \tag{4.68}$$

where $\alpha = S/(6371+400)$, $Z_1 = (6371+400) \times \cos(\alpha)$, and $Y_1 = (6371+400) \times \sin(\alpha)$, and R_{start} is the range from the radar of the mobile anti-aircraft missile system to the satellite located at the design point at the time of the launch of the interceptor missile. It is quite clear that in order to issue a "Start" command to the missile from the command post of the air defense system, it is necessary to first detect and lock onto the satellite for auto-tracking. Based on the fact that for stable auto-tracking of a target, the following condition must be met:

$$10 \log[P_r/P_{min}] \geq B_0, \tag{4.69}$$

when the signal-to-noise ratio is greater than a certain threshold value B_0 [dB] and taking into account that the equality is true:

$$\frac{P_{t1} G_t G_r \lambda^2 \sigma_0}{(4\pi)^3 R_1^4} = \frac{P_{t2} G_t G_r \lambda^2 \sigma_0}{(4\pi)^3 R_2^4}, \tag{4.70}$$

therefore, the required power of the radar station of the mobile anti-aircraft missile system P_{t2} for the orbit height $H_s = 400$ km can be expressed as

$$P_{t2} = \left(\frac{R_2}{R_1}\right)^4 \times P_{t1} = 4.078 \text{ MW}, \tag{4.71}$$

provided that the line-of-sight range of a satellite flying in a circular orbit at an altitude $H_s = 400$ km is $R_2 = 2292.771$ km, the line of sight of the S-200D air defense missile system $R_1 = 377.587$ km, and the power of the emitted signal $P_{t1} = 3{,}000$ W. From (4.71) it follows that it is impossible to provide such a large power with the help of a standard mobile diesel power plant; therefore, the detection and auto-tracking of satellites is possible only with the help of stationary missile defense radars connected to a stationary power plant with a capacity of at least 6 MW. Alternatively, we can consider a power supply scheme consisting of three mobile diesel power plants with a capacity of 2 MW each, operating in parallel

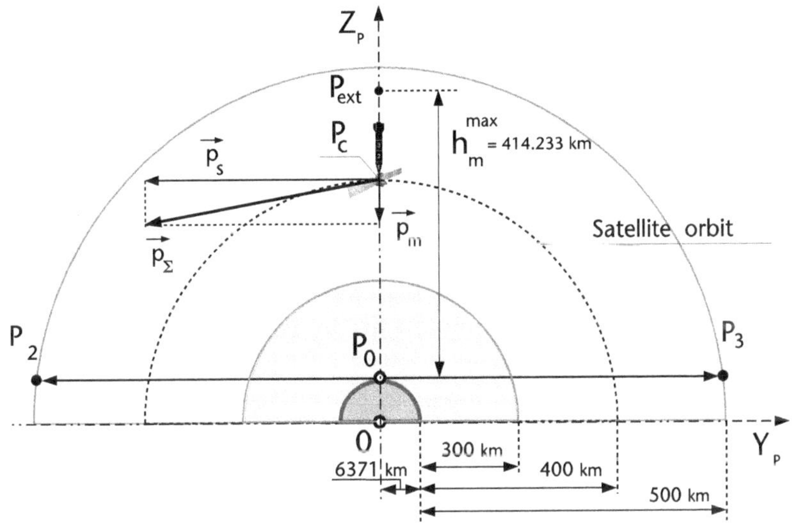

Figure 4.25 Satellites' collision capabilities from their orbits.

in a precise synchronization mode, which provides an equal value of the voltage level, current frequency, and phasing on each generator. When implementing this method, it will also be necessary to upgrade the transmitter and the antenna system, which will ensure operation with such a high power, provided that the specified beamwidth and the existing algorithm for the operation of the antenna array phase shifters are implemented. In the final phase of the missile flight, it is advisable to carry out precise aiming by the method of parallel approach using a homing head operating in an active or semi-active mode.

The use of a high-explosive fragmentation warhead for the purpose of destroying or damaging an enemy's satellites is seen as inexpedient due to the fact that after the detonation of the missile warhead, a large number of damaging elements (fragments) begin moving at a speed greater than the speed of the attacked spacecraft. The presence of an expanding cloud of damaging elements in near-earth space will significantly complicate the operation of other satellites and pose a threat to manned space programs.

In order to push a satellite out of its orbit so that it subsequently burns up in dense layers of the atmosphere, it is necessary that the vector of the momentum of the kinetic impact on the satellite $\vec{p}_m$ would be directed to the center of the Earth (Figure 4.25).

$$\vec{p}_m = m_0 \vec{V}_m = 174\,668.08 \text{ kgm/s}, \tag{4.72}$$

where: $|\vec{V}_m| = g \cdot \sqrt{\frac{2(h_m^{max} - 400 \text{ km})}{g}} = 528.18$ m/s is the speed of the missile near the point of kinetic impact (point of collision); $g = 9.81$ m/s^2 is the acceleration of gravity

$$m_0 = m_{0\_miss} + m_{wh} + m_{hh} = 330.7 \text{ kg} \tag{4.73}$$

is the missile mass before collision point P_c

$$m_{0\_miss} = \rho_{Al} \pi d_m^2 L_m = 123.7 \text{ kg} \tag{4.74}$$

is the missile body mass with completely burnt fuel, $m_{wh} = 180$ kg is the warhead mass, $m_{hh} = 27$ kg is the homing head mass, $\rho_{Al} = 2800$ kg/m^3 is the density of duralumin, $L_m = 7.5$m is the missile length, and $d_m = 0.0005$m is the wall thickness of the missile body.

$$\vec{p}_s = m_s \vec{V}_s = 59\ 250\ 000.00 \text{ kgm/s}, \quad \text{where:} \tag{4.75}$$

$m_s = 7500$ kg is the satellite full mass (e.g., for a satellite with a nuclear power setting), $|\vec{V}_s| = 7900$ m/s is the satellite velocity, 1-st space velocity.

The impulse conservation law for an inelastic impact is

$$m_s \vec{V}_s + m_0 \vec{V}_m = (m_s + m_0)\vec{V}_\Sigma \tag{4.76}$$

Substituting (4.72) and (4.75) in (4.76) we get

$$|\vec{V}_\Sigma| = \frac{m_s \vec{V}_s + m_0 \vec{V}_m}{m_s + m_0} = 7588.68 \text{ m/s}, \tag{4.77}$$

$$|\vec{V}_\Sigma| = 7588.68 \text{ m/s} < |\vec{V}_s| = 7900 \text{ m/s}, \tag{4.78}$$

where (4.78) is the condition for the descent of the satellite from its trajectory and its subsequent combustion in the dense layers of the atmosphere.

Thus, in order to push a satellite out of its orbit when shooting at zenith, it is necessary that the collision point P_c be below the extreme point of the missile trajectory P_{ext} (Figure 4.25), in which the missile speed is $|\vec{V}_m| = 0$.

With a kinetic impact on the satellite from above, velocity modules of the satellite and the interceptor-missile are less than the first cosmic velocity and have components directed toward the center of the Earth, which ensures a gradual decrease in the heights of their orbits and the subsequent combustion of both the satellite and the interceptor-missile in the dense layers of the atmosphere.

References

[1] Kupar, I. I., *Homing of Anti-Aircraft Missiles to the Target,* Military Publishing, 1963.

[2] Vermishev, Y. K., *Missile Control Basics,* Military Publishing, 1968.

[3] Krysenko, G. D., *Missile Control,* Moscow: Military Publishing House of the USSR Ministry of Defense, 1960.

[4] Draft Design of a System of Anti-Aircraft Jet Guided Weapons (S-200), "The Principle of Construction and General Characteristics of the System-200," Moscow: Order of Lenin Design Bureau, Vol. 1, No. 1, 1959, https://docviewer.yandex.ru/view/0/?page=181&*=JN0w6j2sFY79S5X8o8NXnxrJik57InVybCI6InlhLWR\\pc2stcHVibGljOi8vM2cyd0szMXN0amJxd3JHSjFWd1piQlpyOFhUQzZSTDFPakhwcWlOTlp2ND0iLCJ0a\\XRsZSI6ItCt0J8g0KEtMjAwICDRgjEuZGp2dSgxKSIsIm5vaWZyYW1lIjpmYWxzZSwidWlkIjoiMCIsInR\\zIjoxNjA3MTg0NjczNzM1LCJ5dSI6Ijg5NDk0MjI1ODE2MDU5NzkyNDAifQ%3D%3D.

[5] Radar P-14 (Tall King), *Air Defense Bulletin,* http://pvo.guns.ru/rtv/nitel/p14.htm.

[6] Neupokoev, F. K., *Anti-Aircraft Missiles Shooting*, Moscow: Military Publishing House of the USSR Ministry of Defense, 1980.

[7] *Air Defense Officer Handbook,* Moscow: Military Publishing House of the USSR Ministry of Defense, 1987.

[8] Demidov, V. P., and N. S. Kutyev, *Anti-Aircraft Missile Control,* 2nd Edition, Rev. and Add., Moscow: Military Publishing, 1989.

[9] System A, Missile B-1000, GAFFER, Military, Russia, http://militaryrussia.ru/blog/topic-340.html.

[10] R-5/R-5M (SS-3 SHISTER), Military, Russia, http://militaryrussia.ru/blog/topic-268.html.

[11] Peresada, S. A., *Anti-Aircraft Missile Systems,* Moscow: Military Publishing, 1973.

[12] Karpenko, A. V., Military-Technical Collection, Bastion, 2020, http://bastion-karpenko.ru/system-s-75/.

[13] "U.S. Space Command Public Affairs Office Russia Tests Direct-Ascent Anti-Satellite Missile," U.S. Department of Defense, December 16, 2020, https://www.spacecom.mil/News/Article-Display/Article/2448334/russia-tests-direct-ascent-anti-satellite.

[14] "Competition in Space," U.S. Department of Defense, The National Air and Space Intelligence Center, January 16, 2019, https://media.defense.gov/2019/Jan/16/2002080386/-1/-1/1/190115-F-NV711-0002.PDF.

[15] "Russia Tests Anti-Satellite Missile," Voice of America News Service, December 17, 2020, https://www.golosameriki.com/a/russia-tests-anti-satellite-missile/5374230.html.

[16] A-235, Wikipedia, https://ru.wikipedia.org/wiki/%D0%90-235.

[17] Khazbiev, A., "The Secret of Restful Sleep," *Expert* (Special Issue), No. 3 (971), January 18–24, 2016, p. 49, https://expert.ru/expert/2016/03/tajna-spokojnogo-sna/.

CHAPTER 5

Radar

Radar is all about using radio waves to detect the presence of objects and to find their position. The word radar, first used by the U.S. Navy in 1940, is derived from radio detection and ranging, thus conveying these two purposes of detection and location [1]. The principle of radar is that a transmitter sends out a radio signal, which will scatter off anything that it encounters (land, sea, ships, aircraft, missiles), and a small amount of the energy is scattered back to a radio receiver, which is usually, but not always, located near the transmitter. After amplification in the receiver, the signals are processed to sort out the required echoes from the clutter of unwanted echoes. Modern radars use a combination of both electronic signal processing and computer software (data processing). There are many radar applications, but this book focuses on air-borne and military radars. The main functions that these radars perform are

- Navigation and altitude determination for own air forces;
- Detection of own forces or enemy forces;
- Tracking of air, sea, land, or space targets;
- Ensuring sustainable auto-tracking of air and/or space targets by anti-aircraft guided missiles;
- Guidance of own weapons systems for the listed targets.

In order to ensure the unconditional fulfillment of the task of destroying the enemy's means of aerospace attack, anti-aircraft missile systems are used. The main element in these systems are radars. These are ground and airborne radars that are used in anti-aircraft missile systems. Moreover, the air-based radar is the main element in the homing heads of anti-aircraft guided missiles, since the homing head, in addition to the radar, has a guidance system that provides accurate guidance of the anti-aircraft guided missile to the target.

5.1 Basic Principles

The main element in any anti-aircraft missile system is the radar. There are many types of radars, such as detection radar that performs a circular view of the airspace relative to the point of standing of an anti-aircraft missile system, ground-based radar for the control and guidance of anti-aircraft guided missiles on an enemy aircraft or missile, or a radar located under the radome of an anti-aircraft guided missile, which is part of a homing head on a signal that is reflected from the target and carries the homing of the missile to the target. They all use the same principle

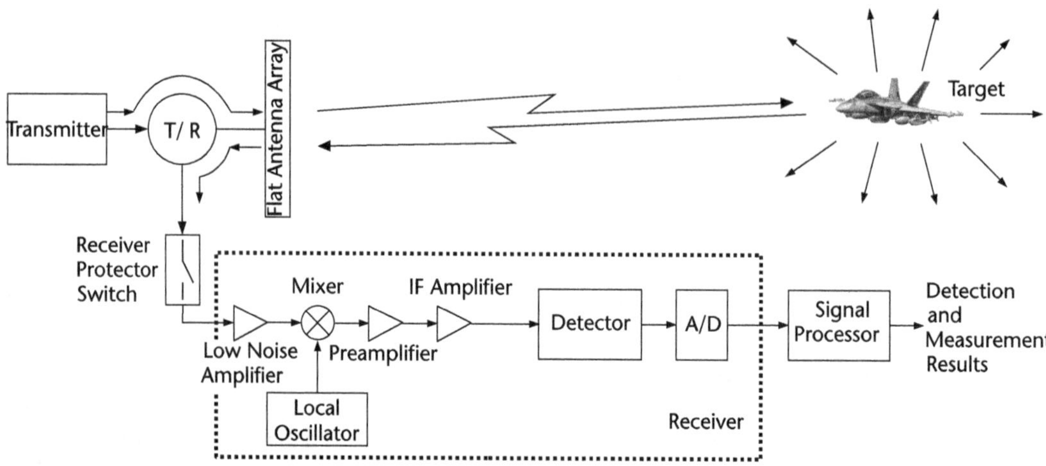

Figure 5.1 The radar transmission/reception process.

in which radar is an electrical system that transmits radiofrequency (RF) electromagnetic (EM) waves toward a region of interest and receives and detects these EM waves when reflected from objects in that region [2]. The transmitted signal propagates through the environment (atmosphere) to the target. The EM wave induces electrical currents on the target metal surfaces, which reradiates the EM wave created by these currents into the environment. Figure 5.1 shows the major elements involved in the process of transmitting a radar signal; propagation of that signal through the atmosphere, reflection of the signal from the target, and receiving the reflected signals. In general, the radar can include a transmitter, antenna, receiver, and signal processor. The transmitter generates the EM waves. The antenna takes as input these EM signals from the transmitter and introduces them into the propagation medium (normally the atmosphere). The transmitter is connected to the antenna through a transmit/receive (T/R) device (usually a circulator or a switch). The T/R device has the function of providing a connection point so that the transmitter and the receiver can both be attached to the antenna simultaneously and at the same time provide isolation between the transmitter and receiver to protect the sensitive receiver components from the high level of transmit signal power. In addition to the desired target, other surfaces on the ground and in the atmosphere—icy snowflakes or raindrops, for example—reradiate the signal. These unintentional and unwanted but legitimate signals are called clutter. Some of the reradiated signal radiates toward the radar receiver antenna to be captured. Propagation effects of the atmosphere and Earth on the waves may alter the strength of the EM waves both at the target and at the receive antenna. The radar receive antenna receives the EM waves that are reflected from the target. This signal may be a target signal, which is of interest, or it may be a clutter signal, which is not of interest. The portion of the signal reflected from the target that propagates back to the radar antenna is captured by the antenna and applied to the receiver circuits as shown in Figure 5.1. The components in the receiver amplify the received signals, convert the RF signal to an intermediate frequency (IF), and subsequently apply the signal to an analog-to-digital converter (ADC) and then to the signal/data processor. The detector is the device that removes the carrier from the modulated target return signal so that target data can be sorted and analyzed by the signal processor. But not all radar systems employ digital signal and data processing. Some systems, such as radars of the previous generation, use analog detection.

The range, R, to a detected target can be determined based on the time, Δt. The EM waves propagate to that target and back at the speed of light. Distance is speed multiplied by the time and the distance the EM wave has to travel to the target and back, which is $2R$, and therefore the range R can be determined as:

$$R = \frac{c\Delta t}{2}, \tag{5.1}$$

Here c is the speed of light in meters per second $c = 3 \times 10^8$ m/s, Δt is the time in seconds for the round-trip travel, and R is the distance in meters to the target.

Note that received target signals exist in the presence of interference. Interference comes in four different forms:

1. Internal and external electronic noise.
2. Reflected EM waves from objects not of interest, often called clutter, such as raindrops or reflections from the underlying surface.
3. Unintentional external EM waves created by other human-made sources, such as electromagnetic interference (EMI). This could be, for example, electromagnetic phenomena in the atmosphere (thunderstorms or lightning creating powerful broadband interference).
4. Intentional jamming from an electronic counter measures (ECM) system, in the form of noise or false targets.

Determining the presence of a target in the presence of noise, clutter, and jamming is a primary function of the radar's signal processor.

EMI is unintentional, as in the case of noise from an engine ignition or electric motor brushes. Jamming signals can take the form of noise, much like internal receiver thermal noise, or false targets, much like a true radar target.

5.2 Radar Types

Radar types can be determined based on their purpose or on the basis of the principle of operation.

While there are hundreds of different types of radars in use, the large majority have three basic functions:

1. Search/detect;
2. Track;
3. Image.

5.2.1 Search and Detect Radars

Almost all radars have to search in a given space volume and detect targets without apriori information regarding the targets' presence or position.

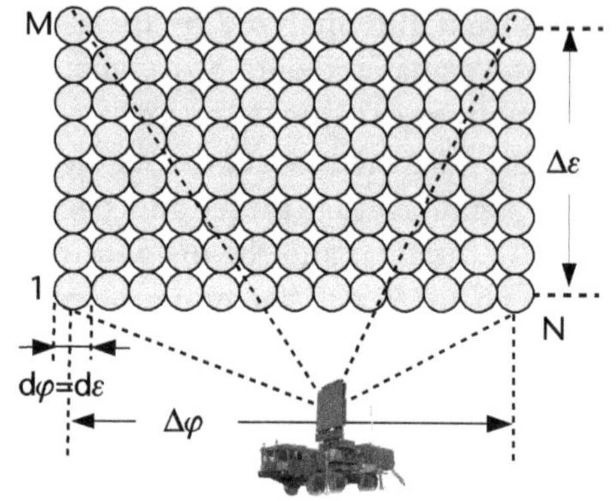

Figure 5.2 Parallel survey principle.

A radar searches a given volume by pointing its antenna in a succession of beam positions that collectively cover the volume of interest. There are two types of airspace survey, which are fundamentally different: a parallel survey and a sequential survey [3, 4]. With a parallel survey (Figure 5.2), the entire solid angle of the radar field of view is densely filled with partial beams intersecting at half-power level with a solid angle of each beam equal to the resolution element on the angle $d\varphi$ and $d\varepsilon$. At the same time, within each partial beam, the target is detected in parallel along the entire range for the time of receiving a burst of reflected pulses: $\tau_b = nT_b$, where n is the number of pulses in a burst, which is determined by a given value of the threshold power at which a decision is made to detect a target signal, and T_b is time interval between pulses in a burst of pulses. Thus, the space survey time can be written as

$$T_{sur} = \tau_b = nT_b \tag{5.2}$$

And therefore, the survey time is equal to the time of receiving one burst of pulses. It is clear that the time of the survey of the region of space $\Delta\varphi \times \Delta\varepsilon$ is reduced due to the introduction of $L = N_{d\varphi} \times M_{d\varepsilon}$ of parallel measuring channels through the formation of a multibeam antenna pattern.

A radar searches a given volume by pointing its antenna in a succession of beam positions that collectively cover the volume of interest. Due to the fact that the beam of the directional pattern of a search and detection radar can have a different shape, this will be a condition determining the number of coordinates that such a radar can measure. In this case, it is necessary to simultaneously have $N \times M$ pencil beams and their corresponding $N \times M$ receiving channels. This greatly complicates the design and greatly increases the cost of such radar. In practical cases, a different scheme is often used (Figure 5.3). Here, M partial beams are created only in the elevation plane, and the radar receives information about target azimuth by the mechanical rotation of the antenna around its H_t-axis with rotation velocity Ω_{sur}. This is similar to 2-D space scanning using fan beam searching (Figure 5.4). In the case when the beam is fan-shaped (e.g., $d\varphi \approx 1^\circ$ in azimuth and $\Delta\varphi \approx$ tens of degrees in elevation), as in Figure 5.4, then only two coordinates can be measured. This is usually the azimuth and range of the target. In this case, as a rule, only circular scanning is carried out in the range of azimuthal angles $[0^\circ, \ldots, 360^\circ]$.

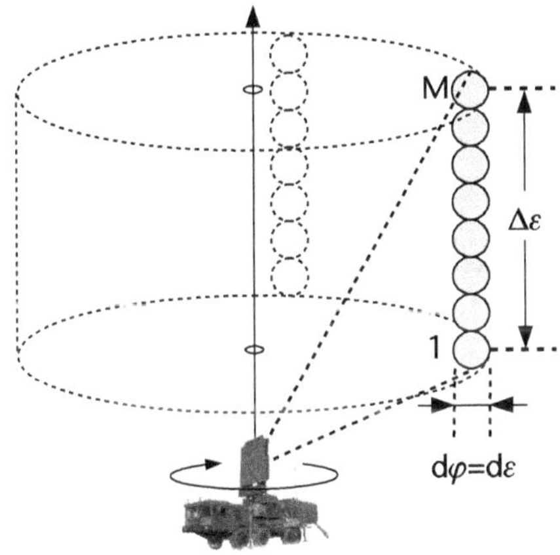

Figure 5.3 Parallel survey in which pencil beams searching a volume providing 3-D target position: $R_t, \varphi_t, \varepsilon_t$.

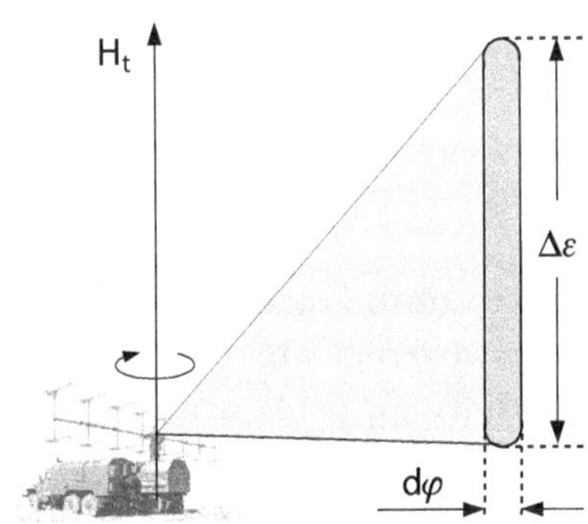

Figure 5.4 P-18 Radar in which fan beam searching a volume providing 2-D target position R_t, φ_t.

In the Soviet anti-aircraft missile forces, such radars were P-18 (Figure 5.4) and P-14 [5,6].

With a sequential survey of space, the beam of the antenna directional pattern moves smoothly or stepwise in a given viewing area, being in each direction of φ_i and ε_j in the course of time $\tau_b = nT_b$, where $i = 1, \ldots N$ and $j = 1, \ldots M$.

With mechanical rotation of the radar antenna, a smooth movement is used, and when using a phased array antenna, only a stepwise movement of the beam is realized. With a smooth movement of the beam, the survey time is determined by the law and the speed of movement of the antenna beam Ω_{sur}. The value of Ω_{sur} is bounded from above by the condition

$$\frac{d\varphi_{0.5}}{\Omega_{sur}} > \tau_b = n \times T_h \quad \Rightarrow \quad \Omega_{sur} < \frac{d\varphi_{0.5}}{n \times T_b} \tag{5.3}$$

Of the large variety of ways to sequentially survey space, it is advisable to consider three main ones: circular, screw, and sector.

With a circular survey (Figure 5.3), the survey area is limited to coordinates $0 \le R \le R_{max}$; $\varepsilon \in [0, \varepsilon^{\circ}_{lim}]$; $0 \le \varphi_t \le 360°$; $d\varphi \in [0, d\varphi_{0.5}]$. Here $d\varphi_{0.5}$ is the width of the radar radiation pattern in the azimuth plane at half power. R_{max} is the maximum range of the radar for a target with radar cross-section σ_t (otherwise known as an effective scattering surface). Taking into account the fact that the speed of the circular survey is given as a rule in the number of revolutions per minute, and the time in radar is set in seconds, we can write for the time required for one turn around the radar location point:

$$T_{1\_turn} = \frac{360°}{\Omega_{sur}} = \frac{360°}{360°/60\ \text{s}} = \frac{360° \times 60\ \text{s}}{n \times 360°} \Rightarrow T_{1\_turn} = \frac{60\ \text{s}}{n}.$$

Taking into account that the speed of circular rotation of the P-18 radar is $\Omega_{sur}^{max} = 6$ degrees per minute [5], consequently, the time of one circular survey cycle will be:

$$T_{1\_turn} = \frac{60\ \text{s}}{n} = \frac{60\ \text{s}}{6} = 10\ \text{s} \tag{5.4}$$

Note that time $T_{1\_turn} = 10\text{s}$ is the radar information update pace. The expression for the survey time T_{sur} will be:

$$T_{sur} = T_{1\_turn} = \frac{360°}{\Omega_{sur}} \geq \frac{360° n T_b}{d\varphi_{0.5}} \tag{5.5}$$

A sequential survey of the radar with a narrow beam with angular velocities ω_β, ω_ε scans the radar coverage area. When using mechanical scanning devices, it will take a long time to see the entire area and there is a high probability of missing the target. Therefore, modern radar technologies use electronically scanned phased array antennas [7] when applicable: screw scanning or spiral scanning (Figure 5.5) and frame-by-frame scanning (Figure 5.6).

A screw survey (Figure 5.5) consists of a circular rotation in the azimuthal plane and a slow change to the beam position in elevation. Spiral survey means that the projection of the beam onto a plane perpendicular to the axis of rotation is in the form of a spiral. Screw and spiral views are used in target search mode, and conical scanning is performed in automatic target tracking mode. With a screw or spiral survey (Figure 5.5), the survey area is limited to the following coordinates:

$0 \leq R \leq R_{max}$, $0 \leq \varphi_t \leq 360°$; $d\varphi \in [0, d\varphi_{0.5}]$ and:

- $\varepsilon \in [0, \varepsilon_{max}^{\circ}]$ is the maximum elevation angle that defines the scan for screw survey, where $\varepsilon_{max}^{\circ}$ area along the ring cut from hemisphere, having radius R_{max};
- $\varepsilon \in [0, 90°]$ for spiral survey, when $\varepsilon_{max}^{\circ} = 90°$ and the radar pencil beam is directed exactly in Zenith.

Let the missile be located from the radar at a distance of $L_m = 350$ km and at an altitude of $H_m = 26$ km. In this case, when finding an antenna phase center at an altitude of $h_a = 6$ m from the Earth's surface, we can find an $\varepsilon_{max}^{\circ}$ for a screw review: $L_m = R_E \cdot \alpha \Rightarrow \alpha = L_m/R_E$=0.05493643, where $R_E = 6371.0$ km is the Earth's radius and $L_m = 350$ km is the arc length, which is situated on the map. $b_1 = (R_E + h_a)\cos(\alpha) = 6361.39453$ km, $b_2 = (R_E + H_m) - b1 = 35.6054668$ km, $b3 = b_2 - H_m = 9.60546677$ km, $\alpha_2 = b_3/L_m = 2.744419 \cdot$

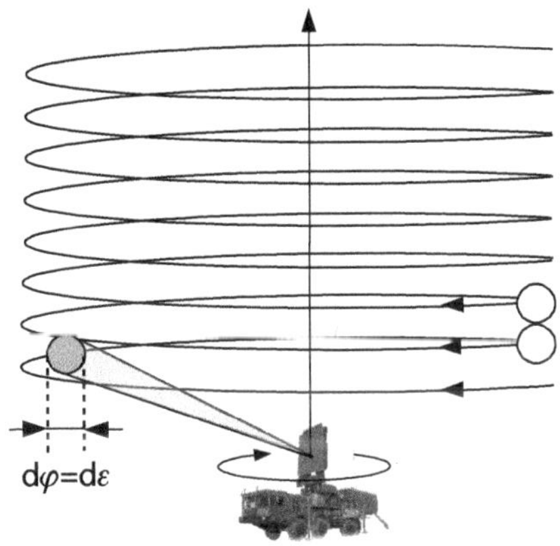

Figure 5.5 Screw survey.

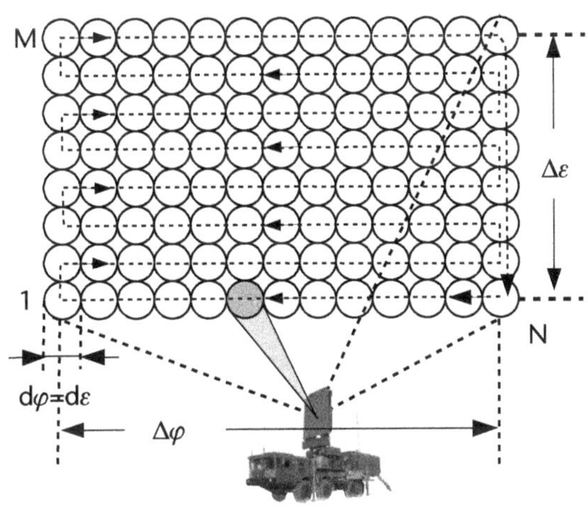

Figure 5.6 Frame survey.

10^{-2} or $= 1.5724363°$, $R = \sqrt{Lm^2 + b_2^2} = 351.806$ km, $\varepsilon_0 = \arccos(L_m/R) = 0.10138113$ $or = 5.80871°$, and therefore we can write:

$$\varepsilon_{max} = \varepsilon_0 - \alpha_2 = 4.23627° \tag{5.6}$$

Thus, $M = \varepsilon_{max}/d\varepsilon$ is the number of concentric rings, having thickness $d\varepsilon$, which is obtained by the following:

$$M = Int\left(\frac{\varepsilon_{max}}{dc}\right) = Int\left(\frac{4.23627°}{1°}\right) = 4, \tag{5.7}$$

where operation "Int" converts a real number to an integer number, rounding it to the closest integer. The time required for one turn around the location point of the radar, which $\varepsilon_i^\circ \approx 0°$ is written similarly to (5.4). The survey time T_{sur} can be determined based on the relationship:

$$T_{sur} = M \cdot T_{1\_turn} \tag{5.8}$$

Frame survey (Figure 5.5) means the oscillatory movement of the beam is in azimuth or elevation with a gradual change in its position along a different angular coordinate. This beam path is beneficial in that it allows you to set the sectors of view independently in azimuth $\Delta\varphi$ and elevation $\Delta\varepsilon$. In this case, given the values of $\Delta\varphi$, $\Delta\varepsilon$, and $d\varphi = d\varepsilon$, it is possible to determine the number of lines M and the number of partial elements N in each line:

$$N = Int\left(\frac{\Delta\varphi}{d\varphi}\right) \qquad M = Int\left(\frac{\Delta\varepsilon}{d\varepsilon}\right) \tag{5.9}$$

This type of scan is similar to the movement of a beam in a cathode ray tube. Here, too, there is a forward path, when the beam moves along the lines in the $\Delta\varphi$ sector (azimuth) with a gradual movement along the elevation angle in the angular sector $\Delta\varepsilon$, after which the beam makes a reverse movement to the starting point of the beginning of the scan. Therefore, the survey time T_{sur} for one frame survey can be determined as follows:

$$T_{sur} = N \times N \times \tau_b \tag{5.10}$$

If the condition is satisfied under which:

$$L = \frac{360^\circ}{\Delta\varphi} \in \mathbb{N} \tag{5.11}$$

Therefore, the survey time T_{sur} for circular frame-by-frame survey can be written as:

$$T_{sur} = N \times N \times L \times \tau_b \tag{5.12}$$

5.2.2 Tracking Radars

Once a target is detected in a given volume, a measurement is made of the target state (i.e., its position in range, azimuth angle, and elevation angle), and often its radial component of velocity. Tracking radars measure target states as a function of time [2]. Tracking implies measuring the position and velocity of a target to an accuracy better than the radar's resolution. A variety of techniques are used to do this. For instance, the azimuth position can be estimated to a fraction of the antenna azimuth beamwidth in a mechanically scanned radar by measuring the detected target strength on several successive pulses as the antenna scans and then computing the centroid of the resulting measurements. Similar concepts can be applied in range and Doppler. Individual measurements are invariably contaminated by measurement noise and other error sources. An improved estimate of the target position over time is obtained by track filtering, which combines multiple measurements with a model of the target dynamics to smooth measurements.

A tracking radar continuously measures the coordinates of a moving target in order to determine its path and to predict where it is going. Tracking can be carried out using range, angle, or Doppler information, but it is the tracking angle that forms the characteristic feature of tracking radars.

To some extent, all types of surveillance radar could be considering as tracking systems since they form an estimate of the target position each time the scanner returns to look in that particular direction—a process known as track-while-scan. Surveillance radars can keep track of many targets simultaneously but the positional accuracy they provide, especially in angle, is not adequate for some purposes.

In contrast to surveillance radars, tracking radars are often used as military systems, designed to automatically track targets and observe them continuously and with great accuracy, usually with the aim of engaging enemy attack aerospace vehicles (aircraft, missiles, and other targets). Some tracking radars have their own search facilities, but the more usual mode of operation is for a main surveillance radar to warn of any targets posing a particular threat and to download the target coordinates to the tracking system, which then searches the immediate area to acquire the target before initiating the tracking process.

The antenna of a tracking radar usually faces the target all the time in order to keep it in the center of the beam and so maximizes the signal-to-noise ratio. If the target moves away from the center of the beam, this produces an error voltage, which is amplified and fed to servo-motors to drive the antenna back onto the target. The azimuth and elevation of the target are then read from angle transducers, such as synchros, mounted on the antenna axes. The methods of producing an error signal from the antenna patterns is one of the main aspects of tracking radar design.

Switching between several beam positions or sequential lobing was one of the earliest methods used to derive angular information about a target. An antenna would have two beams displaced slightly left and right (or up and down) of its centerline and the receiver switched rapidly between them. If the target appeared to grow larger in the left-hand beam, this would drive the antenna to the left until the signal was equal in both beams again. Similarly, two beams displaced vertically could be used to give elevation information.

Around the direction of the target, it is possible to rotate the antenna beam continuously by a conical scanning method. In the case under consideration the effect by pointing the missile velocity vector straight out at a distant object (target) and then begin moving the antenna beam maximum in small circles around the object (the angle between the object and the circle that is describing it) is called the squint angle. If there is a deliberate miss at the object with the center of the scan, it can be seen that once per revolution, the line of the antenna beam maximum comes nearer to the tracking object.

One of the first successful applications of conical scanning was Telefunken's 600 MHz Wrzburg radar used by the German Luftwaffe in World War II for gun laying and ground-controlled interception of Allied aircraft. The technique is still used today as a reliable and low-cost method of tracking in noncritical situations. Conical scan radars usually process 10 or more pulses per revolution. But even so they suffer from the same handicap as sequential lobing in that the radar cross section (RCS) of the target can change between pulses. Also, the repetitive nature of the scanning makes the system vulnerable to electronic countermeasures.

The problem of pulse-to-pulse variations in echo amplitudes can be overcome by using more than one beam simultaneously to measure the angular position of the target on a single pulse. This technique is known as monopulse tracking, a name suggested by Bell Telephone Laboratories in 1946.

Since monopulse radar is typically used in military applications involving guided weapons such as anti-aircraft guided missiles, monopulse radars will be discussed in more detail in the next chapter.

A common application of tracking radar is in the defence of local or point targets from incoming air (and/or space) attack. One of the simplest methods is for the tracking radar to lock on to the target, fire an intercepting missile, and then track both objects with a view to narrowing the gap between them by sending control signals up to the intercepting missile on a communication link. This technique is known as radar command to line of sight (RCLOS) and is used, for example, by the Rapier missile, which distinguished itself during the Falklands crisis. The drawback of this system is that the missile must stay within line of sight from the radar to the target if it is to be tracked accurately, and this may not be the optimum trajectory. For crossing targets, for example, high lateral accelerations are needed by the intercepting missile just before impact—it would be preferable for the missile to travel outside the beam [1]. Once a tracking radar has locked onto a target, there is an invisible beam between the two, and in the late 1940s it was realized that it should be possible to fly a missile up this beam; these weapons have since become known as beamriding missiles [1]. The missile is equipped with a rearward-facing antenna to provide bearing information, which the autopilot uses to keep the missile in the center of the radar beam. The attraction of this approach is that it is simple, and a strong signal-to-noise ratio (SNR) is received by the missile because the signal travels only on a one-way path. This method is used, for example, by antiaircraft systems S-75M3, S-125, and the S-300 family of systems. The disadvantages are that the target must be kept accurately in the center of the beam and, as with RCLOS, the path up the beam may not be the ideal trajectory for a missile to follow. The disadvantages of beamriding can be overcome to some extent through the use of semi-active homing in which the tracking radar acts as a target illuminator. On board the missile a receiver detects the scattered energy from the target, tracks its position, and works out the best trajectory for an intercept. The missile does not have to fly within the illuminating beam and the tracking radar is not required to keep the bore-sight exactly on the target; it merely needs to keep the target illuminated. The most striking application of the homing principle is the Soviet S-200 long-range anti-aircraft missile system. The last modernization of this system, called the S-200D, where I had the honor of taking part in the tests from 1985 to 1989 at the test site Sary-Shagan, is currently capable of defeat targets at a distance of up to 300 km and in the altitude range from 0.03 to 41 km with a probability target defeat up to 0.99 [8]. This system is still in service with the Russian air defense forces in a limited number, but according to other sources, the S-200 system has not been in operation in Russia since 2014 [9].

The main targets for the S-200 system were SR-71 high-speed and high-altitude reconnaissance aircraft, long-range radar patrol aircraft, combat control aircraft (air command posts), and active jamming aircraft, operating from a greater distance but within radar visibility. Before the advent of the S-200 system, these targets were unattainable for other Soviet anti-aircraft missile systems. Active jamming planes served as an ideal target for the S-200 homing missiles. For this reason, for many years, reconnaissance aircraft of the United States and NATO countries, including the SR-71, were forced to make reconnaissance flights only along the borders of the USSR and the Warsaw Pact countries.

The S-200 long-range anti-aircraft missile system has shown its high efficiency during military conflicts as well as in various incidents. On March 22, 1986, the next exercises of the U.S. Navy strike force began in the Mediterranean Sea. The ships maneuvered under the cover of carrier-based aircraft, the actions of which were provided from the loitering zones by the aircraft AWACS E-2C Hawkeye, which had an air strike force control system on board. Established by the official Tripoli, the border of territorial waters 32°30′ north latitude was 145 km from the Libyan coast. On March 24, 1986, over the waters of the Mediterranean Gulf of Sidra, a Libyan S-200 air defense system fired at U.S. Navy carrier-based aircraft [10]. Targets were found consisting of two aircraft heading toward the Sirte object (range, 160 km; azimuth, 12°; altitude, 4.5 km; speed, 160 m/s). Both divisions of the S-200VE group fired single missiles at a rate of 8 seconds for separately observed aircraft of the U.S. Navy (range, 115 km, azimuth, 12°; altitude, 4.5 km; speed, 160 m/s). Two to 30 seconds after the missiles were launched, the targets began a dive maneuver (without turning on the course) and finished it at an altitude of 2–2.5 km. After 95 seconds of flight with an interval of 8 seconds, missiles met with targets at a range of approximately 100 km. Further, the first S-200VE anti-aircraft missile system detected a target at a distance of 120 km and an altitude of 3 km. It was decided to destroy this target with one missile (launch range, 100 km, height, 3 km). From 28–30 seconds after the missile launch, the target carried out a similar dive maneuver to an altitude of 1.5 km. The meeting of the missile with the target occurred at a range of 85 km and an altitude of 1.5 km with a missile flight time of about 85s [10].

The American version of the events that took place is as follows. On March 24, 1986, operating in the immediate vicinity of the Libyan coast, at noon, three American warships crossed the 32°30′ parallel; that is, exactly the line that the leader of the Libyan revolution, Colonel Muammar Gaddafi, called the "Death Line." Libya immediately fired two surface-to-air missiles at the F-14A Tomcat carrier-based fighters of the VF-102 squadron. Both missiles, according to the Americans, fell into the water. A little later, another Libyan air defense missile fired at another American plane. But this missile also missed the target and also fell into the waters of the Gulf of Sidra. Two hours later, the Libyans relaunched the SA-5 missiles, which again flew past American F-14 carrier-based fighters and fell into the water. The fact of any losses on the American side was categorically denied by U.S. president Ronald Reagan [10].

On September 18, 2018, a Russian Il-20 aircraft was shot down by a Soviet S-200 anti-aircraft missile system in service with the Syrian army after being attacked by Israeli F-16 fighters on targets in the Syrian province of Latakia [11]. This happened because the Il-20 plane was on the line "S-200 missile - F-16," so that it was closer to the target illumination radar at the moment the missile was launched. The homing head captured a more powerful signal reflected from the Il-20 aircraft, which had an effective scattering surface (RCS) many times larger than the F-16. Since the homing head of the S-200 system does not select targets by range, the guidance was carried out to a target located closer to the target illumination radar of the S-200 system. As a result, the Il-20 was destroyed, and all 15 crew members were killed.

On October 4, 2001, a Tu-154 of Siberia Airlines was supposed to fly to Novosibirsk from Tel Aviv, but an hour and 45 minutes after takeoff fell into the Black Sea. All 78 people on board were killed, mostly Israeli citizens [12]. According to the conclusion of the investigation of the International Aviation Committee (IAC), which ended in 2004, the Tu-154 was shot down by an S-200 anti-aircraft missile launched by the Ukrainian air defense during the Russian-Ukrainian military exercises on the Crimean Peninsula. The missile exploded 15 meters above the aircraft body. It was found that it was launched from the area of Feodosia. As in the case of the Il-20, the seeker captured a more powerful signal reflected from the Tu-154 aircraft. This could be only in one case, when the Target Illumination Radar captured a Tu-154 civilian plane instead of a target-aircraft or target-missile, and then, everything happened in automatic mode: S-200 system missile's homing head, when it situated on the ground, captured a powerful reflected signal, after which the "Start" command was issued. This tragic combination of circumstances was possible only in the case of an incorrectly issued target designation for the target illumination radar, since this radar is not able to independently search for a target due to a very narrow needle-shaped radiation pattern that is about 1° wide.

The above cases prove the high efficiency of the S-200 system, provided that the target illumination radar only captures a given target for auto-tracking.

5.2.3 Image Radars

Image radars make images of things. The imaging process involves two steps:

1. Developing a high-resolution range profile of the target.
2. Developing a high-resolution cross-range (angular) profile.

The range resolution of the radar ΔR can be present in the form:

$$\Delta R = \frac{c\tau}{2} \tag{5.13}$$

This equation is achieved using a simple, unmodulated pulse of length τ [s], where $c = 3 \times 10^8$ m/s is the speed of light. A pulse width of 1 μs results in a range resolution of $(3 \times 10^8)(10^{-6})/2 = 150$ m, and therefore the range resolution is proportional to the pulse width. A shorter transmitted pulse width will lead to better range resolution (because smaller is better). Improved range resolution can also be achieved by modulation of the signal within the pulse by linear FM, chirp pulse, or by biphase coded pulse.

The Doppler shift is a very important quantity in modern radars Measurement of the Doppler characteristics is used to suppress returns from clutter, to determine the presence of multiple targets at the same range, and to classify and identify moving targets and targets with moving components (e.g., aircraft, helicopters, trucks, and tanks). In a synthetic aperture radar, the measurement of Doppler shift is used to improve the cross-range resolution of the radar. Consider a stationary radar designed to detect moving targets on the ground. The electromagnetic wave return forms a moving target that will have a nonzero Doppler shift, and it can be written in the form [2]:

$$f_d \approx \frac{2V_r}{\lambda} \tag{5.14}$$

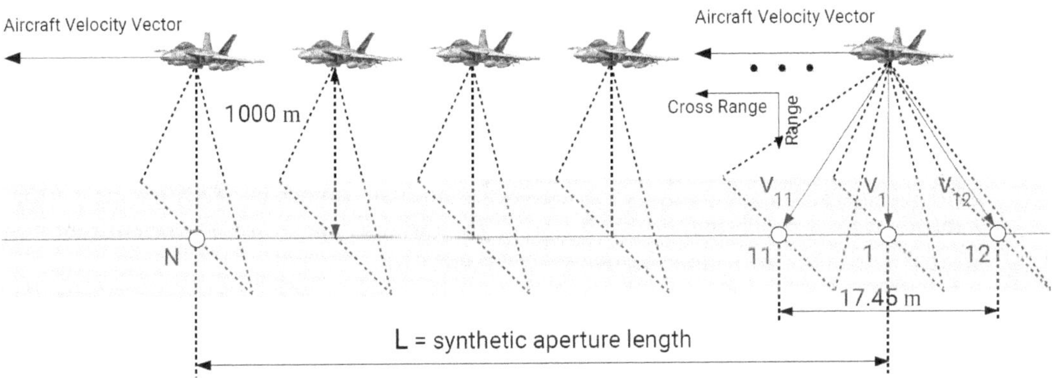

Figure 5.7 Synthetic aperture radar.

Here f_d is the Doppler shift in $[s^{-1}]$, V_r is the target radial velocity [m/s], and λ is the wavelength in [m]. Consider an aircraft with a radar on board pointing out of the side of the aircraft down to the ground (i.e., a sidelooking radar) Figure 5.7, in which v_{11}, v_1, v_{12} are velocity components towards targets $11, 1$ and 12. If the radar's antenna has a 1° beamwidth and the ground is 1000m from the radar, then the antenna beam will be equal 17.45 meters wide in the cross-range dimension (the direction of aircraft motion, which is also perpendicular to the range dimension) on the ground. To implement a narrow radiation pattern in the direction perpendicular to the aircraft speed vector, it is necessary to have the following linear dimensions of the antenna [13]:

$$d_l = \frac{\lambda}{\Delta\varepsilon} = 0.04/0.01745 = 2.292\text{m}, \tag{5.15}$$

therefore, to implement the situation depicted in Figure 5.7, it is necessary to have a linear antenna 2.292m long, which is advisable to be placed along the aircraft fuselage. Doppler shift can be used to resolve objects within the antenna's beam in cross range. As shown in Figure 5.7, objects in the front, middle and back of the beam will have a positive (point 11), zero (point 1), and negative (point 12) Doppler shift, respectively. In fact, each object in the beam with a different cross-range spatial location will have a different Doppler shift. Therefore, if the radar's signal processor is capable of sorting (filtering) the electromagnetic wave returns according to Doppler shift, this is tantamount to sorting the objects in the cross-range dimension, thus resolving objects at different positions.

Synthetic aperture radars from two-dimensional images of an area at resolutions ranging from 100m or more to well under 1m. The first case would typically be used in wide-area imaging from a satellite, while the latter would be used in finely detailed imaging from an airborne (aircraft or unmanned autonomous vehicle (UAV) platform). Here, platform movement is used to electronically simulate an extremely large aperture. With the help of such radars, high-resolution radar images are obtained for remote sensing. The essence of the aperture synthesis method is that the radar data obtained in each separate sounding period is stored together with the platform location data corresponding to these periods. All this data is stored for a certain period of time. The distance L, by which the platform will move during this interval, determines the size of the synthesized aperture (Figure 5.7). At the end of the interval, the received data is processed jointly, as if it was obtained using an antenna array, the elements of which $1, 2, \ldots, N$ correspond to the positions of the platform with radar $1, 2, \ldots, N$ during each of the sounding periods. The processing takes into account the Doppler frequency shifts, which will be different for different relative positions of the radar and the object in each sensing period. In this way, a much higher resolution of the radar image is achieved than that of a radar with the same antenna using a real aperture.

The principle of operation of the synthetic aperture radar (Figure 5.7) is similar to the principle of operation of the phased array radar. The difference lies in the fact that instead of separate antenna elements of the phased array, in this case, only one antenna is used, but with the combination of information from its output at different times, and due to the movement of the platform, at different points in space. The synthetic aperture radar processor stores all echoes (in the form of amplitudes and phases) received during the time interval T, which corresponds to the movement of the platform from point 1 to point N. This makes it possible to generate a signal similar to that which could be obtained using an antenna with an aperture length $L = v \cdot T$, where v is the speed of the platform. Thus, an aperture of considerable length is synthesized, which does not exist in reality. This is where the name for this type of radar comes from. With increasing time T, the synthesized aperture also increases in length, and therefore, an ever higher angular (transverse) resolution is achieved. As soon as the target (for example, a ship) hits the radar beam, the received echoes for each sounding pulse are recorded. The radar platform continues to move and echoes are recorded while the target is in the antenna beam. The point when the platform is located in which the target leaves the radar field of view after some time determines the length of the synthesized (simulated) antenna aperture. The expanding synthesized beamwidth combined with the increased target time in the beam as the track range increases, counterbalance each other so that the resolution remains constant across the entire swath. The potential resolution in the tangential direction does not depend on the range to the target and the wavelength and is determined only by the size of the actual antenna aperture. The mandatory requirements for the implementation of the aperture synthesis method are:

- Stable, fully coherent transmitter;
- Efficient and powerful radar processor;
- Accurate information about the trajectory of the platform and its speed.

The radar antenna alternately transmits and receives pulses at particular microwave wavelengths (in the range of 1 cm to 1 m, which corresponds to a frequency range of about 300 MHz to 30 GHz) and polarizations (waves polarized in a single vertical or horizontal plane). For an imaging radar system, about 1,500 high-power pulses per second are transmitted toward the target or imaging area, with each pulse having a pulse duration (pulse width) of typically 10–50 microseconds (μs). The pulse normally covers a small band of frequencies centered on the frequency selected for the radar. Typical bandwidths for an imaging radar are in the range 10 to 200 MHz. SARs provide unique capabilities as an imaging tool. Because they provide their own illumination (the radar pulses), they can create images at any time of day or night, regardless of sun illumination. And because the radar wavelengths are much longer than those of visible or infrared light, SARs can also see through cloudy and dusty conditions that visible and infrared instruments cannot [14].

5.3 Tactical and Technical Characteristics of the Radar

As is well-known, the three fundamental functions of radar systems are:

- To search for and find (detect) targets;
- To track detected targets;
- And in some cases, make an image of the target.

To implement all these functions, it is necessary to have a transmitter with an output power P_0, a receiver with an appropriate sensitivity P_{min}, and an antenna with an appropriate beamwidth in mutually orthogonal planes $\Delta\varphi,\ \Delta\varepsilon$, which will provide the radar in spatial target selection and the corresponding coefficient directivity antenna (gain) G and antenna working polarization horizontal H, vertical V, or circle polarization C_l or C_r with left or right rotation, respectively. In all of these functions the radar performance is influenced by the strength of the signal coming into the radar receiver from the target and by the strength of the signal that interferes with the target signal. In the special case of receiver thermal noise being the interfering signal, the ratio of target signal to noise power is the SNR; if the interference is in the form of a clutter signal, then the ratio is called the signal-to-clutter ratio (SCR). The ratio of the target signal to the total interfering signal is the signal-to-interference ratio (SIR) [2].

One of the simplest equations in radar theory, which contains all of the radar characteristics listed above, is the radar range equation. One form of the basic radar range equation (also sometimes called the radar equation and the single-pulse signal-to-noise equation, among other names) is [15]:

$$SNR = \frac{E_S}{E_N} = \frac{P_T G_T G_R \lambda^2 \sigma \tau_p}{(4\pi)^3 R^4 k T_s L}, \tag{5.16}$$

where

- SNR denotes the signal-to-noise ratio in units of joules per joule, or J/J. The equivalent units are watt-seconds per watt-seconds, or W-s/W-s.
- E_S denotes the signal energy, in joules (J) or watt-seconds (W-s), at some point in the radar receiver—usually at the output of the matched filter or the signal processor.
- E_N denotes the noise energy, in joules, at the same point that E_S is specified.
- P_T is the peak transmit power, and denotes the average power, in watts, during radar signal transmission. Although P_T can be specified at the output of the transmitter or at some other point, such as the output of the antenna feed, we specify it here as the power at the output of the transmitter.

- G_T denotes the directivity, or directive gain, of the transmit antenna in units of watts per watt.
- G_R denotes the directivity, or directive gain, of the receive antenna in units of watts per watt. In many cases, $G_R = G_T$ when the transmission and reception of a high-frequency signal occurs on one antenna.
- λ denotes the radar wavelength in units of meters.
- σ denotes the average target RCS in units of square meters.
- τ_p is the transmit pulsewidth, in seconds. τ_p is the width of a $rect[x]$ function envelope.
- R denotes the slant range from the radar to the target in units of meters.
- k denotes Boltzmann's constant and is equal to $1.3806503 \times 10^{-23}$ $W/(\text{Hz K})$ or (W-s)/R, although it is often truncated 1.38×10^{-23} $W/(\text{Hz K})$.
- $T_s = T_a + (F_n - 1)T_0$ Kelvin (K) is the system noise temperature, where T_a is an antenna noise temperature used to characterize environment noise.
- T_0 denotes a reference temperature. The Institute of Electrical and Electronics Engineers (IEEE) defines noise figure in terms of a noise temperature of $T_0 = 290\ K$, which results in the rule of thumb approximation of $kT_0 = 4 \times 10^{-21}$ W/Hz.
- F_n denotes the overall radar noise figure and is dimensionless or has units of watts per watt. Note that noise figure and noise factor are often treated as synonyms, although noise figure is used when in logarithmic form (dB), while noise factor is used when in linear form (W/W).
- L denotes all the losses one takes into consideration when using the radar range equation. This term, which accounts for losses that apply to the signal and not the noise, has the units of watts per watt. L accounts for a multitude of factors that degrade radar performance, including those related to the radar itself, the environment in which the radar operates.

The radar range equation of (5.16) is termed the single-pulse radar range equation because the SNR calculation is based on a single transmit pulse. Equation (5.16) is often used to determine the maximum range of the radar when it detects a target that has the RCS σ. This means that if the SNR is a certain value at a given range, it will increase as range decreases. Therefore, we can define the detection range as the range at which it achieve a certain SNR. Expressing the range P from (5.15) through a given SNR, we can write the equation for the target detection range:

$$R = \sqrt[4]{\frac{P_T G_T G_R \lambda^2 \sigma \tau_p}{(4\pi)^3 (SNR) k T_s L}} \tag{5.17}$$

From Table 5.1 it can be seen, that for the S-200 air defense system, the detection range of the B-52 type aircraft is 317.6 km. This is one of the main tactical and technical characteristics of an anti-aircraft missile system. Other characteristics can also be obtained from the analysis of this data. For example, the analysis

Table 5.1 Radar Parameters

Radar Parameter	Value (MKS)	Value (dB)
P_T: transmit power	$1 \times 10^{+3}$ W	30 dBW
G_T: directive gain of the transmit antenna	50000 W/W	46.9897 dB
G_R: directive gain of the receive antenna	800 W/W	29.0308 dB
λ: radar wavelength	0.04 m	
λ^2	0.016 m^2	−27.9588
σ: radar cross section of B-52 aircraft	70 m^2	18.4509 dBsm
Δf: receiver filter bandwidth	400 Hz	
τ_p: duration of one sample	0.0025 s	−26.0205 dB
	$1.12 \cdot 10^{+7}$	70.4921 dB
k: Boltzmann's constant	1.38×10^{-23} W-s/K)	−228.6012 dB
$L = L_t L_{ant} L_{other}$: total losses	2.51 W/W	3.9967 dB
T_s: the system noise temperature	1602 K	32.0466 dB(K)
$(4\pi)^3$	1984.4017	32.9762 dB
SNR	10	10 dB
	$1.1011 \cdot 10^{-15}$	−149.5815 dB
subtotal	$1.0171 \cdot 10^{+22}$	220.0737 dB
B-52 detection range is R	317572.7 m	317572.7 m

of the directivity patterns of the radars, included in the system allows us to make conclusions about its capabilities for target selection in three-dimensional space, as well as its susceptibility to interference, for example, by sidelobes. In the case of the S-200 air defense system target illumination radar, its capabilities for spatial targets selection are 1°, and with a sidelobe level of −41 dB, the setting of noise interference cover for this radar is meaningless.

References

[1] Kingsley, S., and S. Quegan, *Understanding Radar Systems,* New York: SciTech Publishing, 1999.

[2] Richards, M. A., J. A. Scheer, and W. A. Holm, *Principles of Modern Radar,* Vol. I: Basic Principles, New York: SciTech Publishing, 2010.

[3] Bakulev, P. A., *Radar Systems: Textbook for Universities,* Moscow: Radiotekhnika, 2004.

[4] Belocerkovsky, G. B., *Basics of Radar and Radar Devices,* Moscow: Soviet Radio, 1975.

[5] Garin, E. N., et al, *The Device of the Radar Station of the Radio Engineering Forces and the Air Force, Radar Station P-18R,* Military Technical Training, Tutorial in Two Parts, Part 1, Publishing Center of the Library and Publishing Complex of the Siberian Federal University, Krasnoyarsk, 2012.

[6] Radar P-14 (Tall King), *Air Defense Bulletin,* http://pvo.guns.ru/rtv/nitel/p14.htm.

[7] Kvasnov, A. V., and V. B. Polyakov, "Analysis of Airspace Scanning Algorithms with Active Phased Antenna Radar Arrays to Reduce the Search Time for Surface and Aerodynamic Targets," *Radio Industry,* No. 2, 2016, pp. 40–46, UDC 621.396.965.25, https://www.radioprom.org/jour/article/viewFile/129/126.

[8] S-200 Long-Range Anti-Aircraft Missile System (SA-5 Gammon), *Air Defense Bulletin,* http://pvo.guns.ru/s200/s200.htm.

[9] S-200 (Missile), https://en.wikipedia.org/wiki/S-200_(missile).

[10] "Libyan Premiere of the S-200VE 'Vega' Air Defense System," *Aerospace Defense Magazine,* August 13, 2015, http://www.vko.ru/oruzhie/liviyskaya-premera-zrs-s-200ve-vega.

[11] Ahronheim, A., and H. Keinon, "Satellite Images Show Results of Israel's Strike in Syria," *The Jerusalem Post,* September 19, 2018, https://www.jpost.com/Israel-News/Israel-works-to-stem-crisis-with-Moscow-after-downing-of-Russian-plane.

[12] "A History of Fatal Mistakes: Accidentally Shot Down Airliners," BBC Russian Service, January 10, 2020, https://www.bbc.com/russian/news-51064990.

[13] Blake, L. V., and M. W. Long, *Antennas: Fundamentals, Design, Measurement,* Third Edition, Raleigh, NC: SciTech Publishing, 2011.

[14] Freeman T., "What is Imaging Radar?" Jet Propulsion Laboratory, https://airsar.jpl.nasa.gov/documents/genairsar/radar.html.

[15] Budge, M. C. Jr., and S. R. German, *Basic Radar Analysis,* Second Edition, Norwood, MA: Artech House, 2020.

CHAPTER 6

Monopulse Radars

Currently, there is a constant increase in aircraft flight speeds and missile and space technology is actively developing. New methods and technical means of radio countermeasures are constantly being developed. Today there are new aerial vehicles moving at hypersonic speed. In this regard, the requirements for the accuracy of determining the angular coordinates, the speed of processing radar information, and for ensuring the jamming immunity of the radar at a high level are constantly increasing. These requirements have largely found their application in monopulse radar systems. Initially, monopulse radars were understood only as radars that carry out multichannel reception and have the fundamental ability to measure the angular coordinates of a target by one pulse. Subsequently, the principle of simultaneous reception of signals was extended to radar stations with continuous and quasi-continuous radiation modes. Monopulse systems are sometimes referred to as systems having instantaneous signal comparison. However, today the term "monopulse radar" applies to all radars with instantaneous signal comparison.

6.1 Basic Principles

As we know, the direction finding of targets, which consists of determining the direction to the target, is one of the main tasks of radiolocation. Until the 1980s, the most common methods of accurate automatic targeting were the methods of conical, linear (plane) scanning, and sequential switching of radiation patterns. All these methods were implemented in single-channel systems. The most prominent representatives of linear (plane) scanning were the Soviet anti-aircraft S-75 Volkhov and S-125 Neva missile systems. They scanned a three-dimensional space in two mutually orthogonal planes using fan-shaped radiation patterns. The position of the target was determined for each angular coordinate φ_t and ε_t separately relative to the initial angles of the scanning sectors. Obviously, obtaining angular information about the target was accompanied by amplitude modulation of the received signals. This method of calculating an angular error signal requires the reception of a sequence of reflected signals, so it is therefore sensitive to fluctuations in the amplitude of the received signals generated by random changes. So, when tracking targets with the S-75 Volkhov anti-aircraft missile system under certain conditions, time sections could be observed at which there was no signal at all. This phenomenon was called signal fading.

In contrast to the methods for determining the coordinates of the target, using one or another type of scanning, in the monopulse method the reflected signal gives complete information about the angular position of the target. Since in monopulse

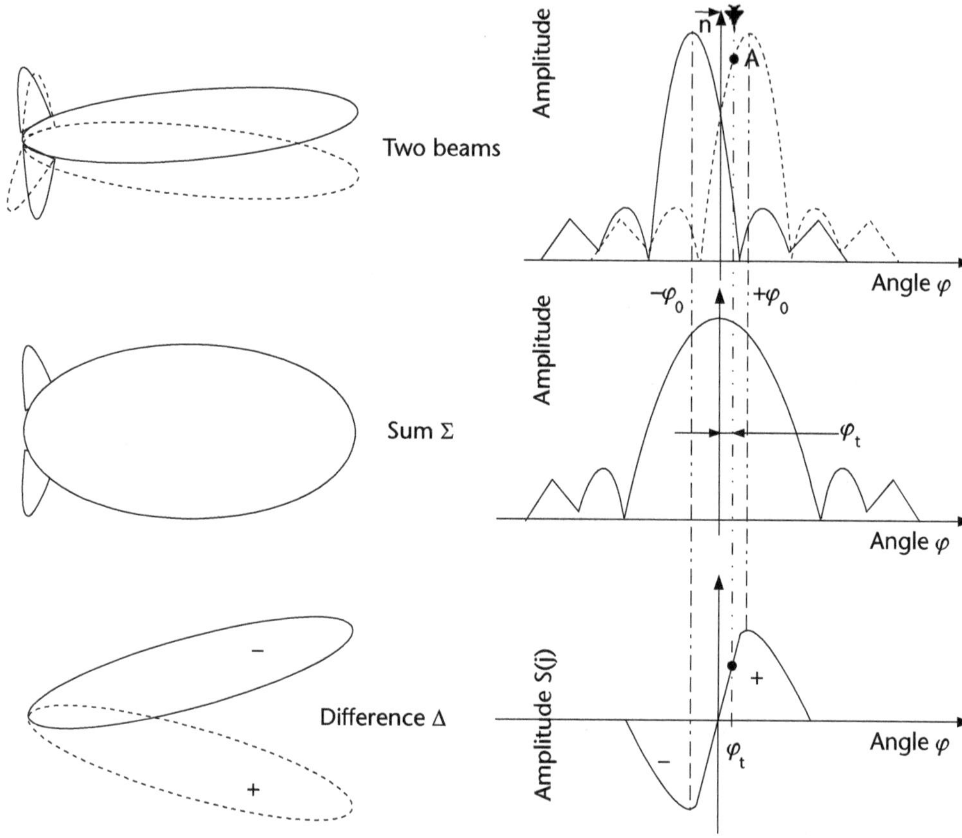

Figure 6.1 The principle of operation of the monopulse method for determining the target angular coordinates.

radars the direction finding of the target is carried out by one pulse, the amplitude fluctuations of the reflected signal do not affect the accuracy of measuring the angular coordinates. This explains the term "monopulse (single-pulse)" direction finding.

As we also know, the original monopulse method was developed for accurate automatic target tracking. But at present, the monopulse method for determining the coordinates of targets is also used in surveillance radars [1].

Depending on the method of extracting the angular information about the target from the received signals, two main methods of obtaining an error signal are distinguished:

- Amplitude method;
- Phase method.

In the amplitude method, to determine the angular coordinate in one plane φ, for example, two intersecting radiation patterns are formed, spaced from each other by an angle of $\pm\,\varphi_0$, relative to the equisignal direction, which is usually the normal $\vec{n}$ of a monopulse antenna.

When the target deviates by an angle of φ_t from the equisignal direction $\vec{n}$ and the target is located at point A (Figure 6.1), the signal received from the right partial

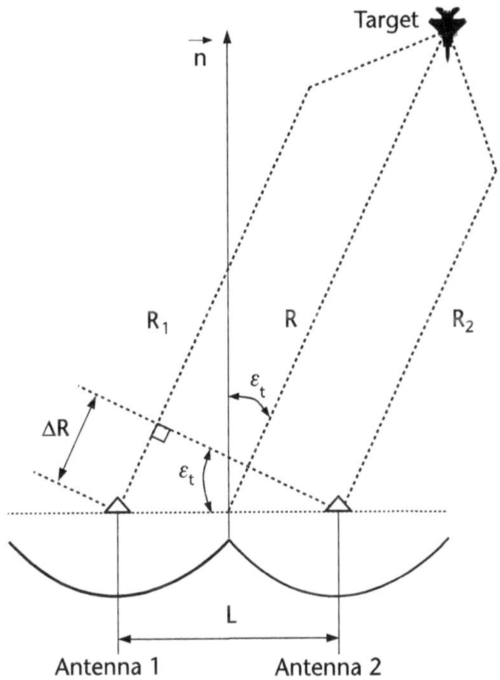

Figure 6.2 Phase method.

radiation pattern of the monopulse antenna is greater than the signal received from the left partial radiation pattern. The difference in the amplitudes of the received signals indicates the amount of deviation of the target from the equisignal direction $\vec{n}$. The sign of this difference indicates the direction of displacement of the equisignal direction relative to the target [2].

When the equal-signal direction is aligned with the target, the amplitudes of the reflected signals received in both partial radiation patterns are equal and their difference becomes zero. Target elevation angle ε_t is determined similarly using two receiving channels and two partial radiation patterns located in the ε plane. In monopulse systems with phase direction finding, the direction to the target in one coordinate plane is determined by comparing the phases of the signals received by two antennas. In Figure 6.2 shows two antennas spaced from each other at a distance L. The line of sight of the target forms an angle of ε_t with the n-axis, which is the normal $\vec{n}$ to the monopulse antenna. The distance R_1 between antenna 1 and the target is

$$R_1 = R + \frac{L}{2} \sin \varepsilon_t \tag{6.1}$$

and the distance R_2 between antenna 2 and the target is

$$R_2 = R - \frac{L}{2} \sin \varepsilon_t \tag{6.2}$$

Therefore, the difference in the distances from the target to the antennas is written in the form

$$\Delta R = R_1 - R_2 = L \cdot \sin(\varepsilon_t) \tag{6.3}$$

and will give an expression for the phase difference:

$$\Delta\varepsilon = \frac{2\pi}{\lambda} \cdot \Delta R = \frac{2\pi L}{\lambda} \cdot \sin(\varepsilon_t), \tag{6.4}$$

where λ is the wavelength. This makes it possible to determine the angle of arrival ε_t from the measured value of the phase shifts of signals, reflected from the target, which are simultaneously received by two antennas, spaced at a distance L as shown in Figure 6.2. It should be noted, that expression (6.4) shows, that the phase shift of the $\Delta\varepsilon$ signals becomes to zero not only at $\varepsilon_t = 0$, but also at other mismatch angles corresponding to the condition:

$$\varepsilon_t = \arcsin\left(\frac{2n\pi}{k_\lambda L}\right), \tag{6.5}$$

where $n = 1, 2, \ldots,$ and $k_\lambda = \frac{2\pi}{\lambda}$ is also known as a wavenumber.

As a result, the direction finding characteristic, which is the dependence of the error signal on the angle of arrival of the signal C (phi), turns out to be sign-alternating and having, along with the main equal-signal direction, many false directions. In this case, there is an ambiguity in measuring the angular coordinates by the phase method. However, this feature of the phase method is not a serious drawback since it manifests itself only in cases where the false equal-signal directions are outside the main lobe of the monopulse antenna pattern. This requires, that the distance between the centers of the receiving antennas does not exceed the diameter of each of them [3–5].

Such a system can be made, for example, in the form of two adjacent antennas. A very original way to solve this problem was found by specialists of the Almaz research and production association, who developed a homing head for the S-200 anti-aircraft missile system. Due to the fact that it is impossible to place two antennas side by side due to the restrictions imposed by the diameter of the rocket body, and it is necessary to make direction finding in the elevation plane, we used the principle of reflection and divided the parabolic antenna by a metal screen so that the reflection plane passed along the line of the antenna diameter and approximately through the phase center of the antenna. As a result, the direction finding scheme in the elevation plane took the form shown in Figure 6.3. This solution made it possible to place the radio fuse antenna in the gap L between the vertical screens and hide its waveguide system in this space.

6.2 Tracking Radars

A tracking radar is a radar that automatically keeps the antenna beam (equisignal direction) axis aligned with a selected target. Any deviation of the target from the beam axis produces a correction signal (usually called an error signal) for each coordinate, approximately proportional to the angular deviation in that coordinate, with a sign or polarity that indicates the sense of the deviation (up or down, left or right). The correction signal is used to drive the antenna axis toward the target [6]. The main function of tracking radars is to constantly receive reliable information about the angular coordinates of the target, if possible, to obtain information about

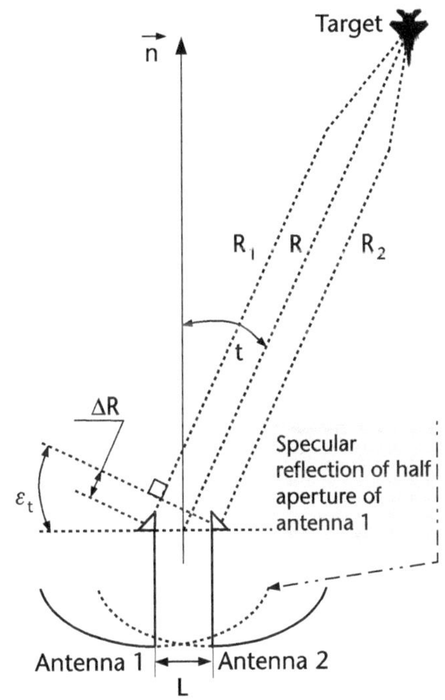

Figure 6.3 Phase method realized in real homing.

the distance to the target. When using tracking radars in the missile homing heads, homing is carried out based on the angular information about the target. The range to the target is necessary for a radio fuse in the final homing phase, when the missile is close to the meeting point with the target and the radio fuse must precisely match the cloud of fragments with the target, so that at the moment of detonation of the warhead the target is in a cloud of fragments of maximum density.

In monopulse systems, information about the angular position of the target is contained in the error signal $S(\varphi)$ or $S(\varepsilon)$ resulting from the comparison of signals simultaneously received by two reception channels, present for each control plane φ, ε. With such a comparison, the voltage at the output of the comparison circuit does not depend on the absolute value of the amplitudes of the received signals, but is determined only by the angle of arrival of these signals. The direction finding characteristic must determine the magnitude and sign of the angle of arrival of the received signal; that is, it must be an odd real function of the angle of arrival of the signal.

The field strength at the receiving point in complex form can be represented as a function of time:

$$\dot{E}(t) = E_m \cdot \exp j(\omega t - \psi_0), \qquad 0 \leq t \leq T, \tag{6.6}$$

where E_m is the amplitude, ω is the frequency, and ψ_0 is the field initial phase. Antenna radiation patterns in complex form can be represented as

$$\begin{aligned} \dot{f}_1(\varphi) &= F_1(\varphi) \cdot \exp[j\psi_1(\varphi)] \\ \dot{f}_2(\varphi) &= F_2(\varphi) \cdot \exp[j\psi_2(\varphi)], \end{aligned} \tag{6.7}$$

where $F_1(\varphi)$ and $F_2(\varphi)$ are the amplitude patterns of 1 and 2 channels, and $\psi_1(\varphi)$, $\psi_2(\varphi)$ are the phase patterns of these channels. It is reasonable to believe that the phase diagrams of the channels are mirror images of each other with respect to the equisignal direction. Amplitude radiation patterns are identical to each other and are displaced by their maxima by an angle $\pm\varphi_0$ relative to the equisignal direction:

$$\left.\begin{aligned} \psi_1(\varphi) &= +\frac{1}{2}\psi(\varphi), & \psi_2(\varphi) &= -\frac{1}{2}\psi(\varphi), \\ F_1(\varphi) &= F(\varphi_0 - \varphi), & F_2(\varphi) &= F(\varphi_0 + \varphi), \end{aligned}\right\} \tag{6.8}$$

where $\psi(\varphi)$ is the difference-phase antenna directivity diagram. The reflected signals received by the diversity antennas can be written as follows:

$$\left.\begin{aligned} \dot{E}_1(t,\varphi) &= \dot{E}(t)\dot{f}_1(\varphi) \\ \dot{E}_2(t,\varphi) &= \dot{E}(t)\dot{f}_2(\varphi) \end{aligned}\right\} \tag{6.9}$$

Therefore, the ratio of the two received signals will be as follows:

$$r_m(\varphi) = \frac{\dot{E}_1(t,\varphi)}{E_2(t,\varphi)} = \frac{\dot{f}_1(\varphi)}{\dot{f}_2(\varphi)} = \frac{F_1(\varphi)\exp[j\psi_1(\varphi)]}{F_2(\varphi)\exp[j\psi_2(\varphi)]} \tag{6.10}$$

and is called the multiplicative ratio.

To form the direction finding characteristic, the ratio of the difference signal to the sum signal, which is called the additive ratio, can also be used:

$$r_a(\varphi) = \frac{\dot{E}_1(t,\varphi) - E_2(t,\varphi)}{\dot{E}_1(t,\varphi) + E_2(t,\varphi)} = \frac{\dot{f}_1(\varphi) - \dot{f}_2(\varphi)}{\dot{f}_1(\varphi) + \dot{f}_2(\varphi)}, \tag{6.11}$$

With amplitude direction finding, the angular information is contained in the amplitude radiation patterns $\dot{f}_1(\varphi)$ and $\dot{f}_2(\varphi)$. In this case, with identical phase diagrams, the multiplicative and additive ratios are expressed only through the amplitude radiation patterns:

$$\left.\begin{aligned} r_m(\varphi) &= \frac{F_1(\varphi)}{F_2(\varphi)} = \rho(\varphi) \\ r_a(\varphi) &= \frac{F_1(\varphi) - F_2(\varphi)}{F_1(\varphi) + F_2(\varphi)} = \frac{\rho(\varphi) - 1}{\rho(\varphi) + 1} \end{aligned}\right\} \tag{6.12}$$

With phase direction finding, in the case when the amplitude radiation patterns are identical, and the angular information is contained in the phase difference, therefore:

$$\psi(\varphi) = \psi_1(\varphi) - \psi_2(\varphi). \tag{6.13}$$

In this case, the multiplicative and additive ratios are expressed only through phase diagrams:

$$\left.\begin{aligned} r_m(\varphi) &= \frac{\exp[j\psi_1(\varphi)]}{\exp[j\psi_2(\varphi)]} = \exp[j\psi(\varphi)] \\ r_a(\varphi) &= \frac{\exp[j\psi_1(\varphi)] - \exp[j\psi_2(\varphi)]}{\exp[j\psi_1(\varphi)] + \exp[j\psi_2(\varphi)]} = j\tan\left(\frac{\psi(\varphi)}{2}\right) \end{aligned}\right\} \quad (6.14)$$

Thus, only three methods of measuring angular coordinates are possible:

1. Amplitude;
2. Phase;
3. Sum-and-difference.

6.2.1 Amplitude Sum-and-Difference System of Automatic Target Tracking

In the amplitude sum-and-difference system, the reception of signal, reflected from the target is carried out using an antenna, which forms two beams in each coordinate plane, deviated from the equisignal direction $\vec{n}$ by an angle $\pm\varphi_0$ and $\pm\varepsilon_0$ accordingly. This system is less demanding on the identity of the characteristics of the receiving channels and therefore is very often used in modern radars and in missile homing heads as well. Here, the signals reflected from the target are received by a monopulse antenna and enter the sum-difference converter (waveguide bridge or stripline ring), where the operations sum and difference of signals are performed at a high frequency $f_0 = 3 \cdot 10^8/\lambda_0 = 7.5$ GHz, provided that the operating wavelength $\lambda_0 = 0.04$m.

From the outputs of the waveguide bridge, the sum and difference high-frequency signals enter the sum and difference receiving channels where they are converted into intermediate frequency signals and amplified to the required value. The amplitude of the difference signal determines the magnitude of the angular error, and the phase difference between the sum and the difference signal determines the sign of the angular error; that is, the direction of deviation of the target from the equisignal direction. A block diagram of this monopulse system is shown in Figure 6.4. The signals at the antenna output of the first and second channels with small deviations of the target from the equisignal direction are determined by the expressions

$$\left.\begin{aligned} \dot{E}_1(t,\varphi) &= E_m F(\varphi_0)(1+\mu\varphi)\exp(j\omega t) \\ \dot{E}_2(t,\varphi) &= E_m F(\varphi_0)(1-\mu\varphi)\exp(j\omega t) \end{aligned}\right\} \quad (6.15)$$

The sum and difference signals at the output of the waveguide bridge, taking into account the power balance, will take the form:

$$\left.\begin{aligned} \dot{E}_\Sigma(t,\varphi) &= \sqrt{2}E_m F(\varphi_0)\exp(j\omega t) \\ \dot{E}_\Delta(t,\varphi) &= \sqrt{2}E_m F(\varphi_0)\mu\varphi\exp(j\omega t) \end{aligned}\right\} \quad (6.16)$$

The dependence of the error signal on the amplitude of the received signals is excluded by the automatic gain control (AGC) system. After conversion to an

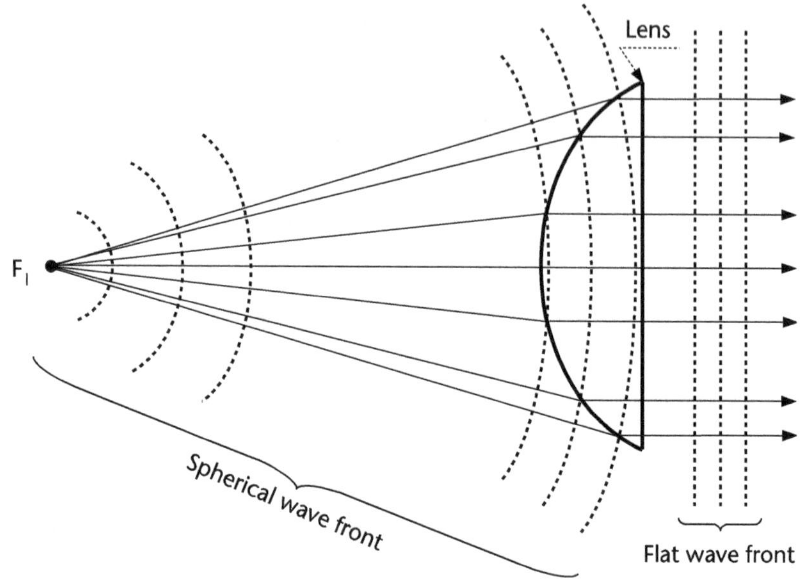

Figure 6.4 Dielectric lens.

intermediate frequency ω_{int} and subsequent amplification, taking into account the operation of the AGC system, at the output of the phase detector, the sum and difference signals can be written in the form:

$$\left.\begin{aligned}\dot{U}_{\Sigma}(t,\varphi) &= \exp[j(\omega_{int}\cdot t+\psi_1)]\\ \dot{U}_{\Delta}(t,\varphi) &= \frac{k_1}{k_2}\mu\varphi\exp[j(\omega_{int}\cdot t+\psi_2)]\end{aligned}\right\} \tag{6.17}$$

where ψ_1 and ψ_2 are phase shifts in channels, and k_1 and k_2 are channel gains. The error signal at the output of the phase detector will be written as

$$S(\varphi) = K_{phd}\cdot\frac{k_2}{k_1}\mu\varphi\cdot\cos(\psi_1-\psi_2). \tag{6.18}$$

Here K_{phd} is phase detector gain. The sum and difference radiation patterns are schematically shown in Figure 6.1, from which it can be seen that the phase of the difference signal at the antenna output changes depending on the direction of deviation of the target relative to the equisignal direction. In the absence of mismatch, the direction to the target coincides with the equal-signal direction of the antenna system, the signals reflected from the target at the input of the receiving channels have equal amplitudes, and therefore the difference signal is zero.

The difference signal is used in tracking radars and missile homing heads, in particular, to control the position of the antenna system in the automatic target tracking process and in the process of guiding the missile to the target.

The sum signal is used in tracking radars and missile homing heads as a reference signal, as well as for target detection, measuring the range to the target and its radial velocity.

6.2.2 Phase Sum-and-Difference System of Automatic Target Tracking

In the phase sum-and-difference system, the reception of signals reflected from the target is carried out using an antenna, which is shown in the Figure 6.3. The signals at the antenna output can be represented as:

$$\left.\begin{aligned}\dot{E}_1(t,\varphi) &= E_m F(\varphi)\exp j\cdot\left(\omega_{int}\cdot t+\frac{\Delta\psi}{2}\right)\\ \dot{E}_2(t,\varphi) &= E_m F(\varphi)\exp j\cdot\left(\omega_{int}\cdot t-\frac{\Delta\psi}{2}\right)\end{aligned}\right\}\qquad(6.19)$$

At the output of the waveguide bridge, the sum and difference signals are written in the form

$$\left.\begin{aligned}\dot{U}_\Sigma(t,\varphi) &= \frac{1}{\sqrt{2}}E_m F(\varphi)\left[\exp j\left(\omega_{int}\cdot t+\frac{\Delta\psi}{2}\right)+\exp j\left(\omega_{int}\cdot t-\frac{\Delta\psi}{2}\right)\right]\\ \dot{U}_\Delta(t,\varphi) &= \frac{1}{\sqrt{2}}E_m F(\varphi)\left[\exp j\left(\omega_{int}\cdot t+\frac{\Delta\psi}{2}\right)-\exp j\left(\omega_{int}\cdot t-\frac{\Delta\psi}{2}\right)\right]\end{aligned}\right\}\qquad(6.20)$$

At the output of the phase detector after conversion to an intermediate frequency ω_{Int} and amplification, taking into account the operation of the AGC system and an additional phase shift in the difference channel at $\frac{\pi}{2}$, we can write

$$S(\varphi)=\Re\left[\frac{\dot{U}_\Delta(t,\varphi)\dot{U}^*_\Delta(t,\varphi)}{\dot{U}_\Sigma(t,\varphi)\dot{U}^*_\Sigma(t,\varphi)}\right]\qquad(6.21)$$

where $\dot{U}^*_\Delta(t,\varphi)$ and $\dot{U}^*_\Sigma(t,\varphi)$ are the complex-conjugate values of the signals at the output of the sum and difference channels. Taking into account (6.20), the expression (6.21) will take the form

$$S(\varphi)=\frac{k_2}{k_1}K_{phd}\times\frac{\sin\Delta\psi}{1+\cos\Delta\psi}=\frac{k_2}{k_1}K_{phd}\times\tan\left(\frac{\Delta\psi}{2}\right)\qquad(6.22)$$

after substitution of the expression $\Delta\psi$ from (6.4), we can write

$$S(\varphi)=\frac{k_2}{k_1}K_{phd}\times\tan\left(\frac{\pi L}{\lambda}\cdot\sin\varphi\right)\qquad(6.23)$$

The error signal $S(\varphi)$ from the output of the phase detector is fed to the input of the antenna control system, which turns toward the target so that the tracking error is minimized, and in the limit, is reduced to zero.

6.3 Monopulse Antenna

It is known that the antenna is the main element of a monopulse radar. A monopulse antenna communicates between a target, emitting a signal and an on board device that calculates the spatial coordinates of this target. The main characteristics of a monopulse radar largely depend on the design of the monopulse antenna. The shape of the antenna directional pattern in a monopulse radar has a significant effect on the ability of a monopulse radar to search for and spatial selection of targets.

Three main categories of antennas used in monopulse radar are lenses, reflectors, and arrays, and each category can be divided into various types. The principal distinction between monopulse antennas and similar antennas used for other purposes is the nature of their feeds, although the feeds are really part of the antennas.

6.3.1 Lens Antennas

The principle of operation of a lens antenna is based on the laws of refraction of rays at the interface between two environments. There are different types of lens antenna surface configurations, which can be divided into two groups based on refraction index. If the index of refraction is less than one ($n < 1$) one or both of the curved surfaces will be concave, whereas for $n > 1$ one or both will be convex. For example, for convex-plane lens: $n > 1$, and for concave-plane lens: $n < 1$. In this case, the convexity or concavity of the surface is considered in relation to the focus of the lens. For convex-plane lens the distance between the first surface (convexity) and the focus should be less than the distance between the focus and the second surface (plane) [7].

It is known from optics, that if a beam falls on a flat interface between two environments with dielectric constants ϵ_1 and ϵ_2, respectively, then the angle of refraction β_1 can be found from the relation [3,4]:

$$\sin\beta_1 = \frac{n_1}{n_2}\sin\beta_2, \tag{6.24}$$

where β_2 is the angle of incidence, and n_1 and n_2 are the refractive indices of the environment. Due to the fact that the refractive index n of an environment is equal to the square root of the dielectric constant ϵ of this environment:

$$n_1 = \sqrt{\epsilon_1}; \qquad n_2 = \sqrt{\epsilon_2}. \tag{6.25}$$

Thus:

$$\sin\beta_1 = \sqrt{\frac{\epsilon_1}{\epsilon_2}}\sin\beta_2 \tag{6.26}$$

Using property (6.26), we can explain the principle of operation of a dielectric lens. Let the source F_l emit a beam of diverging rays and is in the air ($n_1 = 1$). In a dielectric lens, the propagation speed of the wave front is slower than in air, and therefore the path traveled by the wave will be longer at the center of the

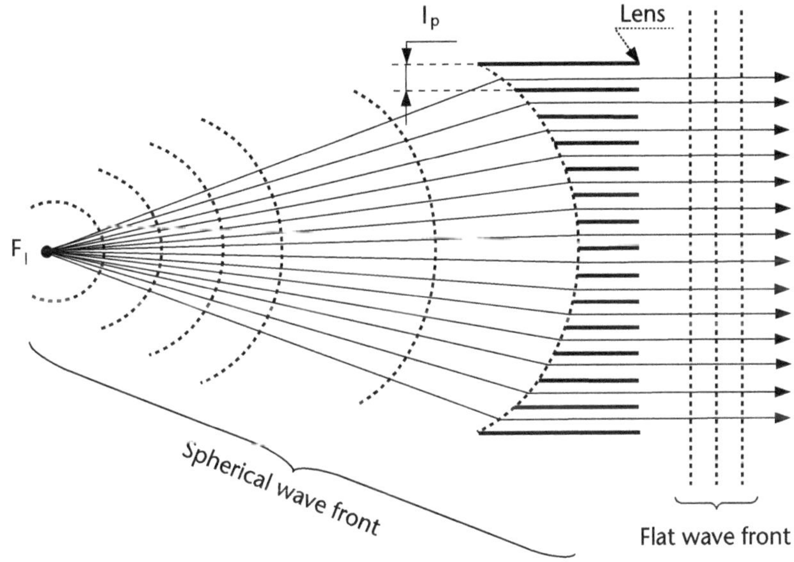

Figure 6.5 Metal-plate lens.

lens and shorter at its edges. Accordingly, the wave front in the central part of the lens propagates more slowly than at the edges of the lens. Thus, as the wave front moves in the direction from the feed to the unlit surface of the lens, it gradually straightens. With the correct choice of the lens profile, a flat wave front appears on its unlit surface. Due to the high cost of dielectric lenses and their high weight, they did not find application in the homing heads of anti-aircraft guided missiles. While dielectric lens focusing technology has now found its way into automotive optics, many automakers are now moving away from automotive optics based on paraboloid-shaped reflectors.

Lens antennas of various types have been built [8, 9]. One type is constructed of dielectric material, in which the phase velocity is less than in air. Such lenses are similar to optical lenses. Since the usual purpose of a microwave lens in reception is to cause a wave from a distant source (essentially a plane wave) to converge at the feed in the focal plane, the dielectric lens is a convex figure of revolution (Figure 6.4). Another type of lens antenna is the metal-plate lens [10,11], illustrated in Figure 6.5, composed of plates parallel to the E-vector and located at a distance of l_p from each other, where $\lambda/2 < l_p < \lambda$. The space between the plates acts as a waveguide, in which the phase velocity is higher than in air, so the refractive index is less than one and can be calculated by the formula:

$$n_2 = \sqrt{1 - \frac{\lambda^2}{4l_P^2}} \tag{6.27}$$

With an appropriate selection of the lens shape, all rays emanating from the F point reach the aperture at the same time, and the field in the aperture will be in phase. The smaller the distance between the metal plates, the longer the lens. But

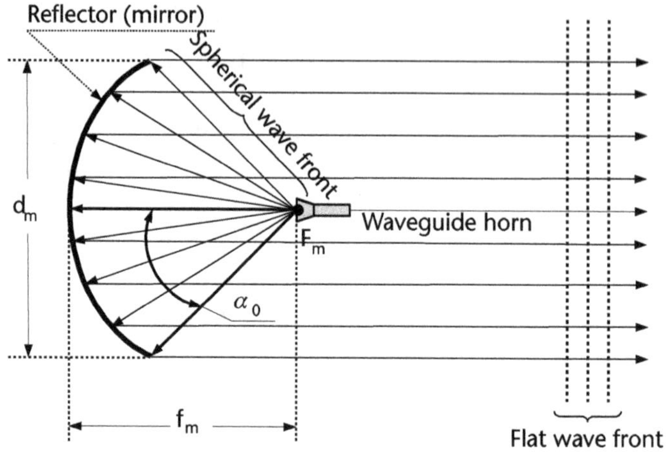

Figure 6.6 Reflector (mirror) antenna.

even with a choice of a refractive index of about 0.5 ($l_p = 0.56\lambda$), the thickness of such a lens is still significant. To produce convergence, the lens must therefore be concave. Although the plates run in only one dimension, the converging effect is two-dimensional if the surface is concave in both dimensions. Because the phase velocity is a strong function of frequency, the useful bandwidth is small. In both Figures 6.6 and 6.7 a single feed horn is shown in order to illustrate the action of the lens. A lens used for amplitude-comparison monopulse would have multiple feed horns in the focal plane displaced from the axis. If lenses were to be used for a phase-comparison monopulse, there would be multiple lenses side by side, each with a single feed horn. To reduce the thickness and weight of dielectric or metal plate lenses, they can be stepped or zoned. Progressing from the thinnest part of the lens, the thickness increases within each zone as in the basic lens, but at the zone boundaries the thickness is decreased in successive steps of one wavelength, causing phase steps of 360° across the aperture. These steps theoretically have no effect, since 360° is equivalent to 0°. However, such lenses are narrowband, since the one-wavelength steps must be matched to a particular frequency. Furthermore, the steps cause some shadowing and discontinuity, resulting in energy loss and increased sidelobes.

Still another type of lens antenna is the constrained metal lens [11], which consists of waveguides or of metal plates perpendicular to the E-vector, connecting the inner and outer faces. The phase velocity is approximately the same as in air, with little frequency sensitivity, permitting operation over a wider frequency band than the metal-plane lens. The converging effect is achieved by arranging the lengths and orientations of the guides so that waves in each guide follow an indirect path between corresponding points on the two faces, with longer length at the center than at the edges.

6.3.2 Reflector Antennas

A diagram of a parabolic antenna is shown in Figure 6.6. The antenna consists of a metal mirror in the form of a paraboloid of revolution and a feed in focus F_m.

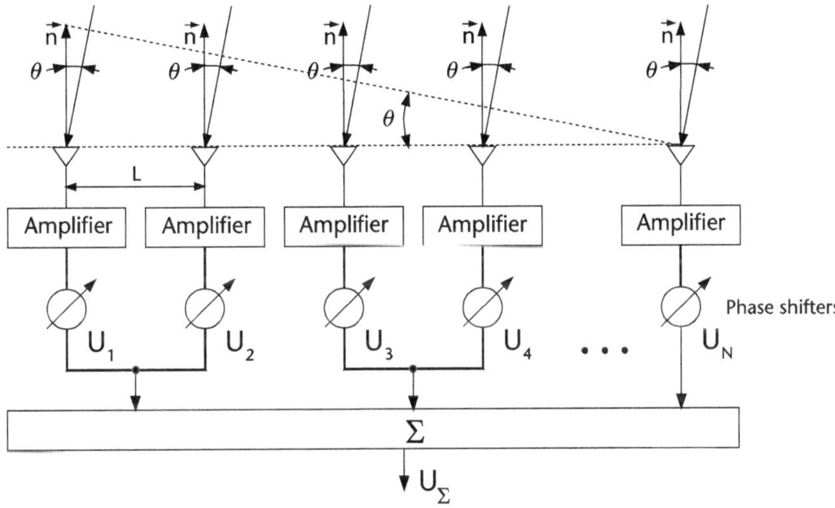

Figure 6.7 Formation reception by means of a phased antenna array.

The main properties of a parabolic (reflector) antenna are as follows:

1. Diverging rays coming from a source located at the focus F_m of the mirror, after reflection from its surface, become parallel.
2. The distance traveled by any ray from the focus to the mirror and after reflection to the plane perpendicular to the paraboloid axis does not depend on the angle at which the ray went out of focus. This results in a flat wave front with a constant phase.

A paraboloid with a larger aperture angle $2\alpha_0 > \pi$ is called short-focus, and a paraboloid of which the inequality $2\alpha_0 < \pi$ is fulfilled is called long-focus. For a short-focus paraboloid, the ratio of the focal length f_m to the mirror diameter d_m is $f_m/d_m < 0.25$, and for a long-focus paraboloid: $f_m/d_m > 0.25$. In practice parabolic antennas have a focal length to diameter ratio in the range: $f_m/d_m = 0.5 \ldots 1.0$. The most common type of feed for monopulse parabolic antennas is the waveguide horn. Microwave energy can be supplied to the horn by means of a waveguide from the side or through the antenna mirror. In any case, the waveguide feeding the horn passes through the mirror, which causes the aperture to be shaded. If the feed is moved from the focus of the parabolic antenna to an angle α_1 around the circle with the center located at the apex of the paraboloid, then the direction of maximum radiation will deviate from the mirror axis by the angle $\alpha_2 < \alpha_1$. In this case, the direction of maximum radiation is rotated in the direction opposite to the movement of the feed. In a monopulse radar with amplitude direction finding, a pair of feeds, displaced symmetrically from the focus, gives symmetrically overlapping amplitude radiation patterns for direction finding in one plane (Figure 6.1). In monopulse phase direction finding radars, the antenna system for direction finding in one plane should consist of two separate parabolic antennas spaced apart at a distance L, which is called the base, with feeds placed at the focus of each of these antennas (Figures 6.2 and 6.3).

Reflector antennas are widely used in monopulse radars and in homing heads of anti-aircraft guided missiles in particular. The homing head of the S-200 anti-aircraft missile system is a classic example of the use of reflector antennas for direction finding targets in two orthogonal planes. This seeker has proven its high efficiency while having a tenfold lower cost compared to seekers with phased array antennas.

6.3.3 Arrays Antennas

Currently, phased antenna arrays are increasingly used in monopulse radars. These antennas have a large number of radiating elements, which are located relative to each other in accordance with the requirement to ensure the uniqueness of the maximum of the radiation pattern of the array, which must be located in the region of real angles [7,12].

Let the antenna array lie in the x, y plane with in-phase excitation of all radiating elements $L = N \times M$, in this case its main beam will be directed along the Z-axis. Thus, a beam is formed that is perpendicular to the plane of the antenna array and collinear to its normal $\vec{n}$. The movement of the beam in space is achieved by a corresponding change in the phases of the signals supplied to each i, jth radiating element:

$$\dot{A}_{i,j} = E_{i,j} \exp(j\varphi_{i,j}), \tag{6.28}$$

where $E_{i,j}$ is excitation amplitude of an antenna array element, $\exp(j\varphi_{i,j})$ is excitation phase of this element. If we provide a linear change in the phase of all elements from 1 element to N element along one linear coordinate along x, for example, then the main beam will deviate from the normal in the $X0Z$ plane by an amount:

$$\Delta\varphi = \varphi_N - \varphi_1 \tag{6.29}$$

Figure 6.7 presents a diagram of a linear equidistant antenna array consisting of H elements spaced from each other at equal distances L. In the mode of receiving signals at the output of the adder, the sum voltage U_sum is formed. Suppose that the elements of the antenna array are isotropic and radiate EM energy uniformly in all directions. The signal of the first element is considered to be a reference with a phase ϕ_0. The phase difference in adjacent elements can be represented as:

$$\Delta\varphi_i = \frac{2\pi L}{\lambda} \sin\theta \tag{6.30}$$

Then the power directional diagram of such a grating will look like:

$$F(\theta) = \frac{\sin^2(N\pi \frac{L}{\lambda} \sin\theta)}{N^2 \sin^2(\pi \frac{L}{\lambda} \sin\theta)}, \tag{6.31}$$

where θ is the antenna main beam position. Due to the fact that it is necessary to consider only the radiation into the front hemisphere, behind the radiating elements of the grating it is necessary to place a metal screen at a distance of a multiple of half the wavelength in order for the direct wave and the reflected from the screen to add up in phase.

When the beam moves in sector $\pm 90°$, the smallest sidelobes are obtained when the distance between the radiating elements is equal to half the wavelength $L = \frac{\lambda}{2}$. In the case when this distance L exceeds the value $\frac{\lambda}{2}$, then diffraction maxima of higher orders appear in the radiation pattern, having an amplitude equal to the amplitude of the main beam. In practice, the distance between the elements is selected based on the condition of obtaining the required width of the directional diagram and ensuring the uniqueness of the main beam in the region of actual scanning angles. If in (6.31) we take $L = \frac{\lambda}{2}$, and the $\sin\theta$ in the denominator is replaced by its argument, then the beam width at the half-power level can be estimated by the expression

$$\Theta_{0.5} = \frac{101.8}{N} \tag{6.32}$$

From (6.32) it follows that the larger the linear dimensions of the antenna array, the narrower its radiation pattern. When using directional radiating elements, the directional diagram of a linear antenna array in terms of power is described by the expression

$$F(\theta) = F_i(\theta) \cdot \frac{\sin^2\left(N\pi\frac{L}{\lambda}\sin\theta\right)}{N^2\sin^2\left(\pi\frac{L}{\lambda}\sin\theta\right)}, \tag{6.33}$$

where $F_i(\theta)$ is the directional diagram of a separate element of the antenna array. It should be noted that (6.33) is valid only in the case of identical radiating elements having the same directional patterns. As emitters of the antenna array, various weakly directional antennas are usually used: vibrators, slots, metal spirals, and dielectric rods. An antenna array forming one beam can be converted into a multi-beam antenna (Figure 6.8) by including additional phase shifters at the input of each element [3–5].

To determine the direction to the target, two beams 1 and 2 are formed simultaneously, spaced from the equisignal direction at an angle $\mp\theta_0$. It is quite clear that the equisignal direction is characterized by the case when the amplitudes of the signals reflected from the target and received by beam 1 and beam 2 are equal. In this case, the output voltage at the output of the error signal amplifier is zero $U_{error}(\theta) = 0$ and monopulse antenna array is directed exactly to the target. The output voltage $U_{\Sigma 1}$ of the beam 1 is formed by the phase shifters $11 \ldots N1$ and the summing device of the first beam (in the diagram on the left). Accordingly, the output voltage $U_{\Sigma 2}$ of beam 2 is formed by phase shifters $12 \ldots N2$ and the summing device of the second beam (in the diagram on the right).

Therefore, if the total number of elements of a flat antenna array is equal to Q, then for the formation of four beams it is necessary to have $M = 4 \times Q$ phase shifters. To implement multipath reception with phase direction finding, it is necessary to coherently add signals (at a high, but more often at an intermediate frequency), provided that the phases of the signals arriving from each quarter of the flat antenna array are preserved.

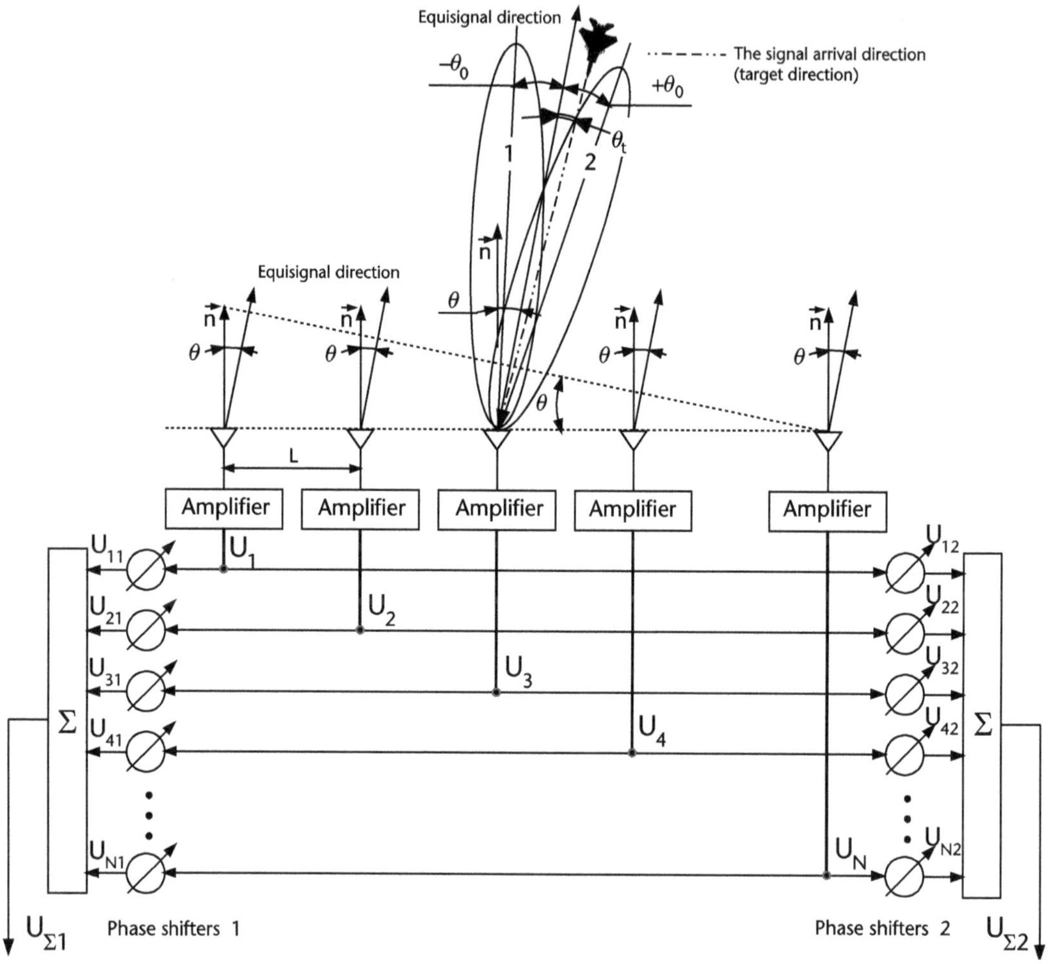

Figure 6.8 Formation of monopulse reception by means of a phased antenna array.

6.4 Cross-Polarization Radiation

In reflector antennas, due to the geometry of the antenna system, as well as the imperfection of the reflecting surface and feeds, displacement of the feed from the focus, and diffraction phenomena, cross-polarization occurs; that is, a wave is emitted or received that is polarized in a different way than required. In reflector antennas, the largest contribution to the formation of cross-polarization radiation is made by the mirror itself; namely, its geometry. Figure 6.9 shows the distribution of the field strength in the aperture of a paraboloid of revolution excited by an electric dipole located parallel to the $0\vec{Z}$ axis. The appearance of cross-polarization radiation, the vector of which is directed along the $0\vec{X}$ axis, is due to the curvature of the mirror surface.

The radiation of a parabolic mirror is formed by two radiation patterns: a radiation pattern for the main (working) polarization and a radiation pattern for cross polarization.

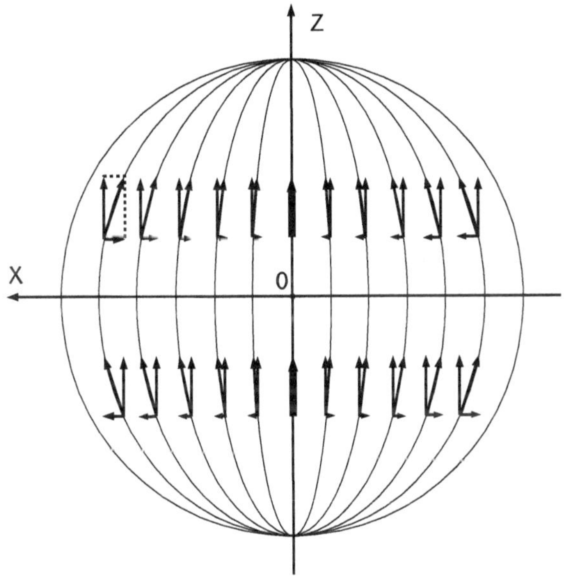

Figure 6.9 Distribution of currents induced on the mirror metal surface.

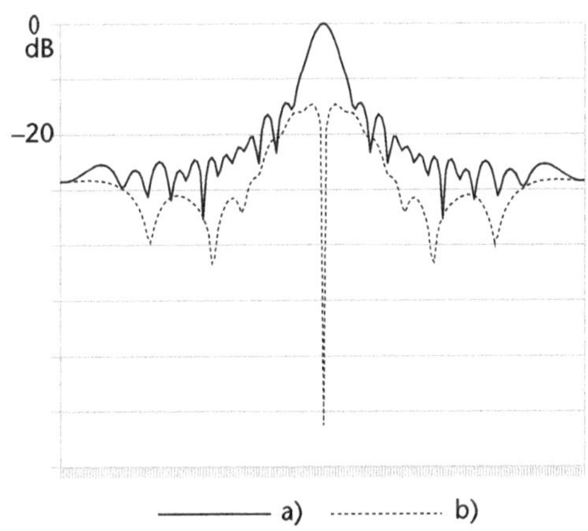

Figure 6.10 Mirror antenna.

Figure 6.10 shows the calculated directional patterns of the reflector antenna for the working and cross polarization: a is $F_{\Sigma}^{w}(\varphi)$ and b is $F_{\Sigma}^{c}(\varphi)$, which are assumption patterns of the monopulse antenna for the sum channel on the working and cross polarization, respectively. It can be seen from these figures that when the mismatch angle is equal to zero, the value of the cross-polarization radiation pattern is estimated at -160 dB. This can be very dangerous if jamming on cross-polarization is organized against the seeker. As shown in [3, 13, 14], for example, in the sum-and-difference scheme of angular tracking, a mutual change of reception channels is observed in the case of exposure to polarization interference: the difference channel becomes the sum channel, and the sum channel becomes the difference channel. Due to the fact that normalization in the normal mode is carried out by the forces of the sum channel, then the normalizing voltage will be zero when exposed to polarization interference, and since this value is in the denominator of the error signal function, the result of such calculation will be infinity, which will inevitably lead to a breakdown of the automatic target tracking. In this case the target will be lost.

Section 7.2 and Chapter 12 of the book are devoted to the cross-polarization interference, as well as methods for dealing with it.

6.5 Radar Resolution on Angular Coordinates

Radar resolution refers to the radar's ability to distinguish between individual targets in a target's group. Quantitatively, the resolution in angular coordinates is estimated by the minimum angle between the directions on the targets at which it is still possible to measure the angular coordinates with the required accuracy of targets that are not resolved in range and speed. It is quite clear that the process of resolving the angular coordinates of two targets is inseparable from the process of their detection. Since in the process of tracking a paired target there will be at least two signals at the output of the receiver, their resolution will depend on the relative energy of these signals and on the degree of their overlap in terms of the resolved parameter; in this case, in angular coordinates.

Obviously, the less the overlap of the output signals in terms of the resolved parameter, the more opportunities for separate tracking of each target, and accordingly, the better the radar resolution.

The resolution can be judged by the correlation function of the input signals carrying information on the angle. Obviously, the resolution will be better the narrower the correlation function for the resolution angle.

For simplicity of reasoning, let us assume that the antenna of a monopulse radar is successively two point targets located at the same distance from the radar and separated in angle by the value $\Delta\theta$, and therefore, the autocorrelation function of the input signals can be represented as [3,4]:

$$\dot{\Psi}(\Delta\theta) = \Re\left[\int_0^T \dot{E}_0(t,\theta)\dot{E}_0^*(t,\theta+\Delta\theta)d\theta\right] \tag{6.34}$$

where $\dot{E}_0(t,\theta)$ is the input signal of the first target, $\dot{E}_0^*(t,\theta+\Delta\theta)$ is the input signal of the second target, while θ and $\theta+\Delta\theta$ are the angular coordinates of the target T is the observation period.

In this case, the input signals are high-frequency signals with an envelope that repeats the shape of the radar radiation pattern.

In [4] it is shown that the correlation function of the input signals as a function of real numbers will have the form:

$$\Psi(\Delta\theta) = 2E_{os}\exp\left(-\frac{\gamma_0^2\cdot\Delta\theta^2}{2}\right) \tag{6.35}$$

where E_{os} is the energy of a high-frequency signal of a given shape, and γ_0 is the coefficient characterizing the width of the radar radiation pattern.

To determine the width of the autocorrelation function at a level of 0.5 relative to the maximum value, it is necessary to substitute the value into (6.35) and solve it with respect to this, we can get:

$$\Delta\theta_{0.5} = \frac{1.17}{\gamma_0} \tag{6.36}$$

Hence, the width of the autocorrelation function at the level of 0.5 will look like:

$$\delta\theta = 2|\Delta\theta_{0.5}| = \frac{2.34}{\gamma_0} \tag{6.37}$$

Expressing the coefficient γ_0 characterizing the width of the radiation pattern of a monopulse radar in terms of the width of the radiation pattern at the level of 0.5, provided that $|F(\theta)| = 0.5$ and $\theta = \theta_{0.5}/2$ = / 2 solving the equation $F(\theta) = \exp(-\gamma_0^2\theta^2)$ with respect to γ_0, we obtain:

$$\gamma_0 = 1.66/\theta_{0.5} \tag{6.38}$$

From this we get the following:

$$\delta\theta = \frac{2.34}{1.66}\theta_{0.5} = 1.4 \cdot \theta_{0.5} \tag{6.39}$$

Expression (6.39) shows that to improve the resolution of a monopulse tracking radar in angles, it is necessary to strive to narrow its directional pattern. It should be noted that this statement is true only when the signal energy is significantly higher than the noise energy of the receiver. In practice, the value of the potential resolving power is the width of the autocorrelation function of the signals of resolved targets, counted at a level of 0.5 from the maximum. In this case, there is no need to take into account the probabilistic approach when assessing the radar resolution.

References

[1] Kingsley, S., and S. Quegan, *Understanding Radar Systems*, SciTech Publishing Inc., 1999.

[2] Richards, M. A., J. A. Scheer, and W. A. Holm, *Principles of Modern Radar, Vol. I: Basic Principles*, SciTech Publishing, 2010.

[3] Leonov, A. I., and K. I. Fomichev, *Monopulse Radar*, Norwood, MA: Artech House, 1986.

[4] Leonov, A. I., and K. I. Fomichev, *Monopulse Radiolocation*, Soviet Radio Publishing House, 1970.

[5] Leonov, A. I., and K. I. Fomichev, *Monopulse Radiolocation*, Second Edition, Revised and Enlarged, Radio and Communication, 1984.

[6] Sherman, S. M., and D. K. Barton, *Monopulse Principles and Techniques*, Second Edition, Norwood, MA: Artech House, 2011.

[7] Blake, L. V., and M. W. Long, *Antennas: Fundamentals, Design, Measurement*, Third Edition, Raleigh, NC: SciTech Publishing, 2011.

[8] Risser, J. R., "Dielectric and Metal-Plate Lenses," Chapter 11 of *Microwave Antenna Theory and Design*, S. Silver (ed.), Vol. 12 of MIT Radiation Laboratory Series, New York: McGraw-Hill, 1947. Reprint, CD-ROM edition, Norwood, MA: Artech House, 1999.

[9] Skolnik, M. I., *Introduction to Radar Systems*, Third Edition, New York: McGraw-Hill, 2001.

[10] Kock, W. E., "Metal-Lens Antennas," *Proc. IRE*, Vol. 34, No. 11, November 1946.

[11] Kock, W. E., "Path-length Microwave Antennas," *Proc. IRE*, Vol. 37, August 1949.

[12] Budge, M. C. Jr., and S. R. German, *Basic Radar Analysis*, Second Edition, Norwood, MA: Artech House, 2020.

[13] Markin, E., "On Interference Immunity of Angle Tracking Systems Under Conditions of Interference Distorting Location Characteristic," *Proceedings of the 2009 IEEE Radar Conference*, Pasadena, CA, 2009, pp. 1–6.

[14] Markin, E., U.S. Patent No. 8,896,483, *Method of Automatic Target Angle Tracking by Monopulse Radar Under Conditions of Interference Distorting Location Characteristic*, 2014, https://www.uspto.gov.

CHAPTER 7

Some Interferences for Monopulse Radar

In modern conditions, it is difficult to imagine the operation of radars that are part of the anti-aircraft missile system without interference. It is quite understandable that air attack weapons are obliged to minimize the likelihood of their destruction by the air defense systems of the defending side. To do this, it is necessary remotely—ideally without entering the affected area of the anti-aircraft missile system—to complicate or make impossible the operation of its radars.

The solution to this problem without the physical destruction of the anti-aircraft missile system (AMS) is possible only by jamming the radars that are part of it. In this regard, the interference immunity of an anti-aircraft missile system is understood as the ability of its radars to qualitatively perform the tasks assigned to them to detect targets, measure their spatial coordinates, and automatically track one or many targets in conditions of interference created. If a radar under interference influence is not able to perform the functions assigned to it, then such a radar is a suppressed by interference and cannot be considered as interference immunity, even having high technical characteristics in the absence of interference.

With regard to the homing heads of anti-aircraft guided missiles, the ability to automatically track a target in coordinates with the required accuracy is the main goal of creating anti-jamming airborne radars that will ensure the unconditional fulfillment of the missile's mission of destroying the enemy.

Before the introduction of the monopulse method for determining the coordinates of the target in space, the methods of conical scanning and sequential switching of the main beam of the antenna directional pattern were used. These methods are vulnerable to organized interferences. Thus, they have low noise immunity from response interference with amplitude modulation at the scanning frequency (switching frequency) of the beam. This is due to the fact that until recently many homing heads of air-to-air and surface-to-air missiles used only these methods of guidance and tracking due to their low cost and ease of implementation. As the jamming stations improved, missiles using these methods of angular tracking increased significantly missing their targets, and they ceased to meet the criteria for target tracking accuracy under the influence of interference, which often led to the missile's failure to destroy the target. Thus, ensuring high interference immunity of an angular measuring channel is a priority task for guidance and automatic tracking radars.

It should be emphasized that the monopulse method is used only for measuring angular coordinates. Range and radial velocity measurements are based on the same principles as in single-channel systems.

It is reasonable to consider as interference any radio signal that, when received, degrades the quality of the radar's operation and violates the reliability of the target information given by this radar. The origin of the interference can be natural or artificial. Natural interference includes interference arising from thunderstorms in the atmosphere, various kinds of precipitation (rain, snow, hail), and other phenomena. Artificial interference includes any deliberate action aimed at reducing the technical characteristics of radars. Artificial interference is usually divided into camouflaging interference and disinforming interference, as well as interference resulting from atomic explosions.

Camouflaging interference is used in order to make it difficult to and ideally prevent the detection of the target and the correct determination of its coordinates. These interferences are divided into two groups:

1. Active interference, which is used by transmitters of noise interference;
2. Passive interference, which are usually dipole reflectors and metallized strips.

Disinforming interference is designed to disorient the radar automatic tracking system or the radar operator when solving problems of target detection and tracking. The disorientation effect is achieved by creating: false signals simulating signals reflected from real targets; jamming signals that form large errors of automatic tracking, for example, in angular coordinates, which, ideally, can lead to disruption (loss) of automatic target tracking; and jamming signals that make it difficult to select true targets from false ones. This problem is especially acute when choosing a target for destruction in an anti-missile defense system, when false targets do not make it possible to select the true target's speed from a certain number of false targets flying in a group with a true target on the outside atmospheric trajectory. Accordingly, the means of creating such interference can be:

1. Electronic warfare stations, which include transmitters of various interference;
2. False targets.

Camouflaging interference is easily determined by the radar operator by the nature of illumination of the tracking radar indicator or the sector in the circular view indicator in the surveillance radar.

Disinforming interference in the case of their use is more dangerous, since they deceive the automatic tracking system or the operator, due to the formation of signals very similar to real ones, which form large errors in determining the coordinates of the target,which often lead to lost of automatic target tracking.

Interference due to high-altitude nuclear explosions.

7.1 Camouflaging Interference

7.1.1 Active Camouflaging Interference

The most versatile camouflaging interference is active noise interference. The classic method of generating noise interference is the emission of a continuous high-frequency signal, modulated in amplitude, frequency, or phase with noise. When the power is sufficient, this causes the screen to fully or partially light up with interference, which in turn causes the blips of the actual targets to be camouflaged. There is selective noise interference and barrage noise interference. Selective jamming is characterized by a narrow spectrum of the noise signal, which is comparable to the passband of the suppressed radar. In this respect, selective jamming has power

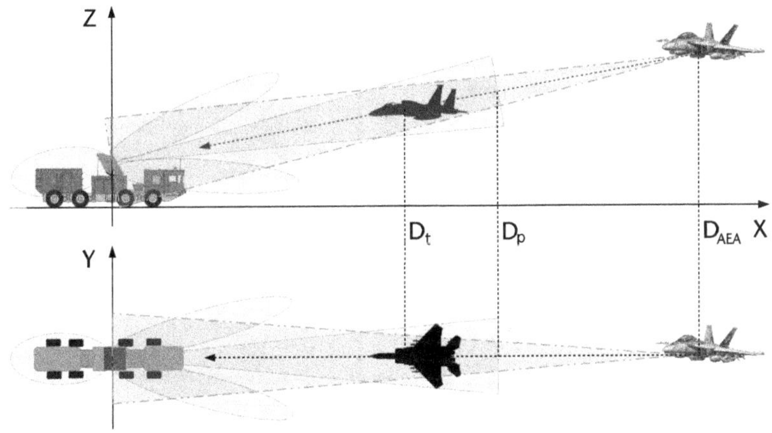

Figure 7.1 An active noise interference of cover by means of a jamming aircraft such as EA-18G Growler.

advantages over barrage jamming. However, selective jamming suppresses only one chosen radar and, with its insufficient mobility, it can be tuned out by changing the working frequency of the radar. Barrage jamming is characterized by a wide spectrum of modulating noises and, in principle, can act simultaneously on several radars. Barrage jamming is characterized by a wide spectrum of noise interference signals. It is quite clear that in order to suppress the selected radar with the same effect, the transmitter of the barrage interference must have a significantly higher power than the transmitter of the aimed interference. The advantage of barrage interference is that, as a rule, the suppressed radar does not have the ability to tune out the interference by retuning the frequency of the signal emitted by the radar [1–3]. Let us consider the case of setting up an active cover interference when the covered aircraft is in the main beam of the suppressed radar pattern, and the jammer is located on an electronic warfare aircraft, for example, a Boing EA-18G Growler (Figure 7.1). The jamming aircraft is at range D_{AEA} outside the range of the anti-aircraft missile system D_p, so that: $D_t < D_p < D_{AEA}$.

With sufficient camouflaging interference power, their effect leads to illumination of the radar screens so that it is visually impossible to distinguish the tracked target against the background of interference. As an example of the impact of aiming interference, the indicators of the I32V unit of the S-75M3 anti-aircraft missile system are given. This is the workstation of the targeting officer [4]. Figure 7.2 shows the case of tracking a single target without interference. There are two screens: the left screen in coordinates "Elevation-Range," and the right screen in coordinates "Azimuth-Range." Each screen shows the target at the intersections of vertical marks (VM) and horizontal marks (HM). Here horizontal marks characterize the distance from the SNR-75 radar to the target, and vertical marks characterize the position of the bisector of the scanning sectors in elevation and azimuth, respectively. At the bottom of each screen (indicator) there are three missile markers (missile packs). In addition, each indicator also has three horizontal marks formed by the APP-75V missile launch time determination device, based on data of angular position, range and speed of the target coming from the cabin coordinate system A: $\mathbf{r}_m$, which is the meeting point, $\mathbf{r}_p$, which is the far border of the affected area when firing with the use of the missile flight passive section, $\mathbf{r}_f$, which is the far border

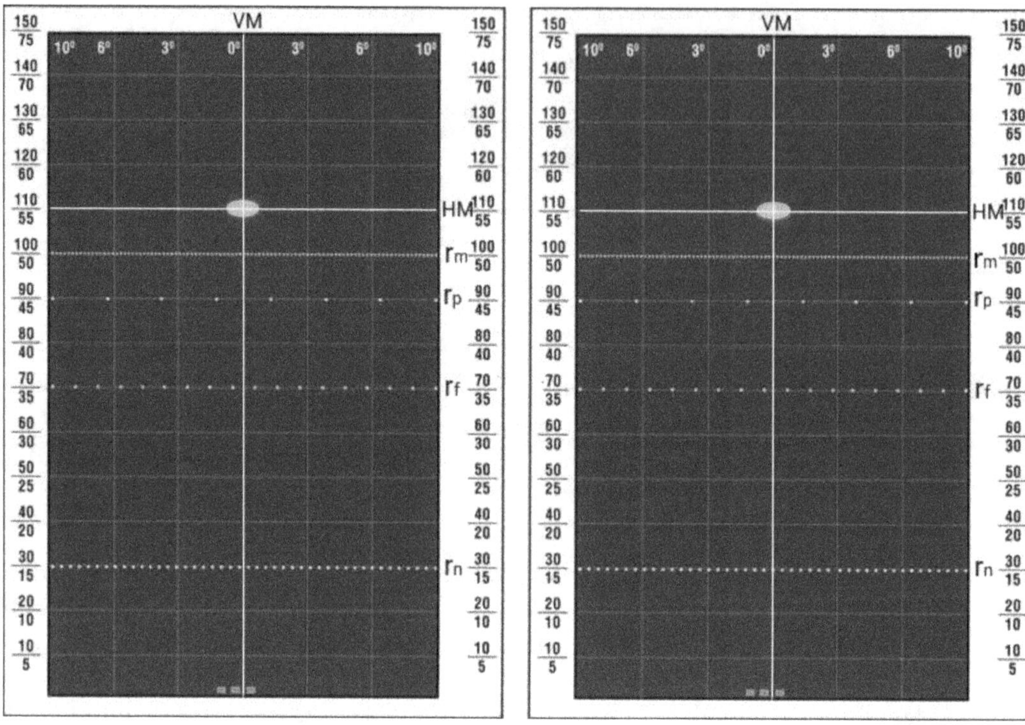

Figure 7.2 Tracking a clean target via the S-75M3.

of the affected area, and $\mathbf{r_n}$, which is the near border of the affected area. Due to the fact that the SNR-75 radar performs sector scanning of space with a frequency of 16 Hz in two mutually orthogonal planes ε and φ: in Wide Beam mode through fan-shaped antenna patterns in the $\Delta_{\varepsilon,\varphi} = \pm 10°$ sector; and in Narrow Beam mode by means of needle antenna patterns in sector $\Delta_{\varepsilon,\varphi} = \pm 3.75°$ relative to the line radar-target. The Narrow Beam antenna patterns have the same width in the ε and φ planes: $F_{-3dB}(\Delta\varepsilon, \Delta\varphi) = F_{-3dB}(0.75°, 0.75°)$. Wide Beam antenna patterns have the following characteristics: in the ε plane (vertical antenna): $F_{-3dB}(\Delta\varepsilon, \Delta\varphi) = F_{-3dB}(2°, \geq 20°)$; and, accordingly, in the φ plane (horizontal antenna): $F_{-3dB}(\Delta\varepsilon, \Delta\varphi) = F_{-3dB}(\geq 20°, 2°)$. As can be seen from Figure 7.2, the mark from the target is clearly distinguishable on the radar screens, and in this case:

$$SNR \gg 1 \tag{7.1}$$

Figure 7.3 shows the case of target tracking under the influence of noise interference along the main beam of the directional pattern of the SNR-75 missile guidance station. For example, the EA-18G Growler is the most advanced airborne electronic attack (AEA) platform and is the only platform produced today [5]. In this case, there is a complete illumination of both indicators, and the bandwidth illumination corresponds to the Wide Beam fan-shaped antenna pattern's width. It is very difficult to visually distinguish a target in interference; therefore, in this case, the following relations are fulfilled:

$$SINR \ll 1 \quad or \quad INSR \gg 1, \tag{7.2}$$

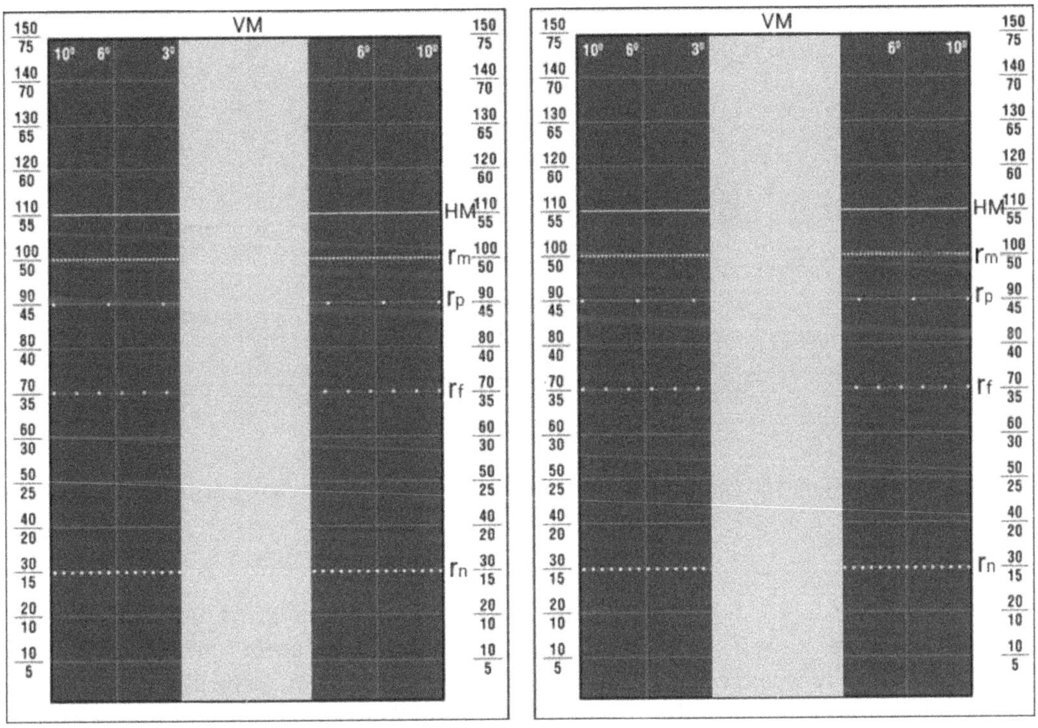

Figure 7.3 Targeted noise interference along the main beam of the antenna pattern S-75M3.

where $SINR$ is the signal-to-interference ratio, and $INSR$ is the interference-to-signal ratio. A similar picture of the illumination indicators (Figure 7.3) of an anti-aircraft guided missile guidance officer develops when the enemy uses self-cover noise interference in the situation shown in Figure 7.4. Here, the attacking aircraft, upon detecting irradiation from the side of the anti-aircraft missile battalion's firing radar, turns on noise interference directed into the front hemisphere in the direction of the attacking aircraft's movement. The onboard equipment of a modern attacking aircraft has a reconnaissance receiver that detects the radar signal that irradiates it and classifies this signal into one of two classes: $R_i \ : \ \forall\, i = \overline{1,2}$, so that the feature space $\{X\}$ would not be contradictory:

$$\begin{cases} X_i \cap X_j = \varnothing & \forall\, i \neq j \\ X_i \cap X_j \equiv X_i & \forall\, i = j, \quad \text{where} \quad \forall i \wedge \forall j = \overline{1,2} \end{cases} \tag{7.3}$$

Let X_1 be the surveillance radar class, and X_2 the missile guidance and tracking radar class. The feature space of the surveillance radars $\{X_1\} = \{x_{1,1}, x_{1,2}\}$ consists of two features:

$$x_{1,1} : \lambda_0 \in \{m\} \qquad or \qquad \lambda_0 \in \{dm\},$$

where the operating wavelength of the signal emitted by the radar belongs, as a rule, to the meter wavelength range, and very high frequency, VHF $\{m\} : \ 1\, m \leq \lambda_0 \leq 10\, m$ or decimeter wavelength range $S : \{dm\} : \ 7.5\, sm \ \leq \lambda_0 \leq 15\, sm$. For P-18 detection radar: $\lambda_0 = 1.2\, m \Rightarrow x_{1,1} = 1$.

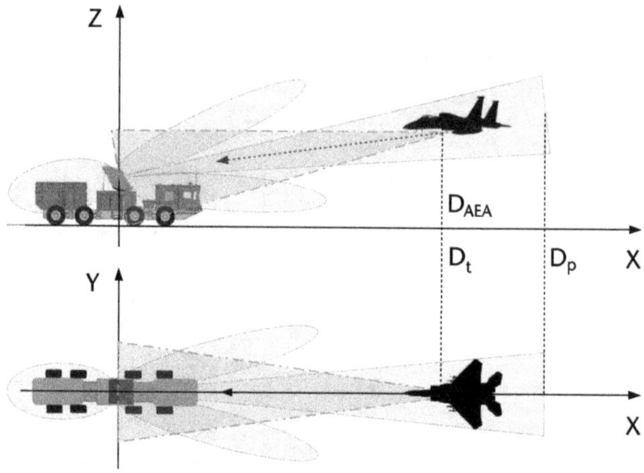

Figure 7.4 Targeted noise interference along the main beam of the antenna pattern S-75M3.

$x_{1,2}$: $t_{irr} << T_{rev}$ is where the target irradiation time t_{irr} is much less than the space survey time T_{rev}; $t_{irr} = N \cdot T_{rep}$, the target's irradiation time, where N is the number of reflected echoes during target irradiation time t_{irr}; and $T_{rep} \geq 2D_{max}/c$ is the repetition period of the probing pulses. Provided that the range is unambiguously measured, this time should be greater than or equal to the maximum delay time of signals reflected from the target located at the maximum detection range D_{max}, where $c = 3 \cdot 10^8$ m/s is the speed of light. $T_{rev} = 360°/\Omega_A$ is for all-round radar with a fan-shaped radiation pattern, where Ω_a is the speed of rotation the main beam of antenna pattern in azimuth [°/s]. Taking into account that for the radar of a circular view $P - 18$: $\Omega_A = 36°/\text{s} \Rightarrow T_{rev} = 10\text{s}$ the time of information update from a given angular direction φ_t, where $t_{irr}(\varphi_t) = 2 \cdot 360 \cdot 10^3/3 \cdot 10^8 = 0.0024\text{s}$ and therefore $x_{1,2} = 1$. Thus, $\{X_1\} = \{1\}$ and it is decided that the signal received by the reconnaissance receiver belongs to the detection radar. In the event that the aircraft's reconnaissance receiver is exposed to radiation from the missile guidance radar (Figure 7.4), which has a wavelength of $1.5 \text{ cm} \leq \lambda_0 \leq 7.5 \text{ cm}$ and belongs to NATO frequency bands $G, \ldots, J \Rightarrow x_{2,1} = 1$. Taking into account that in the case of aircraft escorting by means of missile guidance radar $t_{irr} = T_{rev}$ and $\Rightarrow x_{2,2} = 1$, thus, in this case, a decision is made about the fact that the aircraft is irradiated by the anti-aircraft missile system firing radar $\{X_2\} = \{1\}$. In this case, it is necessary either to perform an anti-missile maneuver and leave the zone of destruction of the anti-aircraft missile system, or to turn on the self-cover interference and continue the combat mission. It is quite clear that the command to turn on the jammer will be issued in the case of $\{X_2\} = \{1\}$.

According to combat work instructions, in case of observing screens similar to that in Figure 7.3, you must immediately switch from the automatic tracking mode AS to the manual tracking mode RS, work which is provided by three military operators $(\varepsilon, \varphi, R_t)$ who manually rotate the corresponding selsyns that control the angular position of the antenna and the range strobe based on visual information received from screens located at the workplace of each operator. Each operator visually carries out the prolongation of the position of the mark from the target

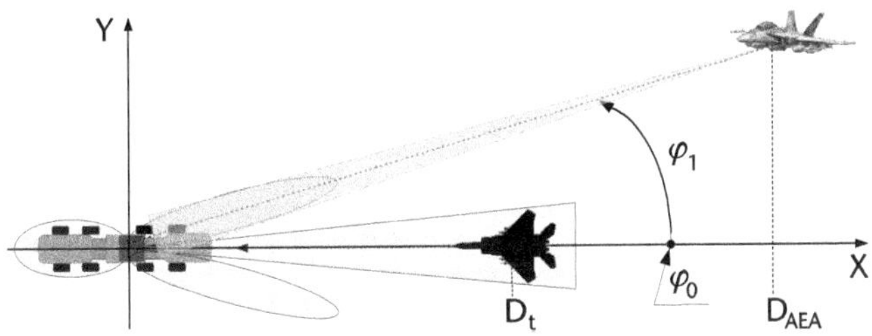

Figure 7.5 Active cover interference on sidelobes of Radar antenna.

on the screen filled with noise. In this case, the effectiveness of the combat mission depends on the coordination and skills of the calculation.

A special place among the active camouflaging interference is occupied by the interference on the sidelobes of the antenna radiation pattern of the suppressed radar. The signal-to-noise ratio in the azimuthal plane can be written as:

$$SINR = \frac{F(\varphi_0) \times P_t}{\sum_{j=1}^{M} F(\varphi_j) \times P_j + N} \tag{7.4}$$

where P_t is the power of the signal reflected from the target, $F(\varphi_0)$ is the position of the target in the sum antenna radiation pattern, when $\varphi_0 = 0°$ and $F(\varphi_0) = max$. There $j = 1, \ldots, M$ is the number of jammers, $F(\varphi_j)$ is the position of the jth jammer in the directivity pattern of a monopulse radar, and N is the receiver noise power. D_{AEA} is the jamming aircraft's loitering zone range, located beyond the far border of the anti-aircraft missile system's engagement zone. For example, the EA-18G Growler flight performance enables performing escort jamming as well as the traditional standoff jamming mission (radar jamming and deception) [5]. D_t is the attacking aircraft range. According to Figure 7.5 the expression (7.4) will be rewritten in the form:

$$SINR = \frac{F(\varphi_0) \times P_t}{F(\varphi_1) \times P_1 + N} \tag{7.5}$$

Taking into account that with the normalized ground radar radiation pattern $F(\varphi_0) = F(0) = 1$, with the level of first side lobes = -40 dB: $k_{lobes} = 10^{(F(\varphi_1)\text{ dB}/10)} = 10^{-40\text{ dB}/10} = 1 \cdot 10^{-4}$.

$$SINR = \frac{P_t}{10^{-4} \times P_1 + N} \tag{7.6}$$

If $N = P_{r_min} = 5 \cdot 10^{-17}$W and $G_t = G_r = 50{,}000$, $\sigma_t = 3\ m^2$, $\lambda = 0.04\ m$, $R_t = \sqrt{(R_E+6)^2 - R_E^2} + \sqrt{(R_E+10000)^2 + R_E^2} = 365.843$ km is the line-of-sight range of a target with $\sigma = 3\ m^2$, which approaches the ground-based radar at an altitude of $H_t = 10$ km, then transmitter power of the ground-based radar can be represented as:

$$P_t = \frac{10 \cdot (4\pi)^3 P_{r\_min} \cdot (R_t)^4}{G_t G_r \sigma_t \lambda^2} = 1500.27\text{W} \tag{7.7}$$

Substituting (7.7) into the expression (7.6) under the condition P_{1_min}: for $SINR = 10$, you can obtain the threshold value of the interference power generated by the aircraft-jammer when exceeding which the considered ground radar will be suppressed:

$$P_1 = \frac{P_t/10 - P_{r\_min}}{10^{-4}} = 1500270\text{W} = 1.5\ \text{MW} \tag{7.8}$$

From the expression (7.8) it follows that to suppress a ground-based radar with a low sidelobe level (≈ -40 dB) at a range of $D_{AEA} = 365.843$ km, you need to have an interference transmitter that would be able to create the power $P_1 \geq 1.5$ MW at the receiving antenna of the ground radar. Taking into account the fact that a pyramidal horn with sides, for example, $5\lambda \times 5\lambda$, is often used as the jammer radiating antenna, therefore $G_1 = 0.5\pi(5\lambda)^2/\lambda^2 = 0.5\pi\ 5^2 = 157.0796327$ or $G_1 = 21.96\ dB$ [6,7]; hence the required power of the onboard jammer P_{AEA} can be represented as follows:

$$P_{AEA} = \frac{P_1}{G_1} = 9.551\ \text{kW} \tag{7.9}$$

From (7.8) and (7.9) it follows that at present it is impossible to create airborne jammers capable of suppressing a ground-based radar with a low level of sidelobes at the far border of the affected area of an anti-aircraft missile system. Thus, lowering the level of sidelobes of the receiving radar antenna is currently the most effective way to combat this interference.

The view of the indicators of the guidance officer of the S-75M3 anti-aircraft missile system when the jammer is working on the sidelobes of the antenna pattern of the SNR-75 missile guidance station is shown in Figure 7.6. Due to the fact that the guidance officer's screens are completely illuminated, it can be concluded that the S-75M3 anti-aircraft missile system is completely suppressed by the interference (completely disabled). This interference is very dangerous for the anti-aircraft missile system, and therefore the most effective means of countering interference along the sidelobes of the radar radiation pattern are measures that reduce the level of the sidelobes of its receiving antenna. So in the S-200D anti-aircraft missile system, the receiving antenna of the target illumination radar has a sidelobe level of ≈ -42 dB. This allows this radar to remain unsuppressed by interference, since at present there are no onboard power sources capable of meeting the condition (7.8).

An intermediate position between the aiming interference and barrage interference is occupied by the interference with the frequency swing (change) of the interference signal. In this case, it becomes possible to concentrate a large interference power in all operating frequency ranges of the suppressed radar. With the correct selection of the frequency swing rate of the interference transmitter, the receiver of the suppressed radar will not have time to restore its full sensitivity in the period between influences.

Camouflaging interference also includes asynchronous impulse noise with low and high frequencies of signal interference sending. This interference creates on

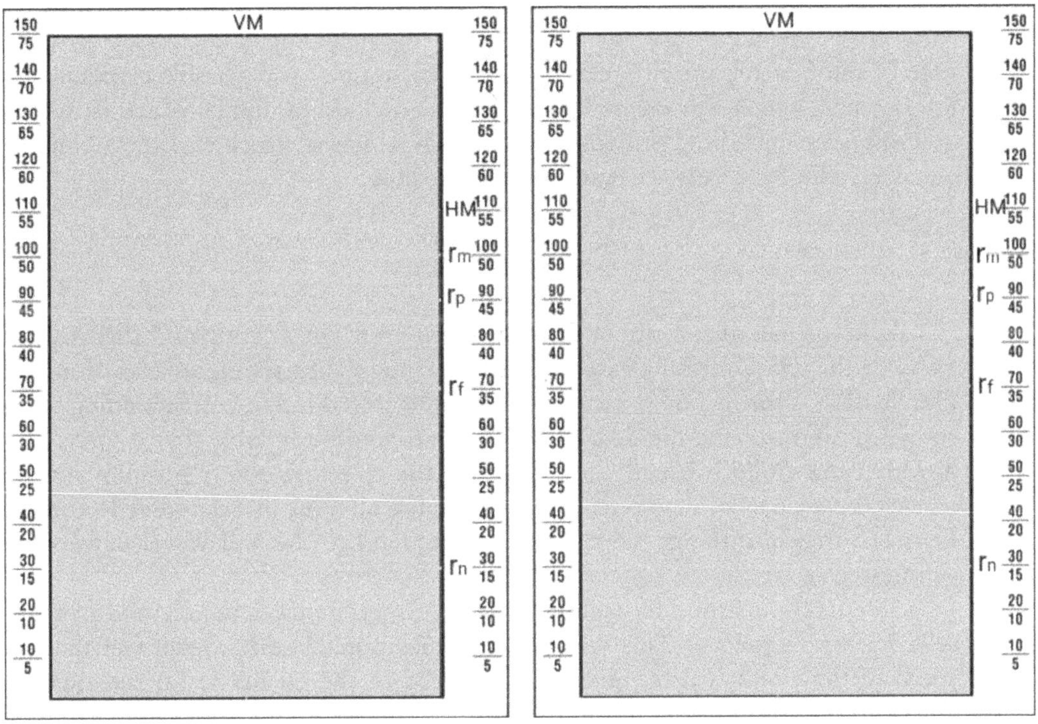

Figure 7.6 Noise interference along antenna pattern sidelobes of the S-75M3 anti-aircraft missile system.

the radar indicators a coarse-grained or fine-grained structure of their illumination or a set of impulse marks moving according to a certain or chaotic law. Active camouflaging interference significantly complicates the detection and auto-tracking of a real target.

There is also an interesting interference in the form of an unmodulated oscillation at a frequency f_0 corresponding to the radar working wavelength λ_0. This interference, with its sufficient power, can reduce the sensitivity of the radar receiver to such an extent that it creates the illusion of a malfunction of the radar as a whole.

The considered active camouflaging interference belongs in the category of power radio countermeasures, since they provide for the suppression of the radar using a sufficiently high interference power.

In addition, it seems promising to develop active camouflaging interference that is emitted by miniature transmitters scattered in the area of responsibility of an anti-aircraft missile system shortly before a missile or missile-space attack. It is reasonable to make these miniature transmitters on solid-state generators capable of generating a jamming signal at a certain frequency for a short period of time sufficient to break through the missile defense system [3, 10].

Thus, active camouflaging interference is still the most dangerous type of interference for anti-aircraft missile systems. With sufficient power of active cover jamming, they are able to bring the radar of an anti-aircraft missile system into an inoperative state, and therefore are one of the main types of jamming that ensure the overcoming of missile defense and air defense systems.

7.1.2 Passive Camouflaging Interference

One of the first organized jamming for radar stations was passive camouflaging interference, which was created by organizing a cloud of dipole reflectors, which, as a rule, were metalized ribbons, the longitudinal size of which was approximately equal to half of the radar operating wavelength, λ_0,

$$l_{mr} \approx \frac{\lambda_0}{2} \tag{7.10}$$

Packs of metallized ribbons are dropped from an aircraft and then carried away by the wind, forming extended spatial clouds. As a result, an area of intense illumination is formed on the screen of an unprotected radar, corresponding to the spatial position of the dipole cloud. It is quite understandable that a target that is within this cloud is usually not visible on the targeting officer's display, as it is masked by a cloud of dipole reflectors. It is quite obvious that dipole reflectors are manufactured in different sizes, which correspond to the half wavelength of the suppressed radar.

An effective method of dealing with such interference is the method of target selection by frequency. This selection becomes possible due to the fact that the speed of the cloud is determined by the speed of the air masses at the spraying height, as well as by their weight and the aerodynamic characteristics of the dipole reflectors. The speed of the cloud of dipole reflectors can change as the dipole reflectors approach the Earth's surface. However, the speed of modern aircraft is much higher than the wind speed, and therefore such a target can be selected by its speed. In the *A* cockpit of the S-75M3 anti-aircraft missile system there were cabinets with moving target selection equipment (SDC), but it was characterized by complex tuning and insufficient efficiency when it was working on clouds of dipole reflectors. Dipole reflectors can be thrown into space in any direction relative to the target using special machines. Ejection of dipole reflectors into the front hemisphere of the protected aircraft allows temporarily disrupting the operation of the automatic target tracking system in range and speed, and in the case of a missile with a homing head flying through a cloud of dipole reflectors, a situation may arise in which the missile's homing head will be reaimed at a cloud of dipole reflectors [3, 8].

In order to increase the efficiency of dipole interference, the illumination of dipole reflectors is sometimes used with the help of special interference transmitters installed on the protected aircraft [3, 9]. In this case, interference signal is emitted either with a frequency shift relative to the operating frequency of the protected radar, which makes it possible to form a false speed of the dipole cloud, or with a time shift, which imitates the dipole cloud movement in range. The method of illuminating dipole reflectors with signals shifted in frequency or time relative to the signal emitted by the radar makes it possible to bring the interference signal closer to the signal reflected from real moving targets, which significantly complicates the selection of real targets from interference.

Currently, there is a tendency to transfer air defense and missile defense radars to the upper part of the decimeter and meter wavelength ranges. This is due to the fact that with an increase in the wavelength at which the radar operates, the geometrical dimensions of the dipole reflectors also increase. Consequently, this tendency to increase the operating wavelength of air defense and missile defense radars can be considered as one of the ways to weaken the interfering effect of passive interference [3,10].

7.2 Active Misinforming Interference

Unlike noise interference, misinforming interference (disinforming interference) provides for the distortion of information about the target through the formation of interference signals that distort the principle of operation of tracking systems. The result of the action of such interference is not a decrease in the signal-to-noise ratio, leading to camouflaging of the signal reflected from the target and, accordingly, reducing the target detection range, but also misleading the radar operators and/or the automatic target tracking system in terms of range, speed, and/or angular coordinates. The creation of such interference requires more information about the principles of operation of suppressed radars, which entails the complication of the jamming equipment in general and the reconnaissance receiver in particular.

Misinforming noises created from one point in space include

- Multiple responding interference;
- Range diversion interference;
- Speed diversion interference;
- Angular coordinate diversion interference.

Multiple responding interference: When creating multiple responding interference (retaliatory multiple interference), the jamming transmitter emits a series of moving false pulses, similar in shape to the received pulse at the same frequency as the pulse of the suppressed radar that was received by the reconnaissance receiver. These false pulses are scattered in range and simulate many targets [3,11] (Figure 7.7). This interference makes it difficult to select a true target from the many false marks, which reduces the effectiveness of the target detection system and can lead to the interception of false targets. The action of this interference leads to the dispersion of the forces and means of air defense, which increases the likelihood of its breakthrough by means of the enemy's aerospace attack. In addition, multiple interference can be used to simulate naval targets. The efficiency of multiple interference increases many times when combined with noise interference. In this way, the targets covered by the jammer are imitated. When creating decoys on the sidelobes of the radiation pattern, the S-75M3 anti-aircraft missile system screens officer of guidance can show the picture seen in Figure 7.8. The creation of false targets on screens of guidance radar of the anti-aircraft missile system is possible with the help of miniature emitters-repeaters, which are ejected from the protected object. Such miniature repeaters are designed on integrated circuits using tunnel diodes designed to capture signal of suppressed radar and then re-emit it with a certain time delay. This method of creating disinforming interference may turn out to be promising in overcoming a missile defense system with ballistic missiles [3,10].

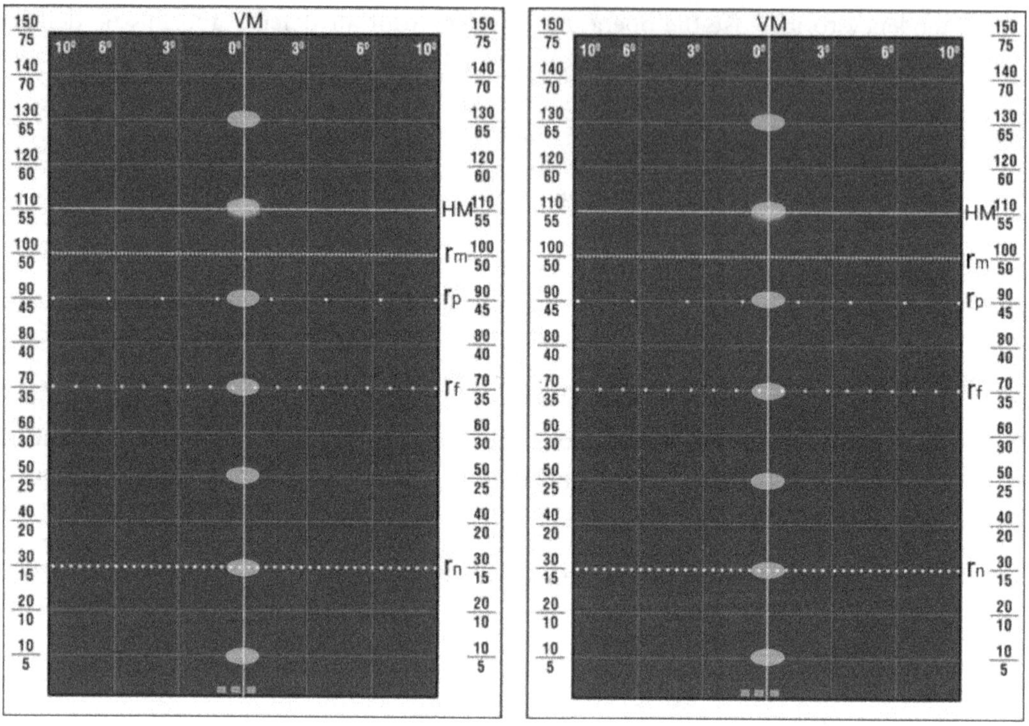

Figure 7.7 Multiple interference for the S-75M3.

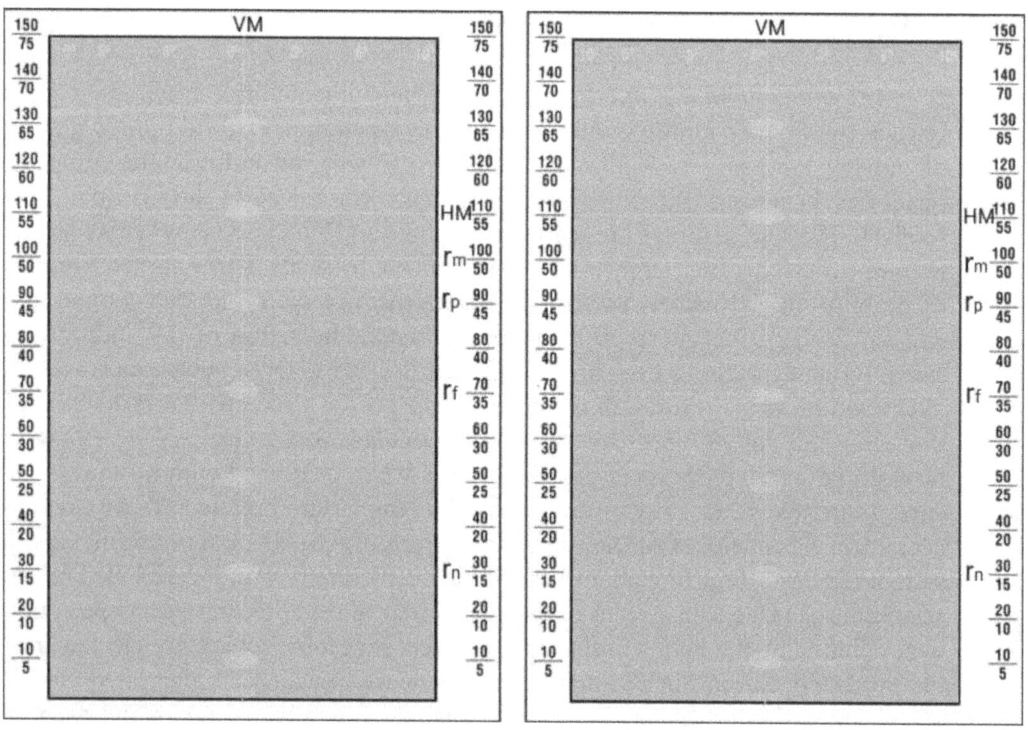

Figure 7.8 Multiple sidelobe interference for S-75M3.

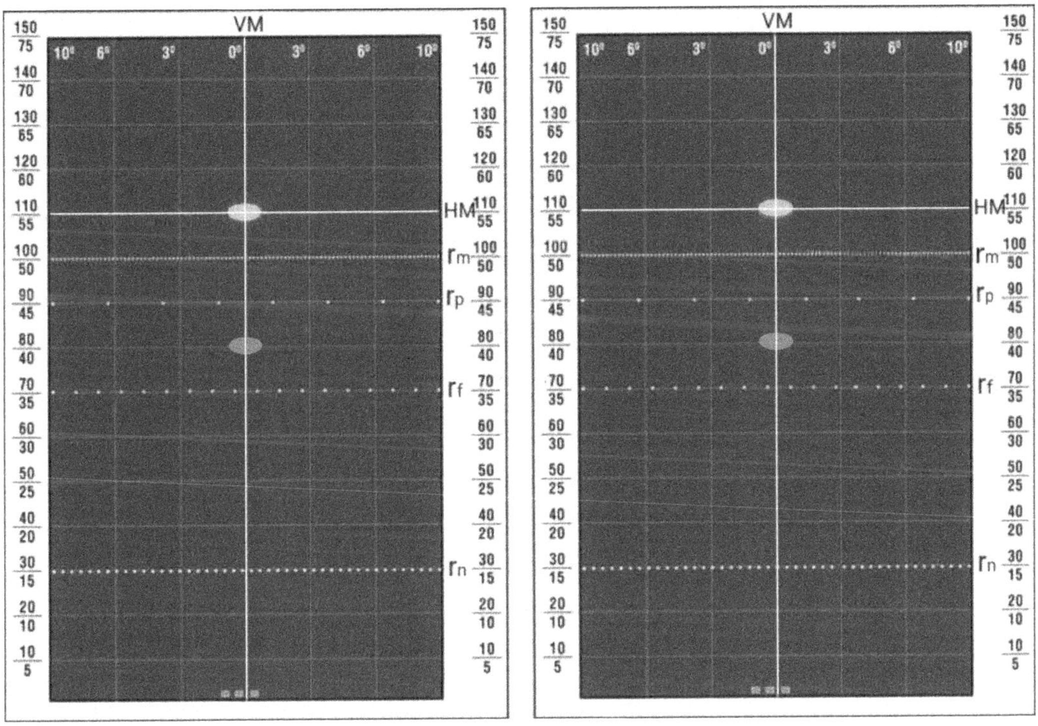

Figure 7.9 Interference leading the tracking system away from a real target for a some range.

Range diversion interference: This interference is created for automatic target tracking systems. The main task of such interference is to force the radar to stop automatic tracking of the target by the reflected signal and switch to automatic tracking of the jamming signal. When generating range diversion interference, the jammer aircraft generates pulsed signals similar to the reflected signal, but with a varying time delay. This simulates the movement of the target at a speed different from the speed of the real target. Figure 7.9 shows a view of the indicator of the guidance officer of the S-75M3 anti-aircraft missile system that is affected by range diversion interference. The real target is located on the indicator at a range of 40 km, and at a range of 55 km there is a false target, which is formed by an interference that leads along the range. Automatic tracking in this case occurs on a false target due to the fact that the power of the jamming signal exceeds the power of the signal reflected from the target. In this case, the strobe of the automatic tracking system will be captured by a jamming pulse and will move away from the real target mark. This will give a false range to the target. The delay of the interfering signal can be decreased or increased until a sufficient amount of range error is provided. After reaching the set error value, the jamming signal is turned off, the automatic tracking strobe loses its mark from the target, and the radar enters the search mode. Thus, the impact on the radar of interference, leading away in range, led to the disruption of automatic target tracking in range.

Speed diversion interference: Speed diversion interference is created for a radar with continuous and quasi-continuous radiation mode; in this case, a smooth variation in interference frequency relative to radar signal frequency occurs. This leads to a smooth variation in Doppler frequency and therefore variation of target speed. With a sufficient surplus of interference power the speed tracking strobe moves according to the noise frequency shift and the servosystem gives out erroneous target speed information. Switching off interference after signal frequency is shifted by a value exceeding the width of the speed strobe leads to loss of target and transition of radar to search mode just as when switching off range diversion interference occurs.

Angular coordinate diversion interference: The effect of angular coordinate diversion interference on the radar antenna will ensure the presence of errors in automatic target tracking in angular coordinates while the real target is on the line of sight radar-target, and the real tracking errors in angular coordinates $\delta_\varepsilon \approx 0°$ and $\delta_\varphi \approx 0°$. A widespread interference in monopulse radars is polarization interference. Due to the fact that often the antenna of a ground-based radar has circular polarization, for example, left-hand rotation, then cross-polarization interference for such an antenna is right-hand rotation. In an airborne radar, linear polarization is usually used, which makes it possible to receive a signal reflected from a target on a working polarization, regardless of the relative position of the missile and the target. This circumstance is decisive in the organization of cross-polarization interference for the homing head of an anti-aircraft guided missile. The direction finding characteristic of the missile seeker at matched polarization when the antenna polarization vector is collinear to the interference linear polarization vector $\alpha = 0°$ is shown in Figure 7.10(a). The direction-finding characteristic of a cross-polarized missile seeker when the antenna polarization vector is perpendicular to the interference linear polarization vector $\alpha = 90°$ is shown in Figure 7.10(b). The essence of the polarization interference lies in the fact that the gain of the direction finding characteristic for the tracking error $\delta_\varphi = (\varphi - \varphi_0) \approx 0$ can be represented as:

$$\begin{aligned} S'(\varphi) &= \left.\frac{S(\varphi) - S(\varphi_0)}{\varphi - \varphi_0}\right|_{(\varphi-\varphi_0)\to 0} \\ &= \begin{cases} K_d, & \forall\, \alpha = 0° - \text{ polarization matched signal} \\ \infty, & \forall\, \alpha = 90° - \text{ polarization interference} \end{cases} \end{aligned} \tag{7.11}$$

It is quite clear that with $S'(\varphi) = \infty$, it is very likely that target auto-tracking will fail. That is why the polarization interference for the missile homing head is the most dangerous interference, since it almost always leads to disruption of the automatic tracking of the jammer in angular coordinates.

It should be noted that active misinforming interference is energetically more advantageous than active camouflaging interference, since it concentrates the maximum power of the jammer in the radar receiver bandwidth. However, from a technical point of view, such interference is more complex, since it requires special and rather sophisticated equipment. So, for example, to effectively stage a polarization interference, it is necessary to know the true position of the cross-polarization vector of the receiving antenna of the seeker. To do this, it is necessary to have a reconnaissance receiver on board the jammer, which has complex algorithms for processing the signal emitted by the missile homing head. At the same time, in the case of semi-active homing, the polarization of the homing head is almost impossible

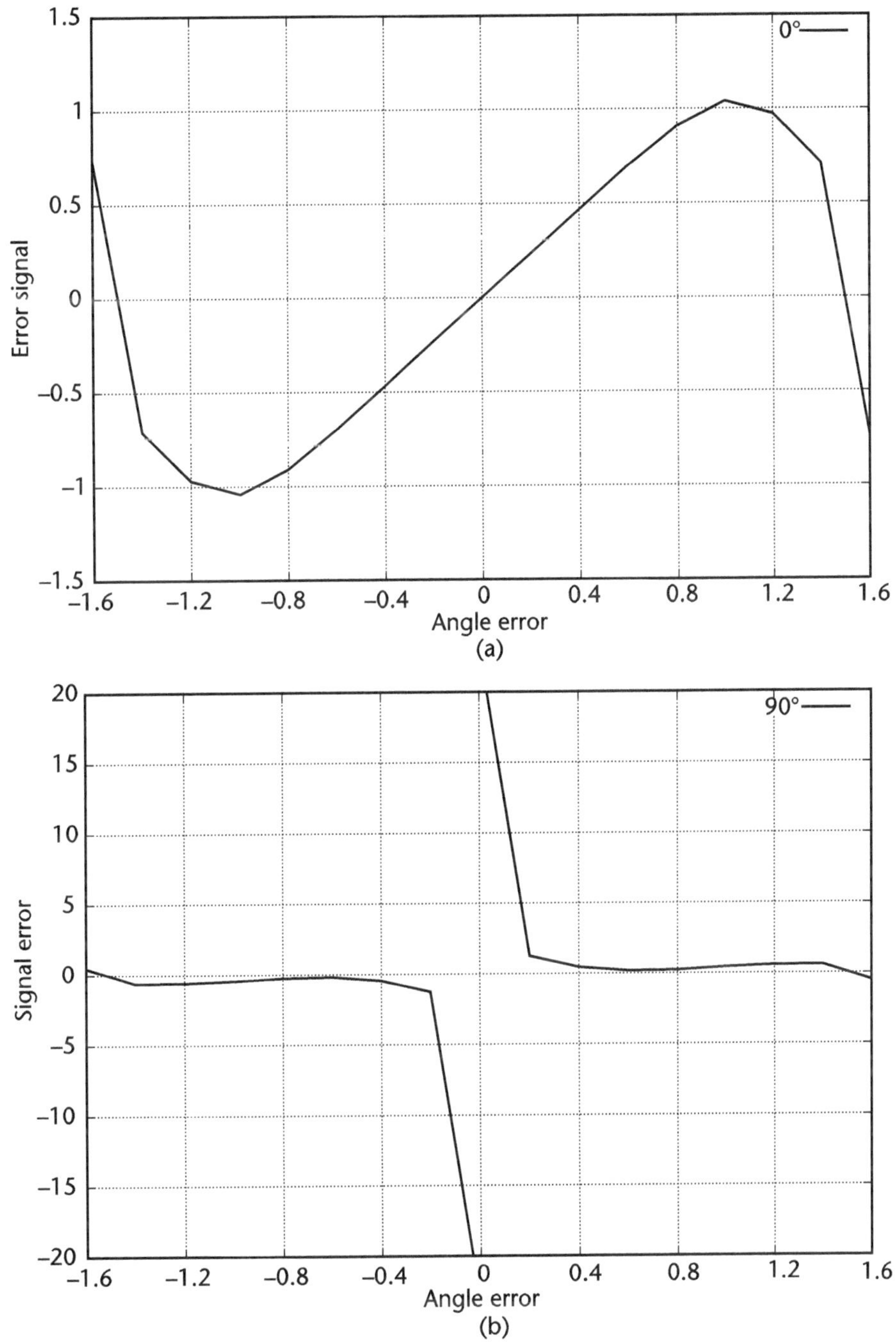

Figure 7.10 Direction-finding characteristics.

to determine. Now decoys have been given considerable attention in the protection of warheads for overcoming antimissile defense systems. Taking into account the specifics of warhead motion, in the middle portions of the flight trajectory, the use of lights decoys is provided (inflated spheres, balloons), and in the final section of the trajectory upon reentry, heavier decoys are provided that are capable of motion in the atmosphere without burning up.

7.3 Interference due to High-Altitude Nuclear Explosions: Electromagnetic Weapons

This interference leads to a violation of the propagation conditions of radio waves and to a deterioration in the quality of the operation of electronic equipment. A damaging factor in this case is the electromagnetic pulse, generated by any explosion of a nuclear weapon. The damaging effect of an electromagnetic pulse is due to the occurrence of induced voltages and currents in various conductors and leads to the failure of electrical and electronic equipment. The most vulnerable are communication and control lines. Breakdown of insulation, damage to transformers, and the most likely failure of semiconductor devices (transistors, diodes, and various microcircuits) can occur. A high-altitude nuclear explosion is capable of creating this interference over very large areas. Therefore, the requirements for the electromagnetic resistance of radio-electronic equipment to the effects of an electromagnetic pulse are included in the list of key indicators for the development of an anti-aircraft missile system. Tests on the effect of an electromagnetic pulse obtained during space and high-altitude nuclear explosions were carried out in the USSR in the period from 1961 to 1962 at the Sary-Shagan test site. A total of five nuclear explosions were made. These experiments were carried out under the letter “K.”

The K1 and K2 explosions were made on October 27, 1961. Ammunition with a capacity of 1.2 KTN was delivered to the center of the Sary-Shagan test site by R-12 (8K63) missiles, which were launched from the Kapustin-Yar test site in the Volgograd region. One charge was detonated at an altitude of 150 km, and the second at an altitude of 300 km above the Sary-Shagan test site. The main purpose of the tests was to test the performance of the “A” missile defense system. In accordance with the test plan, each test involved the sequential launch of two R-12 intercontinental ballistic missiles. The first R-12 missile carried a nuclear charge, and the second missile was equipped with radio measuring equipment to register the damaging effect of a nuclear explosion. The main task during operation “K” was to check the influence of high-altitude and space nuclear explosions on the operation of radio-electronic means of missile attack detection systems and anti-missile defense of the “A” system. Under the conditions of a real nuclear explosion, the second R-12 missile was intercepted by missile B-1000 anti-missile system “A,” equipped with a telemetric warhead (without a warhead).

In October 1962, three other tests were carried out.

The K-3 test took place on October 22, 1962. During this test, at an altitude of $\approx$300 km above the test site Sary-Shagan, or rather 180 km west of the city of Zhezkazgan above the point with coordinates $47°46'48''N$ and $65°19'44''E$, at an altitude of 290 km [14], a nuclear charge with a capacity of $\approx$300 KTN was detonated (Figure 7.11). At this time I was 1 year and 2 months old and I lived with my parents in the city of Zhezkazgan (Kazakhstan) [15]. So, I became a participant in nuclear tests in the USSR without realizing it. When, five years later, I came to my mother with a question about two suns that people saw, my mother scolded me strongly and asked who I heard this from. I replied that children in the schoolyard speak about it. She said that it was not to be discussed and forbade me to tell anyone else about it. My mother at that time worked as the chief editor of a regional television studio and today I understand that she probably signed a nondisclosure agreement with the KGB. In 1962, the Soviet government endangered the civilian population of an entire city. I am sure that the authorities did not have and do not have a complete list of civilians who became unwitting participants in these nuclear tests, since neither my family then, nor I today, received any material compensation. It is just that everyone was silenced about it. This explosion was carried out in order to test the operation of a nuclear charge and its delivery vehicle, and also an assessment was made of the impact of the damaging factors of a nuclear explosion on various types of military equipment, including missiles and military satellites. In addition, the basic principles of the anti-missile defense system Taran proposed by Vladimir N. Chelomey, which provided for the defeat of enemy intercontinental ballistic missiles by a series of nuclear explosions on their way, were tested. On October 20, 1962, an artificial satellite of the Earth, Cosmos-11, was launched from the Kapustin-Yar test site, designed to study the radiation arising from nuclear explosions at high altitudes, in a wide range of energies and efficiencies and also for the development of methods and means of detecting high-altitude nuclear explosions. The reliability of radio communications and radio control equipment was studied, and the accuracy of measuring the parameters of the movement of the warhead and the anti-missile was assessed. The effect of a nuclear explosion on the level of received signals at the input of onboard and ground-based receivers of radio control systems was determined. The main damaging factor of this explosion was an electromagnetic pulse, which led to interference with air defense radars at a distance of about 1,000 km from the epicenter of the explosion. An underground power cable between Almaty and Tselinograd with a length of $\approx$1,000 km, located at a depth of about 1 m, was put out of action. Ceramic insulator breakdowns were noted in ground power transmission lines, causing short circuits; in some areas, the insulators were so damaged that the wires fell to the ground. Also, an electromagnetic pulse caused fires due to short circuits in electrical appliances. One of the fires broke out at the Karaganda thermal power station #3, which was connected to an underground power cable and was located $\approx$423 km in a straight line from the epicenter of the explosion. The straight line distance d in the $WGS-84$ system can be calculated based on the knowledge of the geographical coordinates of two points φ_1, φ_2, latitude of points 1 and 2 in radians, λ_1, λ_2, the longitude of the points 1 and 2 in radians and the radius of the sphere $R_E = 6371$ km, which is the radius of the Earth. Using simple geometric relationships, we can get a formula for calculating the distance d between two points:

$$d = R_E \cdot \tan\left(\frac{(\cos\varphi_2 \sin(\lambda_2 - \lambda_1))^2 + (\cos\varphi_1 \sin\varphi_2 - \sin\varphi_1 \cos\varphi_2 \cos(\lambda_2 - \lambda_1))^2}{\sin\varphi_1 \sin\varphi_2 + \cos\varphi_1 \cos\varphi_2) cos(\lambda_2 - \lambda_1)}\right) \tag{7.12}$$

Taking into account that $\varphi_1 = 47°46'48''N$ and $\lambda_1 = 65°19'44''E$ are the geographic coordinates of the nuclear explosion point, and $\varphi_2 = 47°46'59''N$, $\lambda_2 = 67°46'0''E$ are the coordinates of the city of Zhezkazgan, calculations using the formula (7.12) will give the result:

$$d_{1,2} = 182.139 \text{ km} \tag{7.13}$$

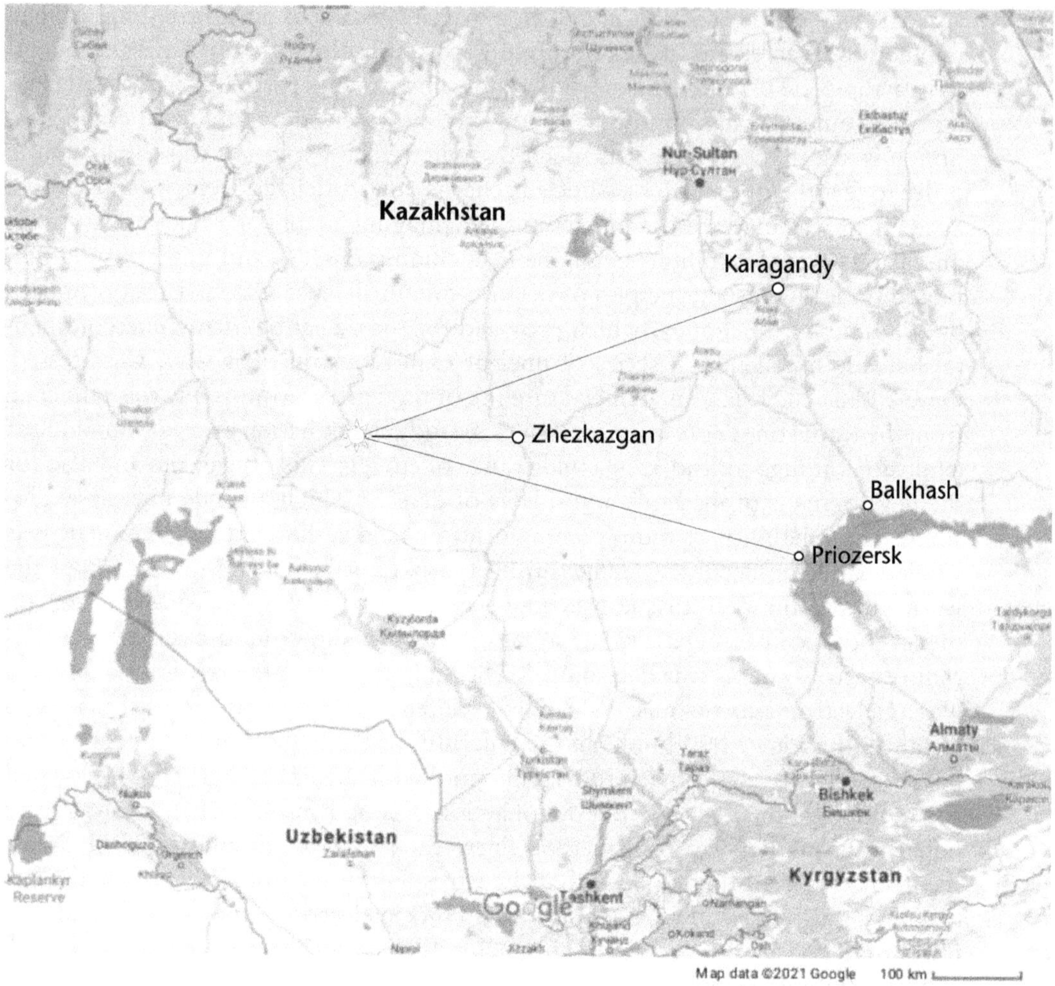

Figure 7.11 Nuclear test K-3, which took place on October 22, 1962.

$d_{1,2}$ is the distance between nuclear explosion (point 1) and the Zhezkazgan city (point 2). The straight line distance between the explosion point and the city of Karagandy (Kazakhstan) $49°49'59''N$ and $73°09'56''E$ was 617.267 km, and the distance from the explosion point to the city of Priozersk (Kazakhstan) was 664.790 km. In Figure 7.11 ☼ is the point of a nuclear explosion. During this atomic test, a 570-kilometer telephone line that ran above the ground was damaged. Telephone lines at that time were aerial and were located on telephone poles exclusively along roads and/or railways. The distance on the map between the cities of Zhezkazgan and Karagandy by road or rail is 536 km. The distance on the road map between Karagandy and Priozersk is 531 km. The road from Priozersk to Karagandy runs north across a flat surface desert. And the road to Almaty goes south, through the mountains, which reduce the impact of the damaging factors of a nuclear explosion. Consequently, it appears that two telephone lines were damaged: between the cities of Zhezkazgan and Karagandy and between Priozersk and Karagandy.

Subsequent analysis showed the presence of a short (≈15 μs) current pulse from 1,500A to 3,400A, caused by the fast, so-called E_1 component of the electromagnetic pulse, caused by synchrotron (magnetic-bremsstrahlung) radiation of electrons, moving from the explosion site in the geomagnetic field, as well as a long (more than 20s) current pulse of 4A induced by the slow E_3 component of the electromagnetic pulse, which is caused by the magnetohydrodynamic interaction of the disturbed region of the atmosphere with the Earth's geomagnetic field. Detectors in the Karagandy region recorded the rate of change in the induction of the Earth's geomagnetic field 1,300 nT/min for 20s after the explosion (E_3 component of the electromagnetic pulse). For comparison, during the Quebec event (a geomagnetic storm on March 13–14, 1989), a change in the Earth's geomagnetic field at a rate of 480 nT/min for 92s turned off the entire power system of the province of Quebec, Canada. In addition, the E_1 component of the high-power electromagnetic pulse is very short in time, hence it generates a very broadband interference signal that affects radar stations of all frequency bands.

Space nuclear explosion K-4 was carried out on October 28, 1962. The detonation of a nuclear charge was carried out at an altitude of 150 km above the Sary-Shagan test site. In all cases, the "A" missile defense system worked without comment.

High-altitude nuclear explosion K-5 was carried out in October 1962 over the Sary-Shagan test site at an altitude of 80 km. Therefore it belongs in the category of high-altitude nuclear explosions. In this explosion, the light flash is transient, as in a cosmic nuclear explosion, but up to 60%–70% of the explosion energy is spent on light radiation. An electromagnetic pulse, which has dangerous parameters for radio engineering during a high-altitude nuclear explosion can spread over hundreds of kilometers.

It is known that if there is a closed conductor in a magnetic field that does not contain a current source, then when the magnetic field changes, an electric current arises in the closed loop. This phenomenon is called electromagnetic induction. The appearance of a current in the circuit indicates the appearance of an electric field in it, which provides a closed movement of electric charges, which is called the electromotive force (EMF). The electric field, which arises when the magnetic field changes, and whose work on the movement of electric charges along a closed circuit is not zero, has closed lines of force and is called a vortex. For a quantitative description of electromagnetic induction, the concept of magnetic flux Φ_B or the flux of the magnetic induction vector $\vec{B}$ passing through a closed loop L, which limits some open surface S (Figure 7.12) [16]:

$$\Phi_B = \oint_S \vec{B} \cdot d\vec{S} = (\vec{,} \vec{S}) = |\vec{B}| \cdot |\vec{S}| \cdot \cos(\alpha) = |\vec{B}| \cdot |\vec{n}| \cdot \cos(\alpha), \tag{7.14}$$

where $\vec{B}$ is the magnitude of the magnetic field (the magnetic flux density) having the unit of Wb/m^2 (tesla), $\vec{S} = \vec{n}$ is the area of the surface, and α is the angle between the magnetic field lines and the normal $\vec{n}$ (perpendicular) to S. A change in the magnetic flux passing through a loop of conductive wire will cause an electromotive force $\mathscr{E}$, and therefore an electric current **I**, in the loop. The relationship is given by Faraday's

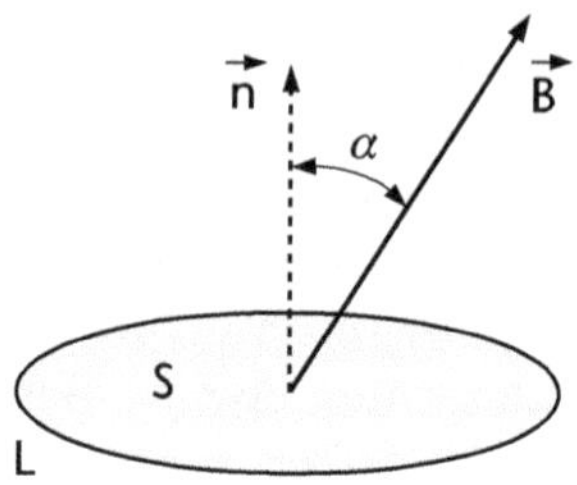

Figure 7.12 Closed conductor in a magnetic field.

law of induction [17]:

$$\mathscr{E} = -\frac{\partial \Phi_B}{\partial t}, \tag{7.15}$$

This law states that an electromotive force of induction $\mathscr{E}$ arises in a circuit covering the time-varying magnetic flux Φ_B, which is proportional to the rate of change of the magnetic flux Φ_B. The minus sign on the right side (7.15) means that the voltage induced in the conductor by the EMF is related to the direction of flux change as the direction of rotation of the left-hand propeller with the direction of its translational motion. EMF can be expressed as a linear integral of the electric field strength $\vec{E}$:

$$\mathscr{E} = \oint_L \vec{E} \cdot d\vec{l}, \tag{7.16}$$

where L is the current of contour, which is the boundary of the open surface S. This surface can usually move and deform, and this is usually a function of time. An electromotive force $\mathscr{E}$ is induced along this boundary and $\vec{l}$ is the direction of the current on an elementary section of the L contour. By definition, the magnetic flux is the flux of the magnetic induction vector $\vec{B}$ through the surface S stretched over the contour L.

$$\Phi_B = \int_S \vec{B} \cdot d\vec{S}, \tag{7.17}$$

therefore (7.15) can be written as an equation:

$$\oint_L \vec{E} \cdot d\vec{l} = -\frac{\partial}{\partial t} \int_S \vec{B} \cdot d\vec{S}, \tag{7.18}$$

which is the second Maxwell equation in integral form. Using the Stokes formula, we pass from an integral over a contour to an integral over a surface:

$$\oint_L \vec{E} \cdot d\vec{l} = \int_S \operatorname{rot} \vec{E} \cdot d\vec{S} \tag{7.19}$$

and therefore, (7.18) and (7.19) can be represented as:

$$\int_S \operatorname{rot} \vec{E} \cdot d\vec{S} = -\frac{\partial}{\partial t} \int_S \vec{B} \cdot d\vec{S} \tag{7.20}$$

The obtained equality is valid for any closed line L and for all surfaces S bounded by this closed line, therefore

$$\mathrm{rot}\,\vec{E} = -\frac{\partial \vec{B}}{\partial t} \tag{7.21}$$

Equation (7.21) is Maxwell's second equation in differential form. According to (7.21) in space, where the magnetic induction changes in time, a vortex electric field appears, of which the direction of the lines of force is associated with a change in magnetic induction by the rule of the left-hand screw (the left-hand rule). According to Lenz's law, the electromotive force of induction EMF creates a current in a conductor in such a direction that the magnetic field of the induction current prevents a change in the magnetic field that caused it, as evidenced by the minus sign in (7.15), (7.18), and (7.21).

It is known that the vector of electrical displacement or the vector of electrical induction is written in the form

$$\vec{D} = \epsilon_a \cdot \vec{E}, \tag{7.22}$$

where ϵ_a is the dielectric constant of the medium or ϵ_0 if the medium is a vacuum. In an isotropic medium, in which the properties of the medium are the same in all directions, the quantity ϵ_a is a scalar, and the vectors $\vec{D}$ and $\vec{E}$ are parallel. In a magnetic field, the linear integral of the magnetic field strength vector along any closed contour L is equal to the total current $\vec{j}_\Sigma$ flowing through any surface S bounded by this contour.

$$\oint_L \vec{H} \cdot dl = \int_S \vec{j}_\Sigma dS \tag{7.23}$$

Equation (7.22) is the total current law or Maxwell's first equation in integral form. The total current $\vec{j}_\Sigma$, which is the sum of the conduction current and the displacement current, can be represented as

$$\vec{j}_\Sigma = \vec{j} + \vec{j}_d = \gamma \vec{E} + \partial \vec{D}/\partial t, \tag{7.24}$$

where γ is the proportionality coefficient, which, according to Ohm's law, has the dimension Ω. Applying Stokes' theorem:

$$\int_S \mathrm{rot}\,\vec{F}(\vec{r}) \cdot d\vec{S} = \oint_L \vec{F}(\vec{r}) \cdot d\vec{r} \tag{7.25}$$

to the left side (7.25):

$$\int_S \mathrm{rot}\,\vec{H} \cdot d\vec{S} = \int_S \left(\vec{j} + \frac{\partial \vec{D}}{\partial t}\right) \cdot d\vec{S}, \tag{7.26}$$

it can be seen that in (7.26) the integrand functions are equal; therefore, the first Maxwell equation in differential form can be written as

$$\text{rot}\,\vec{H} = \vec{j} + \frac{\partial \vec{D}}{\partial t} \tag{7.27}$$

The complete system of Maxwell's equations can be obtained by supplementing the first two Maxwell's equations with three material equations:

$$\vec{D} = \epsilon_a \vec{E}; \qquad \vec{B} = \mu_a \vec{H} \qquad \textit{and} \qquad \vec{j} = \gamma \vec{E}, \tag{7.28}$$

where ϵ_a and μ_a are the dielectric and magnetic conductivity of the environment.

It is known that the law of conservation of energy of an electromagnetic field in differential form has the form

$$\vec{j} \cdot \vec{E}^{ext} = \frac{\partial w}{\partial t} + \frac{j^2}{\gamma} + \text{div}(\vec{E} \times \vec{H}) \tag{7.29}$$

For an arbitrary volume of space, we carry out the integration (7.29) over the volume V and applying the Gauss-Ostrogradsky theorem to the third term, we obtain the power of the source of external forces:

$$P^{ext} = \frac{\partial W}{\partial t} + G + \oint_S (\vec{E} \times \vec{H}) \cdot \vec{n_0} \cdot dS, \tag{7.30}$$

where P^{ext} is the power of the source of external electrical forces; $\partial W/\partial t$ is the energy change of the electromagnetic field in the volume V. G is the power of heat losses; $\oint_S \vec{E} \times \vec{H}) \cdot \vec{n_0} \cdot dS$ is the energy flow through the closed surface S, and $\vec{n_0}$ is the outward normal to the surface S. Wherein, the vector $\vec{S}$:

$$\vec{S} = \vec{E} \times \vec{H} \tag{7.31}$$

is called the Umov-Poynting vector and determines the amount of energy flowing in time through a unit area perpendicular to the direction of propagation of the electromagnetic pulse $\vec{S}$. It is quite clear that in the electromagnetic field there are simultaneously electric $\vec{E}$ and magnetic $\vec{H}$ components, due to the fact that if one of them is equal to zero, then $\vec{S} = 0$ and hence there is no energy emission. Thus, the main property of any current is the ability to excite a magnetic field, and a time-varying magnetic field creates an electric current. According to Maxwell's second equation (7.21) and Faraday's law (7.15), the enormous amperage generated in communication lines and power lines in a nuclear explosion indicates tremendous power released in a very short period of time. The corresponding timing diagrams for the normalized power of the electromagnetic pulse are shown in Figures 7.13 and 7.14 [18]. As can be seen from Figure 7.13, the power rise time from the moment of explosion $t = 0$ to EMP $= max$ was $\tau_1 = 4.875$ ns. The power decay time to 0.5 was $\tau_{0.5} = 124.488$ ns $= 0.124488\ \mu$s, and to 0.01 the power decay

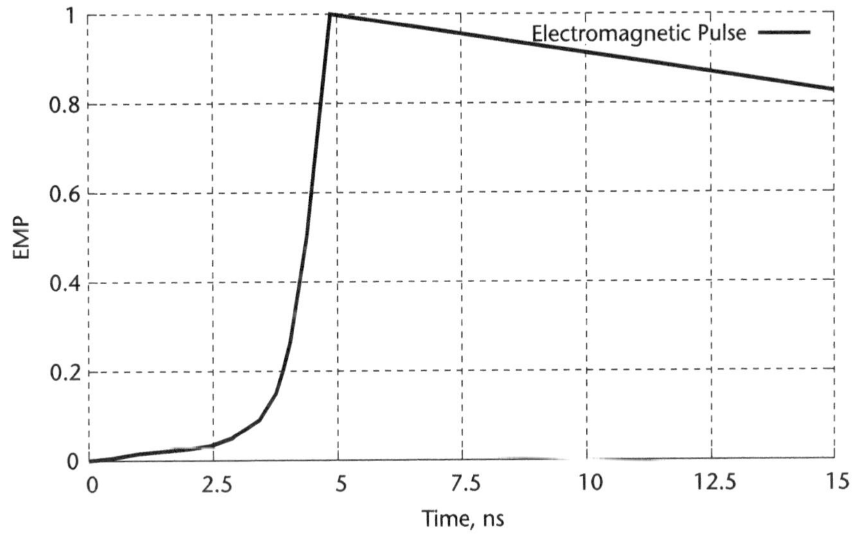

Figure 7.13 EM pulse leading edge.

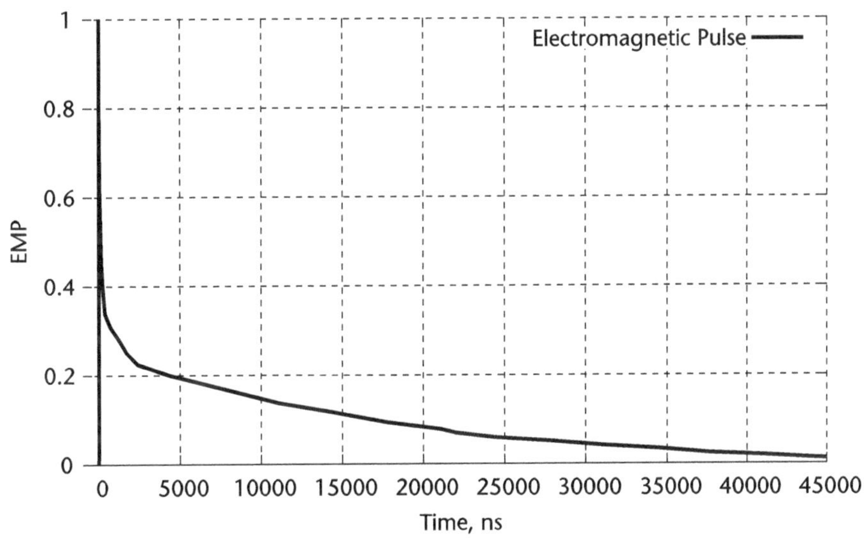

Figure 7.14 EM pulse trailing edge.

time was $\tau_{0.01} = 47.714237\ \mu s$, as a result, the total duration of the electromagnetic pulse at the 0.5 level is $\Delta\tau_{0.5} = 0.1287745\ \mu s$, and at the 0.01 level it is $\Delta\tau_{0.01} = 47.718380\ \mu s$.

Thus, in order to achieve high powers of an electromagnetic pulse, it is not only the power of the explosion that is essential, but also the rate of change of the magnetic flux in time:

$$\mathscr{E} = -\lim_{\partial t \to 0} \left(\frac{\partial \Phi_B}{\partial t} \right) = -\infty \tag{7.32}$$

The principle (7.32) is currently used in high-power explosively pumped lasers. These lasers were successfully tested at the Sary-Shagan test site between 1985 and

1990. Moreover, if a fusible link connected to a power source and a device that automatically replaces it after each electromagnetic pulse is placed in the focus of the parabolic mirror, then with sufficient power of the power source, it is possible to achieve positive results in disabling of predefined enemy radar systems by means of a directional electromagnetic impulse. In addition, in this way it is possible to disable enemy satellites space grouping, as well as the means of telecommunications, communications, and command and control of the enemy's troops, being outside the zone of physical contact with the enemy. At the same time, there is currently no protection against such electromagnetic effects.

References

[1] Vakin, S. A., and L. N. Shustov, "Fundamentals of Radio Countermeasures and Electronic Intelligence," Soviet Radio, Moscow, 1968.

[2] Van Brunt, L. B., *Applied ECM,* E. W. Engineering, 1978.

[3] Leonov, A. I., and K. I. Fomichev, *Monopulse Radar*, Norwood, MA: Artech House, 1986.

[4] *Aerospace Defense Magazine,* No. 5, October 1, 2015, http://www.vko.ru/oruzhie/zrk-s-75-pervyy-sredi-ravnyh-chast-i.

[5] Boeing, EA-18G GROWLER, https://www.boeing.com/defense/ea-18g-growler/.

[6] Blake, L. V., and M. W. Long, *Antennas: Fundamentals, Design, Measurement*, New York: SciTech Publishing, 2009, ISBN 978-1-891121-78-4.

[7] Markov, G. T., and D. M. Sazonov, *Antennas. A Textbook for Students of Radio Engineering Specialties of Universities,* Second Edition, Revised and Enlarged, Energy, 1975.

[8] Skolnik, M., *Introduction to the Technique of Radar Systems*, (Translation from English by K. N. Trofimova), MIR Publishing House, 1965.

[9] Kownacki, S., "Simulation of Radar Range and Doppler Effect by Means of Stationary Target," *IEEE Transactions on Aerospace and Electronics Systems,* 1967, pp. 148–149.

[10] Miller, B., "US Penetration Capability Erodes," *Aviation Week and Space Technology,* 1967, Vol. 17, pp. 90, 95, 98, 99, 105, 108, 109.

[11] Deloraine, E. M., and H. G. Busignils, *Pulse Radar Countermeasure,* U.S. Patent # 2943318, August 5, 1942.

[12] Arkhipov, V. N., V.A. Borisov, A.M. Budkov, et al., *Mechanical Effect of a Nuclear Explosion*, FIZMATLIT, 2003, p. 384, ISBN 5-9221-0261-3.

[13] "Basic Definitions, Terms and Concepts for Military-Technical Training," Section 7.2.2, *Classification and Combat Properties of the Air Defense System,* Institute of Military Education, 2016, http://zrv.ivo.unn.ru/pages/vtp/7/7-2-obshchaya-kharakteristika-zenitnykh-raketnykh-kompleksov.htm.

[14] EIS (Electric Infrastructure Security) Council, *Report: USSR Nuclear EMP Upper Atmosphere Kazakhstan Test*, Dec. 19, 2013, https://www.scribd.com/document/192482647/Saviet-Test-184.

[15] Zhezkazgan, Google Maps, 2021, https://www.google.com/maps/@47.7133743,68.3119193,7z.

[16] Landsberg, G. S. (ed.), *Elementary Physics Textbook,* Vol. 2, Electricity and Magnetism, Moscow: Nauka, 1971.

[17] Krivitsky, B. Kh., and V. N. Dulin (eds.), *Handbook on the Theoretical Foundations of Radio Electronics,* Vol. 1, Energy, 1977.

[18] Electro-Magnetic Compatibility Laboratory, 2021, Moscow: Emctestlab LLC, https://www.emctestlab.ru/post/%D1%8D%D0%BC%D0%B8%D1%8F%D0%B2.

CHAPTER 8

Interference Immunity of Monopulse Radar

In various works the concept of interference immunity is interpreted in different ways, which is largely determined by the meaning of the tasks being solved. However, conceptually, there is a commonality of different approaches of the interpretation of anti-aircraft missile system as a system that performs the functions assigned to it to destroy air and space targets under the influence of interference created by the enemy, in order to suppress the radio-electronic means of an anti-aircraft missile system, as well as its optical means [1].

When assessing the interference immunity of an air defense system, various indicators can be used [2]. In this case, the generalized indicators are made up of the particular indicators of the elements that in turn make up its composition. Note that in this sense the anti-aircraft missile system is unequal; that is, its elements (means) have a different quality of interference immunity. In an air defense system with a semi-active homing channel, it is important to assess the quality of the onboard homing head in conditions of interference [3]. In the general case, the main task of the angular tracking meter in conditions of interference is the task of determining the angular direction of the signal source with minimal errors [3,4]. Due to the fact that the direction finding characteristic of the angular tracking seeker is nonlinear and the parameters of the real system are not invariant in time, the angular tracking seeker in the general case is a nonlinear nonstationary system.

Currently, the main methods for analyzing the accuracy of such systems are based on a single methodological base: linearization [4]. One of its methods are harmonic and statistical linearization methods. Note that the assessment of the quality of nonlinear systems under harmonic impacts is not complete, especially if the influences are random processes [5]. One of the main difficulties of analysis that arises in this case is in the distortion of the laws of distribution of random processes during their nonlinear transformation. In this regard, approximate methods are used based on the approximation of nonlinearity by a linear function. This makes it possible to neglect the distortions of the distribution laws of random input processes. One of these approximate methods is Kazakov's statistical linearization method [6]. The essence of this method is to replace the nonlinear transformation $F(X)$ with a linear one. In this case, the process $Y(t)$ at the output of a nonlinear element can be written as

$$Y(t) = F[X(t)] \approx \phi(t) + k_1(t)[X(t) - \mathsf{M}_x(t)] = \phi(t) + k_1(t)\check{X}(t), \tag{8.1}$$

where $\phi(t)$ is the ordinate of the intersection point of the approximating straight line with the $F(X)$ axis; that is, $\phi(t) = F(X)$, for $\check{X}(t) = 0$;

$\check{X}(t)$ is a random component of the process, and $X(t)$ is a centered random process;

$k_1(t)$ is the coefficient of statistical linearization for a random component;

$M_x(t)$ is the mathematical expectation of the random process $X(t)$.

The choice of the coefficients $\phi(t)$ and $k_1(t)$ is carried out on the basis of minimization in some sense of the linearization error: $\delta(t) = F(X(t) - k_1(t)\check{X}(t)$. There are several methods to minimize the error $\delta(t)$ and determine unknown coefficients [7,8]. One of the known ways of defining $\phi(t)$ and $k_1(t)$ is ensuring zero value of the mathematical expectation and minimum variance of the error approximation. As a result, the expressions for $\phi(t)$ and $k_1(t)$ will take the form

$$k_1(t) = \frac{1}{D_x(t)} \int_{-\infty}^{+\infty} [X(t) - M_x(t)] W(X, t) dX \tag{8.2}$$

$$\phi(t) = k_0(t) M_x(t) = \int_{-\infty}^{+\infty} F(X) W(X, t) dX \tag{8.3}$$

where $\phi(t)$ is the mathematical expectation of the output process $Y(t)$, and $k_0(t)$ is the coefficient of statistical linearization on mathematical expectation.

The expression (8.3) is written for a particular case when $F(X)$ is an odd-symmetric function, while the appearance of the mathematical expectation $M_Y(t)$ at the output of a nonlinear element, which in turn represents the mathematical expectation of the position of the equisignal direction. It is possible only for $M_x(t) \neq 0$, and in this case the expression for $k_0(t)$ will take the form

$$k_0(t) = \frac{I}{M_x(t)} \int_{-\infty}^{+\infty} F(X) W(X, t) dX \tag{8.4}$$

This approach makes it possible to analyze the accuracy of nonlinear systems only with known statistical characteristics of the input random processes. Thus, the process $X(t)$ requires knowledge of its mathematical expectation $M_x(t)$, variance $D_x(t)$ and one-dimensional probability density $W(X, t)$; that is, sections of the probability density function of the process $X(t)$ at any moment of time t. In this case, only the discriminator characteristic (direction finding characteristic) will be considered as nonlinearity. When considering, for example, the elevation plane θ, the input for the discriminator is the angular error $e(t)$ between the true position of the target $X(t)$ and the position of the equisignal direction $Y(t)$ in the receiving basis on angular tracking radar

$$e(t) = X(t) - Y(t) \tag{8.5}$$

The output signal of the discriminator is the error signal $S(\theta, t)$, which is an input impact for the control system antenna of the angular tracking radar.

Often the statistical characteristics of the input random process $X(t)$ are unknown. In addition, $\theta(t)$ and $Y(t)$ depend on the previous values and dynamic properties of the system. And the analysis of the functioning of the system under the influence of interference, when the position and shape of the direction finding characteristic (homing location characteristic) can change, requires an adaptive calculation of the coefficients of statistical linearization, which should minimize the approximation error of the said nonlinearity. Consequently, under conditions of influence on the angular tracking radar of interference, distorting its direction finding characteristic (homing location characteristic), the approximation of the discriminator nonlinearity by a linear function with a constant slope in the region of its definition can lead to incorrect results. In the presence of interference, the main task of angular tracking systems is to determine the direction of the signal source with as few errors as possible [9].

Thus, it becomes necessary to assess the accuracy of the angular tracking of the target in conditions of interference impact for the radar that distorts the direction finding characteristic of its discriminator.

It is known that the main purpose of radar systems is the detection and separation of signals carrying information about an air (space) target from a mixture of signals and interference [3]. The process of forming this mixture is multifactorial, and therefore, signals and interference can be represented as random processes and the problem of separating signals from interference can be solved by methods of mathematical statistics.

Any processing of the received information ends with some result, which can be called decision making. The rules on the basis of which decisions are made will be called decision rules. Usually, the solution to the problems of target signal detection and (or) determination of its angular position can be found in various ways, but it is advisable to choose a solution based on their optimality in a certain sense. When considering signal detection problems, one of the most convenient methods is the method of testing statistical hypotheses [10].

In conditions of interference on the angular tracking radar, the most urgent problem is to ensure the stability of automatic target tracking in angular coordinates. In this regard, there are two strategies for solving: passive and active.

A passive strategy can be characterized by the fact that even before the effect of interference, the angular tracking radar must have certain qualities that provide an unambiguous decrease in the effectiveness of interference during their influence. So, for example, when the level of the first and subsequent sidelobes of the radar receiving antenna is less than minus 48 dB, the cover interference, which acts on the sidelobes of the radar receiving antenna, will be ineffective. However, this level of sidelobes leads to widening of the main beam of the antenna compared to the case of a uniform amplitude-phase distribution of its aperture [11]. The missile diameter body has a fixed size, therefore,it is not possible to provide the required width of the directional pattern of the seeker at a given low level of side lobes by expanding the antenna aperture due to mechanical limitations, imposed by the diameter of the missile body. As a result, the characteristics of the angular tracking radar have to be chosen on the basis of compromise considerations, which in advance allows some deterioration in the quality Q when tracking the target under the influence of each of the interferences.

It is proposed to consider an active strategy as an adaptive procedure (a set of adaptive procedures) that allows maintaining the quality criterion Q at a level not lower than a specified one under the influence of a countable number q of types of interference $H_i, \forall i = \overline{1, q}$. Under these conditions, adaptation can be any (parametric, algorithmic, structural, etc.). It is important that under the influence of the interference H_i the minimum quality criterion is ensured:

$$Q^* = \arg\inf F(H, Q), \tag{8.6}$$

where F is some operator.

It is quite clear that the condition (8.6) can be fulfilled if for the detected interference H_i the most effective countermeasure Z_i is correctly selected. Knowing the types of interference that can affect the angular tracking radar, and the measures leading to compensation for their impact, it is possible to construct decision rules that allow each type of interference H_i to associate a certain set of features $X = \{x\}$, which characterizes it, as well as the compensating impact Z_i. Thus:

$$\begin{cases} H_i = F_i(x_1, x_2, \ldots, x_n) \\ Z_i = L(H_i), i = \overline{1, q} \end{cases} \tag{8.7}$$

L is some linear operator, and q is the total number of considered interference, which is the power of the set of interference types $\{H_i\}$ and, accordingly, the set of compensating influences $\{Z_i\}$. Here, interference types H_i (classes) are described on the set of features $X = \{x\}$ apriori. Let interference types $H = \{H_i\}$ and features $X = \{x\}$ be optimal in the sense of (8.6); that is, such a partition of the feature space in which the optimal value of the quality criterion is attainable.

The task is to find such a rule that will provide the maximum efficiency of interference recognition given the criteria and constraints [12]. This task can be characterized as the problem of interference recognition.

In the general case, the features (interference characteristics) $X = \{x\}$ can have a deterministic, probabilistic, logical, and structural nature [13, 14]. Depending on the type of interference characteristics used, the ratios (8.7) take one form or another.

With regard to deterministic features, the types of interference $H = \{H_i\}$ (classes) can be specified either by the values of signs of objects apriori assigned to each class, or the coordinates of some points in the feature space.

For probabilistic features, descriptions of the types of interference (classes) are probability density functions $f_i(x_1, x_2, \ldots, x_n)$ of feature values $x_1, x_2, \ldots, x_n$, provided that the object belongs to the class H_i, the prior probabilities $P(H_i)$ that the interference, randomly selected from the general population, will belong to the class H_i.

For logical features in the form $X_i = \{1, 0\}, \forall i = \overline{1, q}$, descriptions of interference types (classes) are Boolean relations between these logical features and corresponding interference classes.

Consequently, depending on the type of features that make up the feature space $X = \{x\}$, one can construct certain decision rules that make it possible to recognize the influencing interference from a countable number of possible ones.

The choice of the number of features and their types is the problem of the formation of the feature space, which can be solved only on the basis of the results of the study of the radar operation under the influence of each interference from a given set of types (classes) of interference $H = \{H_i\}$. In addition, the synthesis of decision rules for recognition is possible only after solving the problem of discretization of the feature space, which specifies its division into classes.

Following the terms of reference for the development of the radar, interference types $\{H_i\}$ would be reasonably considered as a countable set of power $q : H_i$, $\forall i = \overline{1, q}$.

It is clear that the feature space $X \subset R^n$ will be developed on the basis of the real capabilities of a particular radar to obtain information. By solving the issue of discretization of the feature space, we mean here a method of partitioning features space of X such that this method is the sum of pairwise disjoint subspaces $\bigcap_{i=1}^{q} X_i = \emptyset$, where $\emptyset$ is an empty set, and the closures $[X_1], [X_2], \ldots, [X_q]$ do not contain isolated points. In this regard, an essential requirement is the agreeing and noncontradiction features, while the feature space is convex and its structure is closed with respect to R^n.

Thus, for $n = q$, each ith hindrance class H_i can be associated with some ith interference detector X_i, which has at the output of this detector a signal $x_i = \{1, 0\}$, $\forall i = \overline{1, q}$. Suppose that at some jth moment of time, the radar was affected by the interference signal H_i, in this case, in the interference detector available on the radar is formed a vector of observations such that $X_j = (x_{j1}, x_{j2}, \ldots, x_{jn})^T \in R^n$. The recognition of the hindrance H_i is supposed to be implemented according to the decision rule, which will have the following form:

$$\forall j, H_i : \sum_{k=1}^{n} (x_{jk} - h_{ik})^2 = 0, \tag{8.8}$$

where $H_i = (h_{i1}, h_{i2}, \ldots, h_{in})$ is the vector of the hindrance image, given apriori.

It can be assumed that the application of this approach to an onboard monopulse homing head will significantly increase its capabilities under interference conditions.

For a given apriori alphabet of noise classes $H_i, \forall i = \overline{1, q}$, the problem arises to find such features $X_i, \forall i = \overline{1, q}$, which uniquely determine the types of interference H_i (classes). The problem of finding such features $X = \{x\}$ can be solved by investigating the operation of an angular tracking radar under the influence of each interference from a given set of interference types $H = \{H_i\}$. This statement is basic in solving the problem of synthesizing angular tracking radars with increased noise immunity.

Thus, an urgent task is to synthesize the structure of an angular tracking radar, optimal in the sense of (8.6) and closed with respect to a countable set of interference classes $H = \{H_i\}$ of power q.

8.1 Interference Immunity Indicators

To analyze the noise immunity of an angular tracking radar, it is necessary to have quantitative estimates characterizing the loss in quality under the influence of various interference in comparison with the case of their absence.

The angular tracking radar solves the problem of determining the unknown angular position of the signal source, which can be considered a random variable with some apriori distribution $P\{\theta_x\}$ [10]. If θ_x is a continuous random variable with a set of values $\theta_x = \{\theta_{xmin}, \ldots, \theta_{xmax}\}$, then $P(\theta_x)$ is a probability density satisfying the normalization condition:

$$\int_{\theta_{min}}^{\theta_{max}} P(\theta_x) d\theta_x = 1 \tag{8.9}$$

or for a discrete random variable $\theta_{xi} = \{1\Delta\theta_x, \ldots, i\Delta\theta_x, \ldots, N\Delta\theta_x\}$, where the events θ_{xi} under $i = 1, 2, \ldots, N$ are incompatible and the only possible ones: they form a complete group of events, the cardinality of the set of values of which is equal to N, so the sum of their probabilities is equal to one:

$$\sum_{i=1}^{N} P_i(\theta_{xi}) = 1 \tag{8.10}$$

It is usually unacceptable to assess quality with a large number of indicators, and therefore, the average risk determined by the [15] ratio can be taken as a generalized quality indicator:

$$R = \iint_{-\infty}^{+\infty} I(\theta_x, \hat{\theta}_x) P(\theta_x) P(\hat{\theta}_x|\theta_x)\, d\theta_x d\hat{\theta}_x \tag{8.11}$$

Here $I(\theta_x, \hat{\theta}_x)$ is a loss function that satisfies the condition:

$$I(\theta_x, \hat{\theta}_x) = \begin{cases} 0, & \theta_x = \hat{\theta}_x \\ 1, & \theta_x \neq \hat{\theta}_x \end{cases} \tag{8.12}$$

where $P(\theta_x)$ is the probability density of the true value of the angular position of the signal source, and $P(\hat{\theta}_x|\theta_x)$ is the probability density of the measured value of the angular position $\hat{\theta}_x$, provided that the true value is θ_x.

From (8.11) it follows that the average risk R is the mathematical expectation of losses: $R = M[I(\theta_x, \hat{\theta}_x]$. Loss function assigns losses corresponding to different combinations of the true value θ_x and the result of its measurement $\hat{\theta}_x$. The most commonly used loss function is a quadratic function:

$$I(\theta_x, \hat{\theta}_x) = (\theta_x - \hat{\theta}_x)^2 = \theta^2, \tag{8.13}$$

wherein the expression (8.11) takes the following form [15]:

$$R = M[\theta^2] = \iint\limits_{-\infty}^{+\infty} (\theta_x - \hat{\theta}_x)^2 P(\theta_x) P(\hat{\theta}_x|\theta_x) d\theta_x d\hat{\theta}_x \tag{8.14}$$

Thus, the average risk coincides with the average value of the square of the tracking (measurement) error. The convenience of the function (8.13) can be explained, on the one hand, by satisfactory compliance with the requirements for a radar system, and on the other hand, it turns out to be convenient for finding the extremum of the quality indicator. In this case, when considering the function of errors in time, it becomes linear after differentiation, and the problem of determining the extremum is reduced to solving a system of linear equations. When using higher powers of p in the loss function $(\theta_x(t) - \hat{\theta}_x(t))^p$, the accuracy of the approximation of the measured angular position function $\hat{\theta}_x(t)$ to the true angular position function $\theta_x(t)$ increases [16], but the problem of finding the extremum of the error function becomes nonlinear, which entails additional difficulties in solving it, and for $p > 4 \ldots 6$ the quality assessment must be carried out using the Chebyshev minimax criterion [17], whose decision functions are not always preferable within the meaning of the problem being solved [18].

In this case, quadratic error (loss) functions will be used. Using a Bayesian approach, the unknown distribution of $P(\theta_x)$ is reasonably considered uniform:

$$P(\theta_x) = \begin{cases} h, & \theta_x = \{\theta_{min}, \ldots, \theta_{max}\} \\ 1, & \theta_x \neq \{\theta_{min}, \ldots, \theta_{max}\}, \end{cases} \tag{8.15}$$

where h is a constant determined from the normalization condition (8.9), that is $h = \frac{1}{(\theta_{max} - \theta_{min})}$. In this case, the expression (8.14) will be written as:

$$M[\theta^2] = h \iint\limits_{\theta min}^{\theta_{max}} (\theta_x - \hat{\theta}_x)^2 P(\hat{\theta}_x|\theta_x) d\theta_x d\hat{\theta}_x \tag{8.16}$$

and the local quality criterion Q will be the minimum of the quadratic loss function, which is determined by the mean value of the squared tracking error: $Q = \min(M[\theta^2])$. It is known that the error in the general case can be represented as:

$$\theta(t) = \theta_s(t) + \theta_f(t), \tag{8.17}$$

where $\theta_s(t)$ is systematic, and $\theta_f(t)$ are random (fluctuation) components of the tracking error.

Let $\theta_s(t)$ is an unconditional systematic error that is, a strictly random component that is an unbiased error; the unconditional expectation of which is zero. In this case, the fluctuation error is the deviation of the total error from its mathematical expectation, and therefore, we can write

$$\theta_f(t) = \theta(t) - \theta_s(t) = \theta(t) - M_\theta(t) \tag{8.18}$$

and, accordingly, the mathematical expectation of the fluctuation error, taking into account the quadratic loss function, is represented as

$$M_{\theta_f^2}(t) = M_{\theta^2}(t) - M_{\theta}^2(t) = D_{\theta}(t), \tag{8.19}$$

that is, the mean of the squared fluctuation component of the auto-tracking error is the variance of that error. In this case, the expression for the mean value of the squared error of automatic tracking, taking into account (8.19), will be written as

$$M_{\theta^2}(t) = M_{\theta}^2(t) + D_{\theta}(t) \tag{8.20}$$

The quality of operation of the angle tracking radar in conditions of interference is often required to be evaluated during the time interval of their exposure $[0, T]$. Taking into account (8.20), the integral quality criterion Q will be written as

$$Q = \int_0^T [M_{\theta}^2(t) + D_{\theta}(t)]dt \tag{8.21}$$

The integrand expression (8.21) represents the dependence of the mean value of the squared error in measuring the angular position of the signal source in time; that is, a function of the local quality criterion.

Thus, the choice of the type of the loss function is dictated by the meaning of the problem and the method of its solution. The choice of the quadratic function leads to automatic linearization of the problem of finding the extremum of the quality indicator. To find this indicator in conditions of interference, it is necessary to determine mathematical expectation and variance of the angular direction measurement error source of the desired signal. Due to the fact that it is always possible to set such a $\theta^* < \infty$, which for any $\theta < \theta^*$ automatic target tracking is disrupted, the problem of determining the interference immunity is reduced to determining the degree of the target auto-tracking stability along the angular coordinates.

Since the input impact in the general case is probabilistic, the meaning of the problem being solved dictates the need for an additional introduction of the auto-tracking failure probability as an indicator of radar interference immunity.

8.2 Criterion for Assessing the Stability of Automatic Target Tracking by Monopulse Radar in Real Time

The composition of the interference immunity indicators was determined earlier:

- The mean value of the squared auto-tracking error;
- The probability of failure of the auto-tracking of the signal source by the radar seeker under the influence of the interference H_i;
- The integral criterion of auto-tracking quality Q.

It is known that the antenna control system of the angular tracking meter minimizes the error signal resulting from processing angular misalignment. Thus, in the process of homing, the error signal is automatically minimized, and the working function is the direction finding characteristic $S(\theta)$. Let the direction-finding characteristic $S(\theta)$ be continuous and differentiable n times in the domain $\theta \in [-\Delta, \Delta]$, with the possible exception of the point $\theta = 0$. Then in a neighborhood of some point $\theta_0 \in [-\Delta, 0 - \circ] \cup [0 + \circ, \Delta]$ it can be represented by Taylor's series. As a first approximation, the function will look like:

$$S(\theta) = S(\theta_0) + (\theta - \theta_0)S'(\theta_0) + \circ(\theta), \tag{8.22}$$

where $S'(\theta_0)$ is the first derivative of the direction-finding characteristic at the point θ_0, and $\circ(\theta)$ is the remainder of the Taylor series. Due to the fact that the process of minimizing the error signal can be represented by the process of solving the equation $S(\theta) = 0$, then from (8.22) we can get $\theta_1 = \theta_0 - (\frac{S(\theta_0)}{S'(\theta_0)})$, and in the general case to find zeros of the direction finding characteristic, we can write

$$\theta_{k+1} = \theta_k - \frac{S(\theta_k)}{S'(\theta_k)}, \qquad \forall k = 0, 1, 2, \ldots \tag{8.23}$$

This is nothing more than a continuous form of Newton's method [19]. Accordingly, the discrete form will be written as

$$\theta_{k+1} = \theta_k - \frac{S(\theta_k)(\theta_k - \theta_{k-1})}{S(\theta_k) - S(\theta_{k-1})}, \qquad \forall k = 0, 1, 2, \ldots \tag{8.24}$$

and using the relations (8.23), (8.24), it is possible to determine the position of stable auto-tracking for the known functions $S(\theta)$ and $S'(\theta)$. Using the relation (8.22) and neglecting the remainder of the Taylor series the expression for the derivative of the direction-finding characteristic existing in the vicinity of the point θ_0 will have the form

$$S'(\theta_0) = \frac{S(\theta) - S(\theta_0)}{\theta - \theta_0} \tag{8.25}$$

Since the function $S(\theta)$ is continuous and n times differentiable in the domain of its definition, then for any point θ_0 belonging to the segment $[-\Delta, \Delta] \setminus 0$, one can find the derivative of the direction-finding characteristic $S(\theta)$, defined on this segment, by the formula (8.25) and therefore construct the function $S'(\theta)$ in the domain of its definition.

To analyze the operation of an onboard radar in conditions of interference, it seems appropriate to use the following model. The time interval $[0, T]$ during which the airborne radar is subjected to interference is divided into N pairwise disjoint half-intervals $[(k-1)h, kh)$ (time quanta), each of length h.

$$[0, T) = \begin{cases} \bigcup\limits_{k=1}^{N} [(k-1)h, kh), \\ \bigcap\limits_{k=1}^{N} [(k-1)h, kh) = \varnothing \end{cases} \tag{8.26}$$

in each of these half-intervals, the automatic angular tracking system is assumed to be linear and invariant in time. Therefore, in the state space $Y : \{\forall y : \|y\| < \infty\}$ and it can be described in each kth time quantum in the form of a system of differential equations with constant coefficients [20]. Assuming that the functions $S(\theta)$ and $S'(\theta)$ are known at each kth time quantum and using the inverse difference formula, the discriminator can be represented as an amplifying link with a gain:

$$K_d(t_k) = S'(\theta_k) = \frac{S(\theta_k) - S(\theta_{k-1})}{\theta_k - \theta_{k-1}}, \tag{8.27}$$

which remains unchanged on the given half-interval $[(k-1)h, kh)$. For any kth time quantum, the equation of phase coordinates in matrix form looks like this:

$$\frac{dY}{dt} = AY, \qquad \forall t \in [(k-1)h, kh) \tag{8.28}$$

If in the state space Y at the initial moment of time $t = 0$ the initial vector $Y_0 : \ Y_0 = [y_{01}, y_{02}, \ldots, y_{0N}]^T$, then finding a solution to the system of differential equations (8.28) with the initial conditions Y_0 is the Cauchy problem [21]. The solution to this problem can be represented as [22]:

$$Y(t) = \exp(At)Y_0, \tag{8.29}$$

where A is the matrix of coefficients of the system of differential equations describing the automatic target tracking system; and Y_0 is the vector of initial conditions.

The solution (8.29) will be stable in the case when the real parts of all roots of the characteristic equation:

$$\det[A - \lambda t] = (-\lambda)^N + p_1(-\lambda)^{N-1} + \ldots + p_{N-1}(-\lambda) + p_N = 0 \tag{8.30}$$

lie in the left half-plane [22–24]. To determine the stability of a system of linear equations, there are various methods, including [22]: direct calculation of the eigenvalues of the matrix A, methods based on the properties of the characteristic polynomial, Nyquist criterion and root locus method (for discrete systems), and Lyapunov's method.

It can be shown that the estimate for the modulus of continuity in the theorem of continuous dependence of the characteristic roots on the elements of the matrix A allows very significant changes in the roots for small perturbations of the matrix such that these perturbations cannot be modeled on the computer [22]. Stability criteria based on the analysis of the location of characteristic roots are not always convenient for practical applications [25]. It is proposed to choose the value æ(A) as a criterion, which is determined through all solutions of the system $\frac{dY(t)}{dt} = AY(t)$ [24]. Each such solution can be uniquely determined by the initial value $Y(0)$ at $t = 0$ of vector $Y(t)$. Moreover, æ(A) is defined as follows:

$$æ(A) = 2\|A\| \sup \left[\frac{\int_0^\infty \|Y(t)\|^2 dt}{\|Y(0)\|^2} \right], \tag{8.31}$$

where $\|A\| = \sup \frac{\|AY\|}{\|Y\|}$ and $\|Y\| = \sqrt{Y_1^2 + Y_2^2 + \ldots + Y_N^2}$. If the matrix A is symmetric, then $\|A\| = \max|\lambda_i|$. Since $\int_0^\infty \left[\exp(-t\|A\|)\|Y(0)\|\right]^2 dt = \frac{\|Y(0)\|^2}{2\|A\|}$, therefore:

$$
\text{æ}(A) = \sup_{Y(0)} \frac{\int_0^\infty \|\exp\{tA\}Y(0)\|^2 dt}{\int_0^\infty \left[\exp\{-t\|A\|\}\|Y(0)\|\right]^2 dt} \tag{8.32}
$$

Using (8.32) and also what $\|Y(t)\| = \|\exp\{At\}Y(0)\| \geq \exp\{-\|A\|t\}\|Y(0)\|$, therefore $\int_0^\infty \|Y(t)\|^2 dt \geq \int_0^\infty \left[\exp\{-t\|A\|\}\|Y(0)\|\right]^2 dt$, whence the inequality follows:

$$
\text{æ}(A) \geq I, \tag{8.33}
$$

where I is the identity matrix.

If zero solution of the system $\frac{dY(t)}{dt} = AY(t)$ is not asymptotically stable, that is, the matrix A has a characteristic root, λ_0, such that $Re[\lambda_0] \geq 0$, then the system under consideration has a solution of the form:

$Y(t) = \exp\{\lambda_0 t\}Y(0) = \exp\{Re[\lambda_0]t\}\|Y(0)\| \geq \|Y(0)\|$,

$\int_0^\infty \|Y(t)\|^2 dt \geq \int_0^\infty \|Y(0)\|^2 dt = \infty$ and, therefore

$\text{æ}(A) = 2\|A\| \sup_{Y(0)} \int_0^\infty \|Y(t)\|^2 dt = \infty$

And if zero solution of the system $\frac{dY(t)}{dt} = AY(t)$ is asymptotically stable; that is, if the matrix A is Hurwitz, then $\text{æ}(A)$ is finite. Moreover, $\text{æ}(A)$ is calculated from ratio:

$$
\text{æ}(A) = 2\|A\|\|H\|, \tag{8.34}
$$

where H is the solution to the Lyapunov equation:

$$
HA + A^*H = I \tag{8.35}
$$

In the case of unique solvability (8.35), the matrix H is strictly positive definite [24]. As a computational procedure (8.35), it is proposed to calculate H in the form of an improper integral [22,24]:

$$
H = \int_0^\infty \exp\{tC^*\}\exp\{tC\}dt, \quad \text{where } C = \frac{A}{\|A\|} \tag{8.36}
$$

and C^* in general is the complex conjugate matrix to C. The integral (8.36) diverges if $\text{æ}(A) = \text{æ}(C) = \infty$. The vector function $Y(t)$ must be measurable on the interval $[0, T)$, and its result of integration on this interval must be finite in order to provide automatic control, therefore, the function $Y(t)$ is defined in the Hilbert space

$L_2[0, T)$, in this case (8.36) is the scalar product in the Hilbert space, and we can write:

$$H_k = \int_0^{2_k} \exp\{tC^*\} \exp\{tC\}, \tag{8.37}$$

The stability criterion takes the form:

$$\lim_{k\to\infty} \|H_k - H_{k-1}\| = 0, \tag{8.38}$$

and in case of instability:

$$\lim_{k\to\infty} \|H_k - H_{k-1}\| = \infty, \tag{8.39}$$

Taking into account the partition of the observation interval $[0, T)$ into N pairwise disjoint half-intervals (8.26), the solution (8.29) can be represented as a finite function:

$$Y(t) = \sum_{i=1}^{N} \exp\{A(t - (i-1)h\}\chi_i(t), \tag{8.40}$$

where $\chi_i(t)$ is the characteristic function of the ith time quantum, having the form

$$\chi_i(t) = \begin{cases} Y_0((i-1)h), & \forall t \in [(i-1)h, ih) \\ 0, & \forall t \not\subset [(i-1)h, ih), \end{cases} \tag{8.41}$$

moreover, its right boundary is the upper limit of integration in (8.36) Thus, the representation of the solution of a system of differential equations in the form (8.40) allows us to find the state vector of the system at any time t, including in the presence of distortions of the direction-finding characteristic due to the influence of interference. Introducing the notation:

$$\begin{aligned} dH_k = \|H_k - H_{k-1}\| = \Bigg\| & \int_{t_{k-1}}^{t_k} \exp\{tC_k^*\} \exp\{tC_k\}dt \\ & - \int_{t_{k-2}}^{t_{k-1}} \exp\{tC_{k-1}^*\} \exp\{tC_{k-1}\}dt \Bigg\| \end{aligned} \tag{8.42}$$

where

$$\begin{cases} C_k, C_k^* = const, & \forall t \in [(k-1)h, kh), \\ C_{k-1}, C_{k-1}^* = const, & \forall t \in [(k-2)h, (k-1)h), \\ t_k = kh, \quad t_{k-1} = (k-1)h, \quad t_{k-2} = (k-2)h \end{cases} \tag{8.43}$$

Therefore, for points on the boundaries of intervals of length h, the following expression can be written:

$$\xi_k = \frac{dH_k}{dt} = \Big\| \exp\{hC_k^*\}\exp\{hC_k\} - \exp\{hC_{k-1}^*\}\exp\{hC_{k-1}\} \Big\| \tag{8.44}$$

Thus, the criterion for assessing the stability of the automatic control system in real time $\forall k \in [1, N]$ can be written as

$$\xi_k = 0 + \epsilon, \tag{8.45}$$

where ϵ is a sufficiently small right neighborhood of zero such that $\epsilon \to 0$, and in this case the instability condition is written as

$$\xi_k = \hat{M} \to \infty, \tag{8.46}$$

where $\hat{M}$ is a large enough number, limited from the top by the bit capacity of the computer. With sufficient speed of the onboard computer, the criterion (8.44) can be used in real time and when a certain threshold Ξ is exceeded

$$\xi_k \geq \Xi \tag{8.47}$$

and after correct detection of the ith interference, the logical command $F_i(x_i)$ should be issued to enable the protection Z_i from the detected interference H_i:

$$Z_i(t_k) = H_i(t_k) \times F_i(x_i), \quad \exists t_k : F_i(x_i) = 1 \tag{8.48}$$

Protection against each ith of the detected interference must be determined before issuing the command $F_i(x_i)$ to protect against this interference.

$$\begin{bmatrix} H_1(t_k) \\ H_2(t_k) \\ \vdots \\ H_3(t_k) \end{bmatrix} \times \begin{bmatrix} F_1(x_1) & 0 & \dots & 0 \\ 0 & F_2(x_2) & \dots & 0 \\ \vdots & \vdots & \ddots & \vdots \\ 0 & 0 & \dots & F_N(x_N) \end{bmatrix} = \begin{bmatrix} Z_1(t_k) \\ Z_2(t_k) \\ \vdots \\ Z_3(t_k) \end{bmatrix}, \forall t \in [0, t_k) \;\&\; t_k = kh \tag{8.49}$$

In this regard, it seems expedient to allocate a separate processor to solve (8.44) in real time. In this case, for one of the orthogonal planes θ or φ, the structural diagram of the onboard homing head under conditions of interference, at which the integral quality criterion of auto-tracking Q is minimized (8.21) will look as shown in Figure 8.1:

Here X is the true position of the target in the seeker antenna receiving base $\{\theta, \varphi\}$;

$Y1$ is the measured value of the angular position of the target;

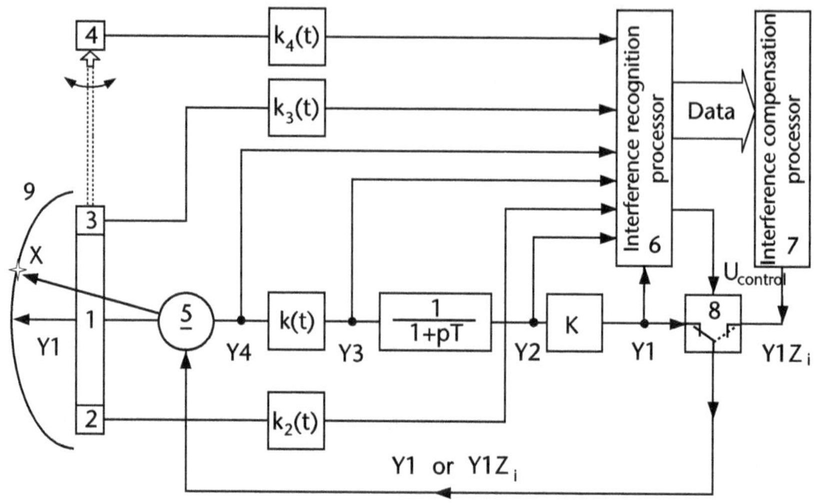

Figure 8.1 Target angle-tracking system simplified structural diagram with interference recognition.

$Y4 = X - Y1$ is an error of target angular tracking;

1 is the antenna of homing head, which is the flat antenna array;

2 is an antenna of an additional channel;

3 is the auxiliary channel antenna;

4 is the antenna array angular position sensor in seeker own coordinate system in $\{\theta\}$ plane or in $\{\varphi\}$ plane, which is mechanically connected to the antenna array 1;

5 is the subtracting device;

6 is the interference recognition processor;

7 is the interference compensation processor;

8 is a switch;

9 is a missile radio-transparent nasal radome.

In this case, the physical quantities that are measured by homing head are X, $Y1$, $Y4$, as well signals from outputs of additional channels 2 and 3 and signal U_{ang} from outputs of antenna array angular position sensor 4, which is proportional to the measured value of the angular position of the target $Y1$ in the receiving base of the seeker. The rest of the signals are formed by means of calculations on a mathematical model (8.29), (8.40), and (8.41), into which on each kth quantum of time the real values of $Y4$ and $Y1$ arrive. Signals $Y4$, $Y1$, and U_{ang} are fed to the interference recognition processor 6, in which they are first converted into digital form by means of an analog-to-digital converter. Signals of additional channels 2, 3, and U_{ang} also pass through the analog-to-digital converter. The remaining signals arriving at the input of the interference recognition processor 6 are presented in digital form, since they are the result of the previous $(k - 1)$ calculations cycle.

In a posteriori analysis, when collecting actual statistics, it is reasonable to offer a relative indicator of interference immunity, at which the accuracy of tracking a target in interference is compared with the case of tracking it without interference. As a result, the following expression can be used as a criterion for interference immunity:

$$Q_j = \ln \sum_{i=1}^{N} e_{ji}^2 - \ln \sum_{i=1}^{N} e_c^2 \tag{8.50}$$

where e_c^2 is taken as equal to the half-width $|\Delta|$ of the direction-finding characteristic on working polarization, N is the number of elementary time samples with duration h, e_{ji} is the auto-tracking error, for $j = 0$ for the case of automatic target tracking in the absence of interference, and $j = 1, 2, 3$ for the case of target tracking in conditions of interference: $j = 1$ on the sidelobes of the radar antenna pattern, $j = 2$ in conditions of interference created from at least two space points, and for $j = 3$ when acting on the radar cross-polarization interference, respectively. It is quite clear that the interference immunity indicator (8.50) will be negative when the system target tracking on angular coordinates will be in a working state, and in the case when the radar is suppressed by interference, the target tracking system on angular coordinates will be inoperative and, therefore, the interference immunity indicator (8.50) will be positive (greater than zero).

References

[1] Basic Definitions, Terms and Concepts for Military-Technical Training (Electronic Reference Book), Section 7.2.2, "Classification and Combat Properties of the Air Defense System," Institute of Military Education, 2016, http://zrv.ivo.unn.ru/pages/vtp/7/7-2-obshchaya-kharakteristika-zenitnykh-raketnykh-kompleksov.htm.

[2] Sharakshane, A. S., and I. G. Zheleznov, *Tests of Complex Systems*, Moscow: Higher School, 1974, p. 182.

[3] Markin, E., "Jamming Detection in Providing for Radar Jamming Immunity," *EUROCON 2009 IEEE*, St. Petersburg, Russia, May 18–23, 2009, pp.1565–1567, DOI: 10.1109/EURCON.2009.5167849.

[4] Markin, E., "On Interference Immunity of Angle Tracking Systems Under Conditions of Interference Distorting Location Characteristic," *2009 IEEE Radar Conference*, Pasadena, CA, May 4–8, 2009, pp. 1–6, DOI: 10.1109/RADAR.2009.4977092.

[5] Marchuk, G. I., *Numerical Methods and Applications*, CRC Press Inc., 1994.

[6] Kazakov, I. E., *Statistic Methods of Control System Design*, Machinostroyeniye, 1969.

[7] Artemyev, V. M., Reference Manual for Study Methods of Electronic Tracking Systems, Minsk:Vysh. Shkola, 1984.

[8] Artemyev, V. M., *Robots Location Systems*, Minsk: Vysh. Shkola, 1989.

[9] Van Brunt, L. B., *Applied ECM*, Vol. 1, Part 4, New York: E. W. Engineering, 1978.

[10] Levin, B. P., *Theoretic Foundations of Statistic Radiotechnics*, Third Amended Edition, Radio i Svyaz, 1984.

[11] Blake, L. V., and Maurice W. Long, *Antennas: Fundamentals, Design, Measurement*, Third Edition, USA: SciTech Publishing, Inc., 2009.

[12] Gorelik, A. L., "General Objective of Object & Phenomenon Recognition," *Kybernetika*, Vol. 6, 1980, pp. 72–75.

[13] Gorelik, A. L., *The Current State of Recognition Problem*, Moscow: Radio and Communication, A. L. Gorelik, I. B. Gurevich, V. A. Skripkin, Higher School, 1985.

[14] Gorelik, A. L., and V. A. Skripkin, *Recognition Methods*, Higher School, 1989.

[15] Gutkin, L. S., *Design of Radio Systems and Radio Devices*, Radio and Communication, 1986.

[16] Dudgeon, D. E., *Multidimensional Digital Signal Processing,* Englewood Cliffs, NJ: Prentice Hall, 1984.

[17] Unruh, P. F., *Chebyshev Approximations,* Kansas State University, 1967, https://core.ac.uk/download/pdf/33364121.pdf.

[18] Ivchenko, G. I., and Y. I. Medvedev, *Mathematical Statistics*, Higher School, 1984.

[19] Stearns, S. D., *Digital Signal Analysis*, Fourth Edition, Rochelle Park, NY: Hayden Book Company, 1979.

[20] Minoux, M., *Mathematical Programming: Theory and Algorithms*, Chichester, UK: John Wiley & Sons, 1986.

[21] Godunov, S. K., *Ordinary Differential Equations with Constant Coefficient* (Translation of Mathematical Monographs), American Mathematical Society, 1987, Alibris ID: 9762292910/.

[22] Godunov, S. K., *Guaranteed Accuracy in Numerical Linear Algebra,* Dordrecht, Netherlands: Springer, 1993.

[23] Ostrem, K., and B. Wittenmark, *Computer Controlled Systems* (Translation from English), MIR, 1987.

[24] Godunov, S. K., *Lyapunov Quadratic Function. Computational Difficulties of Hurwitz Problem and Ways of Their Overcoming. Novosibirks*, Univ. Press, November 1986.

[25] Wilkinson, J. H., *The Algebraic Eigenvalue Problem,* Oxford: Clarendon Press, 1969.

CHAPTER 9

Modeling

To study the main characteristics of monopulse radars, numerical probabilistic (statistical) models [1], also called Monte Carlo models, are the most widely used in practice. A simplified block diagram of amplitude-comparison monopulse radar designed for automatic target tracking along angular coordinates in one plane (φ) is shown in Figure 9.1 [1,2], where:

1 is a monopulse antenna with two outputs: sum Σ and difference Δ;

2 is a microwave combining network, represented by the magic-T hybrid junction or the hybrid ring (rat-race) junction;

3 is a duplexer, which is an antenna switch from transmission to reception and back;

4 is a transmitter;

5 is a mixer of sum channel;

6 is a heterodyne;

7 is a mixer of difference channel;

8 is an intermediate frequency amplifier of the sum channel;

9 is AGC;

10 is an intermediate frequency amplifier of the difference channel;

11 is a phase detector;

12 is a amplitude detector;

13 is an error signal amplifier;

14 is an antenna control system (electric motor), which is mechanically connected to the rotary device of the monopulse antenna 1.

The placement of a monopulse antenna 1 with homing units 2–14 under the radio transparent fairing 15 is shown in Figure 9.2. Here, an asterisk denotes the position of the target in the coordinate system $X_m Y_m Z_m$ associated with the homing head at the time t_k. Point 0 is the phase center of monopulse antenna 1. Vectors $\vec{0Y}_m$ and $\vec{V}_m$ are collinear, since they determine the instantaneous missile direction $\vec{V}_m(t_k)$ at time t_k with velocity $|\vec{V}_m|$. The vector $\vec{n}$ is the direction of normal for homing head monopulse antenna 1 at time t_k, which characterizes the instantaneous position for homing head monopulse antenna $\vec{n}(t_k)$ relative to the instantaneous position of the missile velocity vector $\vec{V}_m(t_k)$. This generates azimuth tracking errors $\Delta_\varphi(t_k) \propto \varphi(t_k)$ and elevation errors, respectively, $\Delta_\varepsilon(t_k) \propto \varepsilon(t_k)$, where $\forall t \in [(k-1)h, kh)$, $\exists k : k = 1, \ldots, N$, and N is the total number of time counts that form the total analysis

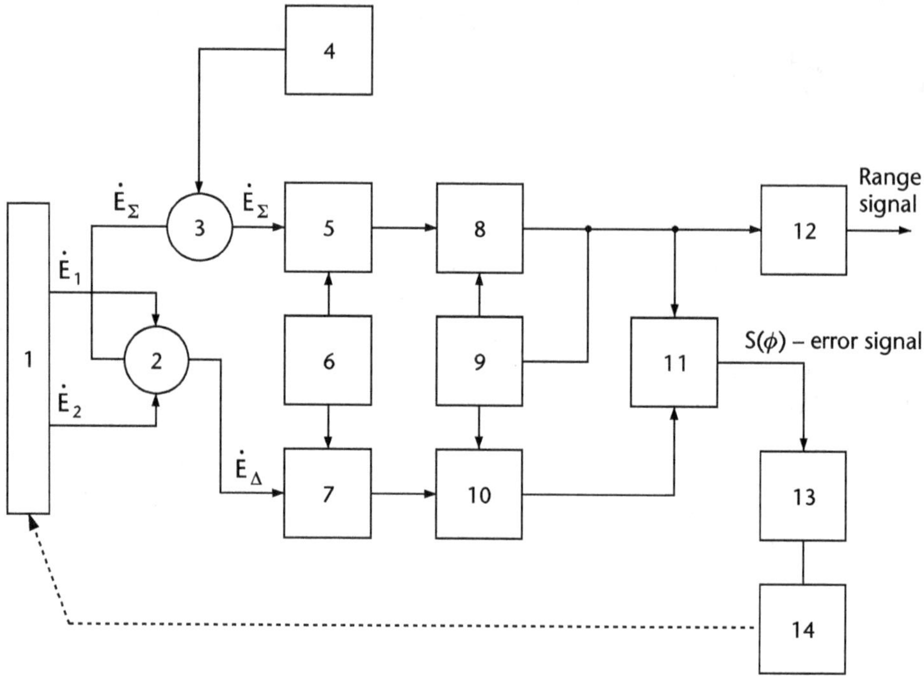

Figure 9.1 Block diagram of amplitude-comparison target tracking monopulse radar on angle coordinate in one plane (sum-difference scheme).

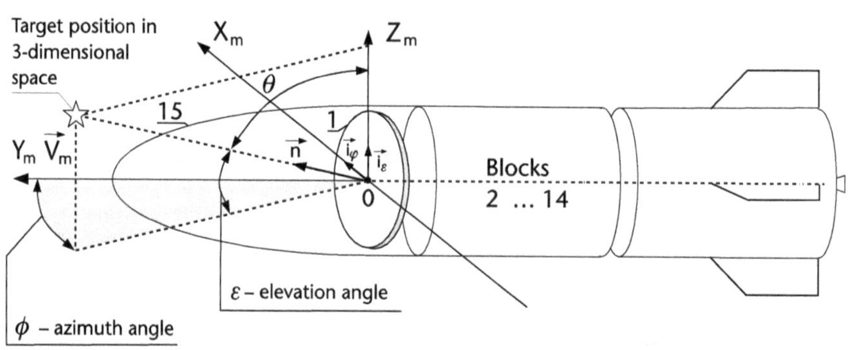

Figure 9.2 Placing a monopulse seeker on board a surface-to-air missile.

time T of homing head operation:

$$T = \sum_{k=1}^{N} [(k-1)h, kh) \tag{9.1}$$

The block diagram of the amplitude sum-and-difference tracking radar model is shown in Figure 9.3.

1. The simulation is started by setting random process realizations counter to the value $m = 1$ and then increasing the counter by 1 up to the value $m = L$. This is an outer loop that is responsible for the formation of realizations of a random

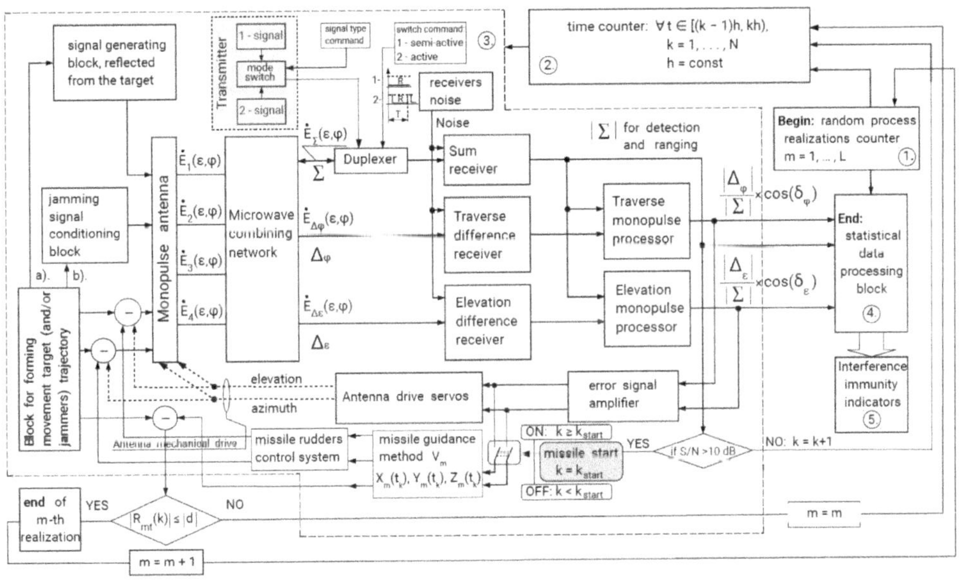

Figure 9.3 Monopulse tracking radar block diagram model.

process from the output of the sum channel Σ and random processes of error signals in angular coordinates from the outputs of the traverse and elevation monopulse processors.

2. All calculations are done sequentially in time on the basis of time samples t_k coming from time counter 2 to all blocks of the homing head model 3 shown in Figure 9.3. According to the partition (9.1) such that

$$\forall k \in \{1, ..., N\} : \; \exists \, t_k \in [(k-1)h, kh), \;\; h = const \tag{9.2}$$

it is advisable to make the following assumption:

instead of variable t_k, it is possible to use the index k to denote time-dependent variables. (9.3)

The block time counter 2 is common to all elements of homing head block diagram 3 model and is an internal loop of the first level. Further, for each kth time quantum, calculations are performed in the following blocks:

3.1: Block for forming movement target (and/or jammers) trajectory

a. By target: $D_{mt}(t_k)$ - range to the target, measured from the point of finding the phase center of the seeker antenna in the horizontal plane X_m0Y_m of the missile-related coordinate system $X_mY_mZ_m$. The range $D_{mt}(t_k)$ is also called "range on the map." $R_{mt}(t_k)$ is the radial range from missile to target (i.e., is the range in the vertical plane Z_m0Y_m of the missile coordinate system $X_mY_mZ_m$), as well as the angular position of the target in the coordinate system associated with the missile $\varepsilon_t(t_k)$, $\varphi_t(t_k)$ and the rate of change in the angular position of the target $\dot{\varepsilon}_t(t_k)$, $\dot{\varphi}_t(t_k)$.

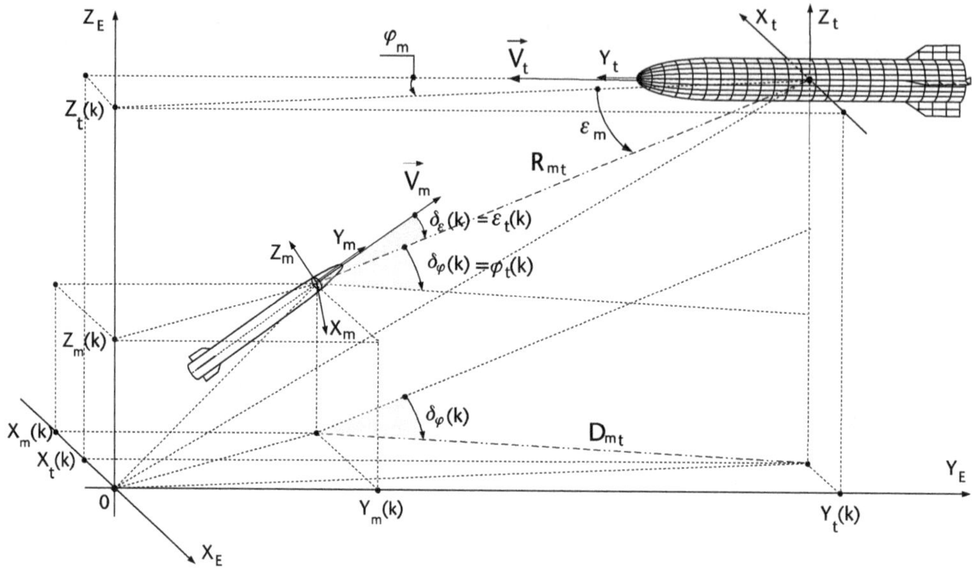

Figure 9.4 Signal reflected from the target.

b. By each jth jammer, such as $\forall j = 1, \ldots, N$: calculated $D_j(t_k)$ - the range to the jth jammer, measured from the point of the air defense system in the horizontal plane $X_E 0 Y_E$ of the earth coordinate system $X_E Y_E Z_E$, which represents is the distance on the map. $R_j(t_k)$ - radial range to the jammer in the $Z_E 0 Y_E$ plane and angular position jth jammer in four directional patterns of a monopulse seeker: $\dot{F}_1[\varepsilon_j(t_k), \varphi_j(t_k)]$, $\dot{F}_2[\varepsilon_j(t_k), \varphi_j(t_k)]$, $\dot{F}_3[\varepsilon_j(t_k), \varphi_j(t_k)]$, $\dot{F}_4[\varepsilon_j(t_k), \varphi_j(t_k)]$, each of which is represented in a complex numbers field.

3.2: Signal generating block reflected from the target. Any target in this consideration is represented by a three-dimensional model in the coordinate system X_t, Y_t, Z_t associated with the target [3], where the vectors $\vec{0Y_t}$ and $\vec{V_t}$ are collinear. The target surface is approximated by flat metal regions, the dimensions of which are $dxdy, dzdy, dxdz \leq \frac{\lambda}{2}$ (Figure 9.4). The signal reflected from the target is formed at a given point in space $\{\varepsilon_m, \varphi_m, R_{mt}\}$, in which the phase center of homing head antenna is located, by summing the complex field amplitudes excited by complex currents on each of the elementary flat surfaces $dxdy, dzdy, dxdz$, which approximated the target.

As a result, for each target, we can pre-calculate the backscatter diagram [3] $\vec{\dot{F}}(\varepsilon_m, \varphi_m)$, for example, for the front hemisphere of the target in question $\varepsilon_m \in [-90°, \ldots, +90°]$ and $\varphi_m \in [-90°, \ldots, +90°]$. For the signal reflected from the target on the operating polarization of receiving antenna of the missile homing head, we can find the value of the signal, received by the antenna seeker from the direction characterized by the relative position of mass centers of the missile and of the target in the terrestrial coordinate system $X_E Y_E Z_E$. According to the assumption (9.3) for each kth time quantum, depending on the relative position of mass centers of the missile x_{mk}, y_{mk}, z_{mk} and of the target x_{tk}, y_{tk}, z_{tk} the projection area of the target onto the plane is calculated perpendicular to missile-target vector $\vec{R}_{mt}(k)$ [3–5], from where the effective scattering area σ_k is calculated for the kth time quantum. Substituting σ_k and missile-to-target radial range $R_{mt}(k)$ in radiolocation formula [3,6]:

$$R_{mt}(k) = \sqrt{(x_{tk} - x_{mk})^2 + (y_{tk} - y_{mk})^2 + (z_{tk} - z_{mk})^2} \tag{9.4}$$

we can find the power $P_r(k)$ (watts) on the kth time interval received from a target by the sum channel of monopulse antenna:

$$P_r(k) = \frac{P_t G_t G_r \lambda^2 \sigma_k}{(4\pi)^3 R_{mt}^4(k)} \tag{9.5}$$

In this expression, P_t is the transmitted power in watts, G_t is the gain of the transmit antenna, G_r is the gain of the receive antenna, equality $G_t - G_r$ is the condition of operation for one antenna "for reception" and "for transmission," which characterizes the homing head operation in active mode, and inequality $G_t \gg G_r$ characterizes seeker operation in the semi-active mode, where G_t is the gain of the target illumination radar antenna, and G_r is the gain of the receiving homing head antenna, λ is the carrier wavelength in meters, $R_{mt}(k)$ is the range "missile-target" in meters, and σ_k is the mean radar cross section (RCS) of the target in square meters on the kth time interval. Denoting homing head receiver sensitivity as N, expressed in watts, it can be noted that when the signal-to-noise ratio (S/N) in the sum channel becomes more than a certain threshold value B_0, for example, $B_0 = 10$ dB:

$$\lg\left(\frac{S_\Sigma(k)}{N}\right) = \lg\left(\frac{P_r(k)}{N}\right) \geq 10\,dB, \tag{9.6}$$

then expression (9.6) is a necessary condition for stable target auto-tracking on angular coordinates, and therefore, sufficient condition for stable target auto-tracking is:

$$\begin{cases} \delta_\varepsilon(k) & \to 0 \\ \delta_\varphi(k) & \to 0 \end{cases} \tag{9.7}$$

When the condition (9.6) is met, the "Start" command is given to the missile launcher and then the missile is guided to the target according to selected guidance method and the corresponding switch is moved to the ON position (Figure 9.3). Guidance errors $\delta_\varepsilon(k)$ and $\delta_\varphi(k)$ are calculated based on equations system solution of guidance contour moments [7–9], assuming vectors $\vec{V}_m$ and $\vec{n}$ collinearity. Assuming that the control system of the missile rudders is inertialess, it can be assumed that the errors in the position of the missile in space relative to the kinematic trajectory of the missile set by the selected guidance method are worked out instantly, and therefore, we can write:

$$\begin{cases} \delta_{\varepsilon,\vec{n}}(k) & = \delta_{\varepsilon,\vec{V}_m}(k) = \delta_\varepsilon(k) \\ \delta_{\varphi,\vec{n}}(k) & = \delta_{\varphi,\vec{V}_m}(k) = \delta_\varphi(k) \end{cases} \tag{9.8}$$

Here $\delta_{\varepsilon,\vec{n}}(k)$ and $\delta_{\varepsilon,\vec{V}_m}(k)$ direction-finding errors and guidance errors in the plane ε, and, respectively, $\delta_{\varphi,\vec{n}}(k)$ and $\delta_{\varphi,\vec{V}_m}(k)$ are direction-finding errors, and in-plane guidance errors φ in kth time quantum. If the missile miss $R_{mt}(k)$ is less than the miss h at which the target will be destroyed with a probability not lower than the specified one, provided that dispersion area of the warhead fragments is coordinated with the target spatial position; then in this case a decision is made to destroy the target. Further computations for this mth realization are stopped, the counter of realizations is increased by 1, and a new computation cycle for $m+1$ realization begins. Otherwise, calculations continue under conditions of the mth implementation, while it is necessary to increase the index k by 1 (i.e., go to a new, $k+1$ time half-interval). Before that, it is necessary to remember tracking errors by elevation angle $\delta_\varepsilon(k)$ and in azimuth $\delta_\varphi(k)$, as well as the value of the signal power $P_r(k)$ received by the sum channel.

3.3: A monopulse antenna forms four partial radiation patterns:

$\dot{F}_1[\varepsilon_t(t_k), \varphi_t(t_k)]$, $\dot{F}_2[\varepsilon_t(t_k), \varphi_t(t_k)]$, $\dot{F}_3[\varepsilon_t(t_k), \varphi_t(t_k)]$, $\dot{F}_4[\varepsilon_t(t_k), \varphi_t(t_k)]$, each of which is represented in a complex field of numbers. The indicated radiation patterns for the antenna used in the homing head should be calculated in advance in the range of angles $\varepsilon_{t,j}$, $\varphi_{t,j} \in [-90.0°, \ldots, +90.0°]$.

3.4. Jamming signal conditioning block where interference influences will be formed in this block:

1. Sidelobe interferences of the missile seeker directional diagram;
2. Polarization interference;
3. Flickering interference from two points in space;
4. Interference, created by a powerful signal, reflected from the Earth's (water) surface.

9.1 Formation Procedure of the Direction-Finding Characteristic

It is known that the vector complex directivity characteristic of the antenna system in the spherical coordinate system $\{\vec{i}_\theta, \vec{i}_\varphi, \vec{i}_{R_t}\}$, provided that the vectors $\vec{i}_\theta$ and $\vec{i}_\varepsilon$ are mutually perpendicular, can be represented as:

$$\vec{\dot{E}}(\varepsilon, \varphi, t) = \vec{i}_\varepsilon \dot{E}(\varepsilon, \varphi, t) \exp(j \arg \dot{E}_\varepsilon) + \vec{i}_\varphi \dot{E}(\varepsilon, \varphi, t) \exp(j \arg \dot{E}_\varphi), \tag{9.9}$$

where $\vec{i}_\varepsilon \dot{E}(\varepsilon, \varphi, t) = |\dot{E}_\varepsilon| = E_w(t)$ is the field component at working (w) polarization and $\vec{i}_\varepsilon$; $\vec{i}_\varphi \dot{E}(\varepsilon, \varphi, t) = |\dot{E}_\varphi| = E_c(t)$ is the cross-polarization (c) field component $\vec{i}_\varphi$.

Assuming in (9.9) the equality of the phases of the orthogonal components of the field, the outputs of homing head channels for direction finding in one plane can be written as follows:

$$\begin{aligned} U_\Sigma(\varepsilon, \varphi) &= E_w(t) F_{\Sigma w}(\varepsilon, \varphi) + E_c(t) F_{\Sigma c}(\varepsilon, \varphi) \\ U_\Delta(\varepsilon, \varphi) &= E_w(t) F_{\Delta w}(\varepsilon, \varphi) + E_c(t) F_{\Delta c}(\varepsilon, \varphi), \end{aligned} \tag{9.10}$$

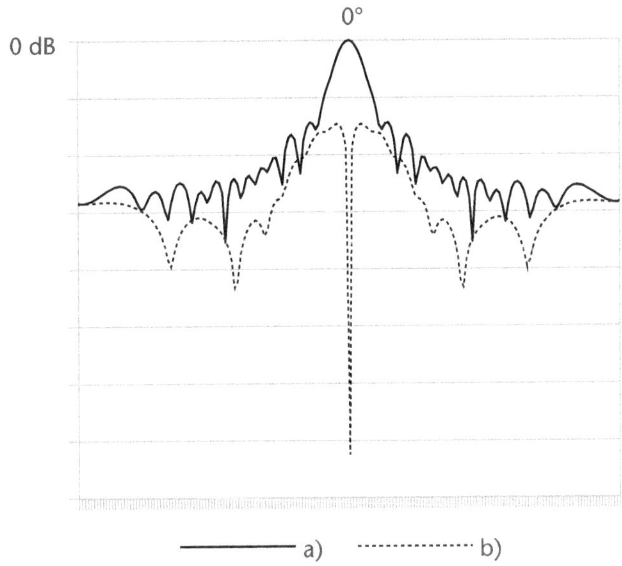

Figure 9.5 Directional patterns of the sum channel: (**a**) $F_{\Sigma w}(\varepsilon)$ on working polarization, and (**b**) $F_{\Sigma c}(\varepsilon)$ on cross-polarization.

where $E_w(t)$ and $E_c(t)$ are components of the signal field in the basis of the homing head receiving antenna on working (w) and on cross-polarization (c) and $F_{\Sigma w}(\varepsilon, \varphi)$, $F_{\Delta w}(\varepsilon, \varphi)$, $F_{\Sigma c}(\varepsilon, \varphi)$, $F_{\Delta c}(\varepsilon, \varphi)$ represent sum Σ and difference Δ radiation patterns on the working (w) and on the cross-polarization (c), respectively. Radiation patterns in the elevation plane $F_{\Sigma w}(\varepsilon)$, $F_{\Sigma c}(\varepsilon)$ of homing head antenna on working polarization (w), when $E_c(t) = 0$, and on cross-polarization (c), when $E_w(t) = 0$, are presented in Figure 9.5. In this case, the signal amplitude $|\dot{\vec{E}}(\varepsilon, \varphi, t)|$ is normalized to one:

$$|\dot{\vec{E}}(\varepsilon, \varphi, t)| = |\dot{\vec{E}}(t)| = \sqrt{E_w^2(t) + E_c^2(t)} = 1 \tag{9.11}$$

Thus, the output voltage of the sum (difference) channel is an additive mixture of the orthogonal components of the signal field, weighted by the corresponding radiation patterns on the working (w) and on the cross-polarization (c). When the homing head tracks the target according to the signal on the working polarization, then in this case its direction-finding characteristic on the working polarization, when the mismatch angle $\alpha = 0°$ between the received signal polarization plane and the vector $\vec{i}_\varepsilon$ of the homing head receiving antenna working polarization will have the form shown in Figure 9.6.

In the case when the received signal polarization plane will be orthogonal to the vector of the operating polarization of the homing head receiving antenna, then in this case $\alpha = 90°$ and the homing head direction-finding characteristic on the cross polarization will have the form shown in Figure 9.7. Taking into account the fact that when the cross-polarization signal is impacted to the homing head ($\alpha = 90°$) the total and difference homing head channels are swapped [1,2], then in this case, for $\varepsilon = 0$, the function $F_{\Sigma c}(\varepsilon)$ vanishes, and the function $F_{\Delta c}(\varepsilon)$ has a maximum value

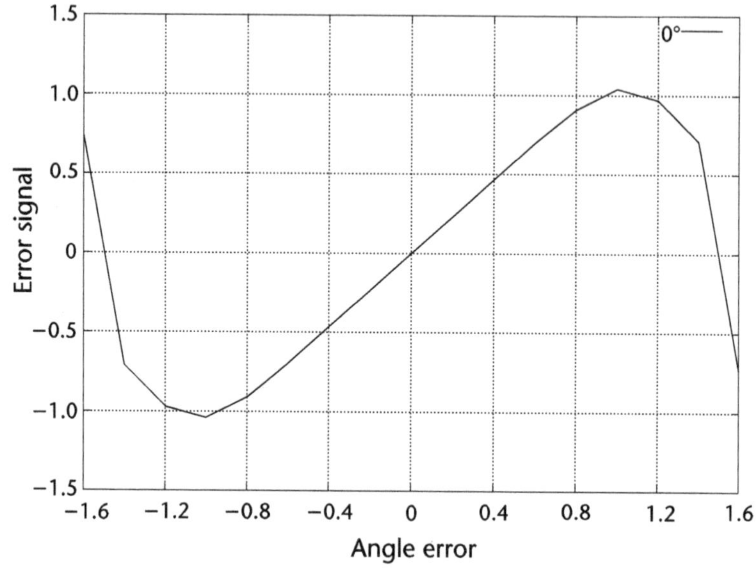

Figure 9.6 $S_w(\varepsilon)$ is the homing head direction-finding characteristic on the working polarization.

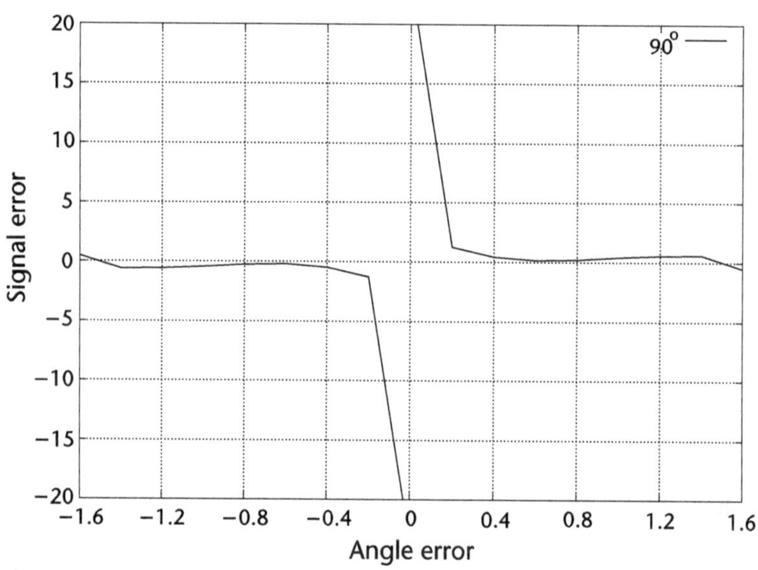

Figure 9.7 $S_c(\varepsilon)$ is the homing head direction-finding characteristic on the cross polarization.

and these functions are within the $[\varepsilon_{\min}, \varepsilon_{\max}]$ can be approximated by expressions:

$$F_{\Sigma c}(\varepsilon) \approx A \sin[\eta(\varepsilon)]; \qquad F_{\Delta c}(\varepsilon) \approx \cos[\eta(\varepsilon)], \tag{9.12}$$

where $\eta(\varepsilon) = \pi d \sin \varepsilon / \lambda$; A is a coefficient normalizing $F_{\Sigma c}(\varepsilon)$ to one; d is the width of the antenna aperture in the plane under consideration; and λ is the operating wavelength.

Therefore, for $E_w(\varepsilon) = 0$, the direction-finding characteristic of the homing head on the cross polarization can be represented as

$$S(\varepsilon) = \frac{F_{\Delta c}(\varepsilon)}{F_{\Sigma c}(\varepsilon)} \approx \frac{\cos[\eta(\varepsilon)]}{\sin[\eta(\varepsilon)]} \quad (9.13)$$

Here, $S(c)$ is an approximation of the direction finding characteristic on cross-polarization. Having experimental cross-polarization radiation patterns $F_{\Sigma c}$ and $F_{\Delta c}$ of a specific real antenna (in the main planes) or radiation patterns of this antenna, calculated by a certain rigorous method, one can always obtain an experimental (calculated) direction-finding characteristic of S_{c0} on cross polarization and subsequently obtain a function of the form

$$S(\varepsilon) = S_{c0}(\varepsilon) K_c(\varepsilon), \quad (9.14)$$

such that

$$\int_{\varepsilon_{min}}^{\varepsilon_{max}} \delta^2(\varepsilon)\, d\varepsilon \to 0 \quad (9.15)$$

where $\delta(\varepsilon) = S_c(\varepsilon) - S(\varepsilon)$ is the approximation error; $S_{c0}(\varepsilon) = 1/\varepsilon$ defines the shape of the direction-finding characteristic on cross polarization in the range of angles $\varepsilon \in [\varepsilon_{min}, \ldots, 0 - \delta[\cup]0 + \delta, \ldots, \varepsilon_{max}]$, where $\delta \to 0$, $0 = \pm\infty$ is the point of uncertainty; and $K_c(\varepsilon)$ is a function of the conversion factor of a monopulse radar by a signal of polarization interference, minimizing the functional $\eta(\varepsilon)$ based on the condition $\frac{\partial \eta(\varepsilon)}{\partial K_c(\varepsilon)} = 0$.

Earlier it was shown that the direction-finding characteristic on the working polarization in the absence of noise interference has the form

$$U_{out}(t) = K_d(\varepsilon) + K_f(\varepsilon), \quad (9.16)$$

where $K_f(\varepsilon) = \sqrt{N(\varepsilon)}\vartheta_0(t)$ is the random component of the error signal; $\sqrt{N(\varepsilon)}$ is the fluctuation characteristic; and $\vartheta_0(t)$ is the Wiener random process (white noise with unit spectral density).

Therefore, if there is no signal on the cross polarization (i.e., $E_c(t) = 0$), then for the direction-finding characteristic on the working polarization, we can write

$$S(\varepsilon) = K_d K_{d0}(\varepsilon) = S_w(\varepsilon) \quad (9.17)$$

where $K_d = \frac{\partial K_d(c)}{\partial \varepsilon} = \mu$ is the angle finder conversion factor, which is the slope of the direction-finding characteristic on the working polarization on its linear section (Figure 9.6); $K_{d0}(\varepsilon)$ is the normalized direction-finding characteristic with a steepness equal to one (determines the form of the direction-finding characteristic on the working polarization).

The value of K_d depends on the signal amplitude and grows with its increase [1]. To eliminate this dependence, as shown above, an automatic gain control system is used, which stabilizes the output of the total channel (average power), but only due to such factors as changing the distance to the object [1]. However, along with this, there are also fast amplitude fluctuations that are not processed by the AGC system. These fluctuations occur when signals are reflected from objects of complex shapes, as well as due to changes in the propagation conditions of the reflected signal [6].

If $E_w(t) \neq 0$ and $E_c(t) \neq 0$, then the expression for the direction-finding characteristic monopulse homing head, taking into account the operation of the automatic gain control system and the given angular velocity of rotation of the plane of polarization of the interference Ω_n, can be written as

$$S(\varepsilon, t) = S_w(\varepsilon)W_w(t) + S_c(\varepsilon)W_c(t), \tag{9.18}$$

where $S_w(\varepsilon) = \frac{F_{\Delta w}(\varepsilon)}{K_{ARU}\bar{U}_\Sigma}$ the direction-finding characteristic at working polarization; $S_c(\varepsilon) = \frac{F_{\Delta c}(\varepsilon)}{K_{ARU}\bar{U}_\Sigma}$ - direction finding characteristic on cross-polarization; $W_w(t) = |\dot{\vec{E}}|\cos(\Omega_n t)$; and $W_c(t) = |\dot{\vec{E}}|\sin(\Omega_n t)$ are the weighting functions for the working and cross-polarization components of the interference signal.

Thus, when exposed to interference with the rotation of the plane of polarization of the interference, the direction-finding characteristic of the missile seeker can be represented by the weighted sum of the direction-finding characteristics on the working and orthogonal polarization of the receiving antenna of the seeker. Consequently, the setting of a polarization interference for the missile homing head leads to a distortion of its direction-finding characteristics and with a sufficiently accurate interference guidance to the seeker ($\varepsilon \approx 0$) can lead to the complete inoperability of the missile homing head due to the fact, that at $\varepsilon = 0 \Rightarrow S(\varepsilon) = \infty$ due to the expression (9.13).

Passing from continuous time t to discrete time counts k with duration h, the expression for the direction-finding characteristic in the kth time quantum can be represented as

$$S(\varepsilon, t) = \Re\left[\frac{\dot{E}_w(\varepsilon, k)\dot{F}_{\Delta w} + \dot{E}_c(\varepsilon, k)\dot{F}_{\Delta c}(\varepsilon)}{\dot{E}_w(\varepsilon, k)\dot{F}_{\Sigma w} + \dot{E}_c(\varepsilon, k)\dot{F}_{\Sigma c}(\varepsilon)}\right], \quad \forall t \in [(k-1)h, kh) \;\&\; k = 1, \ldots, N \tag{9.19}$$

There is no time dependence in (9.19) since the invariance was previously assumed in systems in time during a time quantum of duration h. Further, it is quite clear that the nature of the dependences of $\dot{E}_w$ and $\dot{E}_c$ on time affects the value of h.

9.2 The Initial Impacts Classification

Modern methods of jamming in order to protect an aircraft-bomber, missile, or other carrier of high-precision weapons from anti-aircraft guided missiles of the air defense system, which covers the attacked object, are divided into two groups of protection [10]:

1. Collective protection (interference from a remote point); that is protection of the attacking aircraft with the effect of interference along the sidelobes of the antenna pattern of a ground multifunctional radar and/or the homing head of

an anti-aircraft guided missile. This interference is also called cover interference. Such jamming is usually carried out from the board of an electronic warfare aircraft, also called a jamming plane. At the same time, a certain safety is provided against retargeting anti-aircraft guided missiles. This method requires a high sensitivity of an electronic reconnaissance receiver and a high energy potential of an onboard active jamming station, which makes it possible to use this jamming by electronic warfare aircraft that are at or beyond the reach of the suppressed anti-aircraft missile systems.

2. Personal protection is the effect of interference only on the main beam of the suppressed radar (ground-based radar and/or homing head), which provides savings of 15–35 dB in the potential of the station of active interference in relation to the Doppler-rangefinder concept methods that require capture on automatic tracking of anticipatory action along the sidelobes of the suppressed radar radiation pattern. Due to the fact that this interference affects only the main beam of the suppressed radar, such interference is called sighting interference. Effective protection of aircraft can only be individual. The transition to personal protection provides:
   - Savings of 15–35 dB in the potential of the station of active interference with respect to the sidelobes of the antenna pattern of the suppressed radar;
   - A solution to the problem of lack of sensitivity of receiving paths;
   - A solution to the problem of decoupling the receiving and transmitting antennas;
   - Light weight and smaller size;
   - Low power consumption.

   The modern systemic or integrated approach in electronic warfare, which is the suppression of all radars integrated into the air defense system at the stage of radar detection of aerospace attack weapons, is a consequence of the lack of effective methods of individual protection of missile weapon carriers from homing heads of enemy anti-aircraft guided missiles. For the air forces of the USSR and the United States from the 1980s to the 1990s, this approach, in principle, was acceptable, because one could count on the massive involvement of special aircraft and electronic warfare helicopters for the protection of attack aircraft, which patrolled outside the engagement zones of air defense weapons and in battle formations—the so-called aircraft reinforced with electronic warfare.

At the same time, personal protection of strike aircraft has the following advantages:

1. Reducing the frequency range of suppressed radars. For protection against airborne guided missiles, a range of about 2 GHz in the frequency range 8.5–10.5 GHz is sufficient. To protect against anti-aircraft guided missiles, a frequency range of 6–12 GHz is required. For radar tracking targets of short-range air defense missile systems, the frequency range 12–18 GHz. The decimeter wavelength range imposes increased requirements on the mass-dimensional characteristics of antenna systems capable of providing a given width of the

antenna radiation pattern of aviation jamming radar (electronic warfare). For the same reason, it is difficult to achieve the specified values of the excess of the N_{jammer} noise over the S (N_{jammer}/S) signal. In this regard, it is advisable to exclude from consideration as interference effects the decimeter and meter wavelength ranges, at least for anti-aircraft guided missiles, which is an invaluable advantage of the concept of individual protection of aerospace attack equipment.

2. Reducing the energy potential of the active jamming station to a level that is actually achievable in each class of carriers, even in the case of sidelobe exposure to the radiation patterns of suppressed radars. It is quite understandable that the decisive gain in potential is associated with failure from impact on the sidelobe effect of the radiation patterns of these radars.
3. Elimination of premature unmasking of the carrier of aerospace attack means by jamming until it is detected. For a carrier with measures taken to reduce the effective reflective surface (stealth technology), this is the only way to maintain the benefits achieved in a very costly way. The standard requirements for electronic reconnaissance for modern on board active jamming stations provide for the detection of interference radar signals at a distance of at least $1.2\ times D_{det}$ from the carrier or 1.8 more than the far border of the air defense system affected area, followed by immediate jamming to disrupt (exceptions) detection and capture on an auto-tracking target with an electronic warfare station. Here D_{det} is the detection range of the given carrier by the radar systems of the suppressed anti-aircraft missile system. It is difficult to find the logic in this when the developer of the aircraft achieves a reduction in the effective reflecting surface by an order of magnitude compared to the previous aircraft of this class, and then the developer of the active jamming station for this carrier implements the concept, according to which the suppressed enemy radar will take the direction finding of this aircraft with high reliability by its own jamming signal long before the enemy's air defense systems means are able to detect it by the signal reflected from its glider, in a case when its electronic warfare station is turned off.

It is known that 95% of all weapons control radars are based on monopulse direction finding. A number of firing radars may not have a rangefinder and there may not be a Doppler velocity meter, but modern weapon control radars without a monopulse angular measurement simply do not exist [11].

Of the four radar coordinates—range, velocity, and two angular coordinates—the first two fix the position of the target on the beam (straight line), and the second two on the contracting (expanding) sphere, representing the range and speed of a target by one set of its values, and two angular coordinates of the target by another set of its values. It is quite clear that the cardinality of both sets is the same and is equal to the cardinality of the continuum. However, when applied to the task of guided, and moreover, homing missile weapons, the precise indication of the angular coordinates almost exhaustively solves the problems facing the weapon control system. And in this sense the power of the target angular coordinates measuring system is incomparably higher than the power of the rangefinder-Doppler systems.

Interference, which, at the stage of tracking a target attacking an air defense system and covered by an electronic warfare system, leads to a disruption of the automatic tracking of this target, can be considered as a highly effective means of personal protection of an aircraft that is attacking an air defense system.

Assuming that in the general case at the antenna input there is a large number of N sinusoidal oscillations of the same frequency ω, but with different amplitudes $\dot{E}_i$ and phases ϕ_i. Omitting the time factor $\exp(j\omega t)$, the expressions for the quadrature components of the field in one plane (for example, in the plane of the elevation ε) take the form:

$$\begin{cases} \dot{E}_w(\varepsilon) = \sum_{i=1}^{N} E_i(\varepsilon)\exp(j\phi_i)\cos\alpha_i \\ \dot{E}_c(\varepsilon) = \sum_{i=1}^{N} E_i(\varepsilon)\exp(j\phi_i)\sin\alpha_i, \end{cases} \tag{9.20}$$

where $\dot{E}_w(\varepsilon)$, $\dot{E}_c(\varepsilon)$ are the complex projections of the tension vector of interference signal fields to the unit vectors of the working $\vec{i}_\varepsilon$ and cross-polarization $\vec{i}_\varphi$ directional patterns of the receiving antenna of the missile homing head.

α_i is an angle characterizing the position of the resulting vector $\vec{\dot{E}}_n(\varepsilon)$ in the orthogonal basis $\{\vec{i}_\varepsilon, \vec{i}_\varphi\}$ relative to the vector on operating polarization $\vec{i}_\varepsilon$:

$$|\vec{\dot{E}}_i(\varepsilon)| = |\dot{E}_w(\varepsilon)|\vec{i}_\varepsilon + |\dot{E}_c(\varepsilon)|\vec{i}_\varphi \ , \tag{9.21}$$

where $|\dot{E}_w(\varepsilon)| = |\vec{\dot{E}}_i(\varepsilon)| \times \sin(\alpha_i)$ and $|\dot{E}_c(\varphi)| = |\vec{\dot{E}}_i(\varepsilon)| \times \cos(\alpha_i)$.

1. For cover interference $\vec{\dot{E}}_n(\varepsilon)$, when the sidelobes of the anti-aircraft guided missile homing head antenna are exposed to interference, we can write:

$$\vec{\dot{E}}_n(\varepsilon) = \dot{E}_w(\varepsilon)\vec{i}_\varepsilon + \dot{E}_c(\varepsilon)\vec{i}_\varphi, \quad \forall\varepsilon \in \left[-\frac{\pi}{2}, -\frac{\varepsilon_{mb}}{2}\right[\cup \left]\frac{\varepsilon_{mb}}{2}, \frac{\pi}{2}\right], \tag{9.22}$$

where $\varepsilon_{mb} = \Delta\varepsilon\big|_{0.707}$ - main beam width of homing head antenna directional pattern in voltage (at the level of 0.707).

2. For sighting interference $\vec{\dot{E}}_n(\varepsilon)$, when interference signal affects only on main beam of homing head antenna pattern of the anti-aircraft guided missile, we can write:

$$\vec{\dot{E}}_n(\varepsilon) = \dot{E}_w(\varepsilon)\vec{i}_\varepsilon + \dot{E}_c(\varepsilon)\vec{i}_\varphi, \qquad \forall\varepsilon \in \left[-\frac{\varepsilon_{mb}}{2}, 0\right] \cup \left[0, \frac{\varepsilon_{mb}}{2}\right]. \tag{9.23}$$

in this case, if to divide conditionally the formation of $\dot{E}_w(\varepsilon)$ and $\dot{E}_c(\varepsilon)$ into two levels according to forming the orthogonal pairs method $\{\dot{E}_{wi}(\varepsilon_i), \dot{E}_{ci}(\varepsilon_i)\}$ in the receiving base of the homing head antenna, we get the following tasks:

Task 1.1. There is only one pair $\dot{E}_w = \{\dot{E}_{w1}\}$, $\dot{E}_c = \{\dot{E}_{c1}\}$ whose modules change in time according to apriori given law, and their initial phases $\phi_{1,2}$ - constant.

Procedure:

$$E_w(t) = E^e f(t, \Omega), \qquad \phi_w(t) = \phi_1 = C_1 = const$$
$$E_c(t) = E^e f_\text{T}(t, \Omega), \qquad \phi_c(t) = \phi_2 = C_2 = const$$

Here f is a function that defines the law of amplitude variation $\dot{E}_w$ in time; Ω is the rate of parameter change over time; f_{T} is the function orthogonal to f; $E^e = \sqrt{E_w^2 + E_c^2} = 1$

Task 1.2. There is only one pair $\dot{E}_w = \{\dot{E}_{w1}\}$, $\dot{E}_c = \{\dot{E}_{c1}\}$ whose phases change in time according to apriori given law, and their modules are constant.

Procedure:

$$E_w(t) = A_1, \qquad \phi_w(t) = \phi_w^0 g(t, \Omega)$$

$$E_c(t) = B_1, \qquad \phi_c(t) = \phi_c^0 g_{\mathsf{T}}(t, \Omega),$$

where ϕ_w^0, ϕ_c^0 are the phase initial values at $t = 0$; g is the function that sets the law of phase variation of $\dot{E}_w$ in time; g_{T} is the function orthogonal to g.

Task 1.3. There is only one pair $\dot{E}_w = \{\dot{E}_{w1}\}$, $\dot{E}_c = \{\dot{E}_{c1}\}$ amplitudes and phases of which change in time according to the apriori given law.

Procedure:

$$E_w(t) = E^e f(t, \Omega), \qquad \phi_w(t) = \phi_w^0 g(t, \Omega)$$

$$E_c(t) = E^e f_{\mathsf{T}}(t, \Omega), \qquad \phi_c(t) = \phi_c^0 g_{\mathsf{T}}(t, \Omega)$$

Task 2.1. There are two or more pairs $\dot{E}_w = \{\dot{E}_{wi}\}$, $\dot{E}_c = \{\dot{E}_{ci}\}$. In this case, their amplitudes change in time according to the apriori given law, and the initial phases ϕ_i do not depend on time.

Procedure:

$$E_w(t) = \sum_{i=1}^{N} E_{wi}(t), \qquad \forall i : \dot{E}_{wi} = E^e f_i(t, \Omega_i)$$

$$E_c(t) = \sum_{i=1}^{N} E_{ci}(t), \qquad \forall i : \dot{E}_{ci} = E^e f_{\mathsf{T}i}(t, \Omega_i)$$

$$\phi_{wi} = C_i; \qquad \phi_{ci} = D_i.$$

Task 2.2. There are two or more pairs $\dot{E}_w = \{\dot{E}_{wi}\}$, $\dot{E}_c = \{\dot{E}_{ci}\}$. In this case, their amplitudes are constant, and the phases change in time according to the apriori given law.

Procedure:

$$E_{wi}(t) = A_i, \qquad \phi_{wi}(t) = \phi_{wi}^0 g_i(t, \Omega)$$

$$E_{ci}(t) = B_i, \qquad \phi_{ci}(t) = \phi_{ci}^0 g_{\mathsf{T}i}(t, \Omega), \qquad \forall i = 1, \ldots, N$$

Task 2.3. There are two or more pairs $\dot{E}_w = \{\dot{E}_{wi}\}$, $\dot{E}_c = \{\dot{E}_{ci}\}$. In this case, the amplitudes and phases change in time according to the apriori given law.

Procedure:

The proposed procedure can be applied to describe a signal reflected from a target. Therefore, it will be considered in more detail. The electrodynamic model of the target is a body defined in the space R^3, the surface of which is divided into sufficiently small areas of infinite conductivity, such that each can be considered flat and less than half the wavelength of the irradiated signal. First, according to the known characteristics of the signal emitted by the illumination radar, find the spatial currents on the illuminated part of the target, and then determine the field scattering at the receiving point as a superposition of all elementary fields created by each platform taking into account their amplitudes and phases. The calculation method for this procedure will be discussed below.

Task 3.1. There is only one pair $\dot{E}_w = \{\dot{E}_{w1}\}$, $\dot{E}_c = \{\dot{E}_{c1}\}$ whose modules change in time according to a random law and are random processes $\dot{E}_w(\omega, \varepsilon, t)$ and $\dot{E}_c(\omega, \varepsilon, t)$, and the phases are constant.

Procedure:

For one implementation, we can write:

$$E_w(t) = M[E_{w1}(t)] + \xi(0, \sigma_{w1}), \qquad \phi_{w1} = C_1$$

$$E_c(t) = M[E_{c1}(t)] + \xi(0, \sigma_{c1}), \qquad \phi_{c1} = D_1$$

Task 3.2. There is only one pair $\dot{E}_w = \{\dot{E}_{w1}\}$, $\dot{E}_c = \{\dot{E}_{c1}\}$ whose phases change in time according to a random law and represent random processes $\dot{E}_w(\omega, \varepsilon, t)$ and $\dot{E}_c(\omega, \varepsilon, t)$.

Procedure:

$$E_w(t) = A, \qquad \phi_{w1} = \phi_{w0} + \xi(0, \sigma_{w1})$$

$$E_c(t) = B, \qquad \phi_{c1} = \phi_{c0} + \xi(0, \sigma_{c1})$$

Task 3.3. There is only one pair $\dot{E}_w = \{\dot{E}_{w1}\}$, $\dot{E}_c = \{\dot{E}_{c1}\}$ whose amplitudes and phases change in time according to a random law and are random processes $\dot{E}_w(\omega, \varepsilon, t)$ and $\dot{E}_c(\omega, \varepsilon, t)$.

Procedure:

$$E_w(t) = M[E_{w1}(t)] + \xi(0, \sigma_{w1}), \qquad \phi_{w1} = \phi_{w0} + \xi(0, \sigma_{w1})$$

$$E_c(t) = M[E_{c1}(t)] + \xi(0, \sigma_{c1}), \qquad \phi_{c1} = \phi_{c0} + \xi(0, \sigma_{c1})$$

Task 4.1. There are two or more pairs $\dot{E}_w = \{\dot{E}_{wi}\}$, $\dot{E}_c = \{\dot{E}_{ci}\}$; in this case, the modules change in time in a random manner, provided that their phases have constant values.

Procedure:

$$E_w(t) = \sum_{i=1}^{N}(M[E_{wi}(t)] + \xi(0, \sigma_{wi})), \qquad \phi_{wi} = C_i$$

$$E_c(t) = \sum_{i=1}^{N}(M[E_{ci}(t)] + \xi(0, \sigma_{ci})), \qquad \phi_{ci} = D_i$$

Task 4.2. There are two or more pairs $\dot{E}_w = \{\dot{E}_{wi}\}$, $\dot{E}_c = \{\dot{E}_{ci}\}$. In this case, the phases change in time in random manner, provided that their amplitudes have constant values.

Procedure:

$$E_{wi}(t) = A_i, \qquad \phi_{wi} = \phi_{w0} + \mu(0, \sigma_{wi})$$

$$E_{ci}(t) = B_i, \qquad \phi_{ci} = \phi_{c0} + \mu(0, \sigma_{ci})$$

Task 4.3. There are two or more pairs $\dot{E}_w = \{\dot{E}_{wi}\}$, $\dot{E}_c = \{\dot{E}_{ci}\}$. In this case, both the amplitudes and phases change in time in a random manner.

Procedure:

$$E_w(t) = \sum_{i=1}^{N}(M[E_{wi}(t)] + \xi(0, \sigma_{wi})), \qquad \phi_{wi} = \phi_{w0} + \mu(0, \sigma_{wi})$$

$$E_c(t) = \sum_{i=1}^{N}(M[E_{ci}(t)] + \xi(0, \sigma_{ci})), \qquad \phi_{ci} = \phi_{c0} + \mu(0, \sigma_{ci})$$

In a joint consideration of the described tasks and the formation procedures that ensure their solution, a two-level classification is obtained about which the following considerations can be expressed.

The grouping of tasks by classification levels is carried out:

a. By the form of functions $\dot{E}_w(t)$ $\dot{E}_c(t)$ and corresponds to the differences in the methods of implementation of these procedures:
   - Level 1, when parameters of setting signal (influence) are changes in time in according to an apriori given law;
   - Level 2, when parameters of setting signal (influence) change over time according to a random law and are random processes.

b. According to the approach to the method of forming setting signals (influences). Here, each level of the two-level classification is divided into two sublevels that characterize the differences in the number of signals arriving at the input of the homing head antenna. The horizontal classification characterizes the gradation according to the composition of the time-varying parameters. Tasks and procedures describe methods of jamming in one way or another.

So, Task 1.1 corresponds to the hindrance with variable polarization parameters: circular or sectorial rotation of the polarization plane of a linearly polarized signal, which is created from one space point.

Task 2.1 describes a flickering interference created from two or more points in space, or interference that creates a false target by means of illumination of the Earth's (water) surface when flying a target at low and extremely low altitudes. In this case, with sufficient power of the signal reflected from the ground (water) surface, it is possible to ensure the retargeting of the anti-aircraft guided missile to a decoy target located behind the line of separation of two media: air-ground or air-water.

Task 2.2 characterizes the creation of coherent interference.

Task 2.3 represents a model for determining the probabilistic moments (mathematical expectations) of orthogonal field vectors reflected from the target or the case of coherent hindrance formation.

Task 3.1 reflects the case of setting up hindrance with chaotic polarization when using the uniform law in the procedure for generating $\dot{E}_w(t)$ and $\dot{E}_c(t)$.

Tasks 4.1 and 4.2 can be considered as two options in the implementation of asynchronous flickering interference created by a group of jammers from two or more points in space.

The situations presented by Tasks 1.2, 1.3, 3.2, 3.3, and 4.3 currently seem unlikely for practical application.

9.3 Formation Method of Stochastic Orthogonal Processes

Since disruption probability of automatic target tracking by angular coordinates was previously introduced as one of interference immunity indicators, it would be reasonable to determine the probability measure $P(A_1)$, where the event A_1 characterizes disruption of automatic target tracking. In this regard, it is advisable to define the complete group of events as $A = A_1 \cup A_2$, and $P(A_1) + P(A_2) = 1$. It is natural to assume that in this way a σ-algebra ν is given on some arbitrary set Ω such that $A \subseteq \Omega$. As a result, the probability space (Ω, ν, P) is formed. Due to the fact that any probability measure is a function of time t, therefore, the given probability space will also be measurable in time.

Describing the state of an automatic system with a vector random variable $\mathbf{Y}(\mathbf{t})$ [7] and denoting $\nu(0, T)$ - σ-algebra of events generated by $\mathbf{Y}(\mathbf{t})$, $\forall t \in [0, T]$, behavior of the system in the past and in the future can be described by the σ-algebras of events:

$$\nu^1 = \nu(0, t_0), \quad \nu^2 = \nu(t_0, T) \quad \text{and} \quad ß = ß(t_0 - \epsilon, t_0). \tag{9.24}$$

If, for known outcomes of all events from ß, the events $A^1 \in \nu^1$ does not depend on the events $A^2 \in \nu^2$; that is, the σ-algebra ν^1 and ν^2 are conditionally independent with respect to ß, then the following equality holds:

$$P(A^1A^2|ß) = P(A^1|ß)P(A^2|ß), \tag{9.25}$$

$\forall A^1 \in \nu^1, A^2 \in \nu^2$ and σ-algebra ß is a splitting [12]. It divides the observation interval $[0, T]$ into three segments: $[0, t_0 - \epsilon]$, $]t_0 - \epsilon, t_0[$, $[t_0, T]$ on which, acting respectively, σ-algebras ν^1, ß, ν^2, moreover:

$$\begin{aligned} A^1 &= A_1 \cup A_2, && \forall t \in [0, t_0 - \epsilon]; \\ A^2 &= A_1 \cup A_2, && \forall t \in [t_0, T]. \end{aligned} \tag{9.26}$$

Since $\forall t_0 \in [0, T]$ it is possible to construct a splitting σ-algebra ß and the observation interval can be discretized into N disjoint segments:

$$[0, T] = \bigcup_{k=1}^{N} [(k-1)h, kh - \epsilon]. \tag{9.27}$$

Here h is the duration of the time quantum in the division of the observation interval. In this case, σ-algebras ß will be constructed only in the left ϵ-neighborhood of discretization points. So for the Mth sampling quantum:

$$\begin{aligned} \nu^1 &= \bigcup_{k=1}^{M} \nu((k-1)h, kh) \backslash \nu(Mh - \epsilon, Mh) \\ \nu^2 &= \bigcup_{k=1}^{N} \nu((k-1)h, kh) \end{aligned} \tag{9.28}$$

As a result of the limit passage:

$$\lim_{\epsilon \to 0} \bigcup_{k=1}^{N} [(k-1)h, kh - \epsilon] = \bigcup_{k=1}^{N} [(k-1)h, kh), \tag{9.29}$$

the representation is obtained for the derivative of the direction-finding characteristic of the monopulse homing head:

$$K_d(t_k) = S'(\varepsilon_k) = \frac{S(\varepsilon_k) - S(\varepsilon_{k-1})}{\varepsilon_k - \varepsilon_{k-1}}, \tag{9.30}$$

in a neighborhood of some point ε_0, such that:

$$\varepsilon_0 = \varepsilon_{k-1} + \frac{\varepsilon_k - \varepsilon_{k-1}}{2} \tag{9.31}$$

it is possible to ensure the continuity of the random process realizations $\mathbf{Y}(\omega, t)$.

The following proposition [12] is valid: ß splits σ-algebras ν^1 and ν^2 if and only if the sequence ν^1, ß, ν^2 is Markov sequence [13]; that is

$$P(A|\nu^1 \vee ß) = P(A|ß), \qquad A \in \nu^2 \tag{9.32}$$

and therefore, for a known current state of the automatic control system $\mathbf{Y}(t)$ its future behavior does not depend on the past. As a result, the Markov property will be rewritten as

$$P(A|Y(s), s, t) = P(A|\mathbf{Y}(\mathbf{t})), \qquad \mathbf{A} \in \nu(\mathbf{t}, \mathbf{T}) \tag{9.33}$$

Thus, $\mathbf{Y}(\omega, t)$ is a Markov random process. For $h \to 0$ and the introduction of a stochastic orthogonal measure $\mu(dt) = M|\eta(dt)|^2$ with zero mean, the behavior of the system can be described by the Ito stochastic integral or by a system of differential equations with bounded infinitely differentiable coefficients and generalized random functions [12,14,15]. As a result, the model for analyzing the operation of a monopulse homing head, proposed earlier, seems to be legitimate.

Due to the fact that $\mathbf{Y}(\omega, t)$ is a Markov random process, and interference signal described in Section 9.1 distorts the monopulse radar direction finding characteristic, hence the duration of discretization interval h is determined by the statistical properties of interference signal, which can also be described by a Markov random process.

Thus, the invariance of the automatic control system during a time interval of duration h is determined by interference signal stationarity on this interval. Therefore, for each source of interference signal, it is advisable to determine the measurable space of its states. By numbering the possible states of the automatic control system with integers $i = 0, 1, 2, 3, \ldots$ the process

$$E(t) = \dot{E}(\omega, \varepsilon, t) = \dot{E}_w(\omega, \varepsilon, t)\vec{i}_\varepsilon + \dot{E}_c(\omega, \varepsilon, t)\vec{i}_\varphi \tag{9.34}$$

can be characterized as a process of transitions from the state $E(s)$ to the state $E(t)$ for $t > s$:

$$\begin{gathered} P\left\{E(t) = j|E(s_1) = i_1, \ldots, E(s_n) = i_n, \quad E(s) = i\right\} = \\ P\left\{E(t) = j|E(s) = i\right\} = p_{ij}(s+t), \quad \forall t > s, \quad i, j = 1, 2, \ldots \end{gathered} \tag{9.35}$$

This is a homogeneous Markov process with a countable number of states [13,16]. Here $p_{ij}(s+t)$ is the transition probability from the i state to the j state, which can be represented as

$$p_{ij}(s+t) = \sum_k p_{ik}(s)p_{kj}(t), \quad i, j = 1, 2, 3, \ldots \tag{9.36}$$

and the probability that the system at the time $t > s$ will be in the state j is

$$p_j(t) = \sum_i p_i^0 p_{ij}(t), \quad j = 1, 2, 3, \ldots \tag{9.37}$$

Thus, expressions (9.36) and (9.37) represent a difference scheme for solving differential equations. Therefore, by specifying the initial vector of states of the process $p_i^0 = (p_1^0, p_2^0, \ldots, p_n^0)^T$ and having determined the matrix of transition probabilities P_{ij} it is possible to form a Markov random process $E(t)$ with the desired statistical characteristics.

References

[1] Leonov, A. I., and K. I. Fomichev, *Monopulse Radar,* Norwood, MA: Artech House, 1986.

[2] Richards, M. A., J. A. Scheer, and W. A. Holm (eds.), *Principles of Modern Radar,* Raleigh, NC: SciTech Publishing, 2010.

[3] Markin, E., "Scatter Pattern Calculations and Determination Performance Limitations of Existing Air Defense Systems for Intercepting Hypersonic Vehicles," *2012 IEEE Radar Conference*, Atlanta, GA, May 7–11, 2012, DOI: 10.1109/RADAR.2012.6212189.

[4] Ufimtsev, P. Y., *Theory of Edge Diffraction in Electromagnetics,* Encino, CA: Tech Science Press, 2009.

[5] Ufimtsev, P. Y., *Fundamentals of the Physical Theory of Diffraction,* New York: John Wiley & Sons, 2007.

[6] Shirman, Y. D., et al., *Radar Theoretical Foundations*, Book on Demand Ltd, October 29, 2019.

[7] Markin, E., "On Interference Immunity of Angle Tracking Systems Under Conditions of Interference Distorting Location Characteristic," *2009 IEEE Radar Conference*, Pasadena, CA, May 4–8 2009, pp. 1–6, DOI: 10.1109/RADAR.2009.4977092.

[8] Minoux, M., *Mathematical Programming: Theory and Algorithms,* Chichester, United Kingdom: John Wiley & Sons, 1986.

[9] Godunov, S. K., *Ordinary Differential Equations with Constant Coefficient* Translation of Mathematical Monographs, American Mathematical Society, 1987, Alibris ID: 9762292910/.

[10] Design Concept of Aircraft Individual Protection Against Guided Missile Weapons, Defence Initiatives, 2014, https://defin.by/en/media/publications/21-04-14/.

[11] Budge, M. C., and S. R. German, *Basic Radar Analysis,* Second Edition, Norwood, MA: Artech House, 2020.

[12] Ivchenko, G. I., and Y. I. Medvedev, *Mathematical Statistics,* Higher School, 1984.

[13] Anatol'evich, R. I., *Markov Random Fields*, New York: Springer-Verlag, 1982.

[14] Arato, M., *Stochastic Differential Systems,* Berlin: Springer-Verlag, 1981.

[15] Bakut, P. A., I. A. Bolshakov, B. M. Gerasimov, et al., in *Questions Statistical Radiolocation Theory,* G.P. Tartakovsky (ed.), Soviet Radio, 1964.

[16] Marchuk, G. I., *Numerical Methods and Applications,* CRC Press Inc., 1994.

CHAPTER 10
Interference Detection Problem

The problem with interference detection consists of finding such a rule that would ensure the correct recognition of each interference $H_i, \forall i = \overline{1, q}$ from the set of interferences $\{H\}$ of power q bounded from above, acting on a monopulse radar, based on a set of features $\{X\}$ uniquely characterizing each H_i interference from this countable set of interferences $\{H\}$. In this case, it should be ensured that the quality criterion of target tracking Q is maintained at a level not worse than the specified one. As it was shown in Chapter 8, when each interference H_i from a countable set of interference $\{H\}$ acts on the monopulse homing head, it is necessary to ensure the minimum quality criterion Q:

$$Q^* = \arg\inf F(H, Q), \tag{10.1}$$

where F is a certain operator, and the quality criterion Q is the target tracking error in angular coordinates $e(t)$ over the entire interval $[0, T]$, on which the anti-aircraft guided missile was guided until the time instant T, which characterizes the moment of detonation of its warhead.

$$Q = \int_0^T |e(t)|\, dt \tag{10.2}$$

in the case when the radio fuse gave a command to detonate the warhead, having previously coordinated the spatial position of the scattering area of the fragments with the position of the target in space, and during the missile aiming at the target $[0, T]$ there were no disruptions to the auto-tracking of the target (Figure 10.1), then you can write:

$$Q < C = \sup Q \tag{10.3}$$

Here the value C is the exact upper bound of the set of possible values of $\{Q\}$; that is, it is the smallest element of C that is equal to or greater than all elements of the set $\{Q\} = \{C \in \{C\} | \forall Q \in \{Q\} : Q \leq C\}$.

Introducing the designation $\Delta_{\varepsilon,\varphi}$, which is the width of the direction-finding characteristic of the seeker in the mutually orthogonal planes ε and φ, as shown in Figure 10.2, where α is the angle of deviation of the plane of polarization of the reflected signal or interference from the direction of the working polarization in the receiving base of the homing head antenna [1]. It is clear that in the general case the following inequality can be satisfied: $\frac{1}{2}|\Delta_\varepsilon| \neq \frac{1}{2}|\Delta_\varphi|$.

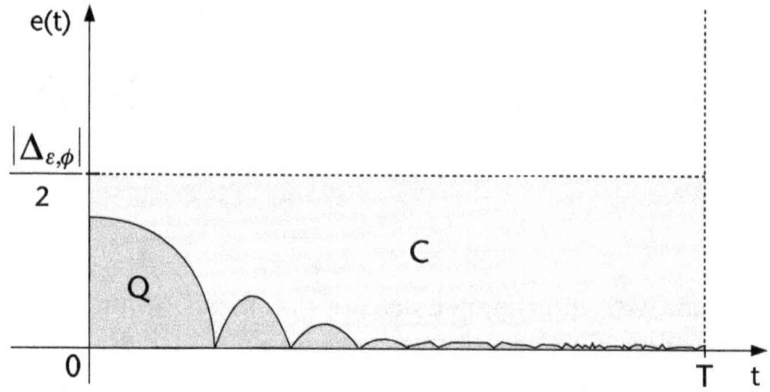

Figure 10.1 Quality criterion Q, when there were no failures of auto-tracking targets.

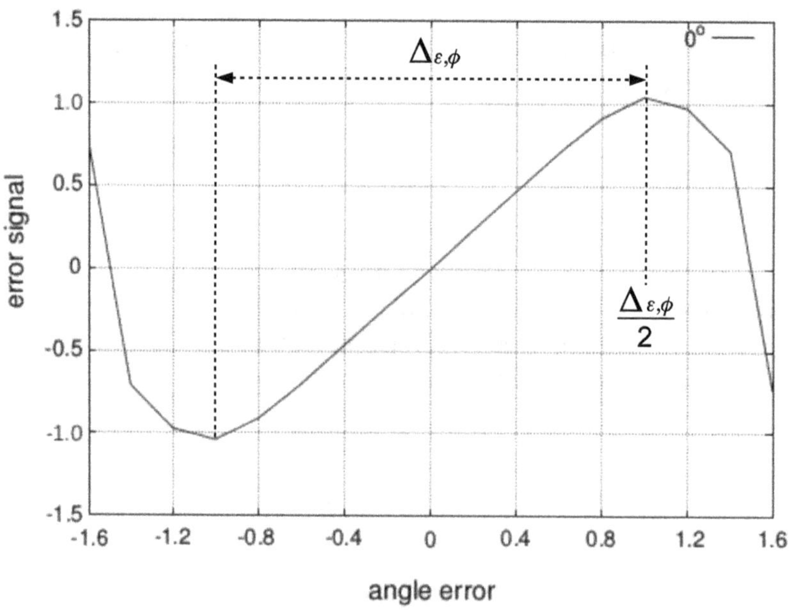

Figure 10.2 Direction-finding characteristics $S(\varepsilon)$ or $S(\varphi)$ of monopulse radar on working polarization $\alpha = 0°$.

It is quite obvious that the condition for disrupting automatic target tracking in angular coordinates (Figure 10.3) will be the expression

$$e(t) > \frac{1}{2}|\Delta_{\varepsilon,\varphi}|, \qquad \forall t \in [T_b, +\infty), \tag{10.4}$$

where T_b is the moment in time at which auto-tracking was disrupted, then for $t \geq T_b$ the error in auto-tracking the target in angular coordinates $e(T_b) = \pm\infty$ and hence in this case

$$C = \pm\infty \tag{10.5}$$

As follows from the form of the direction-finding characteristic on the working polarization shown in Figure 10.2, the "+" or "−" signs in (10.5) are determined by the sign of the error signal $S_{\varepsilon,\varphi}(t)$ at time $t \in [T_b - \delta, T_b] : \delta \to 0$, preceding the

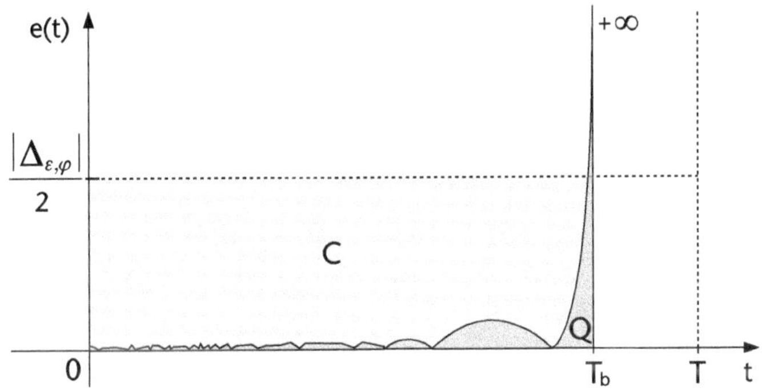

Figure 10.3 Quality criterion Q in the presence of automatic target tracking failures.

auto-tracking failure:

$$\begin{cases} \forall \varepsilon, \varphi > 0 \quad \exists\, S_{\varepsilon,\varphi}(t) > 0 \;\Rightarrow\; C = +\infty, \quad \forall t \in [T_b - \delta, T_b] : \delta \to 0 \\ \forall \varepsilon, \varphi < 0 \quad \exists\, S_{\varepsilon,\varphi}(t) < 0 \;\Rightarrow\; C = -\infty, \quad \forall t \in [T_b - \delta, T_b] : \delta \to 0 \end{cases} \tag{10.6}$$

For the case of stable target auto-tracking, when there were no auto-tracking disruptions during the time $t \in [0, T]$ (Figure 10.1), for the exact upper bound C of the set of possible values $\{Q\}$ can be written as

$$C = \frac{1}{2}\int_0^T \Delta_{\varepsilon,\varphi}\, dt = \frac{T|\Delta_{\varepsilon,\varphi}|}{2} \tag{10.7}$$

It is quite clear that in order to fulfill the condition (10.1), two conditions must be met:

1. It is necessary to correctly recognize the interference H_i from the set of disturbances $\{H\}$ of cardinality q;
2. It is enough to choose such a countermeasure Z_i that will be most effective against this interference H_i, provided that the minimum criterion for the quality of target tracking Q (10.1) is provided under the influence of the interference H_i.

These conditions represent a necessary and sufficient condition for ensuring high jamming immunity of any anti-aircraft missile system radar.

In Chapter 8, it was shown that it would be desirable that the feature space $\{X\}$, which is the union of the feature spaces corresponding to each interference H_i, would be split into q interference classes $H_i : \forall\, i = \overline{1, q}$ so that the feature space $\{X\}$

$$\{X\} = \bigcup_{i=1}^{q} X_i = X_1 \cup X_2 \cup \ldots X_{q-1} \cup X_q \tag{10.8}$$

would not be contradictory:

$$\begin{cases} X_i \cap X_j = \varnothing \quad \forall\, i \neq j \\ X_i \cap X_j \equiv X_i \quad \forall\, i = j, \text{ where } \forall i \wedge \forall j = \overline{1, q}. \end{cases} \tag{10.9}$$

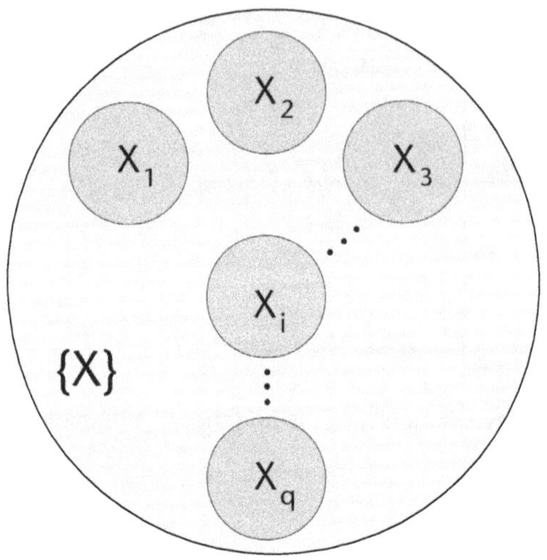

Figure 10.4 Illustration of attribute space {X}.

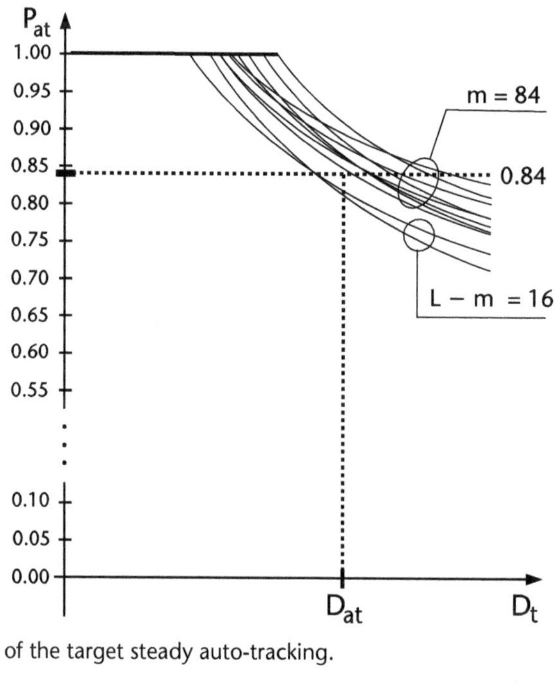

Figure 10.5 Range of the target steady auto-tracking.

In [2] it is proposed to distinguish between two ways of describing a system under the influence of a random process. The first method, called direct description, establishes a connection between the output and input random process in the form of functional dependencies represented by stochastic difference equations or differential equations. The second method, called indirect description, establishes a connection between the probabilistic characteristics at the input and output of the system.

The analysis problem , also called the direct problem, consists of obtaining an indirect description of the system on a known direct description of the system under consideration.

The inverse problem consists in obtaining a direct description of the system according to a given indirect description and is a synthesis problem, which in the general case does not have an unambiguous solution. Probabilistic characteristics can be obtained either on the basis of a series of similar full-scale (flight) experiments or as a result of computer simulation carried out on the basis of a mathematical model of the system verified at specific points. For example, when assessing the characteristics of the combat properties of a long-range air defense system, one of these characteristics is the probability of stable automatic target tracking in angular coordinates, when the missile's homing head is affected by interference created by the jammer on the sidelobes of the seeker's directional pattern from one point in space or interference created by a group of jammers from several points in space, for example, by the forces of several jammers located in one or several loitering zones located at distances exceeding the far border of the anti-aircraft missile system's engagement zone. In this regard, the maximum range radar-target: D_t, at which the probability of stable automatic tracking $P_{at}(D_t)$ of the target by angular coordinates are not less than a certain threshold.

Here the SINR approach can be applied when analyzing the accuracy of target tracking in conditions of interference along the sidelobes of a monopulse radar in one plane (ε for example). In this case, we can write:

$$SINR(\varepsilon_0, D_t) = \frac{F(\varepsilon_0) \times P_t(D_t)}{\sum\limits_{j=1}^{M} F(\varepsilon_j) \times P_j + N} \tag{10.10}$$

where $P_t(D_t)$ is the power of the signal reflected from the target, which is situated on D_t range, $F(\varepsilon_0)$ is the position of the target in the sum antenna radiation pattern, when $\varepsilon_0 = 0°$ and $F(\varepsilon_0) = max$. There $j = 1, \ldots, M$ is the number of jammers, $F(\varepsilon_j)$ is the position of the jth jammer in the directivity pattern of a monopulse radar, and N is the receiver noise power. In this case, the expression (10.11) will be the condition for stable target auto-tracking at a distance of D_t in angular coordinates in one plane (for example, ε):

$$A_{SINR}(D_t) = \lg\left[SINR(\varepsilon_0, D_t)\right] \geq 10 \text{ dB}, \Rightarrow m = m + 1, \quad \forall m \in [1, \ldots, L] \tag{10.11}$$

Accordingly, the condition for disrupting automatic target tracking on angular coordinates at a distance of D_t will be the expression (10.12):

$$\overline{A_{SINR}}(D_t) = \lg\left[SINR(\varepsilon_0, D_t)\right] < 10 \text{ dB}, \Rightarrow m = m, \quad \forall m \in [1, \ldots, L] \tag{10.12}$$

where random events $A_{SINR}(D_t)$ and $\overline{A_{SINR}(D_t)}$, given on the space of elementary outcomes $\Omega = \{1, 0\}$ form a complete group of events such that:

$$P(A_{SINR}(D_t)) + P(\overline{A_{SINR}(D_t)}) = 1. \tag{10.13}$$

In this case, the probability of stable tracking of the target $P_{at}(D_t)$ located at a distance of D_t is calculated as

$$P_{at}\left(A_{SINR}(D_t)\right) = \left.\frac{m}{L}\right|_{D_t} = P_{at}(D_t), \tag{10.14}$$

where m is the number of outcomes when the signal-to-noise ratio is equal to or exceeds the 10 dB level and the condition (10.11) is satisfied, and L is the total number of realizations that must belong to the numeric a series of natural numbers $L \in [50, 100, 150, \ldots, i = i - 1 + 50, \ldots)$, in which $i, L \in \mathbb{N}$. Such a series of numbers provides a threshold equal to 0.84 in the expression (10.14).

If the condition (10.11) is fulfilled, the number of outcomes $A_{SINR}(D_t)$ is incremented by 1: $m = m + 1$, and if not fulfilled (10.11) there is no increase in the number of such outcomes, and therefore $m = m$. As a result, the range of stable auto-tracking of the target D_{at} will be such a distance between the locator and the target, at which $P_{at}(D_t)$ will be equal to or greater than a certain level, for example, the level 0.84:

$$D_{at} = \left. D_t \right|_{P_{at}(D_t) \geq 0.84} \tag{10.15}$$

Then the condition of stable auto-tracking along the angular coordinates of the target located at the range D_t can be written as

$$P_{at}(D_t) \geq 0.84, \quad \forall D_t \in [0, D_{at}], \tag{10.16}$$

and the condition for the loss of the target auto-tracking will have the form:

$$P_{at}(D_t) < 0.84, \quad \forall D_t \in \,]D_{at}, D_{t\,max}] \tag{10.17}$$

Here D_{at} is the maximum range at which the condition (10.16) is already satisfied, and $D_{t\,max}$ is the maximum target detection range with an effective reflecting surface σ_t in the absence of interference on the monopulse radar in general and on the homing head of the anti-aircraft guided missile in particular.

The problem of interference recognition can be formulated as follows: on the basis of some features X_i characteristic of each ith interference H_i from a given set of interference $\{H\}$ of power q:

(a) It is necessary to correctly recognize which particular interference H_i from the given set of interference $\{H\}$ affects the protected radar;

(b) It is enough to enable such countermeasures for the detected interference H_i, which minimize or neutralize its impact on the monopulse radar and can inflict unacceptable damage on the enemy.

The approach described was first implemented by me in 1988, when I served as a test engineer at the Sary-Shagan State Missile Defense Research Site with the rank of captain of the Soviet Army. The center of the Missile Site existed the city of Priozersk, which was located on the shore of Lake Balkhash. The peculiarity of this lake is that it consists of two approximately equal halves: in one half (in the west) there is freshwater, and in the other half (in the east) there is salt (sea) water. Moreover, these two halves essentially do not mix. Despite the fact that the city of Priozersk was located in the fresh part of Lake Balkhash, this freshwater contained a huge amount of different types of salts. For example, in order to make a 200g glass of tea a little sweet, it was necessary to dissolve 5 or 7 teaspoons of sugar in this amount of water, which was 40–56g or 20–28% of water weight in a glass. Due to the fact that this area had severe climatic conditions: in summer the temperature reached +45°C near the lake and much more than +45°C in the desert, where the test sites were located. In winter, temperatures dropped to less than −25°C to −35°C and all this was accompanied by strong winds. The winds were so strong that one frosty winter morning I went to work and suddenly I felt a blow on my cheek like a knife, after which blood came out. It turned out that the wind tore off a leaf from the tree that had remained on it since fall, which cut my cheek. I determined this by the fresh blood on the fallen leaf, which the wind drove away from me. Due to the low temperatures in winter, houses in the city and at test sites were heated with hot-water batteries. Due to the high concentration of salt in the freshwater, the heating pipes became unusable very quickly. They were clogged with salt so that 3–5 years after a major overhaul in an apartment building with the replacement of all cold and hot water pipes in a pipe with a diameter of 20 mm, a nail with a length of 100 mm and a diameter of 4 mm could barely pass through. For this reason, it was cold in the apartments in winter and there were constant accidents with the heating and water supply. To facilitate welding and other repair work, all water supply routes were above ground (i.e., passed along metal supports located above the Earth's surface). To service this entire economy, there was a special military unit at the test site, the main task of which was to ensure uninterrupted supply of cold and hot water to consumers, as well as to ensure uninterrupted water disposal (i.e., to ensure the correct operation of the sewage system). At that time I lived in an apartment on the second floor of a five-story panel building, located in Priozersk on the Boulevard of the Soviet Army. Here, in addition to all the difficulties listed above, one more was added: the apartment on the ground floor began to flood with sewage, which was waste products of human activity, with surprising frequency. My apartment was on the second floor along the same sewer pipe in the same building.

To save my neighbors, and myself too, I began to think about two questions:

1. What exactly is the cause of these periodic sewerage accidents?
2. How can I fight this obstacle that interferes with our normal lives?

Conversations with neighbors living above were unsuccessful. They all swore that everything was fine with them and that they were not draining anything into the toilet.

The answer to the first question was found after another accident in the sewer. I asked the plumbing soldier who came to do the repair to take me with him to fix the accident. We went down to the basement of the house, opened the revision window in the cast-iron sewer pipe, and found a towel, which was acting like a cork and clogging the entire volume of the sewer pipe (Figure 10.6). We drained

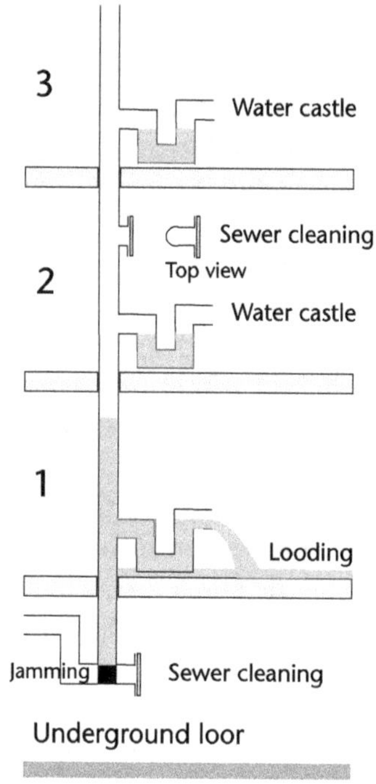

Figure 10.6 Standart situation.

the volume of sewage that had accumulated in the sewer pipe. It became clear that someone was pouring the washed things into the toilet after rinsing. There were no automatic washing machines with dryers in the USSR at that time, at least not in Priozersk. People washed with soap or detergent and rinsed the washed items, usually in a basin. So the reason was established.

The answer to the second question was found based on the fact that I classified this interference and determined the principle of its action. Using the sufficient condition (b) I realized that it is possible to neutralize the effect of this interference with the organization of unacceptable damage to an unknown enemy only if you put some kind of trap, in this above the sewage drainage point of the protected apartment, which would delay all objects inadvertently poured into the toilet, and the sewage drains would pass through it without delay (Figure 10.7). At least technically, the problem was solved by installing three steel pins with a diameter of 8 mm and a length of 170 mm into the hole cover for cleaning the sewer. This length of the pins was determined by the inner diameter of the cast-iron pipe of the sewer riser, which was equal to 150 mm. The flat cast-iron cover was drilled with an electric drill at a minimum of ≤ 400 rpm with the addition of coolant. Next, the steel bars were threaded $M8$ and the three bars were inserted into the holes in the riser cover so that for each metal bar the cover was between two rubber gaskets, two metal washers and two $M8$ nuts. Thus, wastewater and waste products passed without delay, and objects larger than the pitch of the metal rods $d = (150 - 24)$ mm/4 $= 31.5$ mm were retained on the surface of the lattice created by the three metal rods.

With this technical solution, a traffic jam should form above the protected apartment.

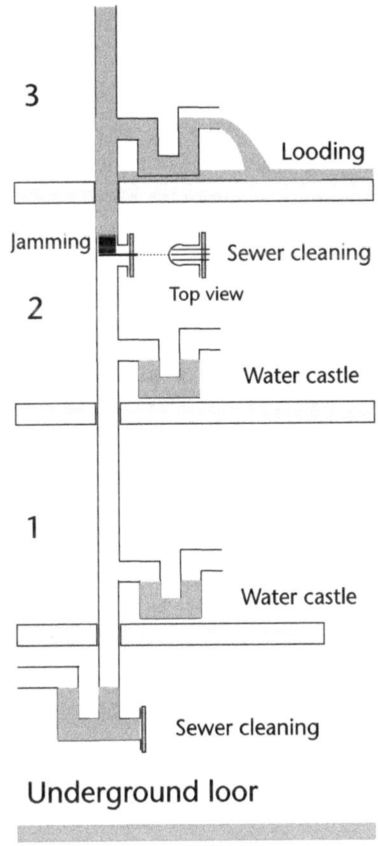

Figure 10.7 Option with protection.

After installing the metal rods to the cover, there was not long to wait. Eventually, my neighbor from the third floor knocked on the door of my apartment and asked if there was flooding by sewerage in my apartment. I answered that everything is dry in my apartment and the sewage system is working fine. I even took her to the bathroom and showed the toilet to work normally. The neighbor was completely perplexed and called the emergency team of plumbers. The plumber soldier who arrived at the scene could not understand why it was dry on the first floor, while the third floor was filled with sewage (Figure 10.7). Then the plumber soldier and I opened the revision window in my bathroom and pulled out a dress that my neighbor from the third floor wore.

After the problem was solved, I took this dress to my neighbor. She had to admit it was her dress. After this incident, the sewage flooding stopped.

It is quite obvious that obtaining a technical solution that will ensure the fulfillment in real time of both the necessary (a) and sufficient conditions (b) will make it possible to create an anti-aircraft guided missile with artificial intelligence, when the missile independently analyzes the interference environment in which it is located and solves the following tasks in real time:

1. The missile independently detects and correctly classifies the H_i interference from a given set of hindrances {H} (i.e., the missile independently solves the problem of recognizing interference);

2. The missile independently decides which technical means of countering Z_i must be switched on in order to minimize, and in the ideal case reduce to zero, the effect of the detected interference H_i on the radar in question.

It is quite clear that in practice these tasks are solved by the homing head and the onboard computer, which should be a multiprocessor computer complex. In this computer complex, ideally, each process should have its own processor. So the problems of controlling the position of the missile in the space R^3 are solved by an autopilot created on one processor or, ideally, on two processors for simultaneous control in two orthogonal planes. This is necessary in order to gain time for carrying out the necessary calculations by excluding writing to memory and reading from memory the data necessary for processing, since the use of switching one processor between two missile flight control processes leads to a delay in the response of the system to input influences. For rapidly changing input processes, this time delay can become critical.

Thus, the problem of correct recognition of the interference affecting the radar is to choose the correct solution based on the feature sample $\mathbf{x} = (x_1, x_2, \ldots, x_n)$, which has a fixed size determined by the number of measuring channels, corresponding to each ith interference feature. The feature vector $\mathbf{x}$ is obtained as a result of observing a specific implementation of the investigated random process.

The result of solving the interference recognition problem is the determination of such a means of combating Z_i with the detected interference H_i, which will ensure the achievement of the minimum of the target tracking error function on angular coordinates $e(t)$ over the entire time interval of the missile guidance to the target $[0, T]$; that is, it will provide the minimum quality criterion (10.1).

10.1 Testing a Simple Hypothesis Against a Simple Alternative

The problem here is that for a given selection of features $\mathbf{x}$, such that it is the correct subset of features $\{X\} : \mathbf{x} = (x_1, x_2, \ldots, x_n) \supset \{X\}$, accept or reject hypothesis A_0, according to which the likelihood function of the sample is set equal to $f(bfx|A_0)$. In this alternative A_1 is assigned some value $f(\mathbf{x}|A_1)$. In this case, it is assumed that:

- A_0 characterizes the situation, when there is no interference effect on the monopulse radar and the missile seeker is aiming at the target according to the signal reflected from the target (i.e., the seeker works on a clean target);
- A_1 is a monopulse radar that operates in conditions of interference that affect the quality of its functioning;
- n is the dimension of the space in which the sample $\mathbf{x}$ of features $\{X\}$ is defined (i.e., it is the number of features involved in making a decision).

For a fixed size of the feature sample $\mathbf{x}$, the observation space or sample space is the n-dimensional Euclidean space $\mathbf{X}^n$, on which two likelihood functions are defined $f(\mathbf{x}|A_0)$ and $f(\mathbf{x}|A_1)$, $\mathbf{x} \in \mathbf{X}^n$. In addition, the probabilities of the hypothesis and the alternative are assumed to form a complete group of events:

$$P\{A_0\} + P\{A_1\} = 1, \tag{10.18}$$

whence, for the probability of the implementation of the accepted hypothesis, we can write

$$p_0 - P\{A_0\} = 1 - P\{A_1\} = 1 - p_1 \tag{10.19}$$

Here the probabilities of realization of the corresponding hypotheses $P\{A_0\}$ and $P\{A_1\}$ are assumed to be known apriori. The solution space Γ consists of two elements: $\Gamma = \{\gamma_0, \gamma_1\}$, where γ_0 is the decision to accept the hypothesis A_0, and γ_1 is the decision to accept the alternative A_1 (i.e., this is a decision to reject hypothesis A_0.)

This implies the division of the feature space $\mathbf{X}^n$ into two disjoint areas: $\mathbf{X}_0 \cap \mathbf{X}_1 = \varnothing$, where $\varnothing$ is an empty set, and $\mathbf{X}_0 \cup \mathbf{X}_1 = R^n$, and the closures $[\mathbf{X}_0]$ and $[\mathbf{X}_1]$ do not contain isolated points. In this case, if the observed sample $\mathbf{x}$ fell into the area $\mathbf{X}_0$ (i.e., when $\mathbf{x} \in \mathbf{X}_0$ and the sample $\mathbf{X}_0$ is realized then the decision γ_0 is made that the monopulse radar is not affected by interference). Otherwise, the decision γ_1 is made that some kind of interference affects the monopulse radar. In this case, the loss matrix will take the form

$$R = \begin{Vmatrix} R_{00} & R_{01} \\ R_{10} & R_{11} \end{Vmatrix}, \qquad \text{where} \quad R_{01} > R_{00} \geq 0, \quad R_{10} > R_{11} \geq 0. \tag{10.20}$$

On the main diagonal of this matrix, the elements R_{00} and R_{11} are the prices of the correct decision (risk), and the elements R_{01} and R_{10} represent the risk of making a mistake of the first kind $\gamma_1 \cap A_0$ and second kind $\gamma_0 \cap A_1$, respectively. Since, when testing a simple hypothesis A_0 against a simple alternative A_1, the likelihood functions $f(\mathbf{x}|A_0)$ and $f(\mathbf{x}|A_1)$ are completely known, the synthesis of the decision-making algorithm by any criterion, it is a synthesis under conditions of complete apriori information.

It is known [2] that the probabilities of errors of the first α and the second kind β can be represented as

$$\alpha = P\{\gamma_1|A_0\} = P\{\mathbf{x} \in \mathbf{X}_1|A_0\} = \int_{X_1} f(\mathbf{x}|A_0)dx \tag{10.21}$$

$$\beta = P\{\gamma_0|A_1\} = P\{\mathbf{x} \in \mathbf{X}_0|A_1\} = \int_{X_0} f(\mathbf{x}|A_1)dx \tag{10.22}$$

Here α is the probability of rejecting the correct hypothesis H_0, which is the level of significance, and the value $(1 - \beta)$ is the probability of rejecting a false hypothesis, which is the power of the decision rule. The prior probabilities of the solutions γ_0 and γ_1 will have the form

$$P\{\gamma_0\} = p_0 P\{\gamma_0|A_0\} + p_1 P\{\gamma_0|A_1\} = p_0(1 - \alpha) + p_1\beta \tag{10.23}$$

$$P\{\gamma_1\} = p_0 P\{\gamma_1|A_0\} + p_1 P\{\gamma_1|A_1\} = p_0\alpha + p_1(1-\beta) \tag{10.24}$$

In (10.23) and (10.24), the products $p_1\beta$ and $p_0\alpha$ are apriori errors probabilities, and the products $p_0(1-\alpha)$ and $p_1(1-\beta)$ are apriori probabilities of correct decisions.

For a given sample dimension **x**, the probabilities α and β cannot be made arbitrarily small at the same time [2]. To develop the decision rule, it is necessary to be guided by the selected quality criterion. It is known that the ratio of the likelihood functions is called the likelihood ratio:

$$\ell(\mathbf{x}) = \ell(x_1, x_2, \ldots, x_n) = \frac{f((x_1, x_2, \ldots, x_n)|A_1)}{f((x_1, x_2, \ldots, x_n)|A_0)} \tag{10.25}$$

$\ell(\mathbf{x})$ is a nonnegative random variable obtained by a functional transformation that maps points of the n-dimensional space of the sample $\mathbf{X^n}$ to the real semiaxis. So, for example, with the Bayesian algorithm for testing a simple hypothesis against a simple alternative, which has the form:

$$\begin{cases} \gamma_0: \ \ell(\mathbf{x}) = \dfrac{f(\mathbf{x}|A_1)}{f(\mathbf{x}|A_0)} < C_b \\ \gamma_1: \ \ell(\mathbf{x}) = \dfrac{f(\mathbf{x}|A_1)}{f(\mathbf{x}|A_0)} \geq C_b, \end{cases} \tag{10.26}$$

where

$$C_b = \frac{R_{01} - R_{00}}{R_{10} - R_{11}} \times \frac{P_0}{P_1} \quad \text{- a certain threshold when} \tag{10.27}$$

the value of $\ell(\mathbf{x})$ is compared with the threshold C_b and if $\ell(\mathbf{x})$ is less than the threshold C_b, then the decision γ_0 and the decision γ_1 is made if the value of $\ell(\mathbf{x})$ is greater than or equal to the threshold C_b (Figure 10.8). Thus, to make a decision, it is sufficient to use the value of one random variable—the statistics of the likelihood ratio $\ell(\mathbf{x})$, and not the values of each element of the sample $x_1, x_2, ..., x_n$ separately. Therefore, the likelihood ratio carries all statistical information about the hypotheses being tested and this statistic $\ell(\mathbf{x})$ is a sufficient. Using the sufficient statistics of the likelihood ratio $\ell(\mathbf{x})$ leads to a mapping of the sample n-dimensional space $\mathbf{X}^n$ onto the real positive semiaxis. The boundary dividing the space $\mathbf{X}^n$ into subspaces $\mathbf{X}_0$ and $\mathbf{X}_1$ is mapped to the point C_b on the axis $\ell(\mathbf{x}) \geq 0$. Moreover, any monotonic transformation $\psi\{\ell(x_1, x_2, \ldots, x_n)\}$ of a sufficient likelihood ratio statistic is also a sufficient statistic [3, 4]. This logarithm is often used as such a transformation. Replacing the likelihood ratio by its logarithm is always advisable when factoring the likelihood function. If the features $\{x_i\}$ that make up the vector $\mathbf{x} = (x_1, x_2, \ldots, x_n)^T$ are independent, then

$$\ln[\ell(\mathbf{x})] = \sum_{i=1}^{n} \ln[\ell(x_i)] = \sum_{i=1}^{n} \ln\left[\frac{f(x_i|A_1)}{f(x_i|A_0)}\right] \tag{10.28}$$

In this case, if conditions are met of the central limit theorem, the sum of independent random variables is asymptotically normal [3,4].

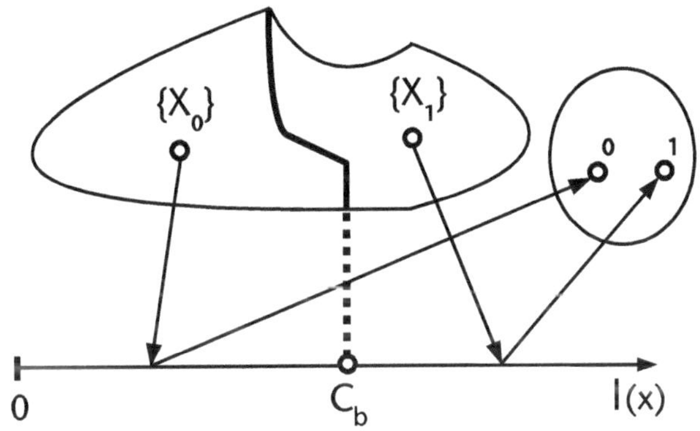

Figure 10.8 Sufficient statistics of likelihood ratio $\ell(\mathbf{x})$.

It is known that the most optimal is the Bayesian decision rule (10.26) [2]; however, all its components are assumed to be known apriori, which often does not correspond to reality and causes difficulties in the practical implementation of this algorithm. The minimax algorithm and the algorithm of maximum a posteriori information require less initial data, but the most acceptable from the point of view of technical implementation are the maximum likelihood algorithms and algorithms that are optimal according to the Neumann-Pearson criterion. In this case, the loss matrix R (10.20), the prior probabilities of the hypothesis A_0, and the alternative A_1 may be unknown. The maximum likelihood algorithm coincides with the optimal algorithm according to the criterion of the maximum posterior probability if the hypothesis and the alternative are equally probable [2]. The optimal algorithm in the sense of the Neumann-Pearson theorem provides the minimum possible probability β of errors of the second kind, provided that the probability of an error of the first kind α does not exceed a given value $o(\alpha)$ [2,5].

An algorithm for checking the sufficient statistics of the likelihood ratio $\ell(\mathbf{x})$ according to the Neumann-Pearson criterion, according to the general decision-making procedure for $\ell(\mathbf{x})$, will be written as

$$\ell(\mathbf{x}) \ \leq \ C(\alpha) = C \tag{10.29}$$

The threshold C is found from the boundary condition:

$$P\{\ell(\mathbf{x})) \ \leq \ C|A_0\} = \alpha \tag{10.30}$$

or

$$P\{\ln \ell(\mathbf{x})) \ \leq \ \ln C|A_0\} = \alpha \tag{10.31}$$

Passing to the probability density function $f(\ln \ell(\mathbf{x}))$, by means of differentiation (10.31), we can write

$$\alpha = \int\limits_{\ln C(\alpha)}^{\infty} f[\ln \ell(\mathbf{x})] d \ln(\mathbf{x}) \tag{10.32}$$

10.2 Testing a Complex Hypothesis Against a Simple Alternative

A complex hypothesis can be considered as a finite, countable, or continuous hypotheses set [2]. Since by virtue of (10.8) and (10.9) the alphabet of classes is a countable set of cardinality q, the alphabet of hypotheses will also be a countable set of cardinality q. As an alternative to $\aleph_0$, any hypothesis is considered for which the estimate of the vector $\hat{\mathbf{x}}$, obtained during the analysis of the sample $\mathbf{x}$, belongs to the space $\aleph$ such that the relation

$$\aleph = R^n \backslash A \tag{10.33}$$

in this case, the set $\aleph$ is closed relatively the operations $\cup$, $\backslash$ and $\cap$:

$$A \cap \aleph = \varnothing \tag{10.34}$$

and has infinite power. Therefore, (10.33) and (10.34) define some set Ω: $A \cup \aleph$ with σ-algebra ν of its subsets: $A \subseteq \Omega$ and $\aleph \subseteq \Omega$. By defining on Ω a positive normalized norm P such that $P(\Omega) = 1$, we can form the probability space (Ω, ν, P). It is necessary to determine the probability measures for the entire σ-algebra ν of subsets of the set Ω using the available sample of observations $\mathbf{x}$ in a given probability space. As a probability measure, we will consider the probability distribution (probability density function) of the random variable x_i, which is a measurable function $x_i = x_i(\omega)$, $\omega \in \Omega$ on the probability space (Ω, ν, P). Assuming that each feature x_i, which by virtue of (10.38) is an element of the vector $\mathbf{x}_j$, is measured by the ith meter with an error distributed according to the normal law with a known variance $\sigma_{x_i}^2$:

$$x_{ij} \sim N(Mx_i, \sigma_{x_i}^2), \quad i = \overline{1, n}, \quad j = \overline{1, m} \tag{10.35}$$

therefore, we can write

$$x_{ij} = Mx_i + \sigma_{x_i} \xi_{ij}, \tag{10.36}$$

where $\xi_{ij} \sim N(0, 1)$ is a normally distributed random variable with zero mathematical expectation and unit variance, and $\xi_{ij} \in R^1$; Mx_i is the unknown mathematical expectation of the ith feature.

The vector $\mathbf{x}_j$, formed during the jth measurement cycle, can be represented as

$$\mathbf{x}_j = MX + \sigma_x \bullet \|\xi_{ij}\|, \tag{10.37}$$

where MX is the unknown mathematical expectation of the vector

$$\mathbf{x}_j = (x_1, x_2, \ldots, x_n)^T \in R^n, \tag{10.38}$$

$\sigma_x = \sigma_{x_1}, \sigma_{x_2}, \ldots, \sigma_{x_n})^T$ is the vector of known standard deviations of the meters of each feature; $\|\xi_{ij}\|$ is a matrix of normally distributed random variables with zero mathematical expectation and unit variance, such that:

$$\|\xi_{ij}\| = (\xi_{1j}, \xi_{2j}, \ldots, \xi_{nj})^T \times (\xi_{1j}, \xi_{2j}, \ldots, \xi_{nj}).$$

Assuming that the channels measuring the ith components of the vector $\mathbf{x}_j$ are independent, the matrix $\|\xi_{ij}\|$ will be diagonal.

10.3 Determination of Estimates of the Maximum Similarities

The assumption about the normality of the error distribution law is based on the fact that a random error is usually caused by a number of reasons and therefore, whatever distribution laws its individual components have, with the same order of their smallness, the resulting error distribution law will be close to normal [2, 5]. Under these conditions, the distribution density of any element x_i of the vector $\mathbf{x}_j$ will be written as [2]

$$f(x_i) = \frac{1}{\sigma_i\sqrt{2\pi}} \times \exp\left\{\frac{-(x_i - Mx_i)^2}{2\sigma_i^2}\right\} \tag{10.39}$$

where Mx_i is the unknown mathematical expectation of the element x_i, and σ_i is a measure of the measurement accuracy of the ith element of the vector $\mathbf{x}_j$, which is the standard deviation of the x_i measurement results from the mathematical expectation Mx_i.

The likelihood function is represented by the distribution density of a random vector (10.38) and is considered for a fixed x_i as a function of the parameter $\omega \in \Omega$ [2]:

$$L(x_{i1}, x_{i2}, ..., x_{im}|\omega_i) = (2\pi)^{-\frac{m}{2}}\sigma_i^{-m} \exp\left\{-\left(\frac{1}{2\sigma_i^2}\right)\sum_{j=1}^{m}(x_{ij} - \omega_i)^2\right\} \tag{10.40}$$

here, as a parametric set Ω a set of certain hypotheses is considered. For example, that ω_i is the mathematical expectation of the value of the ith element of the vector $\mathbf{x}$ for j measurement cycles.

According to the maximum likelihood method [2, 5], the optimal estimate of the maximum likelihood of the parameter ω_i is the point ω_i^* of the parametric set Ω, for which the following condition is satisfied:

$$\omega_i^* = \arg\sup_{\omega_i} L(x_{i1}, x_{i2}, \ldots, x_{im}|\omega_i) \tag{10.41}$$

After taking the logarithm (10.40), the condition (10.41) will be fulfilled in the form

$$\omega_i^* = \arg\inf\left(\sum_{j=1}^{m}(x_{ij} - \omega_i)^2\right) \tag{10.42}$$

The condition (10.42) is, as is known, the formulation of the least squares criterion [2]. Therefore, under the normal distribution of the random variables x_i, the estimates the least squares and maximum likelihoods are the same. Introducing the notation: $\sum_{j=1}^{m}(x_{ij}-\omega_i^*)^2 = Q$, the value of the estimate is ω_i^* is determined from the condition

$$\frac{\partial Q}{\partial \omega_i^*} = -2\sum_{j=1}^{m}(x_{ij}-\omega_i^*) = 0, \tag{10.43}$$

whence follows

$$\omega_i^* = \frac{1}{m}\sum_{j=1}^{m} x_{ij} \tag{10.44}$$

Thus, the estimate of the maximum likelihood of the mathematical expectation of the random variable x_i is its average value $\overline{x}_i$:

$$\overline{x}_i = \frac{1}{m}\sum_{j=1}^{m} x_{ij} \tag{10.45}$$

The estimation of the variance $\hat{D}_i^*$ of the random variable x_i by the maximum likelihood method is determined from the condition

$$\hat{D}_i^* = \arg\sup L(x_{i1}, x_{i2}, \ldots, x_{im} \,|\, \overline{x}_i, D_i),\; D_i = \sigma_i^2 \tag{10.46}$$

After taking the logarithm (10.41), the extremum of the likelihood function is determined from the condition

$$\frac{\partial \ln L(x_{i1}, x_{i2}, \ldots, x_{im})}{\partial D_i} = 0, \tag{10.47}$$

when

$$\hat{D}_i^* = \frac{1}{m}\sum_{j=1}^{m}(x_{ij} - Mx_i)^2 \tag{10.48}$$

Since the true value of Mx_i is unknown in this case, then using its estimate $\overline{x}_i$, the unbiased sample variance estimate will have the form

$$D_i^* = \left(\frac{1}{m-1}\right)\sum_{j=1}^{m}(x_{ij} - \overline{x}_i)^2 \tag{10.49}$$

10.4 Interference Recognition Procedure

In the case of interference recognition, a seeker is considered in which for each interference $H_i \in \{H\}, \forall\, i = 0, \ldots, n$ there is a channel for detecting this interference. Due to the fact that the set of interferences specified in the terms of reference for the development of an anti-aircraft missile system is a countable set of power n, hence the alphabet of hypotheses $\{A\}$ will also be a countable set of power n. Let's pretend that:

- A_0 is a hypothesis characterizing the situation when the seeker has detected a signal reflected from the target and the process of homing the missile at the target is carried out on a signal in working polarization;
- A_1 is a hypothesis corresponding to a situation when a polarization interference affects the seeker;
- A_2 is a hypothesis representing the case of the effect of noise interference along the sidelobes of the directional pattern of the receiving antenna of the homing head;
- A_3 is the hypothesis corresponding to the case of impact on the homing head of flickering interference, which is carried out from two points in space;
- A_4 is a hypothesis that characterizes the operation of the seeker under the influence of interference reflected from the ground or water surface.

In long-range anti-aircraft missile systems such as the S-200D air defense missile system, the missile homing head detects a signal from a target or a jamming signal while the missile is on the launcher and receives the coordinates of the target $(\varepsilon_t, \varphi_t)$ from the target illumination radar; that is, it is oriented in the R^3 space by external target designation. In multi-channel long-range systems such as the S-400 air defense missile system, the missile launches vertically at the command of a multifunctional radar station located on the Earth's surface. After the launch, the missile is displayed by the commands of the ground radar station on the reference trajectory. At this time, the missile seeker does not work. Guidance of the missile on the calculated kinematic trajectory is carried out by commands $\lambda_\varepsilon = \varepsilon_t - \varepsilon_m$, $\lambda_\varphi = \varphi_t - \varphi_m$. Before turning on the homing mode on the missile (active mode AM or semiactive mode SAM), the target spatial coordinates $(\varepsilon_{mt}, \varphi_{mt})$ are transmitted on the missile board via the command radio line radar-missile. The homing head, using obtained target coordinates, searches for a target and captures it for auto-tracking. The target coordinates $(\varepsilon_{mt}, \varphi_{mt})$ transferred to the missile are recalculated by the ground radar for the receiving base of the seeker antenna $\{X_m, 0, Z_m, Y_m\}$, determined by the spatial position of its phase center 0 (Figure 10.9). Here the point C is the center of mass of the missile. At the point A, which is the moment of the beginning of the missile guidance along the trajectory AD, and the point D is the point of detonation of the missile warhead. The point C is spaced from the phase center of the seeker antenna at a distance of δX_{m}. So for the missile S-400 series 40N6E, having a length of 7.5 m [6] at the start $\delta X_{\text{m}} = 3.75\text{m}$. After starting the main engine of the missile and as burnout fuel δX_{m} will decrease. Taking into account the mass of the seeker equal to $\approx$40 kg and the mass of the missile with fully burned fuel equal to $m_0 = 636$ kg, (see Chapter 3, Section 3.2), for the position of the center of gravity C starting from the engine nozzle, we can write

$$x_c = \frac{m_1 x_1 + m_2 x_2}{m_1 + m_2} = 5.743\text{m} \tag{10.50}$$

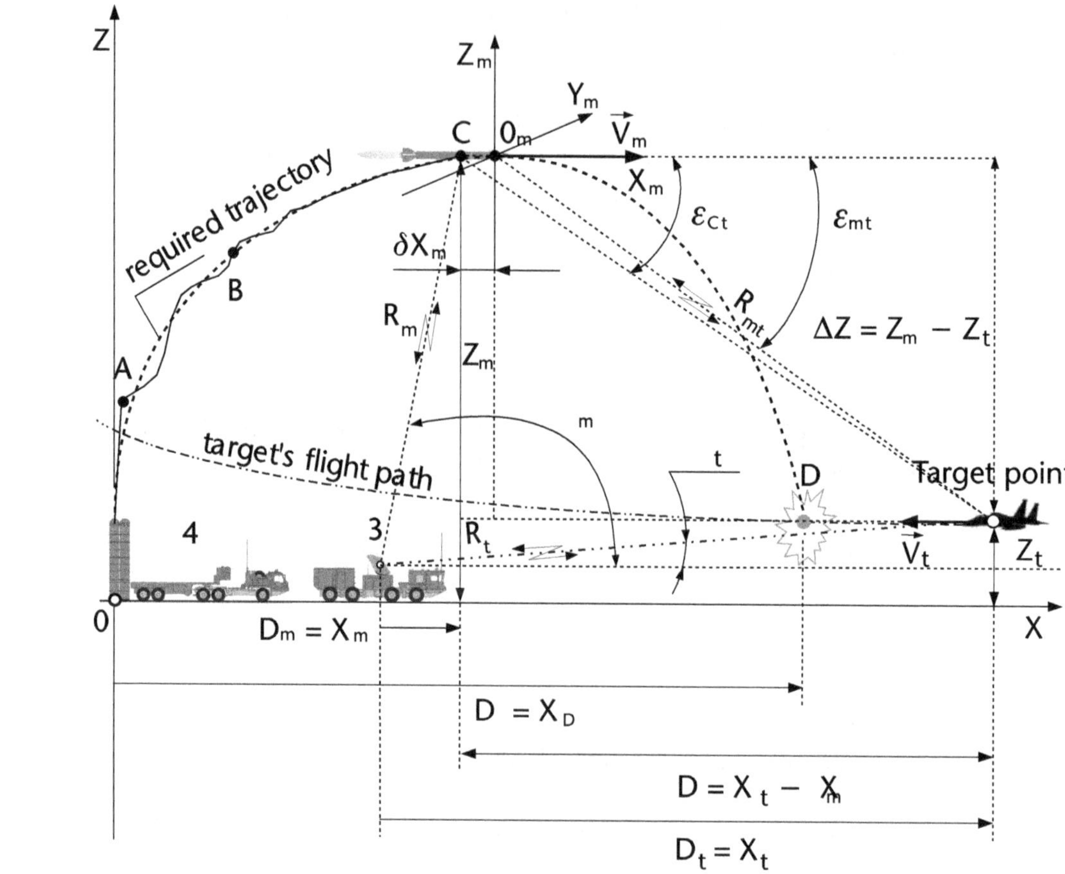

Figure 10.9 Operation of long-range air defense systems against low-altitude targets.

Here $x_1 = 3.75\text{m}, m_1 = 298$ kg, $x_2 = 7.5\text{m}$, and $m_2 = 338$ kg, provided $m1 + m2 = m_0 = 636$ kg. Taking into account the longitudinal size of the fairing $x_R \approx 1\text{m}$, the expression for δX_m becomes

$$\delta X_m = 7,5\text{m} - x_C - x_R = 0.75\text{m}, \tag{10.51}$$

Therefore, for the radial range missile-target $R_{mt} > 100\text{m}$, at which the angular error caused by the displacement of the phase center of the seeker antenna from the center of mass of the missile $\delta\varepsilon_t < 0.429°$, we can write

$$\delta X_m \approx 0, \quad \Rightarrow \quad \delta\varepsilon_t = \varepsilon_{mt} - \varepsilon_{ct} \approx 0° \tag{10.52}$$

Assuming that the target is steadily accompanied by a multifunctional radar, and on the missile after the launch the transmitter has successfully turned on according to which the radar determines the missile spatial coordinates. In a computer located on the ground, in a multifunctional radar, for the elevation plane $X0Z$ located along the azimuth φ_t, coordinates are first calculated that characterize the relative position of the missile and the target based on the known ranges from the multifunctional radar to the R_m missile and to the R_t target, as well as from the known elevation angles of the missile ε_m and the target ε_t:

$$\begin{aligned} \Delta Z = Z_t - Z_m &= R_t \sin \varepsilon_t - R_m \sin \varepsilon_m \\ \Delta D = X_t - X_m &= R_t \cos \varepsilon_t - R_m \cos \varepsilon_m - \delta X_m \\ R_{mt} &= \sqrt{\Delta Z^2 + \Delta D^2} \end{aligned} \tag{10.53}$$

As a result, the target angular coordinates calculated in this radar can be transmitted from the multifunctional radar to the missile in every ith time quantum:

$$\begin{cases} \varepsilon_{mt} = \arcsin\left[\dfrac{\Delta Z}{R_{mt}}\right] \\ \varphi_{mt} = \varphi_t \end{cases} \tag{10.54}$$

Having received the target coordinates ε_{mt} and φ_{mt} from the multifunctional radar, the missile homing head goes into the target search mode by the conical scanning mode is turned on. The task of this mode is to detect the target in a cone formed due to the rotation of the deflected main beam of the sum directional pattern of the seeker antenna at an angle β from the normal of the seeker antenna. As shown earlier (see (3.32) and (3.34)), the optimal value of β_{opt} should be chosen from a compromise between the quality of tracking in angular coordinates and in range, which minimizes the effect of thermal noise of the seeker receiver:

$$\beta_{opt} = 0.39\Theta_B \tag{10.55}$$

Here Θ_B is the width of the scanning beam at half power (at the level -3 dB). Taking into account the fact that the search for a target in the R^3 space is carried out according to the coordinates (ε_{mt}, φ_{mt}), which is given to the homing head by a ground radar, which is already accompanying a certain target. Based on this, in the search volume V_{sv}

$$V_{sv} = \frac{R_{mt}}{3} \pi B_{opt}^2 \tag{10.56}$$

there is either a net target or a jammer aircraft, and therefore, the probability of one of the events occurring is equal to one.

10.4.1 Estimation of the Required Range of a Hypersonic Target Detection by the Missile Homing Head when Firing over the Horizon

Consider the most dangerous variant of a raid on an anti-aircraft missile system, carried out by a hypersonic cruise missile such as the X-51 Waverider, moving at a maximum speed $V_t = 8\text{M} = 2396$ m/s at a minimum height, for example $h_t = 500\text{m} = 0.5$ km Figure 10.10.

The line-of-sight range of such a target by means of a ground multifunctional radar R_{GMR} can be estimated by the expression

$$R_{GMR} = \sqrt{(R_E + h_t)^2 - R_E^2} = 79.82 \text{ km}, \tag{10.57}$$

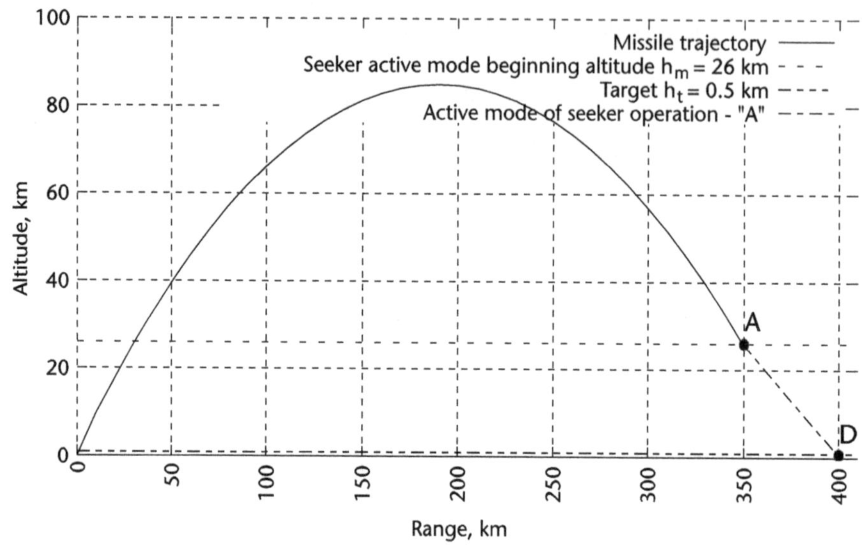

Figure 10.10 Air defense system operation against hypersonic targets.

provided that $R_E = 6371$ km is the Earth radius. It is quite understandable that in this case the ground-based multifunctional radar of a long-range anti-aircraft missile system does not see this target on its own. Therefore, when shooting beyond the horizon, combat work should be carried out according to external target designation coming from the detection radar network, which should be united with this anti-aircraft missile system by a single data transmission system and a single computer system. In this situation, in order to destroy a high-speed low-flying target at the far border of the S-400 anti-aircraft missile system defeat zone according to external target designation data, it is necessary to lock the target at point A, which will be at a distance of $D_m = 350$ km from the multifunctional radar and at an altitude of $h_m = 26$ km from the Earth's surface as shown in Figure 10.10.

For the missile flight altitude $h_m = 26$ km, the line-of-sight range of this missile from the side of the ground multifunctional radar (GMR) according to (10.57) will be $R_{GMR} = 576.16$ km, and therefore, the multifunctional radar has the ability to transmit the target coordinates to the missile, which is located at point A and is accompanied by a ground multifunctional radar. In this case, the missile 40N6E flies to the lead point D along the shortest path AD, which is 56.13 km, the missile flight time will be $t_m = 47.17$s for the missile speed $V_m = 1190$ m/s and as shown in Chapter 3, for the maximum possible flight speed of the 40N6E missile $V_m^{max} = 2417.16$ m/s the time of the controlled flight of this missile in the meeting point of D will be $t_m = 23.22$s. Using formulas given in Chapter 3, for the initial velocity of the missile $V_{m0} = 1980.95$ m/s, we can calculate $V_{mA} = 1856.58$ m/s—the module of the missile velocity vector at the point A, and $V_{mD} = 1976.97$ m/s—the modulus of the missile velocity vector at the meeting point D. Accordingly, the average speed of the missile on the area AD will be equal to $\bar{V}_{AD} = 1916.77$ m/s and the average flight time of the missile from point A to the meeting point D will be $\bar{t}_m = 29.38$s. For a hypersonic target flying at a speed of $V_t = 8M = 2396$ m/s, the range of stable target tracking without interference R_{stt} for a missile homing head should be no worse

$$R_{stt} \geq R_{mt} = \sqrt{(h_m - h_t)^2 + (\bar{V}_{AD}\bar{t}_m)^2} = 74.871 \text{ km} \approx 75 \text{ km}, \tag{10.58}$$

which corresponds to ground range for seeker D_{mt} = 70.395 km or D_t = 420.395 km for the ground radar.

Let us calculate the estimate of directivity $\bar{D}(\varphi, \varepsilon) = \bar{D}_0$ in the direction of the maximum radiation of the seeker antenna $\varphi = 0°$, $\varepsilon = 0°$ by the recursive formula [7–9]:

$$\bar{D}_0 = \frac{(25\ldots37) \times 10^3}{\Delta\varphi_{0.5}\Delta\varepsilon_{0.5}} = \frac{31000}{\Delta_{0.5}^2} = 861.11 \tag{10.59}$$

where $\Delta\varphi_{0.5} = \Delta\varepsilon_{0.5} = \Delta_{0.5} = 6°$ is the width of the seeker antenna directional pattern at half-power level in each of two mutually orthogonal planes. The estimate of the seeker antenna Gain $\bar{G}_A$ can be found from the expression

$$\bar{G}_A = \bar{D}_0\eta_A = 861.11 \times 0.93 = 800.83 \approx 800 \tag{10.60}$$

where $\eta_A < 1$ is the antenna efficiency. It is known that the minimum power at the input of the seeker receiver $\bar{P}_{r\_min}$ by which the signal can be detected is represented by the expression [7]:

$$\bar{P}_{r\_min} = kT_0F_rL\Delta f_r \tag{10.61}$$

The $\bar{P}_{r\_min}$ score is the minimum detectable signal (MDS) used to estimate the ultimate sensitivity of a receiver. This is the minimum received signal level at which a signal can be reliably detected in the presence of white noise. This parameter is mainly used for systems in which signal detection is carried out automatically. Here $k = 1.380649 \times 10^{-23}$ J is the Boltzmann constant, $T_0 = 290°$ K is the receiver noise temperature, F_r is the noise figure referenced to the antenna feed, L is the total losses excluding beamshape loss, and Δf_r is the bandwidth of the seeker receiver filter. Taking into account that $F_r = 5$ dB, $L = 10$ dB, and $\Delta f_r = 400$ Hz, we can write

$$\bar{P}_{r\_min} = 1.380649 \times 10^{-23} \times 290 \times 10^{0.5} \times 10^1 \times 400 = 5.06456 \times 10^{-17}\text{W} \tag{10.62}$$

or

$$\bar{P}_{r\_min}[dB] = 10\log(\bar{P}_{r\_min}) = -162.9545832 \text{ dB}$$
$$\bar{P}_{r\_min}[dB] = -228.5991 + 24.6239 + 5.0 + 10.0 + 26.0205 = -162.9545832 \text{ dB} \tag{10.63}$$

SNR for the case where the radar beam is pointed directly at the target, the radar range equation is written in the following form [7]:

$$SNR = \frac{P_t G_t G_r \sigma \lambda^2}{(4\pi)^3 k T_0 F_r L \Delta f_r R_{mt}^4} = \frac{P_t G_t G_r \sigma \lambda^2}{(4\pi)^3 \ \bar{P}_{r\_min} \ R_{mt}^4} = 10.4169 \tag{10.64}$$

$$SNR_{\,[dB]} = 10\log(10.4169) = 10.3728 \text{ dB}$$

It is known that the S-200D system is capable of detecting a B-52 aircraft at a range of $R_t = 368.390$ km when flying at an altitude of $H_t = 10000$m with an effective reflecting surface $\sigma_1 \approx 100\text{m}^2$ [10]. At the same time, this target is steadily accompanied by the missile homing head when the condition $SNR \geq 10$ dB [10] is fulfilled. Let the characteristics of the seeker when operating in semiactive mode 1 and in active mode 2 are identical. Taking into account that $G_{1r} = G_{2r}$, $P_{1r\_min} = P_{2r\_min}$ and $\lambda_1 = \lambda_2$; hence, we can define power of the missile seeker transmitter P_{2t} from the condition:

$$P_{2t} = P_{1t} \times \frac{G_{1t}\sigma_1 R_2^4}{G_{2r}\sigma_2 R_1^4} \tag{10.65}$$

Here, with the width of the main beam of the target illumination radar at half-power $\Delta_{0.5} = 0.759°$, we can estimate the Gain of the target illumination radar antenna $G_{1t} = 31000/\Delta_{0.5}^2 = 50,000$. As shown in [10–12], the full cross section is

$$\sigma = \sigma^{refl} + \sigma^{sh} = 2S_{sh} \tag{10.66}$$

where σ^{refl} is a scattering cross section of the illuminated part S_0, σ^{sh} is a scattering shadow cross section, $2S_{sh}$ is a cross-sectional area of the shadow (see Figure 1 in [10] and Figure 10.11 in this chapter). In this case the scattering cross section takes the following form:

$$\sigma(\delta\varepsilon_t, \delta\varphi_t) = 2S_{sh} F^2(\delta\varepsilon, \delta\varphi) = 2S_{sh}\Big|_{F^2(\delta\varepsilon,\delta\varphi)=1,\ \delta\varepsilon=0°,\ \delta\varphi=0°} \tag{10.67}$$

Here $\delta\varepsilon$, $\delta\varphi$ are target tracking errors in elevation ε and azimuth φ, respectively, ⊙ are the directions of the $\vec{X}$ and axes $\vec{X}_m$ coordinate systems associated with the missile that are directed toward the reader. $\{XYZ\}$ is a rectangular coordinate system, in which the $\vec{Y}$ axis is parallel to the Earth's surface, and $\{X_m Y_m Z_m\}$ is a coordinate system associated with the missile, in which the $\vec{Y}_m$ axis is directed toward the target. ⊗ - direction $\vec{X}_t$ in the coordinate system $\{X_t Y_t Z_t\}$, which is associated with the target and the axis $\vec{X}_t$ directed from the reader. Due to the fact that when firing to the defeat area far border, the missile always approaches the target from above, therefore the illuminated part of the target has an area of S_0, the projection of which onto the plane $\{Z_m X_m\}$ can be represented as:

$$\begin{aligned} S_{sh} &= Z_{mx} \times B_{mx} \\ Z_{mx} &= L_x \sin(\varepsilon_{mt}) + \frac{H_x}{2}\cos(\varepsilon_{mt}) \end{aligned} \tag{10.68}$$

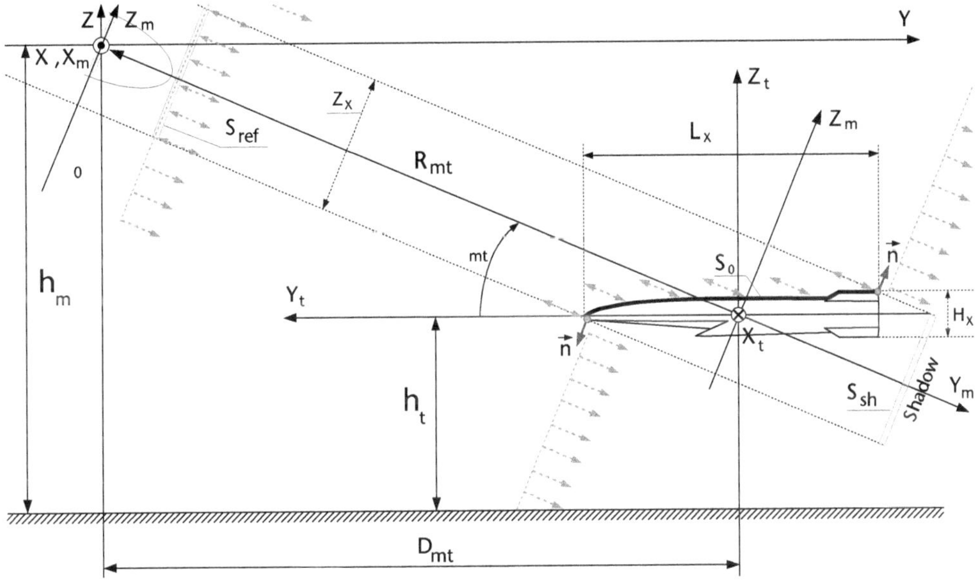

Figure 10.11 Hypersonic target shadow.

where L_x is the longitudinal dimension of the X-51 hypersonic missile along the $\vec{Y}_t$ axis, B_x is the lateral size of the missile along the $\vec{X}_t$ axis, and H_x is the vertical dimension of the X-51 missile along the $\vec{Z}_t$ axis. Taking into account (10.65), the expression for the effective reflective surface of the target can be represented as

$$\sigma_{X-51} = B_x\left(2L_x \sin(\varepsilon_{mt}) + H_x \cos(\varepsilon_{mt})\right) \quad [m^2] \tag{10.69}$$

For the case shown in Figure 10.11, we can find the value of the angle of approach of the missile to the target ε_{mt} at the time of receiving target designation before the target is captured by the seeker for automatic tracking along angular coordinates. In this case calculation the effective reflective target surface σ_{X-51} can be carried out according to (10.68):

$$\begin{aligned}
D_{mt} &= \sqrt{Rmt^2 - (H_m - h_t)^2} = \Big|_{R_{mt} = 75\text{ km};\ H_m = 26\text{ km};\ h_t = 0.5\text{ km}} = \\
&= 70.53190767\text{ km} \\
\cos(\varepsilon_{mt}) &= 0.940425436 = \frac{D_{mt}}{R_{mt}}\Big|_{D_{mt} = 70.53190767\text{ km};\ H_m = 26\text{ km};\ h_t = 0.5\text{ km}} \\
\sin(\varepsilon_{mt}) &= 0.34 = \frac{H_m - h_t}{R_{mt}}\Big|_{R_{mt} = 75\text{ km};\ H_m = 26\text{ km};\ h_t = 0.5\text{ km}} \\
\varepsilon_{mt} &= \arccos\left(\frac{D_{mt}}{R_{mt}}\right) = 0.471615568\text{ rad} \quad \text{or} \quad \varepsilon_{mt} = 19.87687407^\circ \Rightarrow \\
\Rightarrow \sigma_{X-51} &= 3.574091537\ [\text{m}^2]
\end{aligned} \tag{10.70}$$

After receiving target designation, the missile homing head begins to search for a target in an active mode by means of a conical scan. When the SNR exceeds the threshold value equal to 10 dB, a decision is made to detect the signal reflected from the target, when the seeker is working on the clean target (i.e., in the absence of cover interference). The excess of the signal received from the target over the receiver noise by 10 dB ensures target detection with a probability not lower than the specified one, for example, determined by the 0.84 level. The transmitter power of the seeker when it is working on a clean target of the X-51 type when firing beyond the radio horizon can be estimated using the expression (10.64):

$$P_{2t} = 3000\text{W} \times \frac{50000 \cdot 100 \cdot (75000)^4}{800 \cdot 3.574091537 \cdot (368390)^4} = 3000\text{W} \times 3.004194654 = \\ = 9012,583963\text{W}$$

or in dB:

$$\begin{aligned} P_{2t}[dB] &= 10 * \log(9012,583963) = 39.54849324 \\ P_{2t}[dB] &= 34.77121255 + 46.98970004 + 20 + 4 * 48.75061263 - 29.03089987 \\ &\quad - 5.531656712 - 4 * 55.66307833 = 34.77121255 + 4.777280658 = \\ &= 39.54849321 \end{aligned} \tag{10.71}$$

From (10.70) it can be seen that for detection at a range $R_{mt} = 75$ km of a high-speed small-sized target with a radar cross section $\sigma = 3.574091537$ [m^2], it is necessary that the missile on board has a transmitter with a radiation power

$$P_{2t} \geq 9013\text{W}. \tag{10.72}$$

According to open sources [13, 14], the active radar seeker 9B-1103M (diameter 350 mm), which was developed by the Moscow Research Institute AGAT, Zhukovsky, Moscow region, is fully digital. The active radar seeker can operate in a fully autonomous fire-and-forget mode and its onboard computer can be reprogrammed to match the changing circumstances of the engagement. The 9B-1103M active seeker has an acquisition range of targets with RCS $\sigma_{as} = 5\text{m}^2$, not less than $R_{mt} = 40$ km. From (10.63) it is possible to obtain the power emitted by the active seeker 9B-1103M for a target having $\sigma_{as} = 5\text{m}^2$ and located from the seeker at a distance of $R_{mt} = 40$ km:

$$P_{t\_Agat} = \frac{10.4169 \cdot (4\pi)^3 \cdot P_{r\_min} \cdot R_{mt}^4}{G_r^2 \cdot \sigma_{as} \cdot \lambda^2} = 523.45\text{W} \tag{10.73}$$

Due to the fact that a waveguide-slot antenna array is used in the 40N6E missile homing head, in which the transmitter is connected to its sum-channel, as in the 9B-1103M homing head, in the case of using a transmitter from the homing head 9B-1103M in the active seeker 40N6E missile, for $\sigma_{X-51} = 3.574091537\text{m}^2$, we

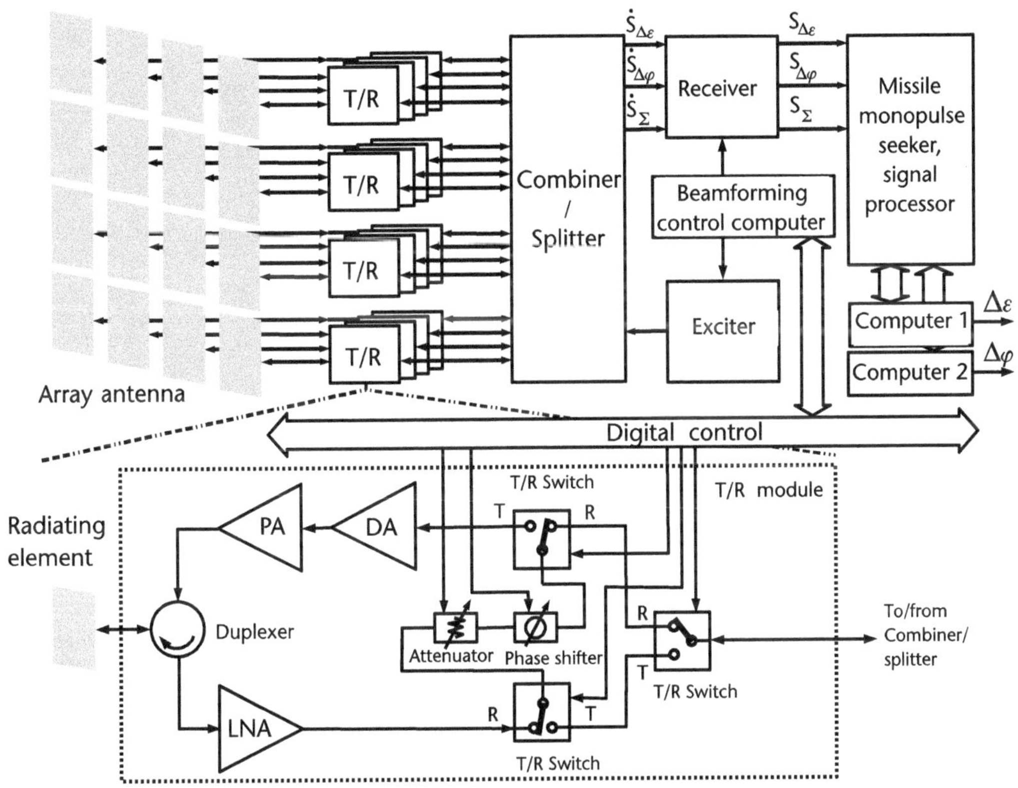

Figure 10.12 Homing head block diagram with antenna array, and T/R module functional block diagram.

can get R_{mt} by expressing it from (10.73):

$$R_{mt} = \left(\frac{P_{t\_Agat} \cdot G_r^2 \cdot \sigma \cdot \lambda^2}{10.4169 \cdot (4\pi)^3 \cdot P_{r\_min}} \right)^{\frac{1}{4}} = 36779.66388329\text{m} \approx 36.78 \text{ km} \quad (10.74)$$

Due to the fact that $R_{mt} = 36.78$ km $\ll R_{stt} = 75$ km based on (10.58), we can conclude that the S-400 anti-aircraft missile system when firing over the horizon is unable to detect a hypersonic target with a small RCS with its active seeker. The reason lies in the insufficient power of the transmitter of the active homing head of the missile and in the outdated design of the seeker antenna in the form of a waveguide-slot antenna array. The solution can be the use in the homing head an active electronically scanned array (AESA) [7, 8, 15], based on transmit/receive modules, for example, as shown in Figure 10.12. Here, PA is the power amplifier, DA is the driver amplifier, LNA is the low noise amplifier, and T/R switch switches the propagation paths of the signal during transmission and when the signal is reflected from the target. Switching is carried out by commands issued by the computer, which controls the formation of the main beam of the antenna pattern of the seeker. The control of the position in space of the main beam of the seeker antenna is carried out by commands coming to the attenuator and phase shifter of each element of the AESA active antenna array. The spatial shape of the beam is determined by the amplitude distribution. $|E_{si}|$ on the antenna surface, formed by

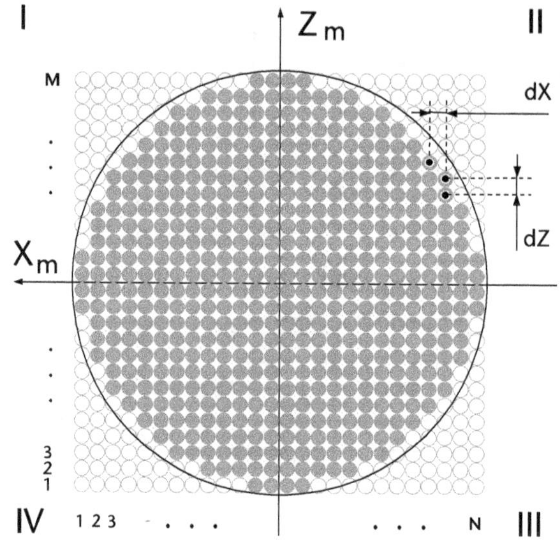

Figure 10.13 Missile homing head antenna array.

the attenuator of each ith element of the array, and the position of the beam in space is characterized by the inclination of the phase plane, which is determined by the phase distribution ψ_{si} over all elements of the antenna array $i = \overline{1, L}$, where for the case of equal lattice length in mutually orthogonal planes $L = M \times N$, N is the number of elements per $\vec{X}_m$, and M is the number of elements per $\vec{Z}_m$ (Figure 10.13). For an anti-aircraft guided missile, it is necessary to fit the antenna array into the cross section of the missile body of radius R_m, while it is clear that the radius of the antenna plane is less than the radius of the inner surface of the missile $R_{ant} < R_m$. An example of such an antenna array is shown in Figure 10.13. A planar antenna array with rectangular packaging is shown here, but triangular packaging with a horizontal or vertical arrangement is also possible. The angles we are using are not the angles traditionally used to develop the radiation pattern for planar arrays [7]. The traditional angles are those associated with a standard spherical coordinate system [8,9]. These angles are θ and φ, where θ is measured from the z-axis, and φ is measured from the x-axis in the x-y plane. In the case under consideration and presented in Figure 10.11, angular coordinates are used in azimuth φ and in elevation ε angles and in the coordinate system of Figure 10.14, which is connected to the missile homing antenna, the field point is located at

$$(S_x, S_y, S_z) = (R_{mt}\cos\varepsilon\sin\varphi,\ R_{mt}\cos\varepsilon\cos\varphi,\ R_{mt}\sin\varepsilon) \tag{10.75}$$

Array element A_{11} is located at the left and bottom of the square antenna array and nmth element is located at (nd_x, md_z), where d_x is the spacing between elements in the x direction, and d_z is the spacing between elements in the z direction. Therefore, the range from the nmth element to the field point in which target situated is

$$R_{nm} = \sqrt{(S_x - nd_x)^2 + (S_y)^2 + (S_z - md_z)^2}$$

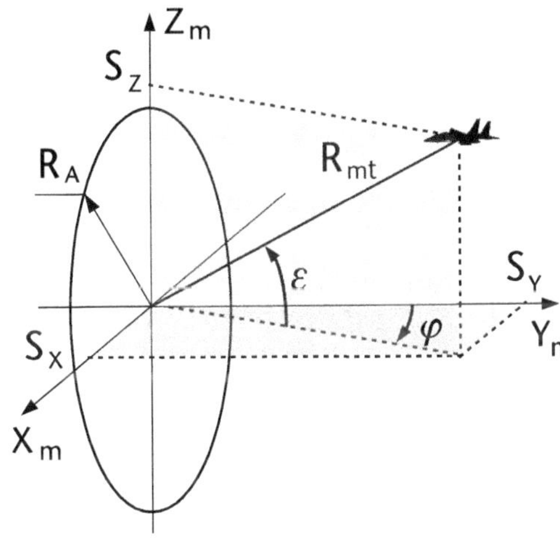

Figure 10.14 Coordinate system associated with the missile homing head.

$$= \sqrt{R_{mt}^2(\cos\varepsilon\sin\varphi)^2 - 2R_{mt}\cos\varepsilon\sin\varphi + (nd_x)^2 + R_{mt}^2(\cos\varepsilon\cos\varphi)^2 + R_{mt}^2(\sin\varepsilon)^2 - 2R_{mt}\sin\varepsilon\cdot md_z + (md_z)^2}$$

$$= \sqrt{-2R_{mt}(\cos\varepsilon\sin\varphi\cdot nd_x + \sin\varepsilon\cdot md_z) + R_{mt}^2[(\cos\varepsilon\sin\varphi)^2 + (\cos\varepsilon\cos\varphi)^2 (\sin\varepsilon)^2] + (nd_x)^2 + (md_z)^2}$$

$$= R_{mt}\sqrt{1 - \frac{2}{R_{mt}}[\cos\varepsilon\sin\varphi\cdot nd_x + \sin\varepsilon\cdot md_z] + (nd_x)^2 + (md_z)^2} \qquad (10.76)$$

Due to the fact that $R_{mt} \gg d_{x,z} \Rightarrow (nd_x)^2 \to 0$ and $(md_z)^2 \to 0$. Since the second term of the radical of (10.76) is small relative to 1, therefore we can use the approximation

$$(1 \pm x)^N \approx 1 \pm Nx \qquad (10.77)$$

and expression (10.76) can be rewritten as

$$R_{nm} = R_{mt} - nd_x\cdot\cos\varepsilon\sin\varphi - md_z\cdot\sin\varepsilon \qquad (10.78)$$

In this case the voltage out of the mnth antenna array element can be written in the form

$$V_{nm}(\varepsilon,\varphi) = V_{mt}a_{nm}\exp\left(\frac{j2\pi R_{nm}}{\lambda}\right) = V_{mt}a_{nm}\exp\left(\frac{j2\pi R_{mt}}{\lambda}\right)\times\exp\left[\frac{-j2\pi}{\lambda}(nd_x\cdot\cos\varepsilon\sin\varphi + md_z\cdot\sin\varepsilon)\right], \qquad (10.79)$$

where a_{mn} is the weight applied to the mnth element. Summing the outputs of all elements gives

$$V(\varepsilon,\varphi)=\sum_{m=1}^{M}\sum_{n=1}^{N}V_{nm}(\varepsilon,\varphi)$$
$$=V_{mt}\exp\left(\frac{j2\pi R_{mt}}{\lambda}\right)\sum_{m=1}^{M}\sum_{n=1}^{N}a_{nm}\exp\left[\frac{-j2\pi}{\lambda}(nd_x\cdot\cos\varepsilon\sin\varphi+md_z\cdot\sin\varepsilon)\right] \tag{10.80}$$

Dividing by V_{mt} and ignoring the first exponential term (which disappears) when from the radiation pattern, we get the array factor

$$A(\varepsilon,\varphi)=a_{nm}\exp\left[\frac{-j2\pi}{\lambda}(nd_x\cdot\cos\varepsilon\sin\varphi+md_z\cdot\sin\varepsilon)\right] \tag{10.81}$$

In the expression (10.79), a_{nm} are weighting factors, assuming the distribution of amplitudes $|a_{nm}|$ and phases Φ_{nm} of each element excitation of the antenna array. They are used to provide a proper taper and steer the beam in space through phase steering.

$$a_{nm}=|a_{nm}|\exp\left[\Phi_{nm}\right]=|a_{nm}|\exp\left[\frac{j2\pi}{\lambda}(nd_x\cdot\cos\varepsilon\sin\varphi+md_z\cdot\sin\varepsilon)\right] \tag{10.82}$$

Note that the array factor $A(\varepsilon,\varphi)$ is the array pattern calculation under the assumption that each array element radiates isotropically. In reality, elements pattern are never isotropic, but they may be described by a function $f(\varepsilon,\varphi)$. If all element patterns were identical in an actual array antenna, the far-field pattern $E(\varepsilon,\varphi)$, because of the principle of multiplication would be [8]

$$E(\varepsilon,\varphi)=f(\varepsilon,\varphi)\times A(\varepsilon,\varphi) \tag{10.83}$$

Often $E(\varepsilon,\varphi)$ is normalized to have a maximum of unity; it provides a practical estimate for actual antenna array patterns. $E(\varepsilon,\varphi)$ and $A(\varepsilon,\varphi)$ are generally complex, and therefore their E-field amplitude patterns are created by plotting their absolute magnitudes.

For the antenna array shown in Figure 10.14 and having a radius R_{ant}, which is inscribed in the missile body with a radius R_m, where $R_{ant}<R_m$, then antenna array emitters coordinates of the centers can be represented as follows:

$$\begin{aligned} x_n &= R_{ant}-(n-1)\frac{\lambda}{2}-\frac{\lambda}{4}, && \forall n=1,\ldots,N \\ z_m &= -R_{ant}+(m-1)\frac{\lambda}{2}+\frac{\lambda}{4}, && \forall m=1,\ldots,M \end{aligned} \tag{10.84}$$

The amplitudes $|a_{mn}|$ from the expression (10.80) in the case of a round aperture can be represented as

$$r_{nm} = \sqrt{x_n^2 + z_m^2}$$
$$r_{nm} < R_{ant} \Rightarrow |a_{nm}| = |A[n, m]|, \quad \Phi_{mn} = \left[\frac{j2\pi}{\lambda}(nd_x \cdot \cos\varepsilon \sin\varphi + md_z \cdot \sin\varepsilon)\right]$$
$$r_{nm} \geq R_{ant} \Rightarrow |a_{nm}| = 0, \qquad \Phi_{mn} = 0. \tag{10.85}$$

Here $|A[n, m]|$ is a discrete amplitude distribution of the excitation voltage or current in the plane of the antenna aperture $\{X_m, Z_m\}$ over the array elements, the position of each is characterized by the indices $m = 1, \ldots, M$ and $n = 1, \ldots, N$.

For an equidistant antenna array (Figure 10.13), the condition for the existence of a single main maximum of the radiation pattern can be written as [8,9]:

$$dx = dz \leq 0.5\lambda \tag{10.86}$$

Let for the 48N6E missile $R_{ant} = 0.256$m, $\lambda = 0.032$m; therefore, in each of the mutually orthogonal planes $\{X_m, Z_m\}$ the maximum number of lattice elements is $N = M = 4R_{ant}/\lambda = 32$. Substituting the found characteristics into (10.84) and (10.85), we can determine the total number of radiators of this antenna array: $L = 812$.

The output power of one element of the transmitting module ranges from 2.5W to 4W when using gallium arsenide (GaAs) transistors, and from 10W to 15W or more when using gallium nitride (GaN) transistors. At present, the technology of integrated circuits based on gallium arsenide [16] is widely used in the output elements of matching circuits. For example, on-board the radar system Zhuk-AU/AE (FGA-29) [17] with an active phased array antenna with an overall diameter of 600 mm (the diameter of the active phased array antenna is about 500 mm), which has 680 transceiver modules, passed a test cycle at the stand of the Fazotron-NIIR Corporation and in the course of flight tests on the MiG-35 aircraft, the detection range of air targets was 130–140 km [18]. The radiating power in this active phased array is contained in each of its elements. In this case, the missile homing head can be represented in the form shown in Figure 10.15. And if the output power of one element of the Zhuk-AUE onboard radar based GaAs transistors is estimated at 8W [19], then when 812 elements are used, the output power of homing head 48N6E with such an antenna will be

$$P_{2t} = 6496\text{W} \tag{10.87}$$

As a result, the detection range R_{mt} of a hypersonic cruise missile similar to X-51 Waverider according to (10.74) will be

$$R_{mt} = 69 \text{ km} \tag{10.88}$$

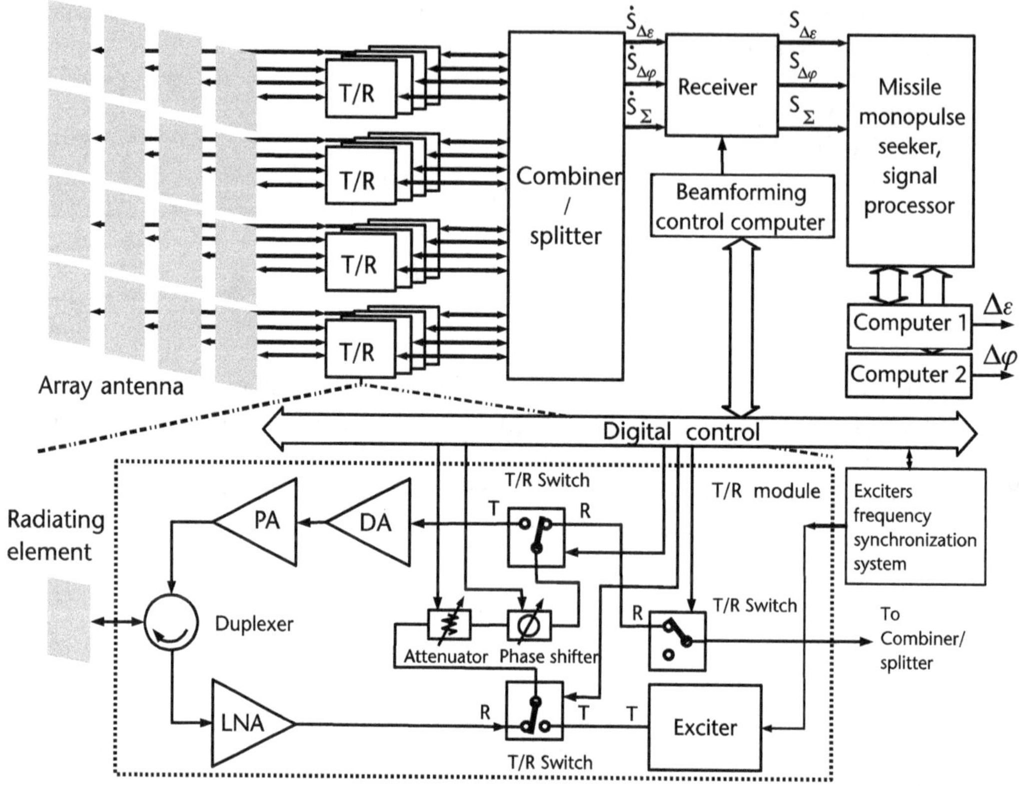

Figure 10.15 Functional block diagram of a missile homing head including AESA with T/R modules.

The obtained detection range is still less than the range of stable auto-tracking of the X-51 target, which is $R_{stt} = 75$ km.

To fulfill the condition (10.58), it is necessary:

1. To use more powerful emitting elements, based on GaAs transistors delivering power $P_t[i] = 10\text{W} \ldots 15\text{W}, \forall i \in [1, \ldots, L]$, where $L = 812$;
2. To use other signals radiated by the seeker.

Evaluation of the first option application of improving of the seeker energy potential can be presented in the form:

$$\begin{aligned}
&1.\ \forall\ P_t[i] = 10\text{W}\ :\ P_{2t} = 8120\text{W}\ \&\ R_{mt} = 72.992\text{ km}\ <\ R_{stt} = 75\text{ km}\\
&2.\ \forall\ P_t[i] = 11.5\text{W}\ :\ P_{2t} = 9054\text{W}\ \&\ R_{mt} = 75.006\text{ km}\ \geq\ R_{stt} = 75\text{ km}\\
&3.\ \forall\ P_t[i] = 15\text{W}\ :\ P_{2t} = 12180\text{W}\ \&\ R_{mt} = 80.779\text{ km}\ >\ R_{stt} = 75\text{ km}
\end{aligned} \tag{10.89}$$

From (10.89) it follows that for the timely detection of a hypersonic small-sized target of the X-51 type by air defense missile, it is necessary to have on board an active antenna array, in which each transmitting module is built on the basis of GaAs transistors and is capable of delivering power $\geq$11.5W each, which corresponds to the result obtained earlier (10.72). According to Baltic State Technical University

VOENMEH, named after D. F. Ustinov [18], the cost of the H010 airborne radar Zhuk-AU/AUE is estimated at 3.5–4 million U.S. dollars. This cost of onboard equipment can be justified when installed on an aircraft, which by definition is a reusable vehicle. The task of an anti-aircraft guided missile is to unconditionally destroy enemy aircraft and missiles. In the process of completing this task, the life cycle of the missile ends, and therefore, the missile is a disposable vehicle by definition. It is clear that the cost of a long-range missile when using a homing head with an active phased antenna array on board increases more than 10 times compared to the cost of existing missiles. In this regard, a reasonable question arises about the economic feasibility of manufacturing such missiles in large quantities. As a result, missiles with a homing head with an active phased antenna array can be manufactured in limited quantities and must be used in the directions of the proposed breakthrough of the air defense, carried out through the use of hypersonic vehicles by the enemy.

As an alternative to active phased arrays, it seems expedient to use other forms of emitting signals. There is a method that makes it possible to increase the radiated power by compressing the emitted pulses. In this case, the internal structure of the probe signal carrier is modulated in phase or in frequency [5, 7]. Consider the option of installing a homing head with a waveguide-slot antenna array and a transmitter from a homing head developed by JSC Moscow Research Institute Agat on a 40N6E missile. As shown previously in this section, the power of this transmitter is estimated at 523.45W. Thus, to increase the energy potential of a pulsed radar, it is necessary to increase the duration of the probe pulse. It is clear that with an increase in the duration of the probing signal, the radar's range resolution will deteriorate. To improve it, as a rule, intrapulse modulation is used, where the probing signal can be represented in the form of separate parts, each of which has its own frequency or phase. Thus, the response from the target can also be divided into N separate parts of τ_0 duration, and the subsequent combination of these parts in time will allow obtaining one short pulse of greater amplitude. The echo-signal compression operation with duration $\tau_p = N\tau_0$ is performed in special filters. As a result, it becomes possible to separate targets that are inside an unresolved volume, the dimensions of which are determined by this long impulse $\tau_p = N\tau_0$. Receiver noise is always broadband and random. At the same time, the spectral density of the noise is rather low in comparison with the spectrum of the echo signal of long duration. Therefore, after the compression filter, the intensity of the noise is significantly reduced. The pulse compression ratio (PCR) is numerically equal to N, which is the number of separate parts of the probe signal with the duration τ_0, also called partial pulses. Consequently, the range resolution is proportional to the duration of one partial pulse of duration τ_0. It is easy to show that the maximum detection range of the radar is increased by an amount equal to the fourth root of the PCR, since according to (10.74) PCR is in the numerator under the fourth root. For example, a binary phase-coded pulses signal (code-phase-shift keyed signal) can be used as a probe signal. In this signal, partial pulses have equal duration, but each is emitted with its own phase determined by the phase code (for example 0 and π or $-\pi/2$ and $+\pi/2$) [7], and hence the phase is binary coded and can be represented as a sequence of only two values, +1 and −1. The code sequence should provide a minimum level of sidelobes that occur after signal compression.

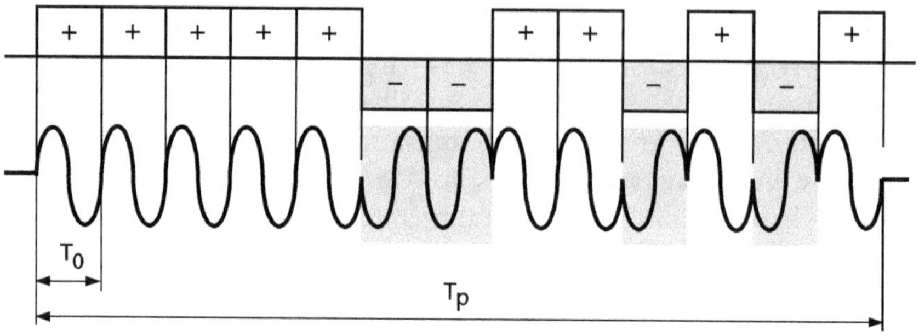

Figure 10.16 Barker-code Waveform.

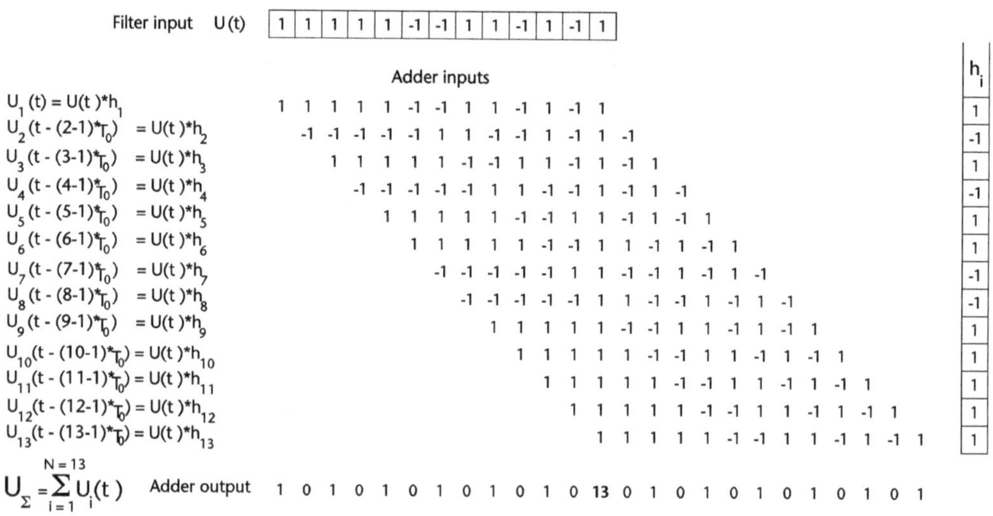

Figure 10.17 Filtering process for a phase-coded waveform.

Barker-codes (Figure 10.16) are optimal in this sense. The maximum Barker code size is $N = 13$. In this case, the level of side voltage peaks at the output of the compression filter is $20\log(1/13) = -22.27$ dB. The filtering process for a phase-coded waveform is presented in Figure 10.17. Here $U(t)$ is the input signal, and $h(t)$ is the impulse response corresponding to the input signal $U(t)$. Figure 10.18(a) shows a diagram of a compression filter based on delay lines with taps, giving a delay of the input signal at each tap by τ_0 and, accordingly, the total filter delay time will be $\tau_{del} = (N-1)\tau_0 = \tau_p - \tau_0$. In Figure 10.18(b), the dependence of the output voltage of the compression filter on time $U_\Sigma(t)$ is presented. According to (10.74), the estimate of the detection range R_{mt} of a hypersonic cruise missile similar to the X-51 Waverider in the case of using a binary phase-coded pulses signal (code-phase-shift keyed signal) will be:

$$
\begin{aligned}
R_{mt} &= (PCR)^{\frac{1}{4}} \times 36.78 \text{ km} \\
&= 69.838 \text{ km}
\end{aligned}
\tag{10.90}
$$

Here, the range of 36.78 km, according to (10.74), corresponds to the case of using the transmitter from the homing head 9B-1103M in the homing head of the 48N6E

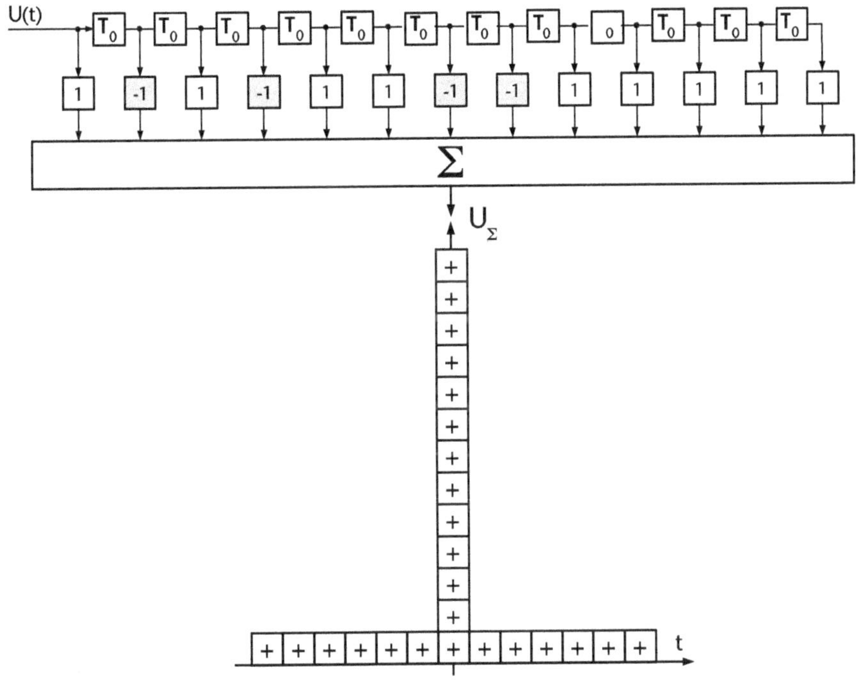

Figure 10.18 Signal compression principle.

anti-aircraft missile, which, according to (10.73), has a power of 523.45W. It is quite clear that in this case the condition (10.57) is not met, since $R_{mt} = 69.84$ km < 75 km. Therefore, it is necessary to use noise-like signals with the largest possible signal base PCR. For binary phase-coded pulses signal (code-phase-shift keyed signal), the maximum value is $PCR = 13$, and as shown above, the radar detection range in this case is limited. To increase the target detection range, it seems expedient to use signals with the largest possible base N in the detection mode, for example, noise-like signals, which, with appropriate processing in the receiver, will allow obtaining the value $SNR >> 23$ dB at the filter output.

10.4.2 Recognition Procedure for Some Interference Types

As shown above, $SNR \geq 10$ dB is a condition under which the seeker makes the decision A_0 when the seeker detects a signal reflected from the target and the target is tracked along the angular coordinates. At the same time, it is reasonable to assume that the signal from the target will either be detected or not. Consequently, the probability of target detection is 0.5, and all other hypotheses account for the remaining 0.5, respectively.

$$\begin{cases} P(A_0) & = 0.5 \\ \bigcup_{i=1}^{q} P(A_i) & = 0.5 \end{cases} \tag{10.91}$$

The main source of information for the seeker is the errors in tracking the signal source in elevation $\delta\varepsilon(t)$ and in azimuth $\delta\varphi(t)$, respectively. Passing from continuous

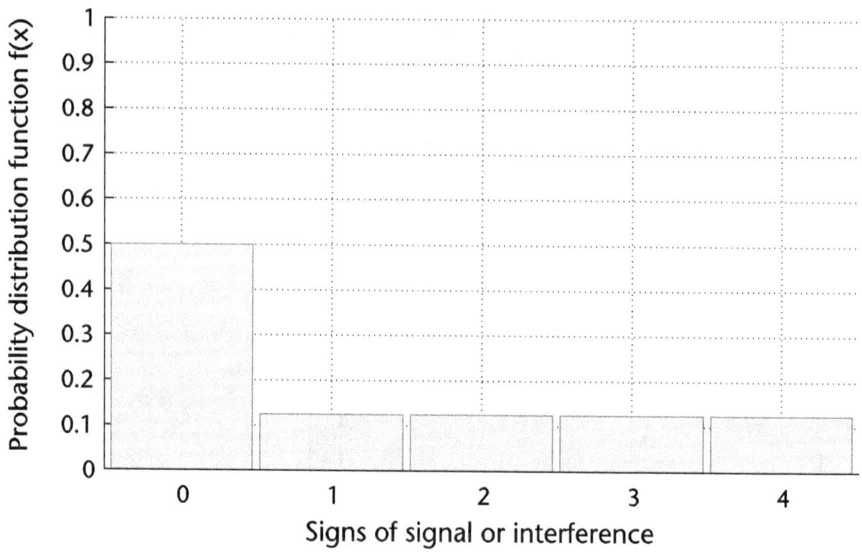

Figure 10.19 Probability distribution function.

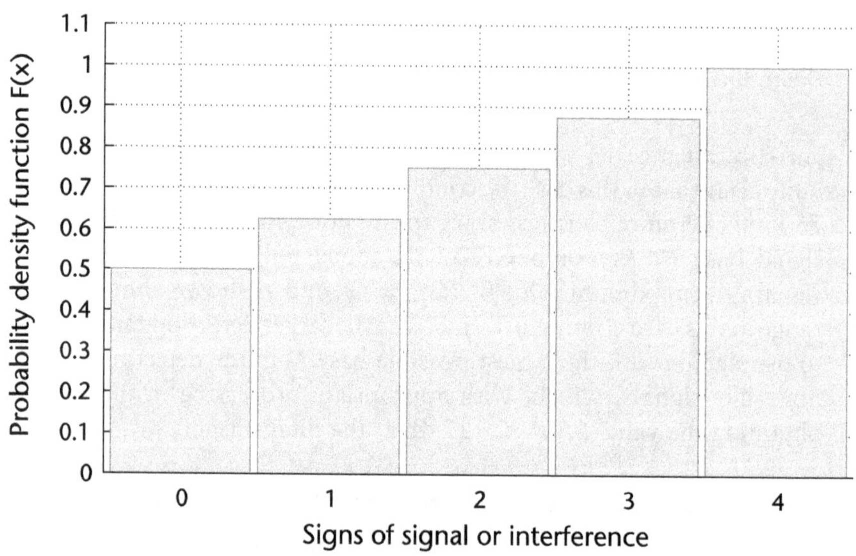

Figure 10.20 Probability density function.

time t to discrete t_j with duration Δt: $t_j = (j-1)\Delta t, \forall j = \overline{1, m}$ we can write numerical sequences on elevation angle $\delta\varepsilon(t_j) = \delta\varepsilon_j$ and in azimuth $\delta\varphi(t_j) = \delta\varphi_j$. Due to the fact that the signs of the polarization interference A_1 and covering interference A_2, acting along the sidelobes of the antenna pattern of the anti-aircraft guided missile, are the outputs of the comparison devices U_{c1} and U_{c2}, which have a set possible values $\{0, 1\}$, these are Boolean signs. A sign of flickering interference A_3 is the alternating signs of automatic tracking errors $\delta\varepsilon_j$ and $\delta\varphi_j$; therefore, it is possible to build a device for determining the effect of flickering interference, the output of which U_{c3} receives values $\{0, 1\}$ and is also a Boolean [10]. The situation of flickering interference is shown in Figure 10.21. Here, jamming is carried out

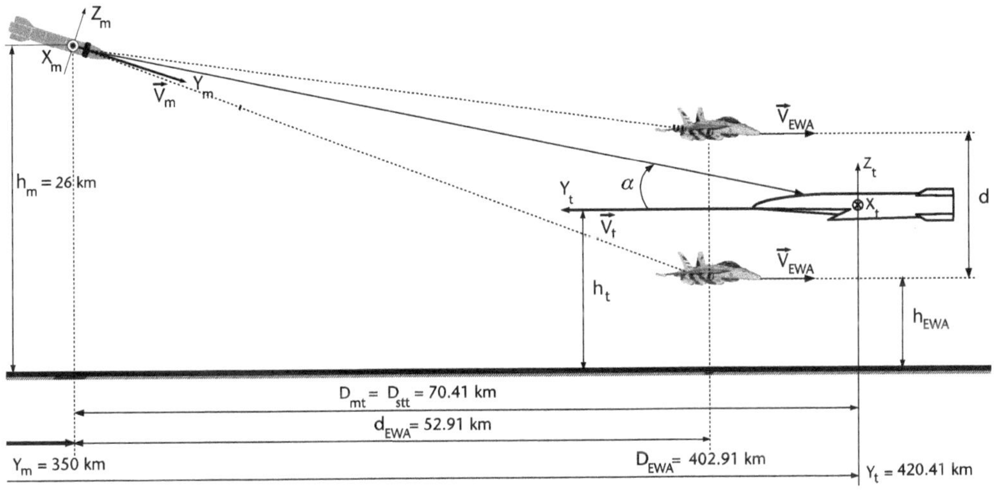

Figure 10.21 Electronic warfire airplanes (EWA).

by two EA-18G Growler electronic warfare unmanned aircraft, controlled via a wireless computer network from the third EA-18G [20, 21] aircraft. The entire group of electronic warfare aircraft is outside the S-400 air defense system defeat zone. Jamming is carried out by means of the tactical jamming system ALQ-99, which is an external carriage Airborne Electronic Attack capability for the EA-18G Growler aircraft used against radar and communications targets for the suppression of enemy integrated air defenses. This system is a digital receiver system, complete communications electronic attack system, and routable network information system [22]. The ALQ-99's jammers are carried in external 950-pound pods, each of which carries typically two 2-kW transmitters having two high band-steerable antennas. Transmitter power is provided by a ram air generator at the front of the pod [23]. In order to reduce the likelihood of electronic warfare aircraft being hit by 48N6E missile, the course of these aircraft $\vec{V}_{ewa}$ is directed in the same direction as the velocity vector of the missile $\vec{V}_m$; hence jamming in this case will be carried out into the rear hemisphere and the speed of approach to the missile will be presented in the form:

$$V_0 = V_m - V_{EWA} = 1856.58 - 1190 = 666.58 \text{ m/s} \tag{10.92}$$

Let the power of one ALQ-99 transmitter $P_t = 1$ kW be radiated by a horn antenna with a gain $G_t = 13$ dB or $G_t = 10^{\frac{13}{10}} = 19.95$. The jamming signal is received by the missile homing antenna with antenna gain $G_r = 800$. Consequently, the range of the interference signal stable tracking, at which the ratio of the interference signal to the receiver noise is more than 10 dB, can be represented as

$$R^2 = \frac{\frac{P_t G_t G_r \lambda^2}{R^2}}{\frac{kTdF10(4\pi)^2}{1}} \Rightarrow R = \left(\frac{P_t G_t G_r \lambda^2}{kTdF10(4\pi)^2}\right)^{\frac{1}{4}} = 56.371 \text{ km} \Rightarrow \tag{10.93}$$
$$\Rightarrow d_{EWA} = R \cdot \cos(\alpha) = 52.91 \text{ km}$$

Here α is the angle of approach of the missile to the target. To recognize the effect of flickering interference, it is necessary to determine the estimates of the mathematical expectation of the measured angular coordinates of the jammers separately for each sign in each measuring plane:

$$\begin{aligned}
\overline{\delta_{\varepsilon}^{+}} &= \frac{1}{k}\sum_{j=1}^{k} \delta^{+}\varepsilon_{j}, \quad k = (j-1)+1 \\
\overline{\delta_{\varepsilon}^{-}} &= \frac{1}{l}\sum_{j=1}^{l} \delta^{-}\varepsilon_{j}, \quad l = (j-1)+1, \quad k+l = m \\
\overline{\delta_{\varphi}^{+}} &= \frac{1}{r}\sum_{j=1}^{r} \delta^{+}\varphi_{j}, \quad r = (j-1)+1 \\
\overline{\delta_{\varphi}^{-}} &= \frac{1}{s}\sum_{j=1}^{s} \delta^{-}\varphi_{j}, \quad s = (j-1)+1, \quad r+s = m
\end{aligned} \tag{10.94}$$

Based on the analysis of the data on the time interval $t \in [0, T_0)$ (see Figure 10.22), the following decision rules can be constructed:

$$U_{c3} = 1 : \begin{cases} \overline{\delta_{\varepsilon}^{+}} > \Delta_{\varepsilon} \times 0.05 \;\; \& \\ \overline{\delta_{\varepsilon}^{-}} < -\Delta_{\varepsilon} \times 0.05 \\ \quad or \;/\; and \\ \overline{\delta_{\varphi}^{+}} > \Delta_{\varphi} \times 0.05 \;\; \& \\ \overline{\delta_{\varphi}^{-}} < -\Delta_{\varphi} \times 0.05 \end{cases} \tag{10.95}$$

$$U_{c3} = 0 : \begin{cases} \overline{\delta_{\varepsilon}^{+}} < \Delta_{\varepsilon} \times 0.05 \;\; \& \\ \overline{\delta_{\varepsilon}^{-}} > -\Delta_{\varepsilon} \times 0.05 \\ \quad and \\ \overline{\delta_{\varphi}^{+}} < \Delta_{\varphi} \times 0.05 \;\; \& \\ \overline{\delta_{\varphi}^{-}} > -\Delta_{\varphi} \times 0.05 \end{cases} \tag{10.96}$$

It is clear that when firing on a collision course, the distance between the missile and the target decreases at a speed of $V_{mt} = V_m + V_t = 1856.58 + 2150 = 4006.58$ m/s, and the distance between the missile and jammers d_{EWA} decreases at a rate of $V_{mE} = V_m - V_{EWA} = 1856.58 - 1190 = 666.58$ m/s. The graph of the angular position of the interference source in the receiving base of the homing head antenna is shown in Figure 10.23. Here $d = 0.368$ km is the distance between jammers, $F = 2$ Hz is the switching frequency of the EA-18G Growler electronic warfare aircraft transmitters, which is set on the leading electronic warfare aircraft and is transmitted within the wireless computer network via a secure communication channel, $dt = 0.001$s is the time sampling interval, and $T_0 = 72.737$s is the analysis time. From Figure 10.23 it can be seen that automatic tracking by angular

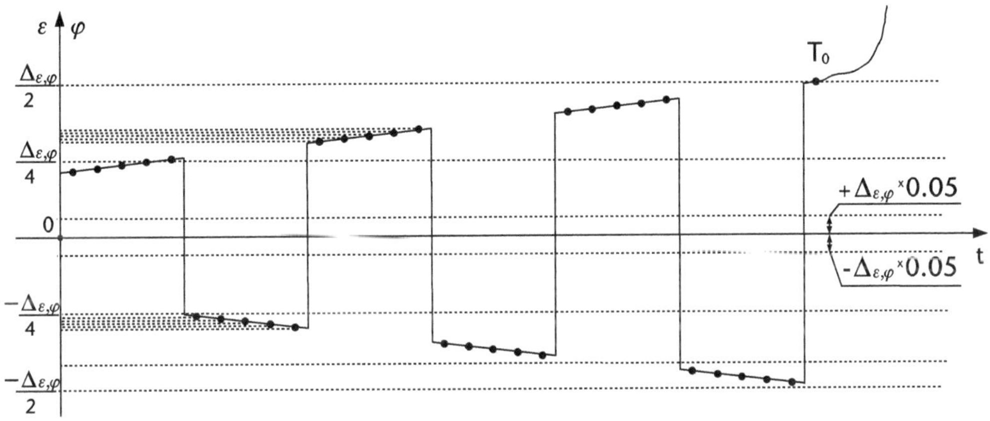

Figure 10.22 Flickering interference recognition principle.

coordinates will fail at a distance of $d_{EWA} \approx 13.5$ km. Interference reflected from the ground (water) surface A_4 is interference that is formed by illuminating an area of the Earth (water) surface, which is a significant region at the radio waves' reflection from the air-ground or air-water interface. This interference is generated by directing the main beam of the jammer's antenna at an angle $\varepsilon_N \approx -30°–90°$ relative to the velocity vector of aircraft, which is masked by this interference. As a rule, this interference is used by an aircraft that overcomes enemy air defenses at low and extremely low altitudes. In the case of such interference, a specular reflection of this aircraft is formed relative to the air-ground or air-water interface, provided that the signal reflected from the Earth (water) surface is exceeded E_N over the signal reflected from the real aircraft E_t:

$$10\log\left(\frac{E_N}{E_t}\right) \gg 10\text{ dB} \tag{10.97}$$

The action of this interference leads to reaiming of the homing head to a false target located in the imaginary region behind the plane of separation of the two media. In this case, there is a sharp change to the main beam position of the directional pattern of the seeker toward negative elevation angles $-\varepsilon_t$. If there is a computer on board the missile, it is easy to calculate the time instant t_a at which the interference is detected:

$$U_{c4} = 1 \; : \; \lim_{dt\to 0}\left[\frac{d\varepsilon_t}{dt}\right] \to -\infty \tag{10.98}$$

or otherwise:

$$U_{c4} = 0 \; : \; \lim_{dt\to 0}\left[\frac{d\varepsilon_t}{dt}\right] \le \Delta_\varepsilon \times 0.05 \tag{10.99}$$

After the moment of time t_a, a false (imaginary) target will be captured for autotracking. In order to prevent this, it is necessary to make a decision about which of the two having targets is reasonable to accompany. It is quite obvious that the decision to track an imaginary target will lead to the fact that the missile will collide with the Earth's (or water's) surface and will not fulfill the combat mission of

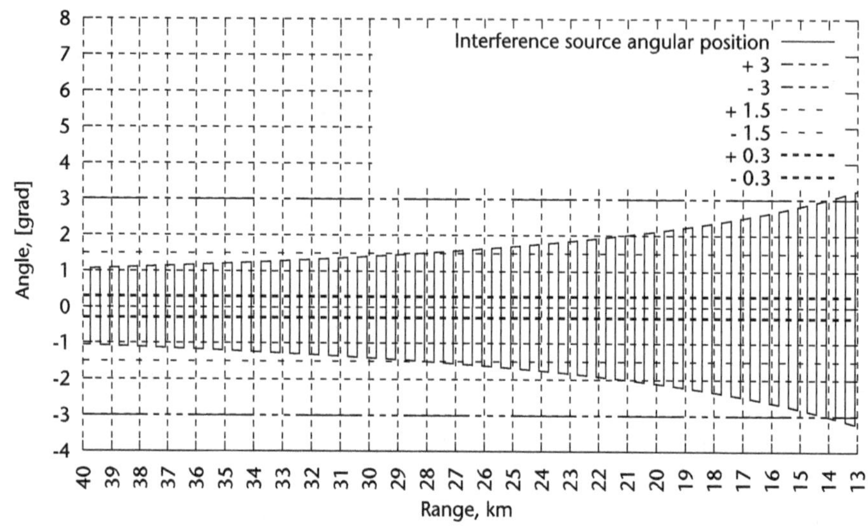

Figure 10.23 Dependence of the angular position of the seeker antenna in time (ranges to jammers).

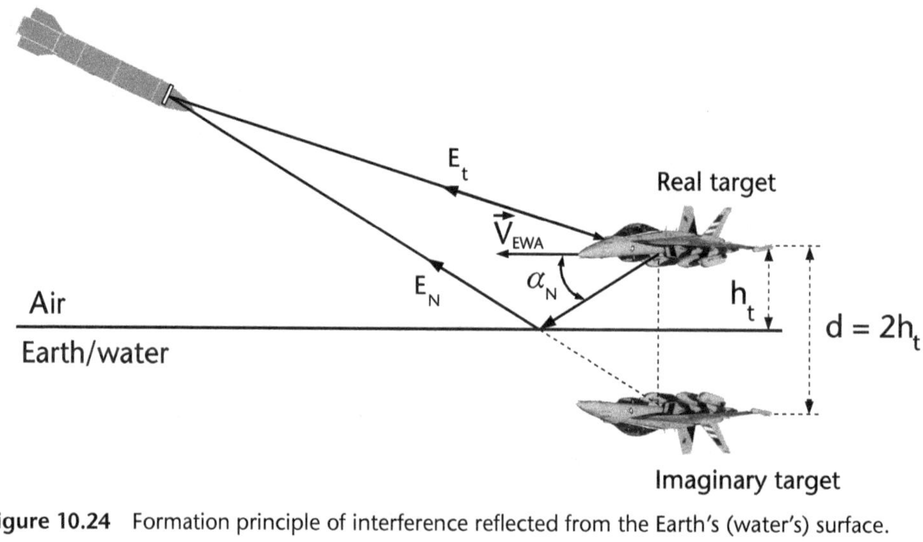

Figure 10.24 Formation principle of interference reflected from the Earth's (water's) surface.

destroying the designated target. Therefore, to prevent an incorrect decision, it is necessary to block the transition of the homing head to an elevation angle that is less than the mathematical expectation of the elevation angle obtained for the previous measurement cycles that preceded the time t_a. Moreover, it is necessary to aim at the interpolation point, which will be calculated by the onboard computer for each time t_j based on the current angular information and information from the onboard computer memory that was received before t_a.

Thus, for each interference H_i it is possible to construct a detector of this interference, the output signal of which takes only two values: $U_{ci} = \{1, 0\}$.

Based on the above, the distribution density of the feature sample $\mathbf{x} = (x_0, x_1, x_2, ..., x_q)$, which is determined by $\forall x_i = \{0, 1\}$, having a fixed size determined by the number of measuring channels corresponding to each ith feature can be represented as

$$f(\mathbf{x}) = \prod_{i=0}^{q} \prod_{j=1}^{m} \frac{1}{\sqrt{2\pi}\sigma_{x_i}} \times \exp\left\{\frac{-(x_{ij} - Mx_i)^2}{2\sigma_{x_i}^2}\right\} \tag{10.100}$$

or

$$f(\mathbf{x}) = \prod_{i=0}^{q} \left(\frac{1}{\sqrt{2\pi\sigma_{x_i}^2}}\right)^m \times \exp\left\{\frac{1}{\sqrt{2\pi\sigma_{x_i}^2}} \sum_{j=1}^{m} -(x_{ij} - Mx_i)^2\right\} \tag{10.101}$$

Passing to the vector form of notation and assuming that vectors $\mathbf{x}_j$ are independent and add up coordinate-wise, the expression for the distribution density of the sample $\mathbf{x}$ takes the form

$$f(\mathbf{x}) = (2\pi)^{-\frac{mq}{2}} (det K_j)^{-\frac{m}{2}} \exp\left\{-\frac{1}{2} \sum_{j=1}^{m} \left(\mathbf{x}_j - \overline{\mathbf{x}}\right)^T K_j^{-1} \left(\mathbf{x}_j - \overline{\mathbf{x}}\right)\right\}, \tag{10.102}$$

where $\overline{\mathbf{x}} = \frac{1}{m} \sum_{j=1}^{m} \mathbf{x}_j = \frac{1}{m} \sum_{j=1}^{m} (x_{1j}, x_{2j}, \ldots, x_{nj})$, K_j^{-1} - inverse covariance matrix, and: $K_j^{-1} = K_{\overline{x}}^{-1} + K_{\sigma_{x_j}}^{-1}$, $K_{\overline{\mathbf{x}}} = cov(\overline{\mathbf{x}}, \overline{\mathbf{x}}^T)$;

Due to the independence of vector elements $\mathbf{x}_j$ the matrix $K_{\sigma_{x_j}}^{-1}$ will be diagonal:

$$K_{\sigma_{x_j}}^{-1} = \begin{bmatrix} \sigma_{x_{1j}}^{-2} & 0 & 0 & \ldots & 0 \\ 0 & \sigma_{x_{2j}}^{-2} & 0 & \ldots & 0 \\ & \ldots & \ldots & \ldots & \\ 0 & 0 & 0 & \ldots & \sigma_{x_{nj}}^{-2} \end{bmatrix} \tag{10.103}$$

Based on the above, we can write an expression for one of the hypotheses of the likelihood function of the sample $\mathbf{x}$ on the space R^q using, for example, deterministic features:

$$f(\mathbf{x}|H_i) = (2\pi)^{-\frac{q}{2}} (det K_j)^{-\frac{m}{2}} \exp\left\{-\frac{1}{2} \sum_{j=1}^{m} \left(\overline{\mathbf{x}} - \mathbf{x}_i\right)^T K_j^{-1} \left(\overline{\mathbf{x}} - \mathbf{x}_i\right)\right\}, \tag{10.104}$$

where $\overline{\mathbf{x}}$ is the estimate of the average value of the feature vector for m measurement cycles, which is a point in the n-dimensional Euclidean space, $\mathbf{x}_i$ is an etalon of the ith class of interference. At the same time, it is assumed that the tasks of the formation and discretization of the feature space have been solved. As a result, after taking the logarithm, the likelihood function ratio will be written as

$$\ln(l(\mathbf{x})) = \ln(f(\mathbf{x}|A_0) - \ln(f(\mathbf{x}|H_i)) \leq \ln C(\alpha) \tag{10.105}$$

In this case, the recognition of the interference affecting on the homing head is supposed to be carried out according to the following decision rule:

$$\ln(l(\mathbf{x})) = \ln(f(\mathbf{x}|A_0) - \ln \sup_i(f(\mathbf{x}|H_i) \leq \ln C(\alpha) \tag{10.106}$$

It can be shown that the distribution (10.106) will obey the χ^2_{q-1} law with (q-1) degrees of freedom [2]. Thus, for each class of interference, a decision rule can be constructed:

$$\begin{cases} x_i[j] - x_i > C_i & \Rightarrow \quad Z_i = 1 \\ x_i[j] - x_i \leq C_i & \Rightarrow \quad Z_i = 0 \end{cases} \tag{10.107}$$

Here x_i is an etalon of the ith class of interference, $x_i[j]$ is the instantaneous value of the ith interference feature, and C_i is the decision-making threshold, the value of which for each ith interference has its own. $Z_i = 1$ is the command "turn on" technical measures to protect the homing head from interference H_i, and $Z_i = 0$ is the command "turn off" for interference H_i.

References

[1] Leonov, A. I., and K. I. Fomichev, *Monopulse Radar*, Norwood, MA: Artech House, 1986.

[2] Levin, B. R., and V. Schwartz, *Probabilistic Models and Methods in Communication and Control Aystems*, Moscow: Radio and Communication, 1985.

[3] Bakut, P. A., I. A. Bolshakov, B. M. Gerasimov, et al., *Questions Statistical Radiolocation Theory*, G. P. Tartakovsky (ed.), Moscow: Soviet Radio, 1964.

[4] Ivchenko, G. I., and Y. I. Medvedev, *Mathematical Statistics*, Moscow: Higher School, 1984.

[5] Shirman, Y. D., et al., *Radar Theoretical Foundations*, Book on Demand Ltd., October 29, 2019, ISBN-13: 978-5458432917.

[6] *Weapons Encyclopedia*, 2014-2021, https://wartools.ru/pvo-rossiya-sssr/zrs-s-400-triumf/#rpoint3.

[7] Budge, M. C., Jr., and S. R. German, *Basic Radar Analysis*, Second Edition, Norwood, MA: Artech House, 2020.

[8] Blake, L. V., and M. W. Long, *Antennas: Fundamentals, Design, Measurement*, New York: SciTech Publishing, 2009.

[9] Markov, G. T., and D. M. Sazonov, *Antennas. A Textbook for Students of Radio Engineering Specialties of Universities*, Second Edition, Revised and Enlarged, Moscow: Energy, 1975.

[10] Markin, E., "Scatter Pattern Calculations and Determination Performance Limitations of Existing Air Defense Systems for Intercepting Hypersonic Vehicles," *2012 IEEE Radar Conference*, May 7-11, 2012, Atlanta, GA, DOI: 10.1109/RADAR.2012.6212189.

[11] Ufimtsev, P. Y., *Theory of Edge Diffraction in Electromagnetics*, Encino, CA: Tech Science Press, 2009.

[12] Ufimtsev, P. Y., *Fundamentals of the Physical Theory of Diffraction*, Hoboken, NJ: John Wiley & Sons, 2007.

[13] Multifunction Monopulse Doppler Active Radar Seeker 9B-1103 (Diameter 350), http://www.missiles.ru/9Bxxx.htm.

[14] JSC Moscow Research Institute Agat, http://siagat.ru/aktivnye-rgs.

[15] Stimson, G. W., *Introduction to Airborne Radar*, Second Edition, New York: SciTech Publishing Inc., 1998.

[16] Garmash, S., A. Kishchinsky, V. Lapin, and M. Rogozhnikova, "Application of Technology of Passive Integrated Circuits on Gallium Arsenide in the Manufacture of Broadband Microwave Power Amplifiers," *Proceedings of the 9th International Crimean Microwave Conference*, CriMiCo, 1999.

[17] Stages of Phazotron, NIIR Work on Active Phased Antenna Arrays, http://www.missiles.ru/AESA_ph_5.htm.

[18] H010 "ZUK," Baltic State Technical University VOENMEH, http://library.voenmeh.ru/jirbis2/files/materials/ifour/book2/book_on_main_page/15.9.htm.

[19] Karpenko, A. V., "The Radar with an Active Phased Antenna Array (APAA) of the 'Zhuk' Military," Technical Collection, April 28, 2021, http://bastion-karpenko.ru/guk-a/.

[20] "Navy Completes F/A-18, AARGM-ER Flight with Separation Test Vehicle Naval Air Systems Command," May 7, 2021, https://www.navair.navy.mil/news/Navy-completes-FA-18-AARGM-ER-flight-separation-test-vehicle/ Fri-05072021-1035.

[21] "Boeing Successfully Tests EA-18G Growler Drones," *Aviation Explorer*, February 5, 2020, https://www.aex.ru/news/2020/2/5/208188/print/.

[22] "ALQ-99 Tactical Jamming System Naval Air Systems Command," https://www.navair.navy.mil/product/ALQ-99-Tactical-Jamming-System.

[23] Kopp, C., "The Anatomy of the TacJammer," *Australian Aviation*, January 1989, http://www.ausairpower.net/TE-Tacjammer.html.

CHAPTER 11

Assessment of an Anti-Aircraft Missile System Operation in Interference Conditions

When evaluating the operation of an anti-aircraft missile system in conditions of interference, the quality indicators of its radar, including the missile homing head, are the potential resolving power: in radial speed, in range, and in angular coordinates. At the same time, the achievement of high interference immunity can be achieved in different ways, and an indicator of the quality of the anti-aircraft missile system as a whole is the probability of hitting a given target with one missile. The fact of defeat is understood to mean the infliction of such damage on the target, which entails the termination of its combat mission. In this regard, it should be emphasized that the modern anti-aircraft missile system must provide an effective fight against low-flying aircraft and cruise missiles, as well as against hypersonic air vehicles. Currently, cruise missiles have a terrain enveloping system, which makes it possible for them to fly at extremely low altitudes. Under these conditions, the detection range, and hence the range of stable auto-tracking of such targets by ground-based radar, is sharply limited by topographic features and the curvature of the Earth's surface. In addition, the visible cross-sectional area (RCS) of the cruise missile head part is very small, and therefore, its effective scattering surface is small, which means that, the radar signal reflected from such a target is very weak. All this contributes to a reduction in the detection range and stable auto-tracking of a target without interference (clean target).

It is reasonable to assume that an increase in the range of destruction of low-flying targets can be achieved:

- Through the use of a radar placed on board a guided missile;
- The introduction of an active mode into the homing head, which will increase the range of destruction of targets flying at low and extremely low altitudes in the zone, located beyond the radio horizon, which is determined by the multifunctional ground radar station;
- The possibility of realizing the work of the active homing head according to the coordinate information of the mixed radar field, introduced on board the missile before its launch, with the subsequent elimination of the target designation and reference trajectory errors by means of angular scanning in $\Omega°$ solid angle;
- Ensuring the operation of the homing head in a semiactive mode, when the target assigned for destruction is tracked by a multifunctional radar station on the ground, and the guidance method through the missile is implemented.

Table 11.1

Target type (class)	H_t	V_t	D_{int}	P_{int}
B-52	10 km	950 km/h	400 km	0.84
jammers	10 km	≤ 950 km/h	370 km	0.84

11.1 Assessment Probability of Target Defeat by One Missile

In the S-400 anti-aircraft missile system, the circumstances of the assessment probability of target defeat by one missile allow, given the missile's flight and ballistic characteristics, to provide a predetermined engagement zone when receiving target designation from external means at the required ranges (Table 11.1). Here H_t, V_t is the target flight altitude and its speed, D_{int} is the target interception range measured from the point of the multifunctional radar (ground radar) location, representing the horizontal range, measured by geodetic maps of the area, $D_{int}|_{B-52} = 400$ km is the destruction range of strategic bombers with a large effective scattering area, which characterizes the very name of the S-400 anti-aircraft missile system, which, apparently, was specified in the tactics-technical characteristics, $D_{int}|_{jammers} = 370$ km is the defeat range of electronic warfare aircraft that interfere with the multifunctional radar and are located in loitering zones. This range is calculated on the assumption that

$$D_{int}|_{jammers} < D_{det}|_{jammers} \tag{11.1}$$

where $D_{det}|_{jammers}$ is the line-of-sight range of the low-altitude detection radar, which is part of the S-400 anti-aircraft missile system, which is calculated by the formula

$$D_{det}|_{jammers} = \sqrt{(R_E + h_{lad_1})^2 - {R_E}^2} + \sqrt{(R_E + H_{jam})^2 - {R_E}^2} = 374.587 \text{ km}, \tag{11.2}$$

where $h_{lad_1} = 23.8$m is the phase center height of the low-altitude detection radar 5N66 that has a universal tower 40V6M, or according to (11.2) $D_{det}|_{jammers} =$ 379.334 km, when used as part of a low-altitude detection radar 5N66 universal tower 40V6MD with height $h_{lad_2} = 38.8$m above the Earth's surface [1]. $H_{jam} = H_t = 10$ km is the flight altitude of an electronic warfare aircraft and a target above the Earth's surface, and P_{int} is a given probability of target defeat by one missile. The probabilities of targets defeat set for the anti-aircraft missile system lead to the need to use the performance characteristics of the radio fuse and the missile warhead. For the missile warhead, as such a characteristic, we can determine the radius of fragments dispersion R_0, which is expedient to characterize as the maximum distance from the point of detonation of the warhead, at which defeat the target with a probability not lower than the specified one. It is rather difficult to give a rigorous mathematical description of this indicator from a probabilistic point of view due to the multifactorial nature of defeat process an aerodynamic target. Some of these parameters are:

- Warhead charge mass;

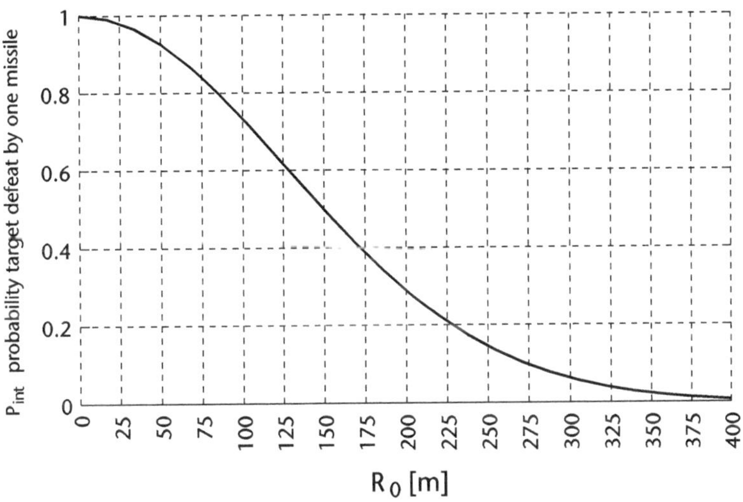

Figure 11.1 $P_{int} = f(R_0)$ is the probability target defeat by one missile.

- Size and geometry of warhead fragments;
- Characteristics of the target's vulnerability;
- Type of warhead explosion (directional, nondirectional);
- The degree of consistency in space of the sighting vector missile-target and the maximum of flux density vector of warhead fragments dispersion in a directional explosion.

It is quite obvious that with an increase in the distance from the point of detonation, the probability of causing damage to the target decreases, in which further performance of its combat mission becomes impossible. This process, for example, for the 5V28M missile of the S-200D anti-aircraft missile system, can be described by the distribution of the modulus of a Gaussian random variable with zero mathematical expectation (Figure 11.1). The steepness of the distribution function is determined by the variance of R_0 as a random variable and is estimated from a sample of data obtained as a result of numerical or nature experiments. The probability density function shown in Figure 11.1 is calculated for the case of firing at Tu-16 aircraft flying at a speed of $V_t = 250$ m/s, in this case, the missile speed was assumed to be equal to $V_m = 1000$ m/s, and its approach angle to the target $\hat{\vartheta}_m = -40.2°$.

The angle of approach to the target $\hat{\vartheta}_m$ in this case is an estimate of the mathematical expectation obtained by averaging over the set of trajectories of missiles aimed at low-altitude targets. The sample size was limited to $\hat{N} = 25$, since for $N > \hat{N}$ the values of the t-distribution change weakly for any confidence level. Their unbiased sample estimates were used as the variances of the random variables R_0 and $\hat{\vartheta}_m$. All results are valid for the case for optimal matching vector maximum of flux density of the scattering of fragments with the target. The function of coordinating the vector fragments scattering with the target and determining the moment of detonation of the warhead is determined by a radio fuse, the operation of which is further assumed to be optimal. Optimal matching is understood

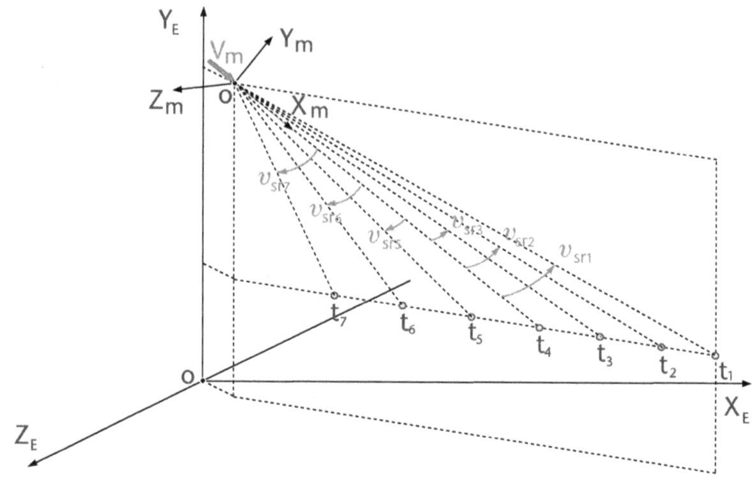

Figure 11.2 Actuation angles ϑ_{sr}.

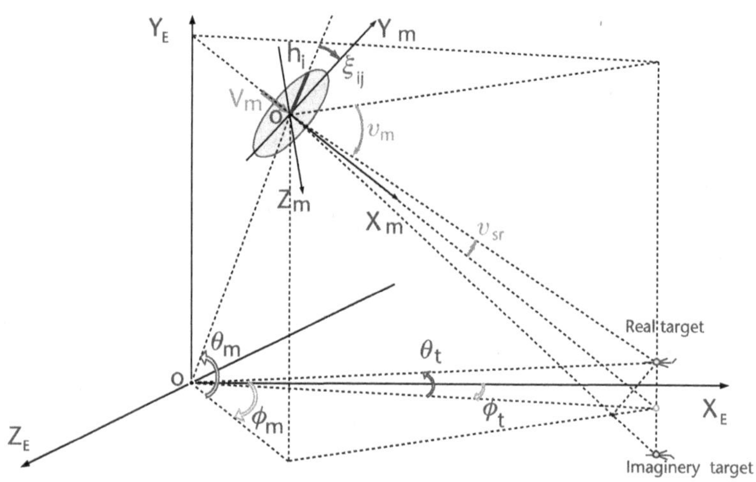

Figure 11.3 h_i is the miss and its phase ξ_{ij}.

in the sense that the coordinate law of hitting the target is calculated based on the given initial data, which makes it possible to determine the optimal firing angle $\hat{\vartheta}^*_{sr}$, at which the target will be defeated with the maximum probability. To find $\hat{\vartheta}^*_{sr}$, estimates of the probability of defeat were calculated for each ith value of the miss $(i = \overline{1, N})$ over all M values of its phase $(j = \overline{1, M})$. These circumstances are illustrated in Figures 11.2 and 11.3, on which the following designations are adopted: $X_E Y_E Z_E$ is the earth coordinate system in the center of which the ground radar is located; $Y_m O Z_m$ is the picture plane of the associated coordinate system (the coordinate system $X_m Y_m Z_m$ associated with the missile); ϑ_m is the angle of approach of the missile and the target; ϑ_{sr} is the angle of operation of the radio fuse; θ_t, φ_t, θ_m, φ_m are coordinates of the spherical coordinate system of the target and missile, respectively; h_i is the ith miss value; and ξ_{ij} is the jth phase value for the ith miss in the picture plane of the coordinate system, associated with the missile. The value of the probability of hitting a target in the situations under consideration can be estimated by the formula

$$\hat{P} = \exp\left[\frac{-h_0^2}{2(1+\sigma_0^2)}\right] \times \frac{1}{1+\sigma_0^2}, \tag{11.3}$$

where $h_0 = \frac{\hat{m}_{h\Sigma}}{R_0}$; $\sigma_0 = \frac{\hat{\sigma}_{h\Sigma}}{R_0}$ which are normalized value estimates of the magnitude of the miss and its variance; $\hat{m}_{h_\Sigma}$ and $\hat{\sigma}_{h_\Sigma}$ represent the expectation estimates and the unbiased estimate of the miss variance, respectively, from a sample of at least 20 realizations; R_0 is the the radius of fragments scattering at which the defeat occurs a given target with a probability of at least 0.7.

11.2 Radomes and Estimation of the Active Seeker Operating Wavelength

It is known that the diameter of the missile is 40N6 $D_m = 0.575$m [2, 3]. The missile fairing has an ogival shape, and the fairing length was adopted earlier as $L_f = 1.0$m, and thus the section of this fairing in the Z_m0Y_m plane is described by the following equations:

$$\begin{cases} Z^+(y_i) = \sqrt{(D_m^2 - y_i^2)} - \frac{D_m}{2} - D_m \times 2.273 \\ Z^-(y_i) = -\left(\sqrt{(D_m^2 - y_i^2)} - \frac{D_m}{2} - D_m \times 2.273\right) \end{cases} \tag{11.4}$$

Let us assume that the thickness of the fairing walls is $S = 2$ mm, and the gap between homing head antenna and its fairing is equal to $\delta s = 1$ mm and this gap is sufficient for mechanical movement of the waveguide-slot antenna array inside the fairing (Figure 11.4). As a result, the diameter of the antenna array can be estimated by the value $D_A = 2 \cdot R_A = 0.568$m. Due to the fact that the waveguide-slot antenna has a narrow bandwidth we can try to estimate the wavelength at which the homing head operates. From the course Microwave Technique [4] it is known that cross-sectional waveguides are widely used in various radars, since the deformation of the waveguide walls has little effect on the structure of the field, there is a high stability of the polarization plane, a high level of power of the emitted signal, and low losses in signal reception are provided that are reflected from the target. In waveguide-slot antenna arrays, rectangular waveguides are usually used. In the cross-sectional plane, the structure of the field in the rectangular waveguide corresponds to the structure of standing waves. The fields in a rectangular waveguide are denoted by the symbols E_{mn} and H_{mn}, where the subscript m denotes the number of field half-periods (half-waves) that fit on the wide wall of the waveguide a, and n characterizes the number of half-waves, which fits on the narrow wall of the waveguide b. Each wave has its own critical wavelength $\lambda_{critical}$ above which the wave does not propagate in the waveguide.

$$\lambda_{critical} = \frac{2\pi}{\sqrt{(\frac{m}{a})^2 + (\frac{n}{b})^2}} \tag{11.5}$$

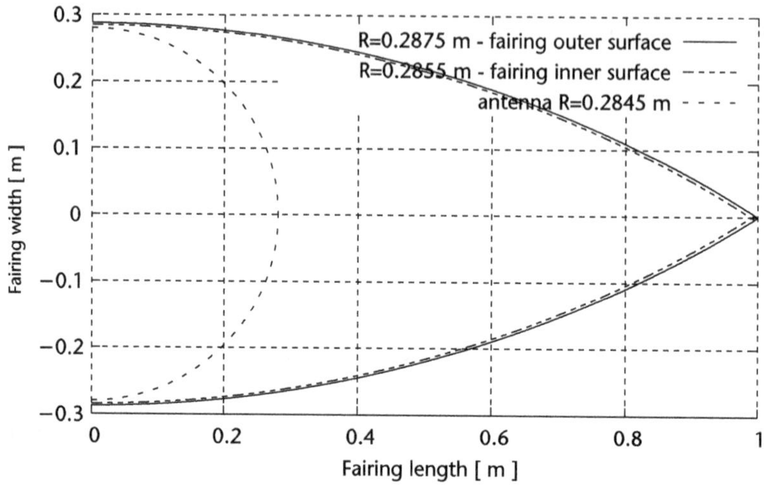

Figure 11.4 Fairing section by plane *ZOY*.

$\lambda_{critical} = max$ for $m = 1$ and $n = 0$; that is, this is the wave that is most often called the operating wavelength. Thus, the main wave in a rectangular waveguide is the H_{10} wave, for which:

$$\lambda_{\text{critical}}^{H_{10}} = 2a \tag{11.6}$$

If the condition is met:

$$\left.\begin{matrix} a \\ 2b \end{matrix}\right\} < \lambda < 2a, \tag{11.7}$$

then a single wave H_{10} will propagate along the waveguide and the right inequality (11.7) is the condition for the uniqueness of the fundamental wave, and the left inequality (11.7) is the condition for the nonpropagation of the fundamental wave in the waveguide. The operating wavelength range of a rectangular waveguide is the range within which the waveguide operates on the fundamental wave and has sufficient power and relatively low attenuation. The practical range is determined by the inequality

$$1.1a \leq \lambda \leq 1.8a, \tag{11.8}$$

where it is not recommended to choose $\lambda \geq 1.8a$, since in this case the power transmitted through the waveguide decreases and the attenuation increases. Option $\lambda \leq 1.1a$ is also not recommended to choose, since waves of higher types H_{20}, H_{01}, E_{11}, and H_{11} will attenuate weakly. It is quite understandable that the choice of the operating wavelength of the homing head of an anti-aircraft guided missile will be based on compromise considerations. On the one hand, the wavelength λ should be such that there is no scattering and reflection of an electromagnetic wave from meteorological precipitation (snow, hail, raindrops) so that the geometric

Table 11.2

Standard			*f, GHz*	*λ, mm*	*Section, mm*	*1.8a, mm*
GOST 51914-2002, Ru			6.85 ... 9.93	43.80 ... 30.21	28.50 × 12.60	51.30
GOST 51914-2002, Ru			8.15 ... 12.05	36.81 ... 24.90	23.00 × 10.00	41.40
GOST 51914-2002, Ru			9.93 ... 14.71	30.21 ... 20.39	19.00 × 9.50	34.20
GOST 51914-2002, Ru			11.55 ... 16.66	25.97 ... 18.01	17.00 × 8.00	30.60
EIA	RCSC	IEC				
WR137	WG14	R70	5.58 ... 8.20	53.76 ... 36.59	34.85 × 15.80	62.73
WR112	WG15	R84	7.05 ... 10.00	42.55 ... 30.00	28.50 × 12.62	51.30
WR90	WG16	R100	8.20 ... 12.40	36.59 ... 24.19	22.90 × 10.16	41.22
WR75	WG17	R120	10.00 ... 15.00	30.00 ... 20.00	19.05 × 9.52	34.29
WR62	WG18	R140	12.40 ... 18.00	24.19 ... 16.67	15.78 × 7.90	28.40

dimensions of raindrops or pieces of hail $\delta_{x,y,z}$ do not exceed a quarter of the emitted wavelength and the following inequality was fulfilled:

$$\delta_{x,y,z} \ll \frac{\lambda}{4} \tag{11.9}$$

Taking into account that $\delta_{x,y,z} \approx 5$ mm, therefore we can write:

$$\lambda > 4\delta_{x,y,z} = 20 \text{ mm} \tag{11.10}$$

The upper limit of the wavelength is determined by the diameter of the missile body on one side and the minimum characteristics for spatial target selection or the required width of the direction-finding characteristic of the seeker on the other side. In any case, the dimensions of rectangular waveguides are strictly determined and are stated in the corresponding [5,6] standards (Table 11.2). Based on (11.10), it is advisable to exclude from consideration $\lambda_4 = 25.97\ldots 18.01$ [mm] [5], and WR62 c $\lambda_5 = 24.19\ldots 16.67$ [mm] [6]. In this case, according to Table 11.2, the average estimate is:

$$\begin{aligned}
\overline{1.8a_i} &= \frac{1.8}{3}\sum_{i=1}^{3} a_i = 42.3 \text{ mm} \\
\overline{\lambda_2} &= \frac{1.1a_2 + 1.8a_2}{2} = \frac{25.30 + 41.40}{2} = 33.35 \text{ mm} \\
\overline{\lambda_2} &= \frac{1}{2}\sum_{i=L}^{H} \lambda_{2i} = \frac{36.81 + 24.90}{2} = 30.85 \text{ mm}
\end{aligned} \tag{11.11}$$

Which gives reason to choose a rectangular waveguide with a cross section of 23.0 × 10.0 mm or 28.50 × 12.60 mm. The real wavelength of the anti-aircraft missile system is unknown, and, as a rule, information about it is secret. Therefore, for simplicity of calculations carried out to evaluate the operation of the homing head of the anti-aircraft missile system, the operating wavelength $\hat{\lambda} = 0.04$m was chosen as the estimate for $\hat{\lambda}$.

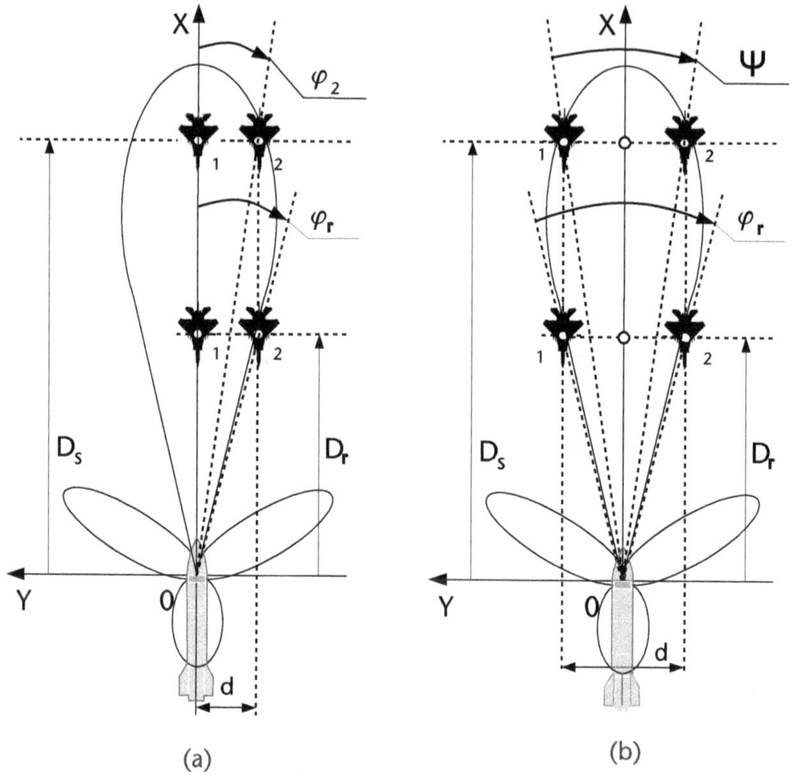

Figure 11.5 Resolution of two purposes: (a) while tracking the first target when $\varphi_1 \approx 0$, and (b) while tracking the energy center of two targets.

11.3 Operation of an Active Homing Head on a Group of Targets-Self-Cover Jammers

To assess the functioning of the active homing head (AHH), in a variety of conditions, a basic situation was selected. The group target is represented by two self-cover jammers emitting noise interference. At the initial segment of the range D_s, the targets are in the unsolvable volume of the total antenna radiation patterns of the active seeker and are located relative to each other at a distance d along the front. In the plane of the course, this situation is shown in Figure 11.5. Note that there are no fundamental differences in the analysis of the operation of the active homing head of an anti-aircraft guided missile and a ground multifunctional radar for a group of targets-self-cover jammers. When assessing the operation of the active seeker in this situation, the following designations are adopted: d is the distance in front between targets [m]; Ψ is the angular base of targets at range D_s; $F_\Sigma(\varphi)$ is the module of the complex antenna directivity pattern active homing head of the sum (Σ) channel in the plane φ; Δ is the half-width of the direction-finding characteristic of the active homing head angular tracking system; φ_r is the angle at which target resolution occurs; $\overrightarrow{OX}$ is the equisignal direction; φ_1 is the angular position of the sight line of the first target; φ_2 is the angular position of the sight line of the second target; h is a miss after resolving targets.

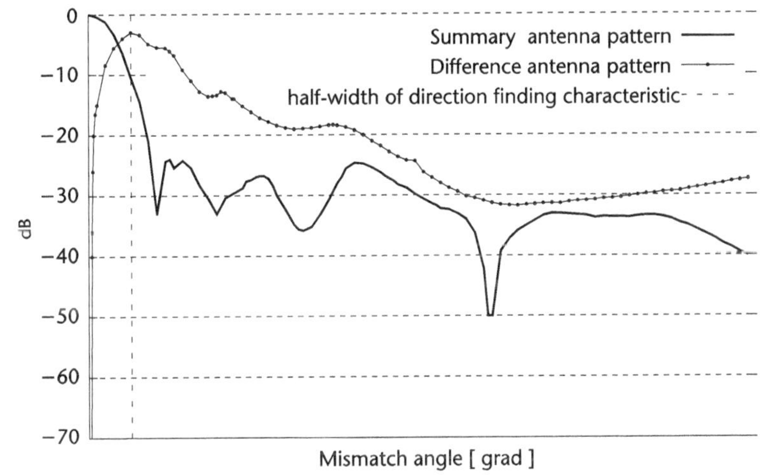

Figure 11.6 Calculated directional patterns of the active homing head in the course plane (plane H).

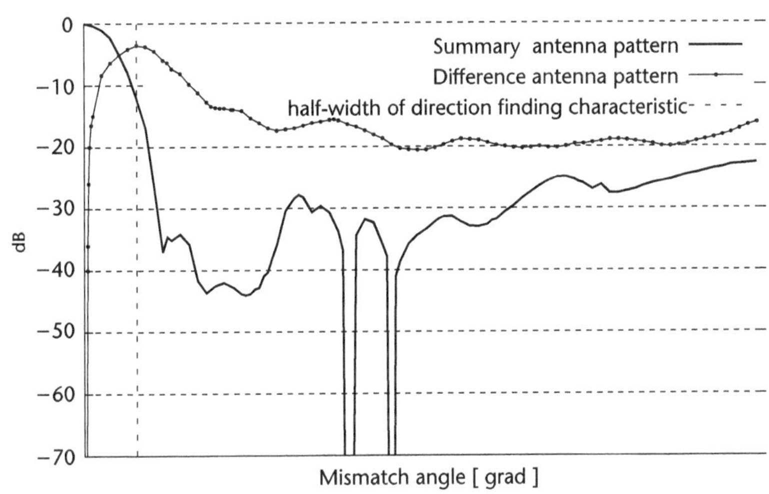

Figure 11.7 Calculated directional patterns of the active homing head in the elevation plane (plane E).

Directional patterns of the missile active homing head antenna for the sum channel and for the difference channels, calculated for the heading H and pitch E planes, are presented in Figures 11.6 and 11.7, respectively.

φ_r is defined as the width of the autocorrelation function $R(\psi)$ module of the complex antenna pattern active homing head of the sum channel $F_\Sigma(\varphi)$:

$$R(\psi) = \frac{\int_0^{\varphi_{max}} F_\Sigma(\varphi) F_\Sigma(\varphi + \psi) d\varphi}{\int_0^{\varphi_{max}} F_\Sigma^2(\varphi) d\varphi} \tag{11.12}$$

In (11.12) ψ is the angular shift relative to the equisignal direction ESD, $\varphi_{max} = 0.84 \times \varphi_{test}$ is the upper limit of integration in (11.12), and φ_{test} is the maximum

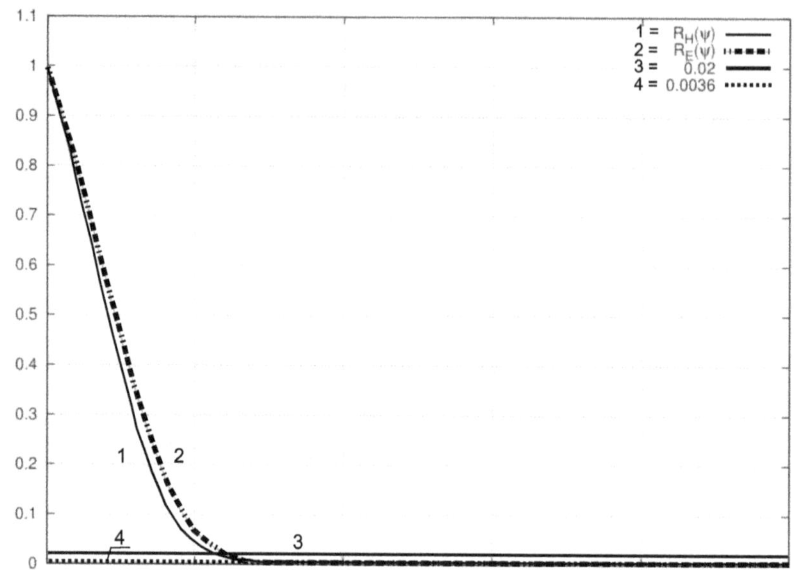

Figure 11.8 Autocorrelation functions: $R_H(\psi)$ in the azimuth plane and $R_E(\psi)$ in the elevation plane.

value of the angular position of the measuring antennas with respect to the same signal direction ESD of the antenna active homing head. Consequently, the capabilities of the active homing head for automatic tracking of targets in the mutually orthogonal planes H and E are characterized by the values of the half-width of the direction-finding characteristic Δ_H and Δ_E, which are a solution to the ψ equation:

$$R_{H,E}(\psi) = 0.02 \tag{11.13}$$

and are determined by the following values:

- For control channel on the course (azimuth), (plane H): $\Delta_H = A^\circ \Rightarrow 2\Delta_H = 2A^\circ$
- For control channel in elevation angle, (plane E): $\Delta_E = 1.081881A^\circ \Rightarrow 2\Delta_E = 2.163763A^\circ$

This is confirmed by the direction-finding characteristics calculated for the missile's active homing head shown in Figures 11.10 and 11.11. From Figure 11.10 it can be seen that the calculated direction-finding characteristics of the active homing head are somewhat different from the ideal ones shown in Figure 11.11. This difference consists of the presence of a dead zone, which is characterized by the fact that in a certain vicinity of the point $\varphi_t \approx \varepsilon_t \approx 0^\circ$ the gain of the direction-finding characteristic is positive, but has a small value. Therefore, for a real active homing head, we can write

$$k_H = \tan\alpha_H = \begin{cases} 0.00606, & \text{if } |\varphi_t| \leq 0.09470483\Delta_H^\circ, \\ 0.17421 & \text{otherwise} \end{cases} \tag{11.14}$$

$$k_E = \tan\alpha_E = \begin{cases} 0.00909, & \text{if } |\varepsilon_t| \leq 0.07788162\Delta_E^\circ, \\ 0.16103 & \text{otherwise} \end{cases} \tag{11.15}$$

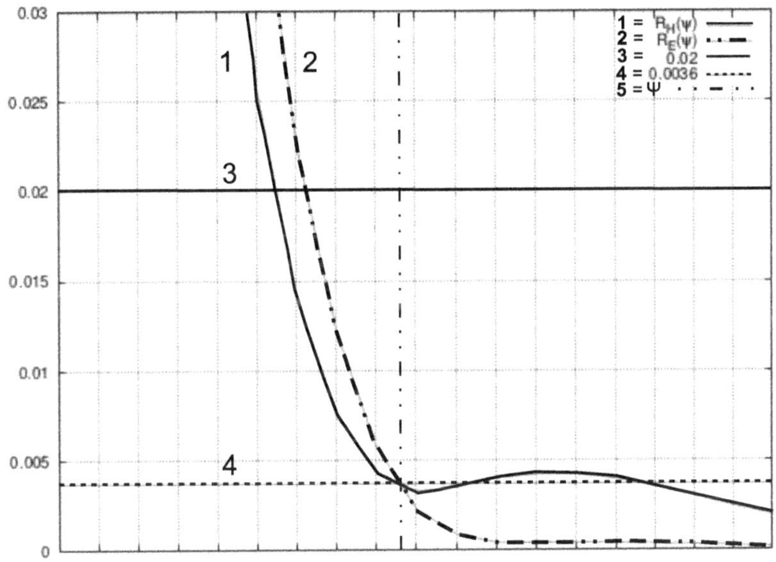

Figure 11.9 Autocorrelation functions: $R_H(\psi)$ in the azimuth plane and $R_E(\psi)$ in the elevation plane.

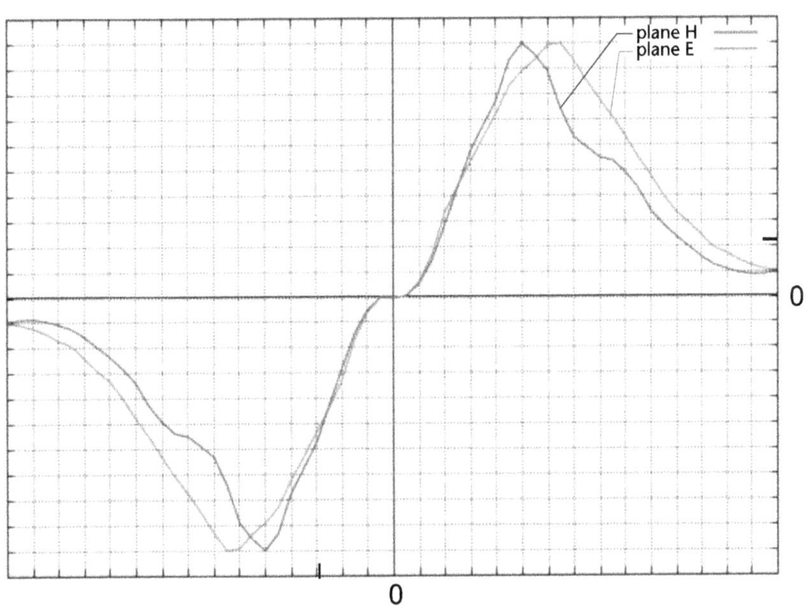

Figure 11.10 Calculated direction-finding characteristics.

In turn, the resolving power of the active homing head in the mutually-orthogonal control channels φ_r is a solution $\psi = \sup \varphi_r(\rho)$ equations:

$$R_H(\psi) = R_E(\psi), \tag{11.16}$$

where the ψ° solution corresponds to the 0.0036 level of both correlation functions, and $\rho = \frac{P_2}{P_1}$ is the ratio of the signal powers of jammers 2 and 1. Therefore, the

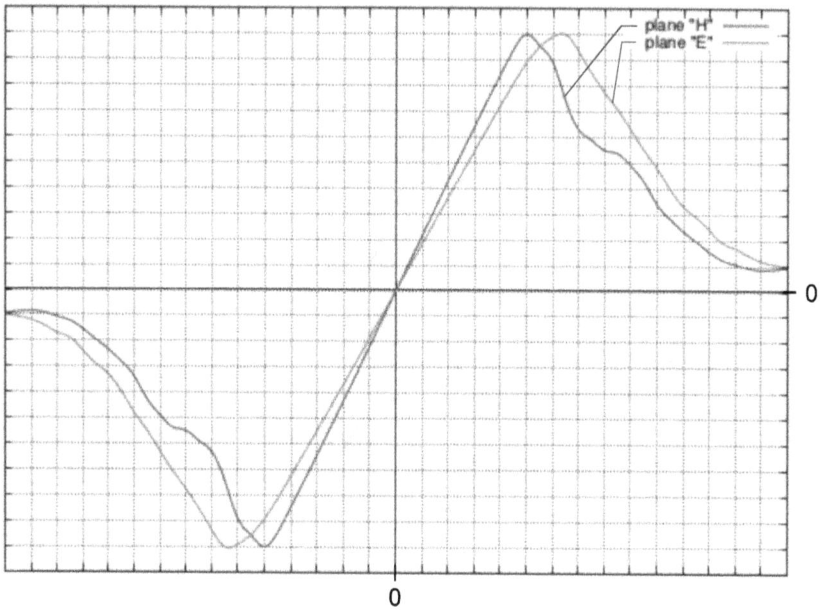

Figure 11.11 Linearized direction-finding characteristics.

resolution of the active homing head in the mutually orthogonal planes $2\varphi_r$ has the following meaning:

$$2\varphi_r = 2\psi^\circ \tag{11.17}$$

This value characterizes the capabilities of the active homing head for spatial selection of targets in the group, as well as the solid angle of the conical scanning during target search. The equations for the approximated direction-finding characteristic as a function of the misalignment angle e in mutually orthogonal planes obtained from the calculated values can be written as

$$S_H(e) = a_1 e - a_2 e^3 \tag{11.18}$$

$$S_E(e) = b_1 e - b_2 e^3 \tag{11.19}$$

The solution of equation

$$\frac{dS_{H,E}(e)}{de} = 0 \tag{11.20}$$

gives the following extrema of functions for the mutually orthogonal planes H and E:

$$\begin{cases} |e|_H = 0.9738675A^\circ \\ |e|_E = 1.1184668A^\circ \end{cases} \tag{11.21}$$

A comparative analysis of two approaches to assessing the capabilities of a missile homing head for spatial selection of targets in a group is given in Table 11.3. Therefore, according to the directional diagram of the sum channel of the seeker, it is possible to estimate its direction-finding capabilities, as well as the required level

Table 11.3

	direction-finding characteristic extremum	*correlation function at level 0.002*
plane H	$0.9738675A°$	$A°$
plane E	$1.1184668A°$	$1.0818815A°$

of interference along the sidelobes, leading the missile homing head to its complete inoperability. In this case, the excess of the jamming signal in relation to the signal from the tracked target should be at least 24.5 dB.

Based on the obtained angles of spatial resolution, the range between the missile and the targets of the group at which their spatial resolution occurs is represented as

$$D_r = \frac{d}{\tan(\varphi_r)} \tag{11.22}$$

The above conditions characterize the most difficult case for an active homing head, when targets emit interference signals of equal power and, therefore, in this case $\rho = 1$.

The difficulty of working on a group target is that the noise interference does not always allow the active homing head to select targets by radial velocity.

Spatial selection of targets in a group is also impossible. The missile is guided by angular coordinates to the equivalent center of jamming signals (in the absence of their coherence). In general, for M jammers, the position of the equivalent energy center in one plane is determined by the equation:

$$S_{\psi,\theta}(e,t) - \sum_{i=1}^{M} \rho_i(t)\bar{F}_{\psi,\theta}(e_i,t) = 0, \tag{11.23}$$

where $\bar{F}_{\psi,\theta}(e_i,t)$ - the value of the directional pattern by the power of the heading channel (yaw ψ) or the pitch channel (θ), normalized by the voltage of the automatic gain control (AGC) of the sum channel for the ith jammer at time t, and e_i is the angular position of the ith jammer in the receiving base of the missile seeker antenna in the ψ and/or θ pitch channel.

$\rho_i(t) = \frac{P_i(t)}{P_1(t)}$ - weight coefficients characterizing the power ratios of jammers' signals $\forall i = 2, \ldots, M$ and $S_{\psi,\theta}(e,t)$ is the error signal of the angular tracking system of the active homing head of the missile at the rate ψ (pitch θ) when operating along the equivalent energy center formed in space by the signals of jammers. In the process of solving (11.23), the tracking error value and the angular coordinates of the equisignal direction (ESD) of homing head antenna are determined. It is assumed that jamming aircraft fly without maneuver in the course and pitch planes, do not maneuver by speed, and that the magnitude of the signals reflected by the jamming aircraft hulls is negligible relative to the jamming signals emitted by them. Unequal emission of jamming signals by jammers leads to the predominant orientation of the equal-signal direction of the active homing head (AHH) to the jammer with higher power. Figure 11.12 shows the case when $P_2(t) > P_1(t)$.

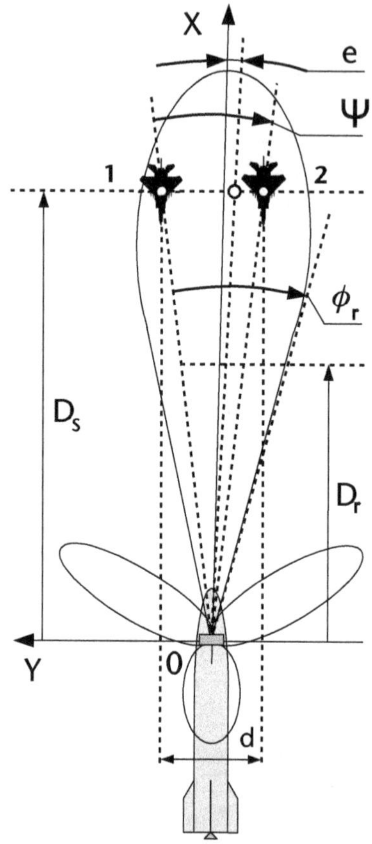

Figure 11.12 Missile homing head operation in conditions of jamming from two points.

In the presence of two sources of interference, which are within the width of the difference directional diagram of the active homing head normalized by its sum directional diagram, the missile homing head will track the direction on the energy center of the interference created by two jammers. The position of this energy center is determined by the expression [7,8]:

$$\frac{e(t)}{\Psi} = \frac{P_1(t) - P_2(t)}{P_1(t) + P_2(t) + P_{s1}(t) + P_{s2}(t)}, \tag{11.24}$$

where Ψ is the angle between the lines of sight: missile, target # 1 and missile-target # 2, and e is the angle between the direction to the energy center of the power emitted by the two jammers and the bisector of the angle Ψ. $P_1(t)$ and $P_2(t)$ are the interference powers emitted by targets # 1 and # 2. $P_{s1}(t)$ and $P_{s2}(t)$ are the powers of the useful signal reflected from targets # 1 and # 2. From (11.24) it follows that the position of the energetic center is determined mainly by the ratio of the power of the interference transmitters and the nature of the change in the interference power over time. When the interference transmitters are switched on alternately, the energetic center will wander within the [1, 2] segment connecting target # 1 and target # 2. When setting a flickering interference, the automatic tracking system in the direction will tend to track one or the other target, the result of which the antenna system of the missile homing head will swing in time with the switching in accordance with the interference signals. This will significantly complicate the determination of the angular coordinates of targets. The movement of the guided missile will occur along a curved trajectory, the amplitude of the oscillations will increase as the missile approaches the targets, and in the case when

$$\Omega > \Delta_H \qquad \text{or/and} \qquad \Omega > \Delta_E, \tag{11.25}$$

the automatic tracking of the target in angular coordinates will be disrupted, and as a result none of the two targets will be destroyed, and therefore, the combat mission of the anti-aircraft missile system will not be completed. This will happen when, in accordance with the missile operation algorithm, after failure of auto-tracking of the target and a negative result for its search, the missile receives the maximum command "up" and after a given time interval the missile self-destructs by detonating its warhead. As the missile approaches the targets, the angle between the lines of sight of the targets increases and the moment comes when the angular position control system of the seeker antenna begins to distinguish targets by angular coordinates. From this moment, the missile will be guided to one of the targets. However, the time to eliminate the initial guidance error on the selected target may not be enough, since the missile's maneuverability is determined by the minimum possible radius of its turn, which, in turn, depends on the speed of movement and the permissible transverse overloads of the missile. This circumstance can lead to a significant miss h when missile guidance at the target. The amount of miss in linear units can be determined from the [7] expression

$$h = \frac{d}{2} - \frac{1}{2} \cdot \frac{ngd^2}{V_{mt}^2 \cdot \varphi_r} \tag{11.26}$$

Here d is the linear distance (base) between two targets, n is the maximum permissible transverse missile overflocking, $g = 9.81$ m/s is the free-fall acceleration, V_{mt} is the relative speed of the missile approaching the target, and φ_r is the critical angle of target resolution in in the group. The switching frequency of $F_{flicker}$ transmitters is limited by the [8] expression:

$$F_{flicker} \leq \frac{\Delta F_{hh}}{2}, \tag{11.27}$$

where ΔF_{hh} is the bandwidth of the tracking system of the homing head. When $F_{flicker} > \frac{\Delta F_{hh}}{2}$, the homing head will track the position of the energy center of the interferers. This case is shown in Figure 11.5(b),where provided that the interference powers $P_1 = P_2$ emitted by jammers 1 and 2 are equal, the energy center will be aligned with the geometric center of these jammers. Usually the value of the flicker frequency $F_{flicker}$ is selected in the range from 0.5 to 10 Hz [8]. Considering the case of unequal radiation of jamming signals by jammers (Figure 11.12) and assuming that φ_1 and φ_2 are tracking errors of the first and second targets, respectively, on the basis of the homing head receiving antenna; $\Delta\varphi_r$ is an error worked out by the missile control loop until the warhead was detonated, provided that the targets were resolved at a range of D_r. Thus, the miss magnitude h can be estimated with equation:

$$h = D_r \left[\tan(\varphi_1) - \tan(\Delta\varphi_r)\right] \tag{11.28}$$

Considering that in practice the values φ_1 and $\Delta\varphi_r$ have small values, the expression (11.28) is represented as an approximate equality:

$$h \approx D_r \left(\varphi_1 - \Delta\varphi_r\right) \tag{11.29}$$

The value of $\Delta\varphi_r$ can be estimated as the value of the angular error that occurred at the time of the encounter (detonation of the warhead) when working out the mismatch φ_1. Obtaining the dependence $\Delta\varphi_r(t)$ analytically is a rather difficult task, since the control loop of the missile operates in a nonlinear and nonstationary mode. To obtain a qualitative dependence of the magnitude of the miss on the missile parameters and the aircraft-jammers order of location in 3-D space, the function $\Delta\varphi_r(t)$ can be approximated by an exponential dependence describing the transient process in the aperiodic link with the time constant T_e:

$$\Delta\varphi_r(t) = \varphi_1 \left(1 - \exp(-t/T_e)\right) \qquad \& \qquad T_e = 0.8 \text{ sec.} \tag{11.30}$$

In this case, the value of the error that has not been worked out by the contour is represented in the form:

$$\varepsilon = \Delta\varphi_1 \exp(-\Delta t_r), \tag{11.31}$$

where $\Delta t_r = (t_b - t_r)$ is the time interval from the moment the targets of the t_r group are resolved to the moment t_b of the missile warhead detonation (to the point where the missile meets the target). Wherein:

$$\Delta t_m = \frac{D_r}{V_{av}}, \tag{11.32}$$

where D_r is the range missile-target after resolving targets in the group and choosing a target for defeat, V_{av} is the average speed of the missile approaching the target selected for defeat from the moment in which the targets in the group are resolved to the meeting point. Assuming, that channels of direction finder sum-and-difference circuit are phased, have equal gains, and are invariant in time, the expression for the error signal can be written as:

$$S(\Delta\varphi) = \mu\varphi_1, \tag{11.33}$$

where μ is the slope of the direction-finding characteristic of the angular tracking system of the active homing head on a linear section. At the time of resolution, the angles between targets will be equal to their angular base α:

$$\alpha = \Delta\varphi_1 + \Delta\varphi_2 \approx \frac{d}{D_r} \tag{11.34}$$

in this case, expressions for tracking errors are written in the form:

$$\begin{cases} \Delta\varphi_1 = \xi_1(\rho)\alpha \\ \Delta\varphi_2 = (1 - \xi_1(\rho))\alpha, \end{cases} \tag{11.35}$$

where $\xi_1(\rho) = \frac{\rho}{1+\rho}$ is the weighting function that determines the signal errors of the first target; $\rho = \frac{P_2}{P_1}$ is the power ratio of jammers signals $\forall\rho \in [0, 1]$.

Thus, at the moment of targets' resolution, at least one target will be on the border of the linear section of the direction-finding characteristic. So, for the second target, we can write:

$$\Delta\varphi_2 = \Delta = (1 - \xi_1(\rho))\alpha, \qquad \text{from where} \quad \alpha = \frac{\Delta}{1 - \xi_1(\rho)} \tag{11.36}$$

and, therefore:

$$\Delta\varphi_1 = \frac{\xi_1(\rho)\Delta}{1 - \xi_1(\rho)} \tag{11.37}$$

Substitution of expressions for $\Delta\varphi_1$ and Δt_r in (11.31) results in an expression for evaluating the target tracking error not worked out by the missile control loop:

$$\varepsilon = \frac{\xi_1(\rho)\Delta}{1 - \xi_1(\rho)} \times \exp\left[-\frac{d(1 - \xi_1(\rho))}{\Delta V_{av} T_e}\right] \tag{11.38}$$

Therefore, the corresponding miss estimate is represented as:

$$h = d\xi_1(\rho) \times \exp\left[-\frac{d(1 - \xi_1(\rho))}{\Delta V_{av} T_e}\right] \tag{11.39}$$

Based on the calculation results presented in Figures 11.13 and 11.14, the following conclusions can be drawn:

1. The most difficult conditions are realized when the active homing head operates by jammers, emitting interference signals of equal power. This achieves the maximum miss value and the minimum defeat probability target from the group by one missile.
2. A decrease in the power ratio of interference signals from 1 to 0 leads to a shift in the energy center of the group interference signal to a target with a higher power, and consequently, to decrease miss and increase likelihood target defeat by one missile.
3. Proceeding from the fact that defeat probability of target with one missile $\hat{P} = 0.84$ corresponds to a miss $h \approx 50$m (Figure 11.14), at a distance along the front $d = 100$m between two jamming planes, both of these planes will be destroyed by one missile with probability $\hat{P} = 0.84$, provided that these planes emit jamming signals of equal power $\rho = 1$. Under the same conditions in terms of the interference power $\rho = 1$ and the probability of two aircraft defeat

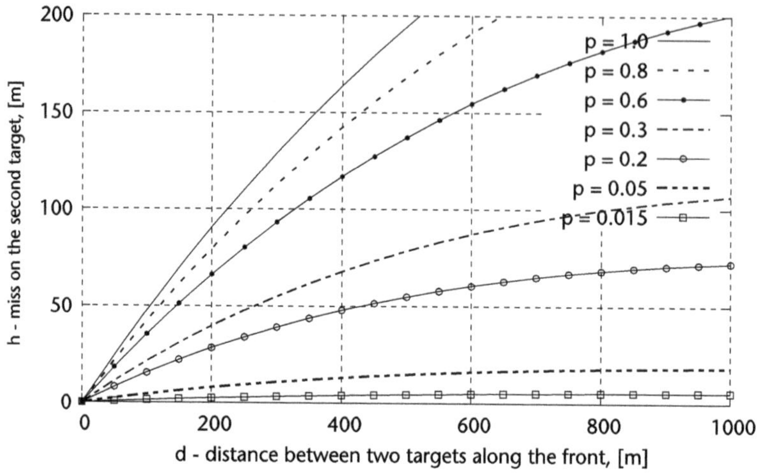

Figure 11.13 Dependence of the miss h on the distance d between targets in the group.

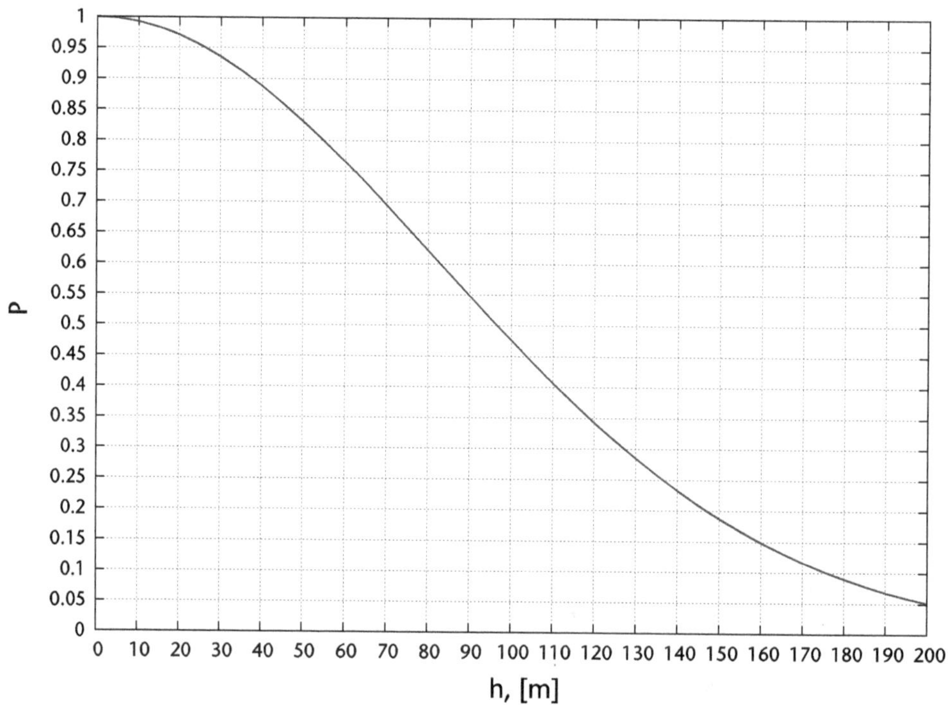

Figure 11.14 Dependence of the estimate of the probability of the target defeat $\hat{P}$ on the miss h.

$\hat{P} = 0.05$ (Figure 11.14), the distance in the front between two jamming aircraft should be at least $d = 400$m, at the same time, at a distance along the front of $d = 250$m, the defeat probabilities of two aircraft $\hat{P} = 0.31 < 0.5$ are realized and can be considered as the minimum safe distance along the front between two jamming aircraft under the condition $\rho = 1$, when the missile is aimed at the energy center of the jamming signals.

11.4 Probability a Target is Defeated by One Missile

The density function of the probability of hitting the target is calculated for the case of firing on a Tu-16 aircraft flying at a speed of $V_t = 250$ m/s. In this case, the missile speed was assumed to be $V_m = 1{,}000$ m/s, and its approach angle to the target was $\hat{\upsilon}_m = -40.2°$. In this case $\hat{\upsilon}_m$ is considered as an estimate of the mathematical expectation, which is the average value of the ensemble of realizations obtained for the subspace of the affected area, the heights of the points of which belong to the set of possible heights {H}.

The sample size is limited to $\overline{N} = 25$, since for $N > 25$ the value of the t-distribution changes weakly for any confidence level. Its unbiased sample estimate was used as the variance. After restricting the set {H} from below $H_t = 20$m and from above $H_t = 100$m minimization of the functional was carried out:

$$Q = \inf F(H_m, H_t, D_{mt}), \tag{11.40}$$

where $Q = \hat{P}$ is an estimate of the probability of defeat Tu-16 aircraft flying at an altitude of H_t, provided that the missile was guided to the mirror mark; H_m is the missile flight altitude at which the target is captured by the active homing head for auto-tracking; D_{mt} is the horizontal distance between missile and target realized after the target is captured; {H} is a limited set of possible target flight altitudes. As a result of calculations, the following was obtained:

$$\forall Q = 0.7, \quad max H_t \in \{\mathbf{H}\}: \quad H_t = 30\text{m} \tag{11.41}$$

This result is valid for the case of optimal matching the flux density vector maximum of the scattering of fragments with a target. For this, in accordance with the given initial data, the coordinate law of target hitting was calculated, which allows to determine the optimal firing angle υ^*, at which the target is defeated with the maximum probability. To find the optimal value of υ_{opt}, estimates of defeat probability were calculated for each ith value of the miss $h_i, \forall i = \overline{1, N}$ for all M values of its phase $j = \overline{1, M}$.

Thus, when firing at targets flying at low and extremely low altitudes, a mirror mark is formed from the target, which lies below the interface between the two media: air-ground and/or air-water. When aiming the missile at the mirror mark from the target, the real target can be defeated with probability $P \geq 0.7$ at altitudes less than 30m under the condition of optimal matching of the vector of the maximum density scattering of warhead fragments with a target.

As a result, to increase the probability of hitting low-altitude targets flying at altitudes above 30m, it is necessary either to increase the charge power, which increases its weight and entails a sharp decrease in the missile's flight-ballistic characteristics, or to increase the resolution of the active homing head in angular coordinates.

11.5 Operating Principles of a Modern Anti-Aircraft Missile System on a Group Target

When the anti-aircraft missile system is operating against a group of self-cover jammers, this group target is accompanied by a multifunctional radar in the automatic tracking mode in angular coordinates AC_{φ}. Depending on the number of targets in the group, it is possible to assign the required number of slave target channels of the multifunctional radar and target firing channels from the group to the same number of missiles, which are required for each dedicated target channel.

In the absence of information about the range to the target in the multifunctional radar, target firing, accompanied in automatic tracking by angular coordinates AC_{φ}, intelligently produce either with coordinate support from the command post of an anti-aircraft missile system or in firing with unknown range FUR mode.

After the missile is launched, during its flight, the multifunctional radar measures the coordinates of the missile and the target, and the missile homing head measures only the coordinates of the target. These measurements are the initial data for calculating the range to the target by the triangulation method. Triangulation range can be used to adjust the filter for statistical processing of target coordinates, for example, every 5 seconds, starting from the moment when the angle between the directions "multifunctional radar-target" and "missile-target" reaches $> \psi°$ (Figure 11.16), and also in the 2 seconds before the meeting point and at the time of the meeting. Shooting at the self-cover jammers from the group in the absence of information about the range to them can be carried out both with single missiles and with a burst. In this case, the range to the first missile, at the time of its meeting with the target, which characterizes the instantaneous value of the range to the target, is reasonable to use to adjust the range of the tracking system of the multifunctional radar. The range determined in this way is entered one time into the tracking system and into missile launch appliance. In this case, it becomes possible to transfer the multifunctional radar from the FUR firing mode to the firing mode with the presence of information about the range to the target.

An active seeker can acquire a target signal approximately 1 second after beginning control. After switching to the homing mode, a continuous calculation of the triangulation range to the target should be made. When the angle between the directions "active homing head-target" and "multifunctional radar-target" of the value $\psi°$ (Figure 11.16) is reached, it is necessary to switch from the homing mode to the command method of guiding the missile to the target. In this case, it is reasonable to use the range calculated by the triangulation method as the multifunctional radar-target range.

It is reasonable to assume that starting from a distance between the missile and the target of several kilometers, there is a transition from the command method of guidance to the homing mode, and from a range of $\approx$1 km, at which for all possible parameters of the group formation, all targets from its composition are practically selected by the active homing head by angular coordinates. In this case it is reasonable to carry out missile guidance to the target only by coordinate information from the active homing head.

To increase the probability of hitting one of the targets in the group a few seconds before the meeting point, it is necessary to deflect the main beam of the multifunctional radar at a small angle, which will provide the primary illumination of one of the targets. It is also advisable to deflect the illumination beam when firing using information about the range, which comes from a multifunctional radar or from target designation data received either from the command post of an anti-aircraft missile system or from a higher command post. It is reasonable to assume that if the difference between the Doppler frequencies of the target measured by the active seeker and the multifunctional radar exceeds a value of several kHz, then it is necessary to retarget the active seeker to the frequency measured by the multifunctional radar.

If the signal reflected from the target disappears in the active homing head, it is necessary to delay the transition to the search by the Doppler frequency by several clock cycles (periods of the clock frequency generator).

11.5.1 The Work of an Active Homing Head on a Group of Targets: Self-Cover Jammers

To assess the functioning of the active homing head in a given variety of conditions, a basic situation has been selected. The group target is represented by two jammers with a spectral power density of 500 W/MHz. An order of aircraft following is Bearing. Bearing is an order of aircraft following, in which they fly parallel courses with at a distance dy, but with acertain distance between them dx along the vector of their speed. See Figure 11.15. At the initial section of the range D_2, the targets are in the unresolved volume of the active homing head antenna pattern and the missile is aimed at the energy center of the jamming signals emitted by jamming aircraft #1 and #2. It is quite understandable that if the computer systems of both jamming airplanes are located in a closed wireless computer network, then in this case flickering interference can be realized by synchronous power switching.

The task being solved is to estimate the probability of hitting one of the targets in the group. In the heading plane, this situation is shown in Figure 11.15. The difficulty of this situation lies in the fact that the use of noise interference by the enemy does not allow the active homing head to select the targets of the group in terms of radial velocity. Spatial selection of targets in a group is also impossible (Figure 11.15) and the missile is guided by angular coordinates to the equivalent energy center of jamming signals in the absence of their coherence.

For M jammers, the position of the equivalent energy center is determined by the equation

$$S^{K,T}(\varphi) - \sum_{i=1}^{M} P_i(t)\bar{F}^{K,T}(\varphi_i, t) = 0, \tag{11.42}$$

where $\bar{F}^{K,T}(\varphi_i, t)$ is the value of the directional pattern for the power of the course (K) or pitch (T) channel, normalized by the voltage of the AGC system of the sum channel for the ith jammer aircraft at time t, and $S^{K,T}(\varphi)$ is the error signal of the active homing head tracking system of the course (K) or pitch (T) channel. In the process of solving (11.42), the value of the tracking error and angular coordinates of the antenna equal-signal direction of the active homing head are determined. It is assumed that jammers fly without maneuver in the heading and pitch planes, and also do not maneuver by speed. It is reasonable to assume that the magnitude

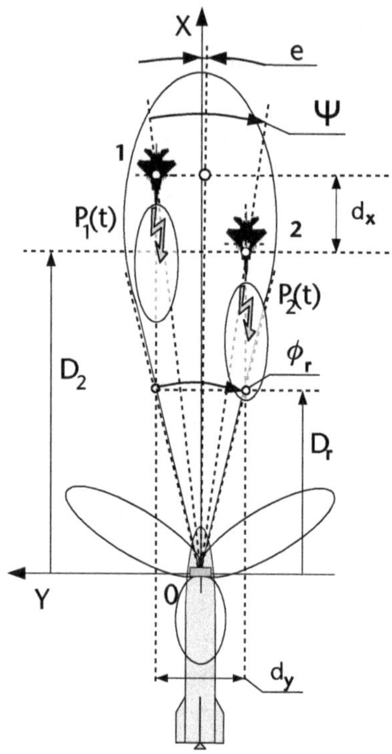

Figure 11.15 Group target consisting of two jamming aircraft.

of the signals reflected by the hulls of the jamming aircraft is negligible relative to the signals emitted by them. Below are the calculation results for group target represented by two jammers. The calculations were carried out for two modes: semi-active mode (SA) and active mode (AM). The position of these goals is characterized by the following offsets: $\Delta_x = 100$m; $\Delta_y = -100$m; $\Delta_z = -100$m;

Estimation of h and φ_r:

- SM:
  - $2\varphi_r = 8°$ for clean target and noise interference;
  - $2\varphi_r = 10°$ upon receipt of target designation for another target.
- AM:
  - $2\varphi_r = 4.4°$ for clean target and noise interference;
  - $2\varphi_r = 8°$ for noise interference.

Operating on two clean targets:

- SM:
  - $\forall\rho = 1 : d_{max0} = 209.5 \ldots 419$ [m] is the distance between 2 targets;
  - $h_{max0} =\ 39.0 \ldots 77.0$ [m] is the deviation from the target (miss);
  - $\forall\rho = 0.5 : d_{max0} = 157.0 \ldots 314.0$ [m] is the distance between 2 targets;
  - $h_{max0} = 20.0 \ldots 39.0$ [m] is the deviation from the target (miss).

Table 11.4

	miss of the first target, [m]				*miss of the second target, [m]*			
	h_x	h_y	h_z	h_Σ	h_x	h_y	h_z	h_Σ
modeling	34.4	77.0	17.0	86.0	57.3	41.0	51.4	87.2

- AM:
    - $\forall \rho = 1 : d_{max0} = 115.0 \ldots 230$ [m] is the distance between 2 targets;
    - $h_{max0} = 21.0 \ldots 42.2$ [m] is the deviation from the target (miss);
    - $\forall \rho = 0.5 : d_{max0} = 86.0 \ldots 173.0$ [m] is the distance between 2 targets;
    - $h_{max0} = 10.0 \ldots 21.0$ [m] is the deviation from the target (miss).

If in the SM mode, target designation is given for another target, which is not selected by the active seeker, then $2\varphi_r = 10°$, therefore: $d_{max0} = 262.0, \ldots, 524$ [m], $h_{max0} = 48.0, \ldots, 96.0$ [m]. By averaging this data with the data for two clean targets in the SM mode, we can write: $d_{max0} = 235.75, \ldots, 471.5$ [m] $h_{max0} = 43.5, \ldots, 86.5$ [m]. This corresponds to the range of probabilities of target defeat with one missile:

For the SM mode: $P \in [0.55, \ldots, 0.09], \forall\ R_0 = 40$ m, $\rho = 1$.

For the AM mode: $P \in [0.87, \ldots, 0.56], \forall\ R_0 = 40$ m, $\rho = 1$.

Knowing the values of the miss in coordinates x, y, z, you can find the total miss:

$$h_\Sigma = \sqrt{{h_x}^2 + {h_y}^2 + {h_z}^2} \tag{11.43}$$

The data in the Table 11.4 allows to carry out studies of guidance accuracy, and consequently, the probability of hitting at least one jammer from the group in active and semiactive modes.

The probability of defeating each target in a group is defined as an estimate of the mathematical expectation of defeat probability by N equiprobable realizations:

$$\hat{P}_i = \frac{1}{N}\sum_{j=1}^{N} \hat{p}_{ij}(\bar{h}_{ij}), \tag{11.44}$$

where $\hat{p}_{ij}(\bar{h}_{ij}) = \int_{\bar{h}_{ij}}^{\infty} \hat{f}(h)dh$ is an estimate of the probability of defeat of the ith jammer in the jth implementation, $\bar{h}_{ij}$ is the total miss, realized for the ith jammer in the jth implementation and obtained by averaging over all phases for the case of the best matching of the vector of the maximum density of the scattering of fragments with the direction to the target, $\hat{f}(h)$ is the estimation of the probability density function of hitting the target for the case of the best match.

The results of calculating the probabilities are shown in Table 11.5. It is assumed that for each jth realization there is a complete group of $M + 1$ events, in

Table 11.5

#	mode	d [m]	clean target	Two jammers: $\rho_2 = 10\lg(P_2/P_1)$ 0 dB	3 dB	4 dB	6 dB	8 dB	10 dB
1.	SM	100 m	0.84	0.920	0.970	0.980	0.990	0.995	1.0
2.	SM	200 m	0.84	0.720	0.950	0.975	0.980	0.990	0.998
3.	SM	300 m	0.84	0.520	0.930	0.960	0.950	0.985	0.991
1.	AM	100 m	0.84	0.925	0.975	0.983	0.991	0.999	1.0
2.	AM	200 m	0.84	0.723	0.951	0.978	0.982	0.993	0.999
3.	AM	300 m	0.84	0.531	0.932	0.963	0.952	0.990	0.995

which each of the M events is the fact of the defeat of the ith jammer ($i = \overline{1, M}$) and one event is the failure to defeat the goals of the jamming group.

As a result, the assessment of the probability of hitting at least one jammer from the group will be written in the form:

$$\hat{P} = 1 - \prod_{i=1}^{M}(1 - \hat{P}_i) \tag{11.45}$$

From Table 11.5 it can be seen that the probability of defeating a clean target remains unchanged, and the probability of destroying the second target in the group decreases with increasing distance d. The most difficult case for the destruction of at least one jammer aircraft from the group is realized in the case of equality of the powers of jamming signals emitted by them, $P_1 = P_2$. Thus, the probability of the second jammer aircraft defeat falls from $\hat{P}_2 = 0.920$ at $d = 100$m up to $\hat{P}_2 = 0.520$ at $d = 300$m when using the semiactive targeting mode SM. In the case when the missile seeker operates in the active mode AM, the probability of the second jammer aircraft defeat is slightly higher than in the semiactive mode and increases as the distance d between the jammer aircraft decreases.

11.6 Operation of the Active Homing Head When Auto-Tracking a Low-Attitude Clean Target and Using an Antipode Jammer

In this section, we look at how the active homing head works when auto-tracking a clean target, which creates a specular reflection from the Earth's (water's) surface, and when using an antipode type of jammer.

The specular reflection problem can be illustrated in Figure 11.16, where H_m and H_t are the absolute heights of the missile and target, respectively, and D_t is the horizontal target range relative to the missile's active homing head. A signal that propagates in a straight line from the active homing head of the missile to the target and back is a useful (direct) signal. In addition, another signal propagation is possible: from the active missile homing head to the target in a straight line, and in

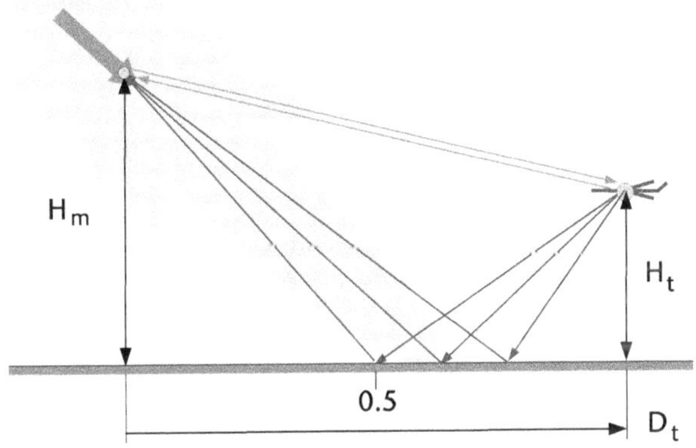

Figure 11.16 Low-altitude target that creates a specular reflection from the ground (water) surface.

the opposite direction-after re-reflection from the Earth's (water's) surface. This is a mirror signal of the first order. A second-order specular signal in both directions propagates like re-reflection from the Earth's (Water's) surface. Since the mirror signal of the second order in comparison with the mirror signal of the first order will be weaker in power and significantly delayed in time, we can restrict ourselves to considering only the first-order mirror signals. In this case, it is necessary to make a few general comments:

1. It is reasonable to assume that during autonomous flight of a missile and the capture of a low-altitude target by an active homing head for auto-tracking, the following condition is fulfilled:

$$H_m > H_t, \tag{11.46}$$

   where H_m and H_t are the heights of the missile and the target in the Earth's coordinate system (absolute heights).
2. Only those first-order mirror signals are taken into account, which are received along the main lobe of the directional pattern of the receiving antenna of the active homing head.
3. The horizontal range between the missile and the target D_t, which is the projection of the line of sight missile-target on flat Earth, is limited from above to 75 km in active mode for a target with an effective surface scattering also called the radar cross section (RCS) $\sigma \approx 20$ m^2.

As shown earlier, the estimate of the mathematical expectation of the angle of approach of the missile to the target $\hat{\vartheta}_t$, when it is displayed along the reference kinematic trajectory, is $-40.2°$ at a range of 30 km from the target. At the same time, in the conditions of shooting at low-altitude targets, there is always an excess in height $H_m > H_t$ (11.46). Therefore, in the general case, the situation presented in Figure 11.17 will be considered. As shown earlier, after the end of the missile movement to a given point in space on the command of the inertial guidance system enable the AM mode, the active seeker switches to the radiation mode and searches

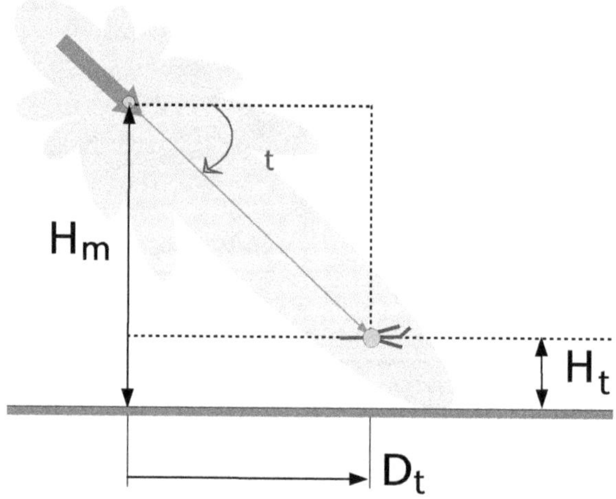

Figure 11.17 The missile approach angle of to the low-altitude target.

for the target by speed, and if necessary, it performs an additional target search by angular coordinates by means of conical scanning in the solid angle ±15°.

The assessment of the functioning of the active homing head when locating low-altitude targets can be carried out on the basis of the Huygens-Fresnel principle (Figure 11.18). According to this principle, the electromagnetic field at an arbitrary point A', created by real sources located in the volume V, can be represented as the sum of the fields that would create at this point fictitious sources continuously distributed over the surface S. The field at the point A' can be defined through the electric hertz vector $\dot{\vec{Z}}(A')$ as a field of primary sources:

$$\dot{\vec{Z}}(A') = -\frac{j}{4\pi\omega\varepsilon}\int_V \frac{\vec{J}}{\rho}\exp[-jk\rho]dV, \tag{11.47}$$

where $\vec{J}$ is a function of the density of electric current vectors, or as the total field of secondary sources, continuously located on a closed surface S, covering the primary sources:

$$\dot{\vec{Z}}(A') = -\frac{1}{4\pi}\int_S \left(\dot{\vec{Z}}_e\frac{\partial\varphi}{\partial\vec{n}} - \varphi\frac{\partial\dot{\vec{Z}}_e}{\partial\vec{n}}\right)dS, \tag{11.48}$$

Here $\dot{\vec{Z}}_e$ is the electric hertz vector created by the primary sources on the element dS of the closed surface S. This is used in the integro-differential equation (11.48) as the basis of Green's function, such that

$$\varphi = \frac{1}{r}\exp[-jkr] - \frac{1}{\rho}\exp[-jk\rho], \tag{11.49}$$

vanishes on the surface S and its derivative $\frac{\partial\varphi}{\partial n}$ also vanishes. Here r is the distance from the radiation point P to some point in the region S, ρ is the distance from point A' to the same point in area S. Representing the derivative of the Green's

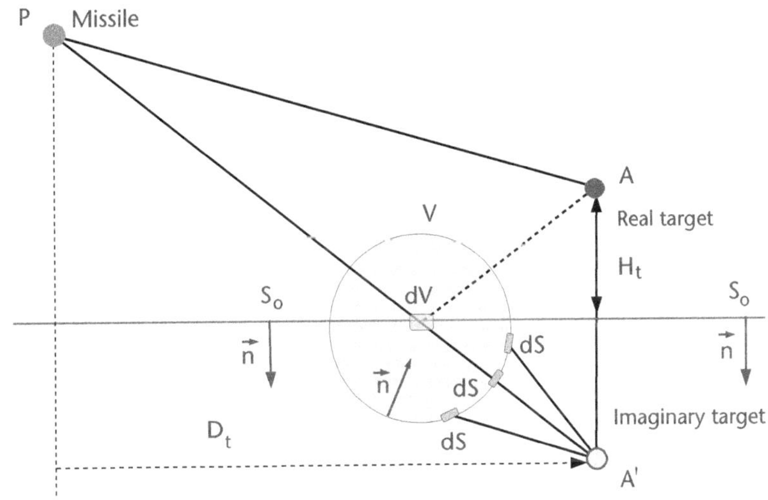

Figure 11.18 Illustration of the Huygens-Fresnel principle.

function in the form:

$$\frac{\partial \varphi}{\partial n} = 2\frac{\partial}{\partial r}\left(\frac{\exp[-jkr]}{r}\right)\cos(\vec{n},\vec{r}) \tag{11.50}$$

and delivering it to (11.48) for an electric dipole oriented parallel to the plane S_0, the electric hertz vector at the point A' after normalization in amplitude will be written as:

$$\vec{Z}(A') = \frac{j}{\lambda}\int\limits_{S_0} \frac{\exp[-jk(r+\rho]}{r\rho}\cos(\vec{n},\vec{r})dS \tag{11.51}$$

Thus, from the expression (11.50) it follows that the field at the point A' is formed as a superposition of the fields created by the elementary surfaces dS.

Dividing the area of integration S_0 into elementary regions S_n, the areas of which are multiples of half the wavelength of the high-frequency signal emitted by the active seeker, which is used to locate the target. You can go to Fresnel zones in three-dimensional space, which are represented by ellipsoids of revolution with focuses at the points A' and P. The intersection of the ellipsoids of revolution with the Earth's surface S_0 gives an area that is significant for the reflection of radio waves (Figures 11.20 and 11.21).

Let us consider the operating case of an active seeker on a low-flying target (Figure 11.19). Suppose that the missile at the start of homing is at a distance of $D_t =$ 75 km from the target and is at an altitude of $H_m = 26$ km from the Earth's surface. In this case, the initial angle of approach of the missile to the target is $\vartheta_{mt} = \Theta$. Obviously, the direct signal propagation path is $r_{mt} = \sqrt{(H_m - h_t)^2 + D_t^2} = 79.057$ km. The phase center of active homing head antenna and the imaginary target are in the foci of the ellipsoid of revolution P and A'; therefore, the distance between two foci of the ellipsoid of revolution can be found: $2C = \sqrt{(H_m + h_t)^2 + D_t^2} = 79.711$ km,

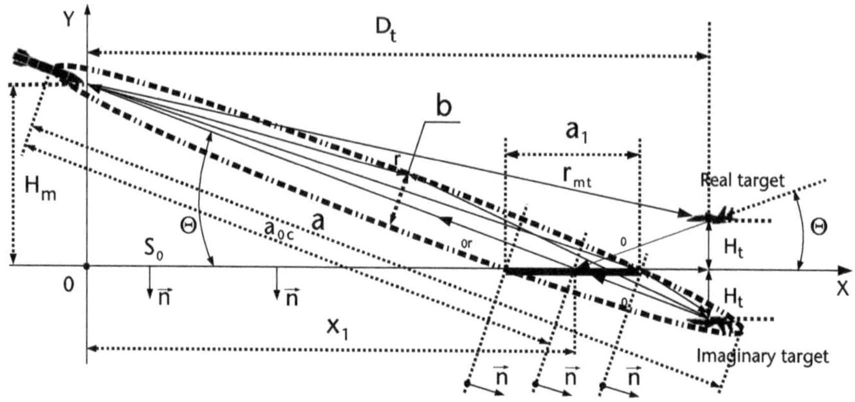

Figure 11.19 Reflections from the ground (water) surface.

and accordingly, $C = 79.711/2 = 39.855$ km. In addition, the sum of the distances from a certain point of the ellipse to its foci is equal to the doubled semimajor axis of the ellipse $2a = \rho + r$. For $|\rho| = |r|$ it is easy to get $\rho = a/2$. Taking into account the fact that in neighboring Fresnel zones the oscillations are in antiphase,

$$r_{mt} - (\rho_o + \rho_{or}) = n\frac{\lambda}{2}, \tag{11.52}$$

therefore, $\rho + r - C - C = \lambda/2$ and we can find the maximum radius of the cross section of the ellipsoid of revolution:

$$R_n = \sqrt{(n\lambda C \cdot C)/(C + C)} = \sqrt{n\lambda \cdot C/2}, \tag{11.53}$$

which, in turn, for $\lambda \approx 0.04$m allows us to find the length of the semiminor axis of the ellipsoid of revolution: $b = R_1 = \cdot\sqrt{(1 \cdot 0.04 \cdot 39855.5)/2} = 28.223$m. It is quite clear that the equation of the straight line connecting the phase center of the homing head antenna and an imaginary target, which is a reflection of a real target, which is below the boundary of the sections of two environment S_0, can be easily represented as $y_r = 26 - x\left(\frac{H_m + h_t}{D_t}\right)$. From the equality $y_r = 0$ it is easy to obtain

$$x_1 = (H_m \cdot D_t)/(H_m + h_t) = 72.222 \text{ km}, \tag{11.54}$$

as well as the distance between the phase center of the seeker antenna and the point of intersection of the direction to the imaginary target with the interface of two environment $\rho_{0r} = \sqrt{H_m^2 + x_1^2} = 76.759$ km. The canonical equation of an ellipse can be written as

$$y = \pm\frac{b}{a}\sqrt{\left(\frac{a}{2}\right)^2 - x^2} \tag{11.55}$$

Here $\frac{a}{2} = 39856.01077$m is the semimajor axis of the ellipse located on the $0X$ axis, and $\frac{b}{2} = 28.223$m is the semiminor axis of the ellipse located on the $0Y$ axis. The + sign is responsible for the part of the graph of the ellipse function for which

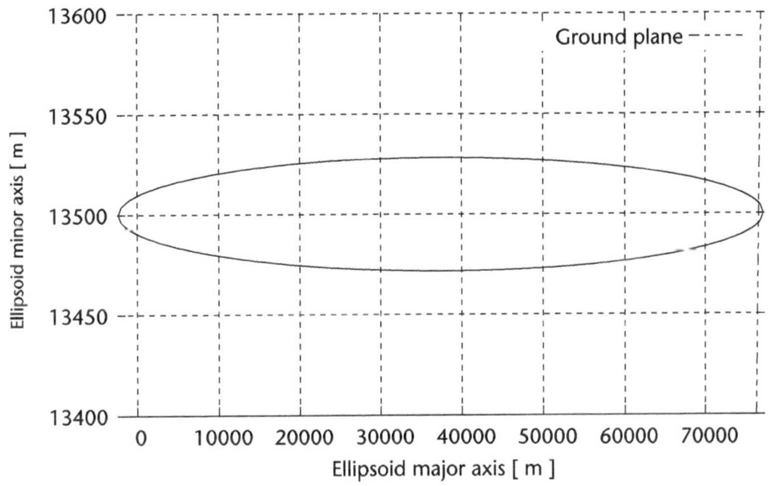

Figure 11.20 Fresnel zone as ellipsoids of revolution.

$y > 0$, and the – sign is responsible for the other part of the graph for which $y < 0$. The value of a can be found from the condition

$$\rho + r = 2a \tag{11.56}$$

By virtue of the equality $2a = 2\rho$, we can write

$$a = \rho \tag{11.57}$$

Due to the fact that the distance from the phase center of the seeker antenna to the imaginary target is $R_{mit} = \sqrt{27^2 + 75^2} = 79.711$ km, respectively, the distance to the point of intersection with the Earth's surface a_{0C}, measured along the major axis of the ellipse, can be represented as

$$a_{0C} = \frac{x_1 R_{mit}}{D_t} = \frac{72.222 \times 79.711}{75} = 76.758 \text{ km} \tag{11.58}$$

By placing the origin of the ellipse coordinates at the point (37500, 13500)m, it can be shown that the equation of the straight line characterizing the Earth's surface can be represented as

$$y_E = (x + 37500) \times 0.36 - 27630 \text{ [m]} \tag{11.59}$$

The joint solution of the equations of the positive part of the ellipse and the straight line (11.59) will give the intersection point $x_1 = 76{,}672$m and for the negative part of the ellipse the intersection point with the Earth's surface (11.59) will have coordinates $x_2 = 76{,}828$m. Thus, the area that is essential for the reflection of radio waves is the section of the ellipsoid of revolution, which is an ellipse, whose major axis has the size $da_F = x_2 - x_1 = 156$m, and the minor axis is $db_F = y_2 - y_1 = (13528.08 - 13471.92)$m $= 56.16$m.

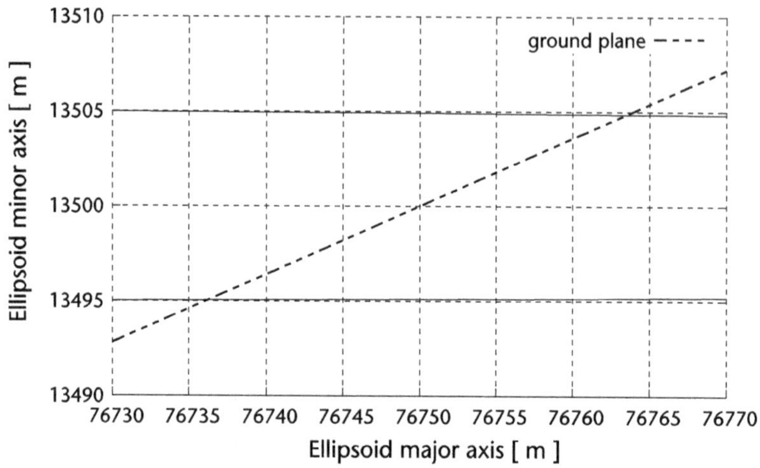

Figure 11.21 Projection of Fresnel zones onto the Earth's (water's) surface.

Due to the fact that only first-order specular signals are calculated (Figure 11.22), the field at the phase center of the receiving antenna of the active homing head can be represented as:

$$\dot{\vec{E}}(P) = \dot{\vec{E}}_1 + \dot{\vec{E}}_2, \tag{11.60}$$

where $\dot{\vec{E}}_1$ is the direct wave field; $\dot{\vec{E}}_2(\psi)$ is the field of the wave, reflected from the Earth's (water's) surface; $\dot{\vec{E}}_r(\psi)$ is the field of the wave, reflected from the target in the direction of the Earth's (water's) surface; ψ is the angle of incidence (reflection) of an electromagnetic wave from the interface of two environment S_0; υ is the approach angle of the missile to the imaginary target.

$\dot{\vec{E}}_1$ is defined similarly to the case of free space, and to find $\dot{\vec{E}}_2$, a reflecting interpretation will be used, according to which the reflected field is represented as:

$$\dot{\vec{E}}_2 = \dot{\vec{E}}(A') \times R \tag{11.61}$$

where $\dot{\vec{E}}(A')$ is the field of the emitter, identical with the real one, placed at the point of specular reflection; R is the wave reflection coefficient of the corresponding polarization in the case of a plane wave incidence on a plane interface between two environment S_0.

The rigor of the method is beyond doubt for the case of a perfectly conducting Earth surface ($\sigma = \infty$). For arbitrary parameters of the Earth, the restrictions are due to the assumption on the plane front of the incident and reflected waves. Due to the fact that the missile under consideration provides for stabilization in two coordinates and effective command processing is achieved due to roll control, the radiation field of the antenna of the active homing head is considered in the general case for an arbitrary orientation of the electric field vector $\dot{\vec{E}}$:
for the incident wave $\dot{\vec{E}}_{down}$:

$$\dot{\vec{E}}_{down} = \dot{\vec{E}}_{\rightarrow} + \dot{\vec{E}}_{\uparrow} \tag{11.62}$$

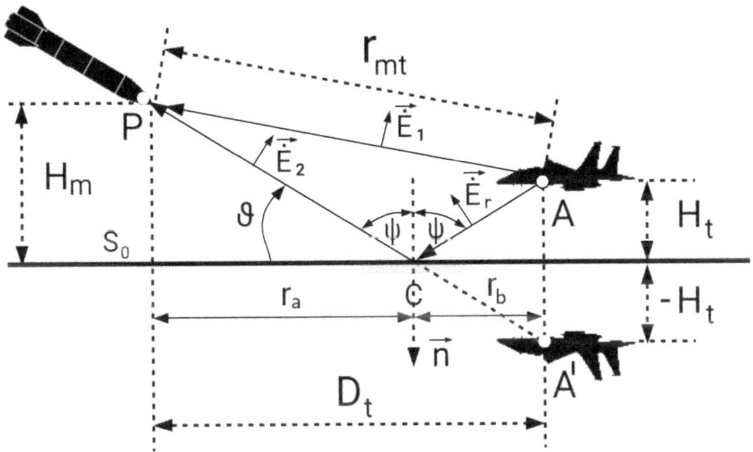

Figure 11.22 Mirror signals of first order.

for reflected wave $\vec{\dot{E}}_{ref}$:

$$\vec{\dot{E}}_{ref} = \vec{\dot{E}}_{ref\rightarrow} + \vec{\dot{E}}_{ref\uparrow} = \vec{\dot{E}}_{\rightarrow}(A')|R_{\rightarrow}|\exp[-j\beta_{\rightarrow}] + \vec{\dot{E}}_{\uparrow}(A')|R_{\uparrow}|\exp[-j\beta_{\uparrow}] \tag{11.63}$$

where $|R_{\rightarrow,\uparrow}|$ and $\beta_{\rightarrow,\uparrow}$ are real numbers representing the modulus and phase of the complex reflection coefficient for horizontal $\rightarrow$ and vertical $\uparrow$ polarization. The quantities $|R_{\rightarrow,\uparrow}|$ and $\beta_{\rightarrow,\uparrow}$ can be determined strictly by solving the boundary value problem for given parameters of anisotropic environment (dielectric constants ε_r and conductivities σ). This approach significantly complicates the solution of the problem with a slight increase in the accuracy of approximating the calculated estimates to the experimental results.

In practical calculations, the following estimates will be used:

$$|\dot{R}_{\uparrow}| = |\bar{R}_{\uparrow}| + N(0, 1) \tag{11.64}$$

$$\beta_{\uparrow} = \bar{\beta}_{\uparrow} + N(0, 1) \tag{11.65}$$

where:

$$|\bar{R}_{\uparrow}| = \frac{1}{MN}\sum_{i=1}^{M}\sum_{j=1}^{N} f(\vartheta, \varepsilon_{rj}, \sigma_i) \tag{11.66}$$

$$\bar{\beta}_{\uparrow} = \frac{1}{MN}\sum_{i=1}^{M}\sum_{j=1}^{N} g(\vartheta, \varepsilon_{rj}, \sigma_i) \tag{11.67}$$

and the following functions are used as functions $f(\vartheta)$ and $g(\vartheta)$ (Figure 11.23):

$$|R_{\uparrow i,j}| = f(\vartheta, \varepsilon_{rj}, \sigma_i) = \frac{\varepsilon_{rE}\cdot\sin(\vartheta) - \sqrt{\varepsilon_{rE} - \cos^2(\vartheta)}}{\varepsilon_{rE}\cdot\sin(\vartheta) + \sqrt{\varepsilon_{rE} - \cos^2(\vartheta)}} \tag{11.68}$$

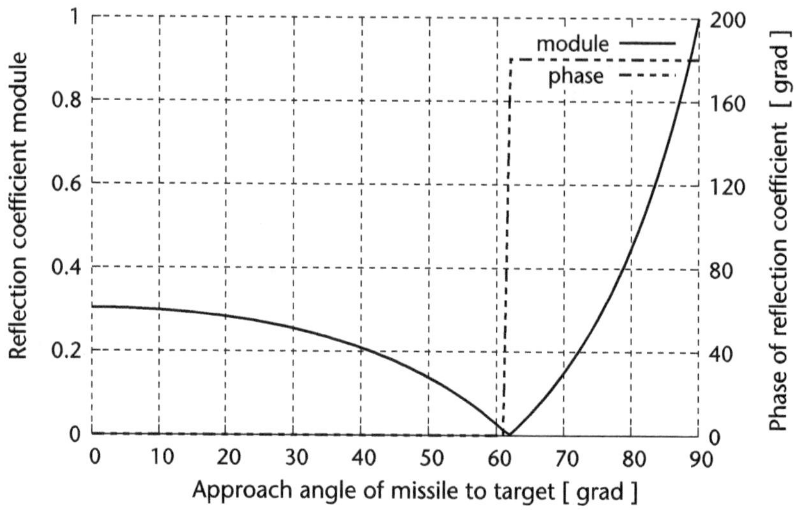

Figure 11.23 Reflection coefficient module and phase.

$$\beta_{\uparrow} = g(\vartheta, \varepsilon_{rj}, \sigma_i) = \begin{cases} 0^\circ, & \forall\vartheta < \vartheta^\circ \\ 180^\circ, & \forall\vartheta \geqslant \vartheta^\circ \end{cases} \tag{11.69}$$

Here ε_{r_E} is an estimate of the average value of the Earth's dielectric constant for temperature $t = 20^\circ$ and dry soil: $\varepsilon_{r_E} = (2+5)/2 = 3.5$; σ_i is an estimate of the average conductivity of the Earth's surface $\sigma_i = (1 \cdot 10^{-1} + 2 \cdot 10^{-1})/2 = 1.5 \cdot 10^{-1}$ S/m at the same temperature. When reflecting from the Earth's (water's) surface, its unevenness is also essential. So, a reflective surface can be considered flat if the maximum irregularity height h_δ meets the Rayleigh [9] criterion (Figure 11.22):

$$h_\delta \ll \frac{\lambda}{8 \cdot \cos(\psi)} \tag{11.70}$$

This circumstance, to one degree or another, leads to the scattering of the energy of the reflected wave, and consequently, to a decrease in the modulus of the reflection coefficient. Taking this phenomenon into account, the field of the signal reflected from the Earth's (water's) surface was calculated as follows.

1. The region S_0, significant in the reflection of radio waves, is represented by a rectangle with sides $a_1 = 2a_F$ and $b_1 = 2b_F$, into which is inscribed the first Fresnel zone, defined by the following formulas:

$$a_F = \sqrt{\frac{r_a \cdot r_b}{r_a + r_b}} \cdot \lambda \tag{11.71}$$

$$b_F = \frac{a_F(r_a + r_b)}{\sqrt{4a_F^2 + (r_a + r_b)^2 \cdot \cot\psi}}, \tag{11.72}$$

where r_a and r_b are the distances from the projections onto the reflecting surface of the points of the phase center of the missile homing head antenna and the target to the point of geometric reflection C, and ψ is the angle of incidence of the wave (Figure 11.22).

2. S_0 splits into $N \times M$ rectangular flat areas dS_i, such that $\forall dS_i << \sigma_t$ (target RCS), the normal vector $\vec{n}_i$ that is oriented according to the normal law $N(0, 1)$.
3. $\forall dS_i : \sigma \rightarrow \infty$ (i.e., elements of the region, essential for the reflection of electromagnetic waves, have ideal conductivity).
4. The assumptions about the incidence of a plane electromagnetic wave on a plane ideal conducting surface dS_i allow us to determine the electric current density on the surface element dS:

$$\vec{J}_{S_i} = 2\left[\vec{n}, \dot{\vec{H}}_{pad}\right] \tag{11.73}$$

Here, $\vec{J}_{S_i}$ is the surface current density vector at the center of the elementary area dS_i; $\vec{n}$ is the unit vector of the normal dS_i in the orthonormal basis $(\vec{i}, \vec{j}, \vec{k})$ of the receiving antenna of the active homing head; $\vec{H}_{pad}$ is the vector of the intensity of the magnetic field emitted by the antenna of the active homing head near the surface dS_i. It should be noted that for the considered conditions the formula (11.73) is exact from the electrodynamic point of view.

5. Due to the fact that the receiving point P for the S_0 region is in the far zone, the expression for the hertz vector potential (11.51), taking into account (11.73), will be rewritten as

$$\dot{\vec{Z}}_p = \frac{1}{j4\pi\omega\varepsilon R} \times \exp[-jkR] \times \dot{\vec{F}}(\theta, \varphi) \tag{11.74}$$

where

$$\dot{\vec{F}}(\theta, \varphi) = \int_S \dot{\vec{J}}(x, y) \exp[-jk(\rho + r_i)] \times \frac{1}{\rho + r_i} dS \tag{11.75}$$

can be characterized as a vector complex scattering diagram of the reflecting surface S. The expression for the Hertzian magnetic vector in the far field will have a similar form. In the general case, electric and magnetic currents will simultaneously exist on the dS area, but the assumption of infinite conductivity allows us to restrict the expression for the electric field strength at the point P to only the electric component:

$$\dot{\vec{E}}(P) = \frac{\omega\mu}{j4\pi} \times \frac{\exp[-jkR]}{R}\left(F_\theta \vec{i}_\theta + F_\varphi \vec{i}_\varphi\right), \tag{11.76}$$

where

$$F_\theta = (F_x \cos(\varphi) + F_y \sin(\varphi) \cos(\theta) - F_z \sin(\theta) \tag{11.77}$$

$$F_\varphi = -F_x \sin(\varphi) + F_y \cos(\varphi) \tag{11.78}$$

and that F_θ and F_φ are defined on the basis $(\vec{i}_\theta, \vec{i}_\varphi)$ receiving antenna of the active homing head. The components F_θ and F_φ are defined according to (11.75) where $\dot{J}(x, y) = [\vec{n}_i, \dot{\vec{H}}_{pad}]$.

6. In general, the complex antenna pattern of the active homing head $\dot{f}\theta, \varphi) = F(\theta, \varphi) \exp[j \arg(\dot{f}(\theta, \varphi))]$ is approximated by radiation pattern with zero-phase$F(\theta, \varphi)$ and it is assumed that an electromagnetic field of linear polarization with a vertical orientation of the electric field strength vector $\vec{\dot{E}}$ is created. In this case, $\vec{\dot{H}}_S$ can be written as

$$\vec{\dot{H}}_S = \sqrt{\frac{\varepsilon}{\mu}}[\vec{i}_\rho\ , \vec{\dot{E}}_S] \tag{11.79}$$

and, therefore

$$\vec{\dot{E}}_{S_i} = \frac{\exp[-jk\rho]}{\rho}\left[\sqrt{\frac{\mu}{\varepsilon}}[\vec{F}_{ma}\ , \vec{i}_\rho] + [\vec{F}_{ma_\mu}\ , \vec{i}_\rho]\right], \tag{11.80}$$

where $\vec{i}_\rho$ - unit vector $\vec{\rho}$. For a rectangular basis of the active seeker $(\vec{i}, \vec{j}, \vec{k})$ we can write

$$\begin{cases} \vec{\dot{F}}_{ma_\varepsilon} = \ \ F(\theta, \varphi)\vec{k} \\ \vec{\dot{F}}_{ma_\mu} = -F(\theta, \varphi)\vec{i} \end{cases} \tag{11.81}$$

After substituting (11.81) in (11.80):

$$\vec{\dot{E}}_S = \frac{F(\theta, \varphi)\exp[-jk\rho]}{\rho}\left\{\sqrt{\frac{\mu}{\varepsilon}}[\vec{i}_\rho, [\vec{k}, \vec{i} - \rho]] - [\vec{i}, \vec{i}_\rho]\right\} \tag{11.82}$$

Performing vector operations in curly braces and substituting the result in (11.79) results in the following expression:

$$\vec{\dot{H}}_S = \sqrt{\frac{\varepsilon}{\mu}} \times \frac{F(\theta, \varphi)}{\rho} \times \left\{H_x\vec{i} + H_y\vec{j} + H_z\vec{k}\right\} \tag{11.83}$$

Substitution (11.83) into (11.73) will allow finding the electric current density on the elementary surface dS.

The transition to the two-dimensional direct Fourier transform with integration, according to (11.75) over the region S_0, allows us to find the vector of the field $\vec{\dot{E}}_2(\theta, \varphi)$, reflected from ground (water) surface under any direction (θ, φ) in the receiving antenna base of the missile's active homing head.

7. The influence of the dielectric properties of the Earth's (water's surface) is taken into account in accordance with (11.68), (11.69) and is normalized to the value of the signal reflected by a rectangular, flat, metal sheet with an area of 1 m^2.

The aerodynamic target is also divided into N flat ideally conducting plates, in each of which the electric current density vector is determined. Note that the currents are only on the illuminated part of the target; that is, that its part that is directly irradiated by the active homing head. The reflected signal is also determined through the direct Fourier transform and the effective scattering surface of the target is normalized for the realized angles θ, φ. The electric field strength at the input of the active homing head antenna is presented in the form

$$\vec{E}(\theta,\varphi) = \left(\dot{E}_{1\theta}(\theta,\varphi) + \dot{E}_{2\theta}(\theta,\varphi)\right)\vec{i}_{\theta} + \left(\dot{E}_{1\varphi}(\theta,\varphi) + \dot{E}_{2\varphi}(\theta,\varphi)\right)\vec{i}_{\varphi} \tag{11.84}$$

The use of a flat waveguide slotted antenna array in the active homing head makes it possible to form a directional pattern with practically no cross-polarization component. However, the use of a nose fairing leads to a significant increase in the cross-polarization component and, accordingly, to the appearance of direction pattern of the antenna-fairing system on cross polarization.

11.6.1 Signal Processing and Decision Rules

After receiving the target signal or interference by the antenna system and lowering it to an intermediate frequency, for all channels we can write

$$\dot{U}^{\Sigma,K,T}(\theta,\varphi,t) = \dot{U}^{\Sigma,K,T}(t)\omega(t) = U^{\Sigma,K,T}(t)\exp[j\Phi_0(t)] \times \omega(t), \tag{11.85}$$

where

$$U^{\Sigma,K,T}(t) = \sqrt{(E_{1,\theta}^{\Sigma,K,T}(\theta,\varphi,t) + E_{2,\theta}^{\Sigma,K,T}(\theta,\varphi,t))^2 + (E_{1,\varphi}^{\Sigma,K,T}(\theta,\varphi,t) + E_{2,\varphi}^{\Sigma,K,T}(\theta,\varphi,t))^2} \tag{11.86}$$

is the signal envelope.

After the analog-to-digital transformation of the function (11.85) with the sampling frequency $f_d = B$ Hz of each channel of the active homing head into the time series, the direct Fourier transform of the real function $\Re[\dot{U}(t)] = X(t)$ is carried out over N_d points. Note that for the sum channel of the active seeker, operation $\Re[\dot{U}(t)] = X(t)$ is identical to the voltage $\overline{U}_{agc}(t)$ of the AGC, which normalizes the amplifiers of the intermediate frequency of the course channel φ (K) and pitch θ (T) channel, which in turn normalizes the amplifiers of the intermediate frequency of the course channel φ (K channel) and pitch channel θ (T channel):

$$X^{\Sigma}(t) = \Re e[\dot{U}^{\Sigma}(t)] \equiv \overline{U}_{agc}(t) \tag{11.87}$$

Based on the results of N soundings, the result of filtering in the frequency domain can be represented in the form

$$Y^{\Sigma,K,T}(f_k) = |H(f_k)| \times \frac{1}{N}\sum_{i=1}^{3} X_i^{\Sigma,K,T}(f_k) \tag{11.88}$$

Spectrum factorization (11.87) for course (K) and pitch (T) channels leads to the following expression:

$$G^{K,T}(f_k) = 10\log\left[Y^{K,T}(f_k) - Y^{\Sigma}(f_k)\right] \tag{11.89}$$

The selection of the received signals can be carried out using the decision function, which is a discrimination function:

$$Dec^{K,T}(\Delta k) = \begin{cases} -1, & \forall\, \Delta k = \Delta G_{-3\text{dB}}(f_k) = 1\ldots 2 \\ 0, & \forall\, \Delta k = \Delta G_{-3\text{dB}}(f_k) = 3\ldots 5 \\ 1, & \forall\, \Delta k = \Delta G_{-3\text{dB}}(f_k) > 5 \end{cases} \tag{11.90}$$

where $Dec^{K,T}(\Delta k) = -1$, harmonic signal; $Dec^{K,T}(\Delta k) = 0$, narrowband interference; $Dec^{K,T}(\Delta k) = 1$, broadband interference; and Δk is the number of frequency filters occupied by the reflected signal or interference.

Target capturing for auto-tracking by speed and angular coordinates can be carried out only in the first case, when the ratio signal/weighted noise exceeds the threshold $G_{lim} = 10$ dB. Consequently, the condition of locking the target for automatic tracking in the speed and angular coordinates is written in the form

$$G^{K,T}(f_k) > G_{lim}, \quad \forall \Delta k : \max G^{K,T}(f_k), \quad Dec(\Delta k) = -1 \tag{11.91}$$

Jammers tracking on angular coordinates is carried out according to the number of the maximum component of the sum channel. Under these conditions, the correlation integral for N sounding cycles has the form

$$Z = \int_0^{T_r} \dot{U}(t)\dot{U}^*(t)\omega(t)dt, \tag{11.92}$$

where T_r is the integration period equal to N sounding cycles.

The decisive detection function $Dec(Z)$ in accordance with the accepted algorithm can be written as

$$Dec(Z) = \begin{cases} 1\,, & \forall\, Z > Z_{lim} \\ 0\,, & \forall\, Z \le Z_{lim} \end{cases} \tag{11.93}$$

where the threshold Z_{lim} is determined by the requirements for the probability of a false alarm.

The above formulas and algorithms were included in the program for the computer, and therefore, the assessment of the functioning of the active homing head in the active mode under the influence of interference signals was carried out in the form of a computational experiment.

Values to be estimated are

- Estimation of the mathematical expectation of the range of capture and stable tracking of the target $\hat{D}_t$ in accordance to probability level 0.84;
- Estimation of the probability of tracking $\hat{P}$ real target after its angular resolution based on the results of M realizations.

This composition of estimates assumes two options for solving the problem, but in both cases, the Monte Carlo method was used to reveal the probabilistic picture.

Table 11.6

$H_t\,[m]$	25	50	75	100	125	150	175	200	225	270
$\hat{D}_t\,[km]$	64.40	66.45	68.34	70.32	71.95	73.14	73.81	74.53	74.89	75
$\hat{P}$	0.49	0.50	0.51	0.54	0.58	0.68	0.69	0.76	0.84	1.00

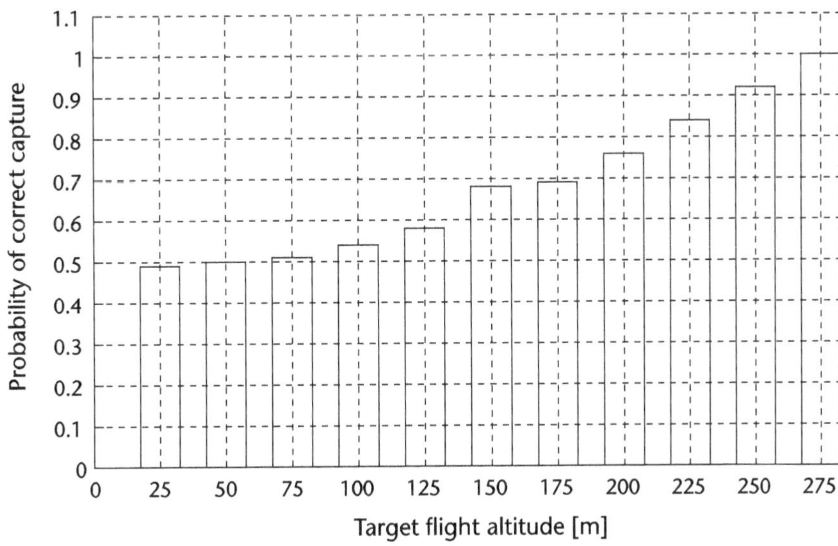

Figure 11.24 Estimates of the probability of correct capture for tracking a real target after its angular resolution.

11.6.2 Operation of the Active Seeker on the Clean Target, Which Creates a Mirror Image from the Earth's (Water's) Surface

The calculations were carried out for the case when a Tu-16 aircraft was used as a target, which performs a horizontal flight over an average (11.70) Earth's surface with a speed $V_t = 250$ m/s, in a range of altitudes (25–300m). The missile speed was assumed to be $V_m = 1000$ m/s. The results of the computational experiment are presented in Table 11.6, and Figures 11.24 and 11.25.

The data in Table 11.6 indicate that as the real altitude of the target increases, the range $\hat{D}_t$ of its stable auto-tracking and the probability $\hat{P}$ correct tracking real targets are increasing.

The spectral characteristics are shown in Figure 11.26.

During the rapprochement of the missile for the purpose of velocity differences in behavior, the signal from the target and its specular reflection were not detected.

Conclusion: Amplitude and velocity selection of signals from the target and its specular reflection is difficult.

11.6.3 The Work of the Active Seeker on a Target, that Creates Interference such as Antipode

In the case of an active seeker that creates interference, it was assumed that a repeater was installed on board the target, which rereflects the signal from the transmitter of

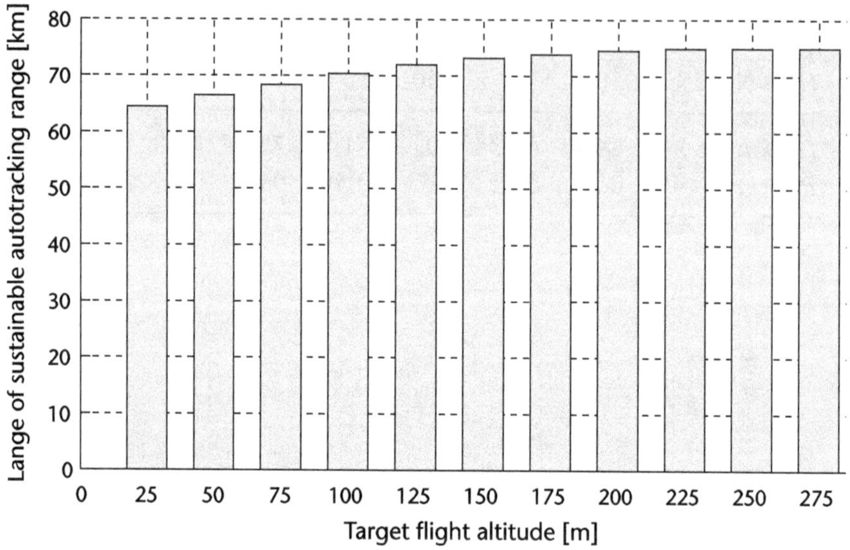

Figure 11.25 Estimates of the range of sustainable auto-tracking a real target.

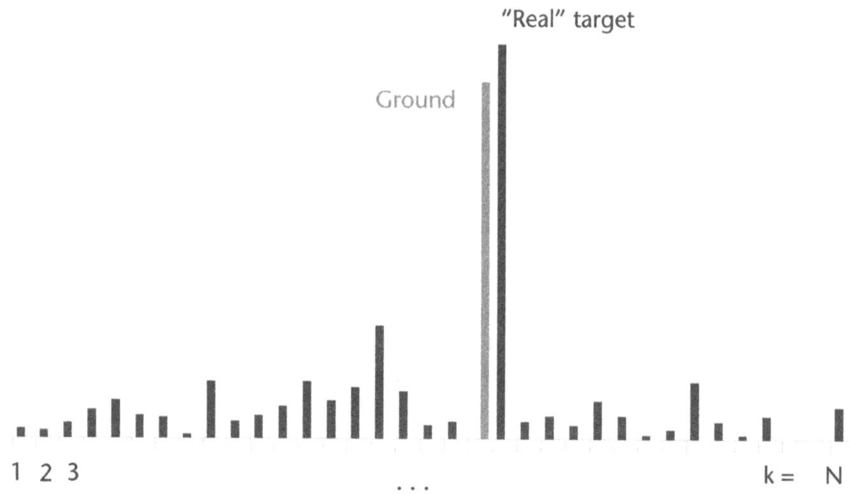

Figure 11.26 Spectral characteristics.

the active homing head to its front hemisphere with a gain $\geq$ 20 dB and some delay τ. Thus, it is possible to artificially form a mirror image of the target; that is, to form its antipode. In this case, the width of the radiation pattern of the transmitting antenna of the repeater accepted 15°.

When evaluating the operation of the active homing head under these conditions, it is assumed that the source of the signal reflected from the Earth's (water's) surface is the phase center of the first Fresnel zone determined by the position of the jammer above the Earth's (water's) surface and the relative position of the target and the equal-signal direction of the antenna of the active head homing. When performing the simulation, it was assumed that the power of the signal reflected from the ground (water) surface—a signal of the antipode type—is 10 dB less than the power of the interference signal. The calculation results for the estimation of the

Table 11.7

#	mode	H_t [m]	clean target	mirror and direct signals: $\rho_2 = 10\lg(P_2/P_1)$ 0 dB	−3 dB	−4 dB	−6 dB	−8 dB	−10 dB
1.	SM	25	0.84	0.96	0.98	0.99	0.995	0.999	1.00
2.	SM	50	0.84	0.9231	0.97	0.98	0.99	0.995	1.00
3.	SM	100	0.84	0.720	0.950	0.975	0.980	0.990	1.00
4.	SM	150	0.84	0.520	0.93	0.96	0.97	0.985	0.991
5.	SM	200	0.84	0.340	0.90	0.94	0.97	0.989	0.993
1.	AM	25	0.84	0.97	0.99	0.995	0.997	1.00	1.00
2.	AM	50	0.84	0.933	0.980	0.990	0.998	1.00	1.00
3.	AM	100	0.84	0.730	0.960	0.985	0.990	1.00	1.00
4.	AM	150	0.84	0.53	0.94	0.97	0.98	0.995	1.00
5.	AM	200	0.84	0.35	0.91	0.95	0.98	0.999	1.00

Table 11.8

H_t [m]	25	50	75	100	125	150	175	200	225	270
$\hat{D}_t$ [km]	75.0	75.0	75.0	75.0	75.0	75.0	75.0	75.0	75.0	75.0
$\hat{P}$	0.00	0.00	0.00	0.00	0.00	0.00	0.00	0.01	0.04	0.08

probability of the jammer defeat for the active AM and semiactive SM modes are presented in Table 11.7.

From Table 11.7, it follows that the most difficult case for the seeker is the case of equality of the amplitudes of the interference signal emitted by the jammer's aircraft and the signal, reflected from the ground (water) surface. With an increase in the flight altitude of the jammer's aircraft, the probability of its destruction decreases. In the case when the power of the interference signal is greater than the power of the signal reflected from the Earth's (water's) surface by 10 dB, the probability of its defeat tends to 1.0. As a result, we can conclude that if the jammer's aircraft provides illumination of the underlying surface so that the inequality is fulfilled:

$$10\log\left(\frac{P_{int}}{P_E}\right) \geq 10\text{ dB}, \tag{11.94}$$

where P_{int} is the power of the interference signal, and P_E is the power of the signal reflected from the Earth's (water's) surface. Then in this case, when the target is flying at low altitudes $H_t = \{25\ldots200\}$m, a false target will be hit with a probability close to one.

At the previously indicated flight speeds of the missile and the Tu-16 aircraft, the results of calculations to determine the defeat probability of a real target are presented in Table 11.8.

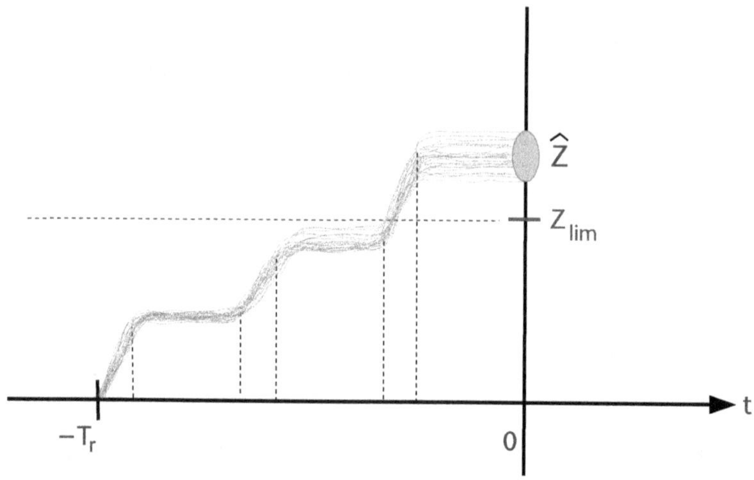

Figure 11.27 Correlation integral.

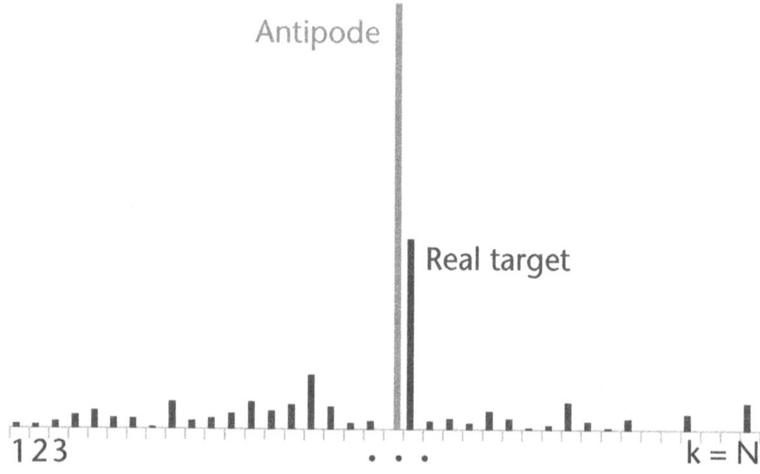

Figure 11.28 Spectral characteristics.

The spectral characteristics and the correlation integral are shown in Figures 11.28 and 11.29.

The given data indicate a stable capture and tracking signal "Antipode." This is due to the organization on the Earth's surface of an area with a powerful interfering signal and the synchronous movement of this spot with the target.

Conclusion: Amplitude and high-speed selection of signals reflected from the target is impossible.

In addition, it is theoretically possible to control the guidance error of an anti-aircraft guided missile by organizing the spot movement in the direction away from the target. However, in this sense, the enemy needs feedback with the missile. At the same time, it is difficult to achieve good quality of such feedback due to the limited time interval during which the missile approaches the target.

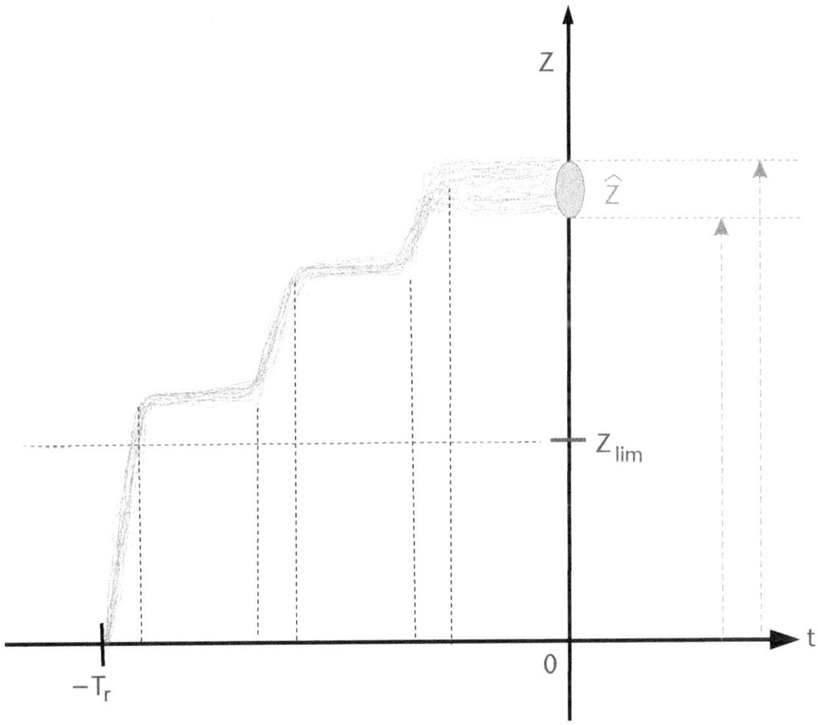

Figure 11.29 Correlation integral.

References

[1] "S-300P System," *Technics and Armament Magazine*, No. 11, 2004, https://military.wikireading.ru/55473.

[2] 40N6-Long-Range Anti-Aircraft Missile of the S-400/S-500 Systems, http://youinf.ru/40n6-zenitnaya-upravlyaemaya-raketa-sverxbolshoj-dalnosti/.

[3] 40N6-Ultra-Long-Range Anti-Aircraft Guided Missile (AGM), Over 400 km, With the Ability to Intercept Targets in Near Space for the S-400/S-500 Air Defense Systems, https://stankotec.ru/raznoe/40-n-40.html#i-2.

[4] Yurtsev, O.A., and Y. S. Grossman, *Microwave Technique*, Lecture Notes, Vol. 1-7, MVIRTU, 1966.

[5] GOST RV 51914-2002, *State Military Standard of the Russian Federation, Elements for Connecting Microwave Paths of Electronic Measuring Devices, Connecting Dimensions*, Official Edition, Moscow: Publishing House of Standards, 2002, http://antic-r.ru/ntd/gost_r/gost_rv51914-2002.pdf.

[6] Rectangular Waveguide Sizes: EIA, RCSC, IEC Standards, https://www.everythingrf.com/tech-resources/waveguides-sizes.

[7] Vakin, S. A., and L. N. Shustov, *Fundamentals of Radio Countermeasures and Electronic Intelligence*, Moscow: Soviet Radio, 1968.

[8] Nikolsky, B. A., *Methods and Means of Electronic Protection of Aircraft, Part 1, Study Guide*, Samara State Aerospace University, Samara, 2004.

[9] Krivitsky, B. Kh., V. N. Doolin, Handbook on the Theoretical Foundations of Radio Electronics, Vol. 1, Moscow: Energia,1977.

CHAPTER 12

Synthesis of Solutions on Improving Interference Immunity Against the Impact of Polarization Interference

Development of air assault weapons and mass-scale deployment of measures and aids to suppress the enemy air defence in air combat operations highlights the demand for creating jammer-resistant radar systems. A vital issue for the onboard radar of anti-aircraft guided missiles is to provide for stable automatic target tracking by angular coordinates, especially under the conditions of interferences affecting the missile homing process. As was shown in Chapters 7 and 8, the airborne radar of an anti-aircraft guided missile is the weakest among anti-aircraft missile systems with semiactive and active homing channels. Polarization interference is the most dangerous because it can result in the complete failure of the monopulse direction-finding system (i.e., full loss of automatic target tracking) [1–4].

It is commonly known that the reason is because the presence of antenna cross-polarization radiation leads to reduction of direction-finding accuracy; it can result in the complete failure of the monopulse radar (i.e., automatic tracking loss) [2] (Chapters 6 and 8) [3]. This phenomenon occurs during direction finding of the targets with marked depolarization effect, which the majority of real aerodynamic targets possess. But this problem is most important when so-called polarization interference is used as electronic countermeasures means [2, section 8.5.2, 3, 4]. In order to reduce the cost of monopulse onboard equipment for guided missiles and unmanned aerial vehicles, it is reasonable to use mirror antennas. For example, a mirror monopulse antenna is used in the homing head of the Soviet Air defense system S-200 or SA-5 Gammon in NATO classification.

A photo of this homing head is shown in Figure 12.1 [5]. Because the sum-and-difference scheme of monopulse direction finding is the most common in practice, it is reasonable to consider an analysis of the effect of polarization interference on the airborne direction finder on the basis of this scheme.

The flow diagram of the sum-and-difference monopulse radio direction-finder (monopulse radar) is shown in Figure 12.2. In the figure and in the text, the following designations are used:

1 is the monopulse antenna;

F_1, F_2 are the feeds, displaced symmetrically from the mirror antenna focus;

2 is the stripline ring;

3 is the mixer of the sum channel;

4 is the mixer of the difference channel;

Figure 12.1 Homing head 5G24 of the Soviet Air defense system S-200.

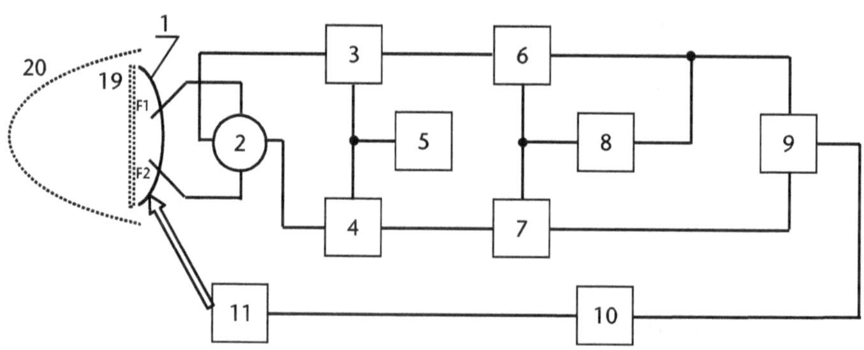

Figure 12.2 Sum-and-difference monopulse direction-finding system.

5 is the heterodyne;

6,7 are the intermediate-frequency amplifiers of the sum and difference channels;

8 is the automatic gain control system of the sum channel;

9 is the phase detector (phase-sensitive detector or dot-product detector);

10 is the error-signal amplifier;

11 is the monopulse antenna drive (mechanism);

19 is the polarization filter;

20 is the radome.

A method of target angle tracking by the sum-and-difference monopulse radar is known, in which reception of signals from the target in the sum-and-difference channels on two orthogonal (cross) polarizations is used to decrease tracking errors [2, p. 249]. The described radars possess the possibility to operate on the group of reception channels that have polarization most closely coinciding with the polarization of electromagnetic waves coming from the target.

However, the drawback of the above-mentioned method is the necessity of doubling the number of monopulse radar reception channels (six instead of three), which makes this method virtually unacceptable for usage in, for example, the airborne equipment of aerial vehicles, including unmanned aerial vehicles (UAVs) and the like, due to weight and size restrictions.

There is a method of target angle tracking based on the usage of polarization filtering of electromagnetic waves coming from the target in the monopulse radio direction-finder [2, pp. 69–71, 168–169]. In this case polarization filtering is performed with the help of the polarization array mounted in the monopulse antenna mouth that allows to weaken an adverse effect of signals on cross polarization on the target direction-finding accuracy.

However, the presence of the diffraction effect on the edges of the polarization array doesn't allow getting a cross-polarization level less than minus 35 dB [2, pp. 165–169] with the help of polarization filtering, which is insufficient to protect from modern polarization interferences that create interference exceeding the signal by 40 dB and more [2, p. 224]. As well, this mode is often inefficient when the monopulse direction-finder antenna is located under the radome 20 or under the blister (for example, in airplanes, missiles, or unmanned aerial vehicles). The radome owing to the curvilinearity of its surface considerably (up to minus 30 dB to minus 15 dB) increases the cross-polarization level of the receiving antenna with a polarization filter, increases the susceptibility of the direction finder to the influence of interference polarization, and leads to the degradation of target tracking accuracy [2, p. 158; 4].

Thus, it is the actual problem of designing technical solutions that would eliminate the negative impact of cross-polarization signals to the accuracy of direction-finding on target, subject to significant restrictions on size and weight for onboard radars.

12.1 Calculation of Direction-Finding Characteristics

Based on the flow diagram of the sum-and-difference monopulse direction-finder cited in Figure 12.2, a scheme of impact of the polarization interference on the monopulse direction-finder is compiled and presented in Figure 12.3.

In accordance with the IEEE Standard definition, a cross-polarization is an orthogonal polarization to a reference polarization [6]. For circular polarization this is adequate, but for linear or elliptical polarization the direction of the reference polarization must still be defined [7]. Let the monopulse antenna 1 be a paraboloid of revolution with a pair of feeds F_1 and F_2 and its reference polarization is vertical one, and a polarization filter 19 is mounted in the mouth of monopulse antenna 1 (see Figure 12.4(a)), the principle of operation of which is shown in Figure 12.4(b). A polarization filter 19 is a system of metal strings stretched parallel to the electric field vector with cross-polarization. The distance between the strings is chosen so

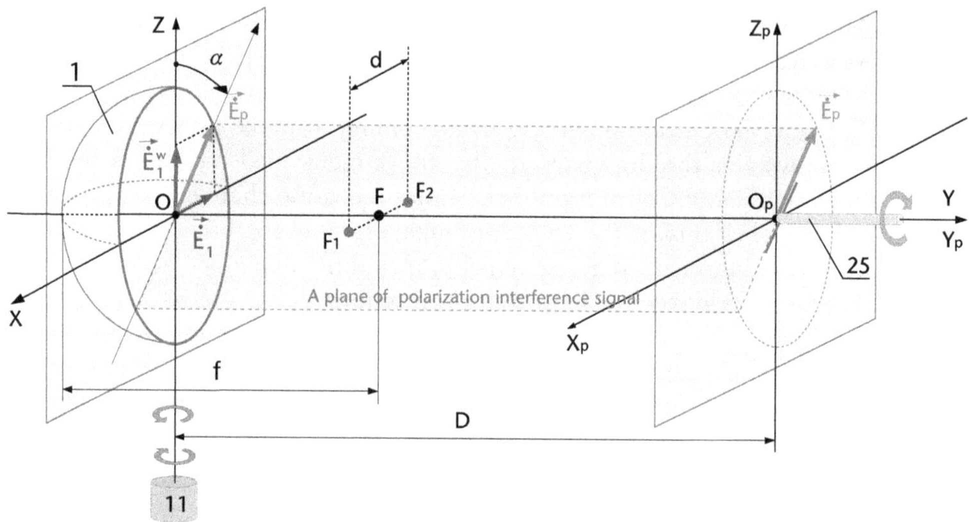

Figure 12.3 Polarization interference impact on a monopulse antenna.

that an incident wave with horizontal polarization (wave on cross-polarization) is reflected, and a wave with vertical polarization (wave on working or main polarization) passes almost without loss to the receiving irradiators $F1$ and $F2$ of monopulse antenna 1.

It is known that any antenna receives and radiates the electromagnetic field on the reference polarization and on the orthogonal polarization, which is called cross-polarization; therefore, any antenna has two directional patterns on the reference and cross-polarization simultaneously. In accordance with the fact that the reference polarization for antenna 1 is a vertical one; that is, the vector of the electric field strength on the reference polarization $\dot{\vec{E}}_1^w$ is parallel to the axis $\vec{OZ}$. The electric field vector of the monopulse antenna 1 on cross-polarization $\dot{\vec{E}}_1^c$ is orthogonal to the electric field strength vector on the reference polarization $\dot{\vec{E}}_1^w$ and it is parallel to the axis $\vec{OX}$, respectively. Due to the fact that the monopulse direction finder carries out auto-tracking of a target signal on reference polarization and it loses the target signal on the cross-polarization [4]; hence, in this case a working capacity of the monopulse direction finder is impossible. Therefore, in the case when a monopulse direction finder receives target signals on the reference polarization, this polarization will be called the working polarization, working, main polarization, or main, and the orthogonal polarization to it will be called cross-polarization or cross. The term "working polarization" emphasizes the fact that with this polarization of the received signals, the monopulse direction finder has maximum operability.

As shown in Figure 12.3, the monopulse antenna 1 and the polarization interference antenna 25 are located on a single straight line that passes through the center of the XOZ coordinate system and the focus F of the mirror antenna. The polarization interference antenna is separated from the monopulse antenna by a distance D so that it is in its far-field zone, where the principle of superposition of the electromagnetic fields created by feeds F_1 and F_2 are satisfied.

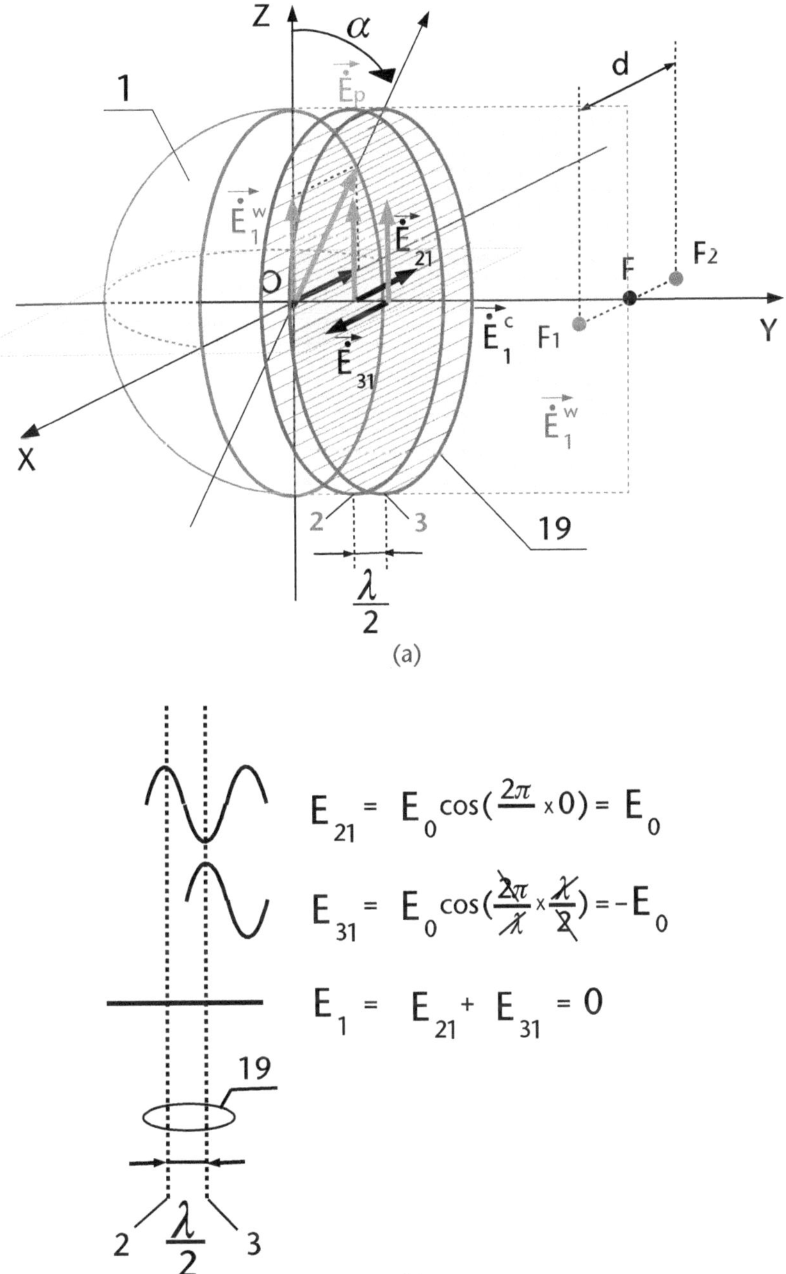

Figure 12.4 A polarization filter 19, mounted in the mouth of monopulse antenna 1.

It is known that in monopulse systems with amplitude direction finding, the pair of feeds F_1 and F_2 displaced symmetrically from the focus gives symmetrically overlapping amplitude radiation patterns for direction finding in one plane. In this case the antenna's directional patterns on working polarization were calculated in the azimuth plane in the range of angles $\varphi \in [-90; +90]$ degrees at a zero tilt angle ($\theta = 0$ degrees) and are shown in Figure 12.5. Here, the angle φ is the angle

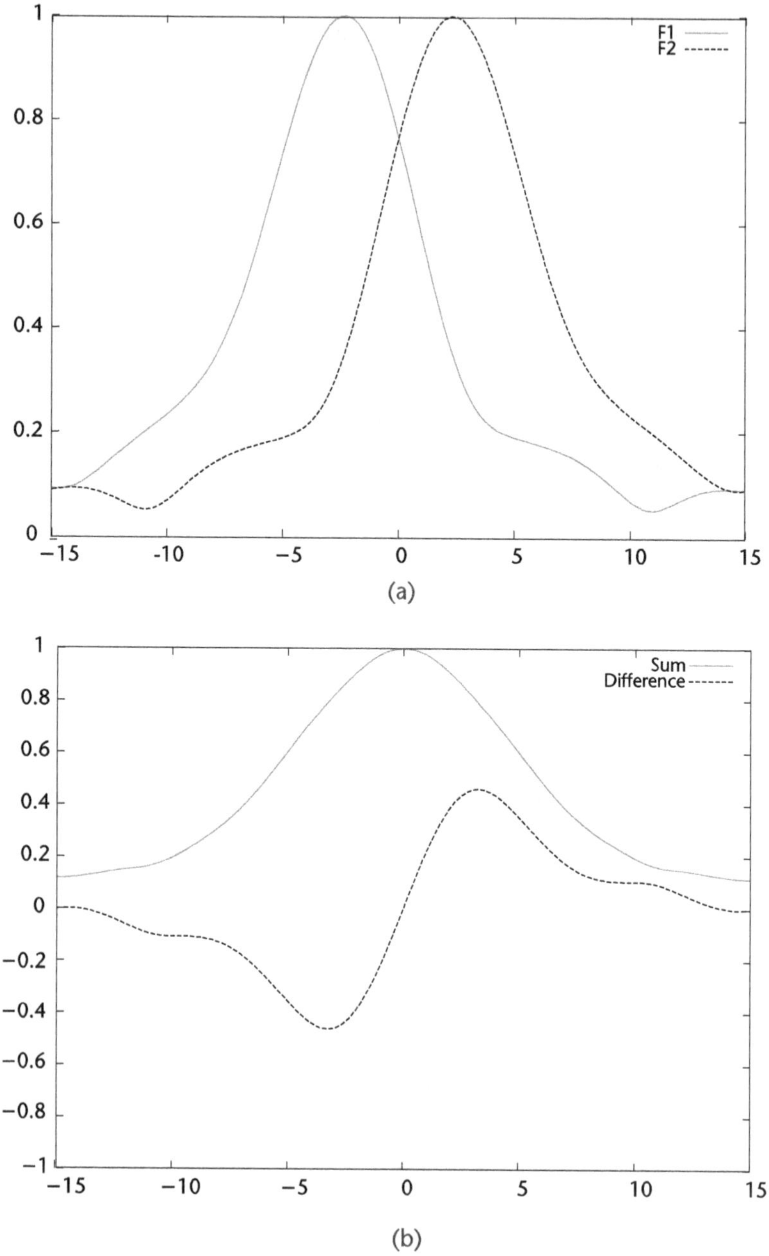

Figure 12.5 Partial, sum, and difference patterns of monopulse antenna on the vertical (working) polarization in the plane φ. (a) Partial directional patterns $F_1(\varphi)$ and $F_2(\varphi)$, and (b) sum and difference antenna patterns $F_\Sigma(\varphi)$ and $F_\Delta(\varphi)$.

measured from a boresight of monopulse antenna in the azimuth plane and if $\varphi = 0$, a target is on the boresight of the monopulse direction finder.

Monopulse antenna 1 consists of two antenna feeds and parabolic mirror. Each feed emits signals on the working and cross polarization and leads to the surface of a metallic mirror spatial currents on working and cross polarization (see Figure 12.6(a)). Moreover, the cross-polarization currents are formed additionally

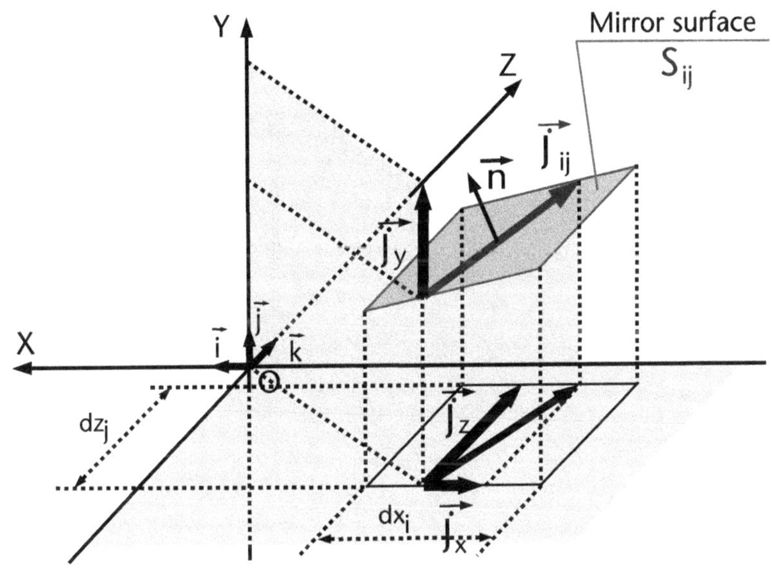

(a)

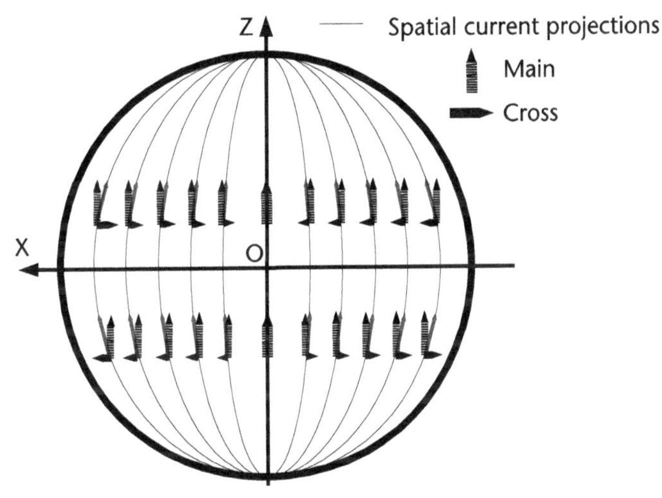

(b)

Figure 12.6 Spatial currents on mirror surface: (a) a spatial current $\vec{J}_{ij}$ and its projections, and (b) main and cross projections on mirror surface.

due of the spatial curvature of the mirror (see Figure 12.6(b)). The field on working polarization and cross polarization, as well as the spatial currents induced on the mirror, have the amplitude and phase, and hence are complex quantities. The antenna pattern induced by the complex currents will vary depending on the focal length of a mirror antenna. Let the focus of mirror antenna F be located in the center of a rectangular coordinate system so that the aperture mirror is in the plane XOZ, and antenna feeds F_1 and F_2 are located at $(-d/2, f, 0)$ and $(+d/2, f, 0)$, where f is the focal length mirror, and d is the distance between monopulse antenna feeds.

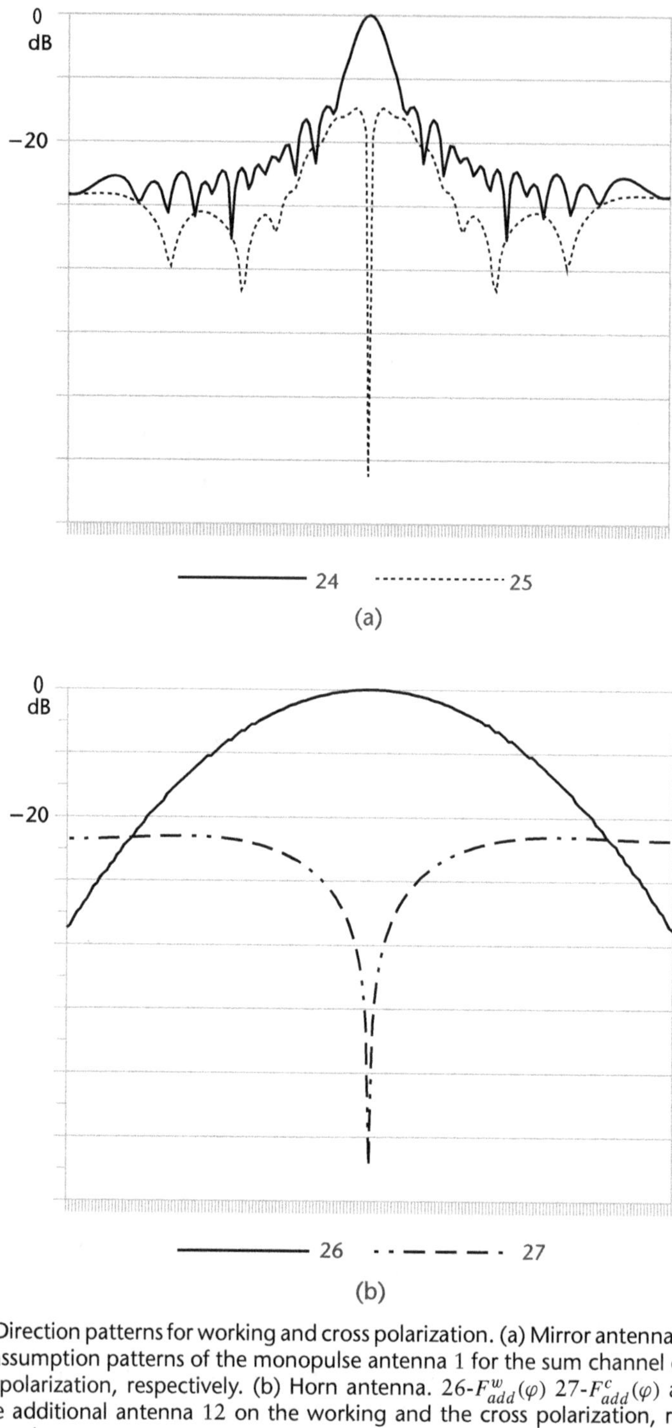

Figure 12.7 Direction patterns for working and cross polarization. (a) Mirror antenna. 24-$F^{w}_{\Sigma}(\varphi)$ and 25-$F^{c}_{\Sigma}(\varphi)$ are assumption patterns of the monopulse antenna 1 for the sum channel on the working and the cross polarization, respectively. (b) Horn antenna. 26-$F^{w}_{add}(\varphi)$ 27-$F^{c}_{add}(\varphi)$ are assumption patterns of the additional antenna 12 on the working and the cross polarization, respectively, for additional channel.

After splitting the mirror in the plane XOZ such that the following conditions:

$$\begin{cases} dx_i, dz_j << \lambda, & \forall i = 1, \ldots, N \\ dx_i, dz_j << R_m, & \forall j = 1, \ldots, M \end{cases} \tag{12.1}$$

Here dx_i, dz_j are the dimensions of an elementary area $S_{ij} = dx_i \times dz_j$ on coordinates x and z, respectively, λ is the wavelength, and R_m is the radius of curvature of the mirror. In this case, the field excited by induced complex currents $\vec{J}_{ij}$ can be considered flat. At the same time the spatial current $\vec{J}_{ij}$ on elementary area S_{ij} in the rectangular basis of the monopulse antenna can be written

$$\vec{J}_{ij} = \dot{J}_x\vec{i} + \dot{J}_y\vec{j} + \dot{J}_z\vec{k} \tag{12.2}$$

where $\vec{i}, \vec{j}, \vec{k}$ are the unit vectors of the rectangular basis XOZ of the monopulse antenna 1, and $\dot{J}_x, \dot{J}_y, \dot{J}_z$ are the corresponding projections of the complex current $\vec{J}_{ij}$ of the elementary area S_{ij} in this rectangular basis. Strictly speaking, x-, y-, and z-projections of the $\vec{J}_{ij}$ vector must also be written with ij-indices each, since they refer to the considered elementary area S_{ij}. However, in this case, in the designation of the x-, y-, and z-projections of the complex current $\vec{J}_{ij}$, cited ij-indices were not used, so as not to clutter up the text and drawing (see Figure 12.6(a)).

The field of antenna feed on working polarization excites the complex currents on the mirror surface on the working and cross polarization, and the field at the cross polarization excited on the surface mirrors the complex currents in cross polarization and working polarization, respectively. It is obvious that the spatial components of these currents must form in accordance with the rules of addition of complex numbers. Furthermore, the sum-and-difference antenna feed pattern monopulse antennas, for example in the plane of course, are formed simultaneously by two antenna feeds. Hence, at each elementary area partitioning antenna reflector S_{ij} shape spatial currents, excited four-bedded partial fields: field antenna feed 1 on the working polarization $\vec{\dot{E}}_{1w}$, the field of the antenna feed 1 on cross polarization $\vec{\dot{E}}_{1c}$, field antenna feed 2 on working polarization $\vec{\dot{E}}_{2w}$, the field of the antenna feed 2 to cross polarization $\vec{\dot{E}}_{2c}$.

$$\vec{J}_{ij} = F\left[\vec{\dot{E}}_{1w} + \vec{\dot{E}}_{1c} + \vec{\dot{E}}_{2w} + \vec{\dot{E}}_{2c}\right], \tag{12.3}$$

Here F is a function that provides the solution of internal problem of electrodynamics according to which the known characteristics of the incident field are spatial currents $\vec{J}_{ij}$ on the irradiated surface of the mirror. Further, from the spatial currents induced on the mirror surface is determined by the electromagnetic field. In this book, it is shown that the problem of finding the field in the far zone at working and cross-polarization has a solution. It is known that under interference conditions, the primary objective of angle tracking systems is to determine the angle direction of a signal source with as few errors as possible [4]. In the process of angle tracking, the radar control system automatically minimizes an error signal obtained as a result of angular misalignment processing. With that, the work function is represented by a direction-finding characteristic that has the following expression for a sum-and-difference pattern [2]:

$$S(\varphi) = \Re\left[\frac{\dot{F}_\Delta(\varphi)}{\dot{F}_\Sigma(\varphi)}\right] \tag{12.4}$$

Here $\dot{F}_\Delta(\varphi)$ and $\dot{F}_\Sigma(\varphi)$ are the complex difference-and-sum direction patterns of monopulse radar antenna, and $\varphi = \varphi_c - \hat{\varphi}$ is the angle misalignment between true and measured target bearing. Here φ_c is the target true position while $\hat{\varphi}$ is the measured track angle of the target on the basis of the monopulse radar receiver antenna (its boresight). The above expression for the direction-finding characteristics can be written as:

$$S(\varphi) = \Re\left[\frac{\vec{i}_w(\dot{F}_1(\varphi) - \dot{F}_2(\varphi))\cos\alpha + \vec{i}_c(\dot{F}_1(\varphi) - \dot{F}_2(\varphi))\sin\alpha}{\vec{i}_w(\dot{F}_1(\varphi) + \dot{F}_2(\varphi))\cos\alpha + \vec{i}_c(\dot{F}_1(\varphi) + \dot{F}_2(\varphi))\sin\alpha}\right] \tag{12.5}$$

Here, $\dot{F}_1(\varphi)$ and $\dot{F}_2(\varphi)$ are complex partial radiation patterns of the monopulse antenna, formed by feeds 1 and 2, respectively, and α is inclination angle of the signal plane of polarization interference in the receiving basis of the monopulse antenna, where $\vec{i}_w$ and $\vec{i}_c$ are unit vectors of working and cross polarization, such that:

$$\begin{cases} \vec{i}_w = \vec{k}, & |\vec{i}_w| = |\vec{i}_k| = 1 \\ \vec{i}_c = -\vec{x}, & |\vec{i}_c| = |\vec{i}_x| = 1 \end{cases} \tag{12.6}$$

Based on the above considerations, calculated direction-finding characteristics were carried for different values of the angle of the plane of the polarization interference monopulse basis in the receive antenna. In Figure 12.7 shows the directional patterns, which were calculated for a mirror antenna (sum channel) and for a horn antenna. Here, curve 24,26 is the work polarization, and curve 25,27 is the cross polarization for the mirror antenna and horn antenna, respectively. The results of calculations for the direction-finding characteristics for different values of the angle α are presented in Figures 12.8 and 12.9. The calculation result for angle value $\alpha = 0°$ is shown in Figure 12.8(a). In this case, automatic target tracking is carried out on the working polarization. As shown in Figure 12.8(b), for values of $\alpha > 80°$ the condition of continuity of the direction-finding function is disrupted and the direction-finding characteristic has a gap in the zero values of the error signal. As the value of the angle α on the orthogonal position is 90°, the direction-finding characteristic is maximally distorted and in the point $\varphi = 0°$ is not defined. In this case the normalization condition (12.4) is violated. Error signal module in a certain ϵ-neighborhood of the point $\varphi = 0°$ will have a maximum value due to division by zero in (12.4), therefore an error signal value tends to infinity and the monopulse direction finder is completely inoperative. The results of calculations for the values of the angle $\alpha = [88°; ...; 89, 9°]$ are presented in Figure 12.9(a), and in Figure 12.9(b) direction-finding characteristic on cross polarization are presented.

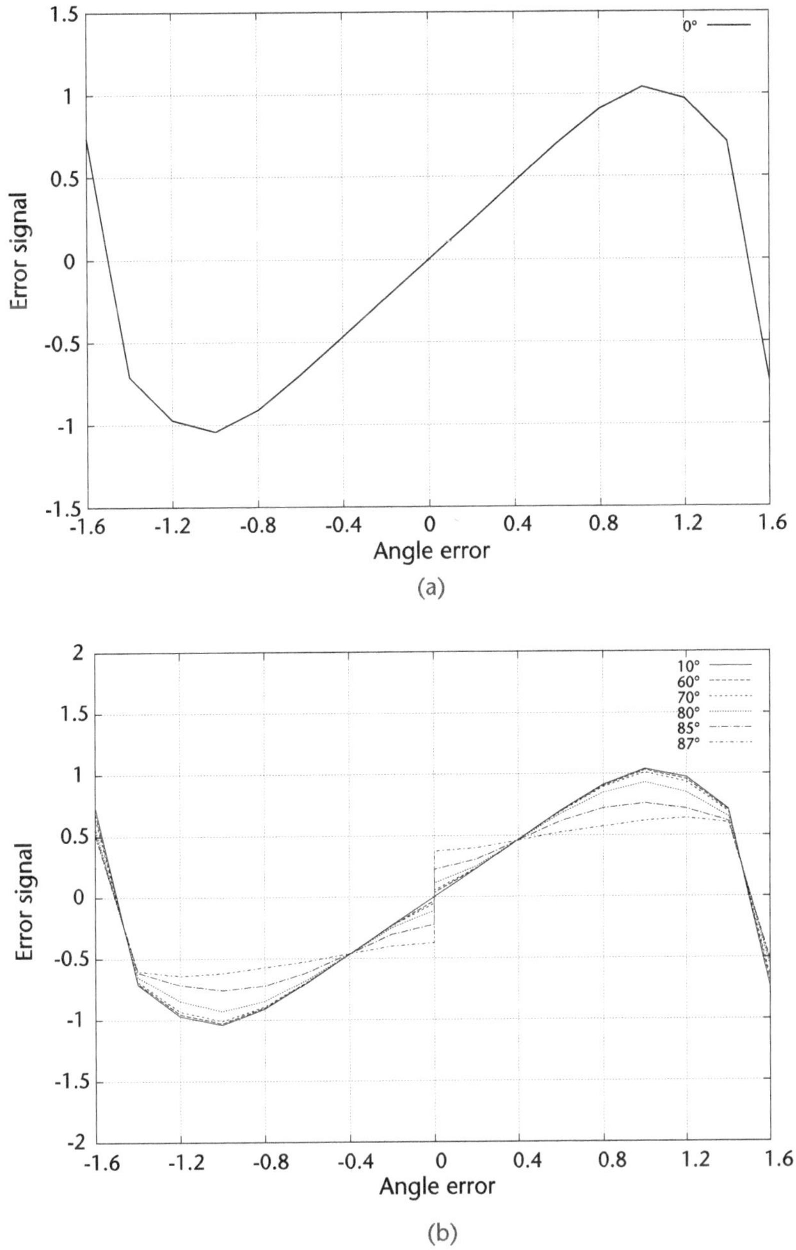

Figure 12.8 Direction-finding characteristics. (a) $\alpha = 0°$, and (b) $\alpha = \{10°;\ldots;87°\}$.

12.2 Assessing Accuracy of an Angular Auto-Tracking System of the Target under Conditions of Polarization Interference

Let us assume the function of direction-finding characteristic $S(\varphi)$ is continuous and n times differentiable in the definition domain $[-\Delta, \Delta]$, except, perhaps, a point $\varphi = 0$. Then, in the vicinity of some point $\varphi_o \in [-\Delta, 0 - o] \cup [0 + o, \Delta]$ it can be represented by Taylor series. In the first approximation, the function $S(\varphi)$ has the

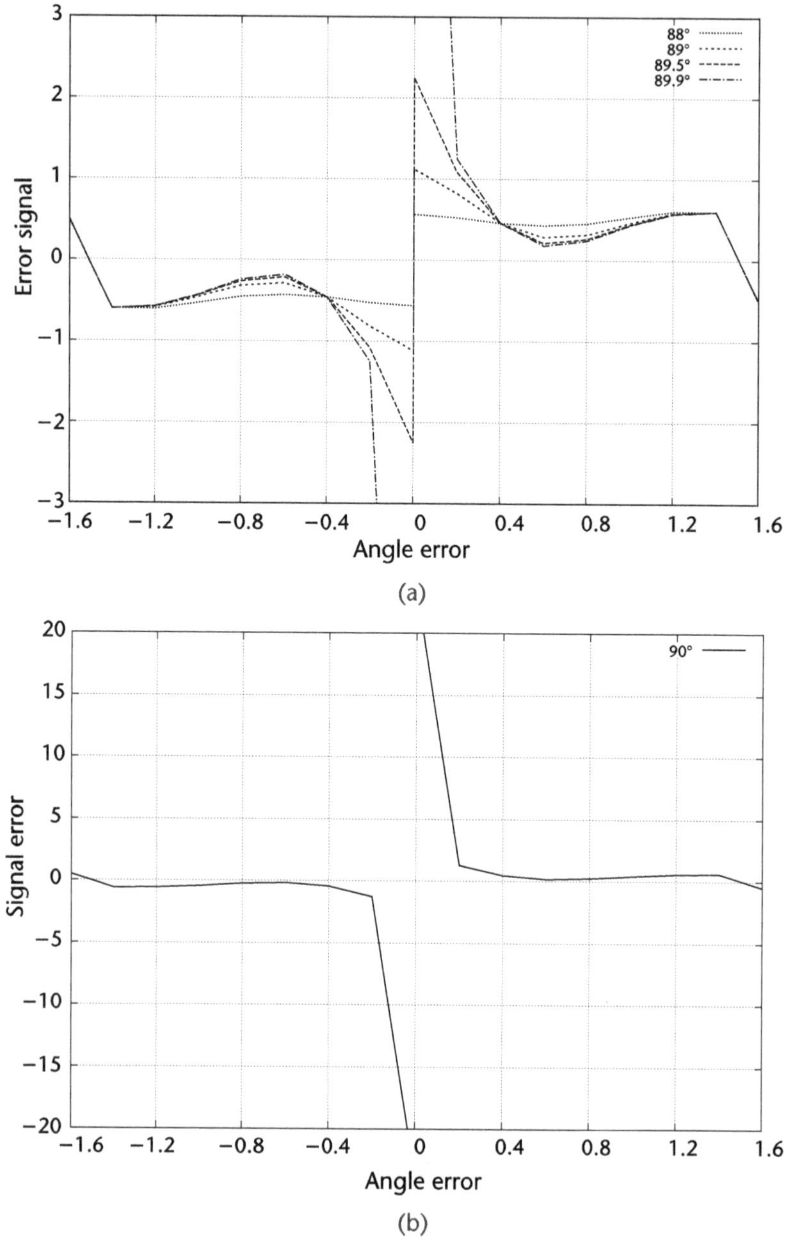

Figure 12.9 Direction-finding characteristics. (a) $\alpha = \{88°; \ldots; 89, 9°\}$, and (b) $\alpha = 90°$.

following appearance:

$$S(\varphi) = S(\varphi_0) + (\varphi - \varphi_0)S'(\varphi) + o(\varphi) \quad (12.7)$$

where $S'(\varphi)$ is the first derivative of direction-finding characteristic in point φ_0, and $o(\varphi)$ is residual member in the series. Since the processes of error signal minimization can be represented as an equation-solving process $S(\varphi) = 0$, then it is possible to obtain an expression for iterative process for finding nulls of the direction-finding characteristic from (12.7):

$$S_{k+1} = S_k - (S(\varphi_k)/S'(\varphi_k)) \tag{12.8}$$

It is clear that (12.7) is a continuous form of Newton's method [12], and by using (12.8) it would be possible to determine the stable autotracking position for known functions $S(\varphi)$ and $S'(\varphi)$. With the residual member of Taylor series in (12.7) neglected, the expression for derivative direction-finding characteristic in the vicinity of point φ_0 will have the following appearance:

$$S'(\varphi) = \frac{S(\varphi) - S(\varphi_0)}{\varphi - \varphi_0} \tag{12.9}$$

Since the function $S(\varphi)$ is continuous and n times differentiable in its definition domain, then for any point φ_0 within the segment $[-\Delta, \Delta]\backslash 0$ it is possible to find a derivative of direction-finding characteristic $S(\varphi)$, determined over this segment per formula (12.9), and hence, to plot the function $S'(\varphi)$ in the domain of its definition. As a result, the following model of angle tracking system operation seems reasonable. The observation interval $[0, T]$ is split into N pairwise disjoint half-intervals $[(k-1)h, kh)$ of duration h each:

$$[0, T) = \bigcup_{k=1}^{N} [(k-1)h, kh), \tag{12.10}$$

where $\bigcap_{k=1}^{N} [(k-1)h, kh) = \emptyset$ at each of these half-intervals, the automatic tracking systems is assumed to be linear and invariant in time. Therefore, within the state space $\{Y\}$, it can be described on each k-unit of time as a differential equation system with fixed factors.

Assuming the functions $S(\varphi)$ and $S'(\varphi)$ are known on each k unit of time and applying the reciprocal difference expression, it is possible to represent the discriminator of the angle tracking system as an amplifying circuit with the gain factor:

$$k(t_k) = S'(\varphi_k) = \frac{S(\varphi_k) - S(\varphi_{k-1})}{\varphi_k - \varphi_{k-1}}, \qquad \forall k = 1, 2, \ldots \tag{12.11}$$

remaining stable in each half-interval $[(k-1)h, kh)$. For each k-unit of time, it is possible to write an expression of state coordinates in the matrix form:

$$\frac{dY}{dt} = AY, \qquad \forall t \in [(k-1)h, kh) \tag{12.12}$$

If the state space $\mathbf{Y}$ has the initial vector $Y_0 = [y_{01}, y_{02}, \ldots, y_{0N}]^T$ set at the starting point of time $t = 0$, the solution of differential equation systems (12.12) with fixed factors and initial conditions Y_0 can be represented as [14]:

$$Y(t) = \exp^{At} Y_0, \qquad \forall t \in [0, h], \tag{12.13}$$

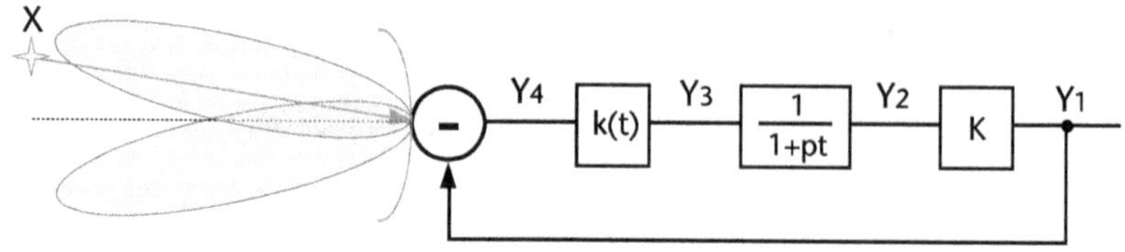

Figure 12.10 A simplified structural diagram of an angle-tracking system.

where **A** is the factor's matrix for a differential equation system, which describes the automatic tracking system, while Y_0 is the initial conditions vector. Thus, the solution (12.13), with account for breakdown of observation intervals $[0, T)$ into N pairwise disjoint half-intervals, can be represented as the finite function:

$$Y(t) = \sum_{k=1}^{N} \exp^{[A(t-(k-1)h]} \chi_k(t), \qquad \forall k = 1, 2, \ldots, N \tag{12.14}$$

where $\chi_k(t) = \begin{cases} Y((k-1)h), & \forall t \in [(k-1)h, kh) \\ 0, & \forall t \notin [(k-1)h, kh) \end{cases}$ - characteristic function kth time unit. Targets auto-tracking system on angular coordinates can be represented by a simplified structural diagram presented in Figure 12.10, where $Y_4 = X - Y_1$ is an error signal in which X is a real target coordinate and Y_1 is the coordinate of target that has been measured by the monopulse radar. Figure 12.11 presents error functions in tracking of linear-polarized interference signal source in static mode, at different angular rotation rates Ω_c of the interference signal polarization plane in the receiving basis of the monopulse antenna. It results from presented dependencies that the tracking of the linear-polarized source of the signal with the time-variable polarization plane is implemented error-free until the signal polarization plane attains position 90°, which corresponds to conditions of polarization interference impact. Clearly, it is possible to match such rotation speed Ω_c of the signal polarization plane at which the loss in automatic tracking of the signal source is achieved with probability close to 1.

It is necessary to note that calculation of the values $k(t_k)$, $\forall k = 1, 2, \ldots$ were calculated according to (12.11). In this case, the values of the direction-finding characteristics of the discriminator $S(\varphi_k)$ and $S(\varphi_{k-1})$ were found from (12.4), which were received by solution of the electrodynamics problem in each kth time moment. The surface of the mirror antenna situated in a rectangular coordinate system {x,y,z} was split in the XOZ plane into M elementary segments, such that each side of each elementary segment is less than half the wavelength λ, in accordance to condition (12.1). Complex vector of the current $\dot{J}_{x,y,z}$ was defined on the each elementary segment. On found distribution complex current $\dot{J}_{x,y,z}$ on surfaces of the mirror were calculated complex sum-and-difference direction patterns of monopulse radar antenna $\dot{F}_{\Sigma}(\varphi)$ and $\dot{F}_{\Delta}(\varphi)$ for operation on working polarization and for cross polarization on each kth quantum of time and for each importance of the corner of plane of polarization of the signal in the base of the monopulse radar antenna. The direction-finding characteristics and discriminator gain were calculated in (12.4) and (12.11) accordingly.

12.3 Stability of Target Auto-Tracking in Conditions of Polarization Interference

Since cross-polarization interference is one of most serious and dangerous interference types to distort the direction-finding characteristic of monopulse radar, and is capable of rendering the radar totally inoperative [2–4] it is worthwhile to consider the stability of angle-tracking systems against impacts of such interference.

According to the method of study advanced above for the structural diagram of the angle-tracking system (12.9), at each k-time unit of breakdown (12.10), there is differential equation system (12.14) solved where:

$$A = \begin{bmatrix} 0 & 0 & 0 & K \\ 0 & -k(t_k)/T_s & k(t_k)/T_s & -1/T_s \\ -k(t_k) & 0 & 0 & k(t_k)K \\ -1 & 0 & 0 & K \end{bmatrix} \tag{12.15}$$

K is the gain factor of the integrating element, T_s is the time constant of control loop and $k(t_k)$ is the discriminator gain factor at kth time unit. Solution (12.14) will be stable when actual parts of all roots in the characteristic polynomial:

$$\begin{aligned} \det[A - \lambda I] = (-\lambda)^N + P_1(-\lambda)^{N-1} + \ldots \\ + P_{N-1}(-\lambda) + P_N = 0 \end{aligned} \tag{12.16}$$

stay in the left semi-plane [14–16]. Here I is the identity matrix of N-order. It follows from the theorem on asymptotical stability of the linear system [14] that it is possible to run direct testing of a specified system by computing the range of matrix eigenvalues (12.15). There are good numerical algorithms to accomplish this task [17]. Yet it is difficult to calculate eigenvalues for systems with an order above two; besides, this method is not applicable when the matrix has parameters in proper factors. Sometimes it is simpler to compute a characteristic polynomial. The characteristic polynomial for matrix (12.15) has the following appearance:

$$(\lambda k(t_k)/T_s + \lambda^2)(\lambda^2 - K\lambda - K) = 0 \tag{12.17}$$

It is quite clear that one of solutions of (12.17) will be the radical:

$$\lambda = -\frac{k(t_k)}{T_s}, \tag{12.18}$$

this being in conformance with the theorem on continuous dependence of characteristic roots on matrix elements [17]. Hence, the relation of the discriminator gain factor, which has the minus sign, to the constant of auto-tracking time is identically equal to one of **A** matrix eigenvalues. Thus, if a discriminator gain factor is negative, then any autotracking system will become unstable due to one of eigenvalues of system matrix factors. It is possible to calculate the function of derivative per (12.11) for the discriminator cross-polarization direction-finding characteristic given in

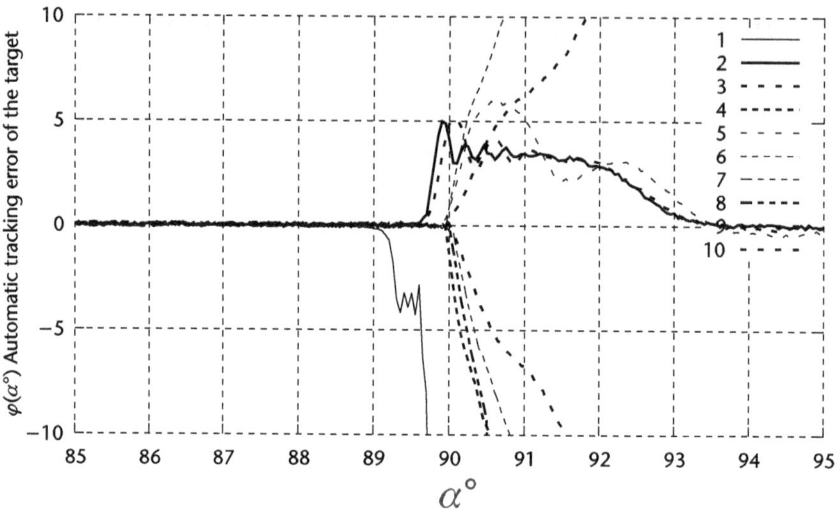

Figure 12.11 Error functions.

Figure 12.9(b), whence it follows that as the locator comes under impact of polarization interference, its discriminator gain factor will be negative due to negativeness of difference $S(\varphi_k) - S(\varphi_{k-1})$ in (12.11). Therefore, $k(t_k) < 0, \ \forall t \in [(k-1)h, kh)$ and due to (12.18) one of **A** matrix eigenvalues will be positive.

Figures 12.8(a) and 12.9(b) are the present direction-finding characteristics for work polarization mode ($\alpha = 0°$) and on cross-polarization mode ($\alpha = 90°$), respectively, while Figure 12.12 shows diagrams of change in the discriminator gain coefficient.

A comparison of Figures 12.11 and 12.12 shows that in realizations 1,4,6,7, and 10 in Figure 12.12(b) and in realization 8, and 9 in Figure 12.12(c) the monopulse radar loses the target.

Figures 12.11 and 12.12(a) show realizations 2,3, and 5 where the monopulse radar has automatic tracking of the target with a large error on the angular coordinates and with a subsequent transient process that returns the system to its normal state when $\alpha = 0°$. It is obvious that the duration and shape of this transient process depends on the rotational speed Ω_c of the linear polarization plane of the interference signal, which has an impact on the monopulse antenna. It is evident from Figure 12.12 that the impact of polarization interference, which unambiguously results in loss of automatic target tracking, is characterized, as a rule, by a large numerical value of gain factor, and necessarily, by the negative sign.

So, at an angular velocity of rotation of the polarization plane of an interfering signal Ω_c = 1°/sec (realization 1), an oscillatory process is observed, after which is the loss of the automatic tracking of the target along the angular coordinates. The oscillatory process, having three minimums and two maximums, respectively, for $\alpha = 89.35°$ the angular tracking error $\varphi = -4, 19°$, for $\alpha = 89.40°$: $\varphi = -3.21°$, for $\alpha = 89.45°$: $\varphi = -3.91°$, for $\alpha = 89.50°$: $\varphi = -3.23°$, for $\alpha = 89.55°$: $\varphi = -4.29°$, and after $\alpha = 89.60°$: $\varphi = -2.83°$ there is a breakdown in the automatic tracking of the source of polarization interference. The described oscillatory process is characterized by the fact that in the interval $\alpha \in [89.35; 89.55]$ the value of the discriminator gain does not exceed units and is of a sign-alternating character. And

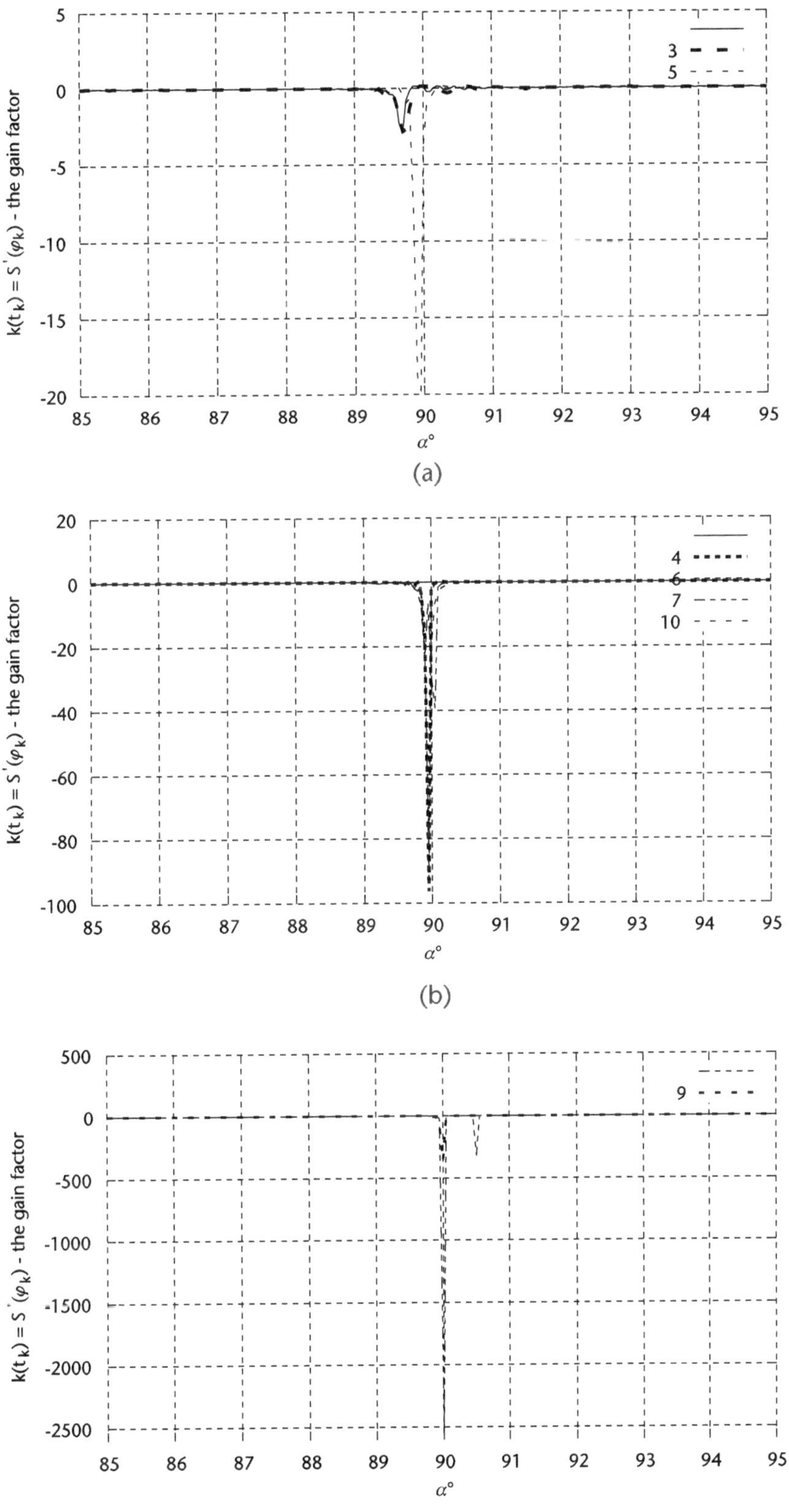

Figure 12.12 Discriminator gain.

at $\alpha = 89.60°$: $\varphi = -2.83°$, the discriminator gain becomes negative, up to the moment of loss of the automatic tracking of the source of interference. As can be seen from Figure 12.12(a), implementations 2,3,5, when the monopulse radar has automatic target tracking with a large error in angular coordinates, are given for the angular velocities of rotation of the polarization plane of interference for $\Omega_c = 2°/\text{sec}$, $\Omega_c = 3°/\text{sec}$, and $\Omega_c = 5°/\text{sec}$, respectively. The discriminator gain in the range of angles $\alpha \in [87.95; 90.50]°$ does not exceed the value $S'(\varphi) = -2.91$ at for $\Omega_c = 2°/\text{sec}$ and for $\Omega_c = 3°/\text{sec}$ respectively, and does not exceed the value $S'(\varphi) = -19.73$ for $\Omega_c = 5°/\text{sec}$.

In the case when the monopulse direction finder is aimed at an interference source with a certain angular error (approximately 0.3 deg.) and with a small value of the discriminator gain (up to -20), its equal-signal direction is deviated on $\varphi = 5.98°$, at $\alpha = 90.65°$, under these conditions the gain of the discriminator has a positive sign, and the angular tracking of the source of interference returns to normal mode. Therefore, with inaccurate homing at the source of polarization interference in the interval $\alpha \in [87.95; 90.05]°$, the minimum gain value does not exceed -20.

Figures 12.11, 12.12(b), and 12.12(c) show realizations of 1,4,6,7,8 and 9,10 for $\Omega_c = 1°/\text{sec}$, $\Omega_c = 4°/\text{sec}$, $\Omega_c = 6°/\text{sec}$, $\Omega_c = 7°/\text{sec}$, $\Omega_c = 8°/\text{sec}$ and $\Omega_c = 9°/\text{sec}$, $\Omega_c = 10°/\text{sec}$, respectively. As we can see in Figures 12.11 and 12.12(b), realizations 4,6,7,8 and 9,10 describe the loss of auto-tracking of the polarization interference source. Moreover, for implementation 4, the loss of auto-tracking occurs when $\alpha = 89.95°$ and $S'(\varphi) = -96.33$. Moreover, the loss of auto tracking occurs for realization 6, when $\alpha = 89.95°$ and $S'(\varphi) = -84.11$, and, respectively, for realization 7 when $\alpha = 90.05°$ and $S'(\varphi) = -39.26$, for realization 10 when $\alpha = 89.95°$ and $S'(\varphi) = -93.15$, for realization 8 when $\alpha = 90.00°$ and $S'(\varphi) = -2615.82$, and for realization 9 when $\alpha = 90.00°$ and $S'(\varphi) = -361.10$.

Thus, impact of polarization interference leads to negative definiteness of function $S'(\varphi)$ on the segment of its definition domain, and therefore, to shift of eigenvalues range of $\mathbf{A}$ matrix factor to the right semiplane, resulting in instability of the auto-tracking system and subsequent loss of automatic tracking of the target.

12.4 The Decision Synthesis, Providing Auto-Tracking Stability of the Polarization Interference Source

If the auto-tracking system is controllable and observable [13, 15, 19, 20], then, according to Bellman's optimality principle [19], there is the optimal control $U(t)$, a solution to the Bellman equation when the $U(t)$ function is linearly integrated into the model while the minimizing functional is quadratically integrated. In this connection, the synthesis task is reduced to the Riccati equation [20], whose solution exists and is unique if the system is controllable and completely observable [19,20]. If the angle-tracking system operates from the source of polarization interference, the linear relationship between the tracking error φ and its measured value $S(\varphi)$ is disrupted. In this case, one of Y vector coordinates ceases to be observable, and therefore, it is not possible to develop a vector-function outside the system and optimal after Bellman, which would enable to stabilize the solution of the $Y(t)$ perturbed system as required. There possibly exists, however, some transformation F that would make it possible, under these conditions, to affect control with minimal errors. Let us assume that $\hat{Y}(t)$ is a solution (12.14) for a system unsuppressed by interference while $Y(t)$ is a solution for a system with a distorted direction-finding characteristic. Both automatic systems are presumed to be identical except for their discriminators with direction-finding characteristics $\hat{S}(\varphi)$ and $S(\varphi)$, respectively. Consequently, it is necessary to find the operator F that operates in space L_2 so $\hat{Y}(t) = FY(t)$. In this case, the following requirements are set to the operator.

1. Norm minimum:

$$\min \left\| Y(t) - \hat{Y}(y) \right\|^2 \tag{12.19}$$

2. Boundedness of solution components $Y(t) : \left\| Y(t) \right\| < C < \infty$, and hence, $\int_0^T \left\| Y(t) \right\|^2 dt < M < \infty$, where C and M are sufficiently large numbers that characterize the upper limits of possible values (closure of state space $\{Y\}$).
3. Continuity of realization $Y(t)$; that is $\forall t_0 \in [0, T] \quad \exists \lim\limits_{t_0 - 0} Y(t) = \lim\limits_{t_0 + 0} Y(t) = Y(t_0)$.

For open (12.18) we can use the quadratically form [21]

$$A(Y, Y) = \sum_{ik=1}^{n} a_{ik} Y_i Y_k = Y^T A Y, \tag{12.20}$$

where $A = \|a_{ik}\|^n$ is the symmetrical matrix (weighting matrix) and for identity weighting matrix I the minimizing functional $Q\{Y\}$ will assume the shape

$$\begin{aligned} Q\{Y\} = & tr\left[\int_0^T Y(t) I Y^T(t) dt \right] + \\ & tr\left[\int_0^T \hat{Y}(t) I \hat{Y}^T(t) dt - 2 \int_0^T \hat{Y}(t) I Y(t) dt \right] \end{aligned} \tag{12.21}$$

Let the polarization interference act against the angle tracking system in the time interval $[0, T]$ and system (12.12) be linear and invariant in time over the given interval. It is necessary to find a solution that would provide the functional (12.21) with extremum; that is, to find such $Y(t)$ that

$$\forall \delta Y(t) \not\equiv 0 : \ \delta Q\{Y\} = 0 \tag{12.22}$$

The equation for functional variation would be written as:

$$\delta Q\{Y\} = tr\left[\int_0^T \left\{Y(t) - \delta Y(t)I(Y^T(t) + \delta Y^T(t)) + \hat{Y}(t)I\hat{Y}^T(t) - 2(\hat{Y}(t)I(Y(t) + \delta Y(t))) - Y(t)IY^T(t) - \hat{Y}(t)I\hat{Y}^T(t) + 2\hat{Y}(t)IY(t)\right\}dt\right] = 0 \quad (12.23)$$

After terms grouping and cancellation, it is possible to write, keeping the main part of functional variation in a shape:

$$\delta Q\{Y\} = \delta Y^T(t)[3Y(t)I] + \delta Y(t)[2IY^T(t)] = 0 \quad (12.24)$$

Therefore, functional variation $\delta Q\{Y\}$ will reach the global minimum at:

$$Y(t) = 0 \quad (12.25)$$

Since the only control parameter in the locator is the error signal $S(\varphi, t)$, consequently, shaping the weighting matrix A^* such that $a_{ij} = 0, \forall i \neq j; a_{ij} \to 0, a_{44} = 1$ by substitution of such weighting matrix A^* in (12.20), we affect the transfer to scalar control, and hence, the condition of applying (12.25) is identical to the disconnection of system on error signal.

Thus, when an angle-tracking system becomes affected by polarization interference, the optimal reaction to this would be the disconnection of the tracking system for the period of interference impact.

Let the polarization interference act on the angle tracking system and let an outside observer detect this impact, disconnect the tracking system on error circuit, and then reconnect it once the interference impact is over.

Suppose that in this case an interference detector is realized in which the procedure for recognizing polarization interference gives a true result. See Chapter 10.

Then, in the case when $d\varphi/dt = 0$, the solution to the differential equation system (12.12) is identical to null over the whole half-interval $[0, T)$. When $d\varphi/dt \neq 0$, system solution (12.12) may significantly differ from null $\forall t \in [T, T_1]$. If, in this case the error signal modulus is larger than the half-width of direction-finding characteristic on the operating polarization ($|\varphi| \geq |\Delta|$), then $S'(\varphi) < 0$, and consequently, we will have loss of stability and, most likely, loss of target automatic tracking.

Thus, in the case of tracking system disconnection, tracking errors induced by target dynamics are not taken into account. Such errors emerge on completion of interference polarization impact, and theoretically can lead to loss of automatic angle tracking of the target. Hence, on the interval greater than the interference impact time, disconnection of the tracking system on the error signal may turn out to be a nonoptimal solution in providing for protection of the angle auto-tracking system against interference impacts that distort its direction-finding characteristic.

Therefore, it is necessary to analyze the operation of the angle auto-tracking system after interference termination for some time interval $[0, T_1] = [0, T] \cup [T, T_1]$. As a result, the minimizing functional will appear as:

$$\begin{aligned}\delta Q\{Y\} &= tr\left[Y(T+\epsilon)IY^T(T+\epsilon) - \hat{Y}(T+\epsilon)I\hat{Y}^T(T+\epsilon)\right] \\ &+ tr\left[\int_0^T \left\{Y(t)IY^T(t) + \hat{Y}(t)I\hat{Y}^T(t) - 2\hat{Y}(t)IY(t)\right\}dt\right] \\ &+ tr\left[\int_T^{T_1} \left\{Y(t)A^*Y^T(t) - 2Y(t)A^*\hat{Y}^T(t) + \hat{Y}(t)A^*\hat{Y}^T(t)\right\}dt\right]\end{aligned} \quad (12.26)$$

On completing actions similar to the first case, the terms for functional variation (12.26) turning null at nonzero variation of solution $Y(t)$ will be written as:

$$Y(t) = \left[C\right] \times \left[2A^*\hat{Y}(t)\right]^{-1} \quad (12.27)$$

Since $Y(t) \neq 0$ and $\hat{Y}(t) \neq 0$, consequently $[C]$ represents a matrix of size $n \times n$:

$$Y(t)2A^*\hat{Y}^T(t) = [C], \quad (12.28)$$

where:

$$\begin{aligned}[C] &= \delta Y^T(T+\epsilon)IY(T+\epsilon) + \delta Y(T+\epsilon)IY^T(T+\epsilon) \\ &= E\left[Y(T+\epsilon)\right]\end{aligned} \quad (12.29)$$

and characterizes the state vector evaluation $Y(t)$ of the automatic control system in some sufficiently small ϵ-vicinity of polarization interference termination point. Correspondingly, for $\hat{Y}(t)$:

$$\hat{Y}(t) = \left[C\right] \times \left[2A^*Y(t)\right]^{-1} \quad (12.30)$$

Thus, to provide for optimal reaction of the angle auto-tracking system to the impact of polarization interference, it is necessary to inversely transform the product of system state vectors by weighting matrix, and subsequently, normalize results through the C-matrix.

With provision for the weighting matrix **A**, actual implementation of this solution was affected by applying error signal inverse function.

It can be shown that the solution to the task of providing monopulse radar interference immunity against polarization interference impact exists (12.30), and as shown in (12.31), is the only solution and is unique.

Let there be another constraint, continuous and n times differentiated solution $Y_2(t)$ at the same Cauchy condition: $\hat{Y}(t_0) = Y_0$, where $\|Y_2(t)\| < C < \infty, \forall t \in$

$[0, T] \supset [-\infty, +\infty]$, then it is possible to build up the function $Z(t) = Y_2(t) - \hat{Y}(t)$ and write the matrix equation as:

$$\begin{cases} dZ(t)/dt = AZ(t), \\ Z(t) = 0, \quad \forall t \in [0, T] \end{cases} \tag{12.31}$$

Therefore, as per uniqueness theorem [14]: $\forall t_0 \in [0, T]$ the solution of system (12.31) is identically zero over the whole interval $[0, T]$ of its existence; that is, $Z(t) = 0, \forall t \in [0, T]$, and therefore, the solution $\hat{Y}(t)$ is unique.

12.5 Evaluation of the Stability of an Angular Auto-Tracking System of a Target Under Conditions of Polarization Interference

As is known, the expression for the derivative of the direction-finding characteristic in the neighborhood of point φ_0 will have the following appearance (12.9):

$$S'(\varphi) = \frac{S(\varphi) - S(\varphi_0)}{\varphi - \varphi_0} \tag{12.32}$$

Since the function $S(\varphi)$ is continuous and n times differentiated in its definition domain, then for any point φ_0 within the segment $[-\Delta, \Delta]\backslash 0$ it is possible to find a derivative of direction-finding characteristic $S(\varphi)$, determined over this segment per formula (12.32), and hence, to construct the function $S'(\varphi)$ in the domain of its definition. In practice, target auto-tracking failure probability is frequently used as one of the automatic target tracking system jamming resistance indexes. Hence, it is useful to define a probability measure $P(A_1)$ such that the event A_1 characterizes automatic target tracking failure. Therefore it is necessary to determine exhaustive events as $A = A_1 \cup A_2$, where $P(A_1) + P(A_2) = 1$. It would appear reasonable that hence σ-algebra $\mathfrak{U}$ is given over some set Ω such that $A \subseteq \Omega$. As a result, the probability space $(\Omega, \mathfrak{U}, P)$ is set. Because any probability measure is a function of time t, hence, the defined probability space will also be measurable in time. As the automatic system state is described by vector random variable $Y(t)$ and $\mathfrak{U}(0, T)$, σ-algebra of events generated by quantities $Y(t), \forall t \in [0, T]$ is denoted by $U(0, T)$ system behavior in the past and the future can be described by the σ-algebra of events $\mathfrak{U}^{(1)} = \mathfrak{U}(0, t_0)$, $\mathfrak{U}^{(2)} = \mathfrak{U}(t_0, T)$ and $\mathcal{B} = \mathcal{B}(t_0 - \epsilon, t_0)$. If events $A_1 \in \mathfrak{U}^{(1)}$ do not depend on the events $A^{(2)} \in \mathfrak{U}^{(2)}$ with known outcome of all events of $\mathcal{B}$, the following equality is right:

$$P(A^{(1)}A^{(2)}|\mathcal{B}) = P(A^{(1)}|\mathcal{B})P(A^{(2)}|\mathcal{B}), \tag{12.33}$$

$\forall A^{(1)} \in \mathfrak{U}^{(1)}$, $\forall A^{(2)} \in \mathfrak{U}^{(2)}$, and σ-algebra $\mathcal{B}$ is splitting [22]. It divides the observation interval $[0, T]$ in three segments: $[0, t_0 - \epsilon],]t_0 - \epsilon, t[, [t_0, T]$, where σ-algebras $\mathfrak{U}^{(1)}, \mathcal{B}, \mathfrak{U}^{(2)}$, act respectively with $A^{(1)} = A_1 \cup A_2, \forall t \in [0, t_0 - \epsilon]$; $A^{(2)} = A_1 \cup A_2, \forall t \in [t_0, T]$. By virtue of the fact that $\forall t_0 \in [0, T]$, one can construct a splitting of σ-algebra $\mathcal{B}$ such that the observation interval can be sampled in N

nonintersected segments:

$$[0, T] = \bigcup_{k=1}^{N} [(k-1)h, kh - \epsilon], \tag{12.34}$$

Here, h is quantum duration of the observation interval partition. In this case σ-algebras $\mathfrak{U}^{(1)}$ and $\mathfrak{U}^{(2)}$ will depend on the time and cleave σ-algebra $\mathcal{B}$ will be constructed only in the left ϵ-neighborhoods of the sampling points. So for the Nth sampling quantum, the following expression can be written:

$$\begin{aligned} \mathfrak{U}^{(1)} &= \bigcup_{k=1}^{N} \mathfrak{U}((k-1)h, kh)|\mathcal{B}(Mh - \epsilon, Mh)), \\ \mathfrak{U}^{(2)} &= \bigcup_{k=1}^{N} \mathfrak{U}((k-1)h, kh) \end{aligned} \tag{12.35}$$

By the result of the passage of the limit:

$$[0, T) = \lim_{\epsilon \to 0} \bigcup_{k=1}^{N} [(k-1)h, kh - \epsilon] = \bigcup_{k=1}^{N} [(k-1)h, kh), \tag{12.36}$$

random process is $Y(t)$ implementation continuity can be ensured. The following sentence is true [22]: $\mathcal{B}$ splits σ-algebras $\mathfrak{U}^{(1)}$ and $\mathfrak{U}^{(2)}$ if and only if the sequence $\mathfrak{U}^{(1)}, \mathcal{B}, \mathfrak{U}^{(2)}$ is a Markovian sequence; that is, $P(A|\mathfrak{U}^{(1)} \vee \mathcal{B}) = P(A|\mathcal{B})$, $A \in \mathfrak{U}(2)$ and therefore at the known current state of the system $Y(t)$ its behavior in the future does not depend on the past. As a result, the Markov property can be rewritten as:

$$P(A|Y(s), s < t) = P(A|Y(t)), \quad A \in \mathfrak{U}(t, T) \tag{12.37}$$

Thus $Y(t)$ can be approximated by a Markov random process. With $h \to 0$ and entering of an orthogonal stochastic measure $\mu(dt) = M|\eta(dt)|^2$ with zero mean automatic control system behavior can be described by a stochastic Ito integral or a differential equation system with bounded indefinitely differentiable coefficients and generalized random functions. Due to the fact that $Y(t)$ is a Markovian process and jamming influencing the direction-finder distorts the direction-finding characteristic, duration of the sampling interval h is determined by statistical properties of the interfering signal, which can also be described by Markovian random process. Thus, invariance of an automatic control system during the time interval of duration h is determined by interference stationarity on this interval.

Let us suppose that the interference signal can be represented by a certain sequence of its stationary states of length h each. As a result, the following model of angle-tracking system operation seems reasonable: time interval $[0, T]$ where a vehicle-borne direction-finder goes through interference attack is split into N

pairwise disjoint half-intervals $[(k1)h, kh)$ (quantums) of duration h each:

$$[0, T) = \begin{cases} \bigcup_{k=1}^{N} [(k-1)h, kh), \\ \bigcap_{k=1}^{N} [(k-1)h, kh) = \oslash \end{cases} \qquad (12.38)$$

At each of these half-intervals the automatic directional tracking systems is assumed to be linear and stable by this time. Therefore, it can be presented by differential equation system with fixed factors on each hth quantum within state space $\{Y\}$: $\{\forall y : \|y\| < \infty\}$. Assuming that the functions $S(\varphi)$ and $S'(\varphi)$ are known on each kth quantum and applying the reciprocal difference formula, it is possible to represent the discriminator as an amplifying circuit with the following amplifier gain:

$$k(t_k) = S'(\varphi_k) = \frac{S(\varphi_k) - S(\varphi_{k-1})}{\varphi_k - \varphi_{k-1}}, \qquad \forall k = 1, 2, \ldots \qquad (12.39)$$

remaining stable in each k-half-interval $[(k-1)h, kh)$. It is possible to write phase coordinates equation in the matrix form for each kth quantum:

$$\frac{dY}{dt} = AY, \qquad \forall t \in [(k-1)h, kh) \qquad (12.40)$$

If the initial vector $Y_0 = [y_{01}, y_{02}, \ldots, y_{0N}]^T$ is given in the state space Y at the start time $t = 0$, then finding the solution of differential equation systems (12.40) with factors fixed on h-sized interval and initial conditions Y_0 represents a Cauchi problem [19]. This problem solution can be represented as [19]:

$$Y(t) = \exp^{At} Y_0, \qquad \forall t \in [0, h], \qquad (12.41)$$

where A is the coefficient matrix of the differential equation system that describes the automatic tracking system while Y_0 is the initial condition vector. Solution (12.41) will be stable in the case when real parts of all characteristic equation roots:

$$\begin{aligned} \det[A - \lambda t] = (-\lambda)^N + p_1(-\lambda)^{(N-1)} + \ldots \\ \ldots + p_{(N-1)}(-\lambda) + p_N = 0 \end{aligned} \qquad (12.42)$$

are located in the left semiplane [19]. There are different methods to determine the stability of the linear equation system, including [19] direct calculation of eigenvalues for matrix A, methods based on the properties of the characteristic polynomial, Nyquist criterion and the method of the root locus (for discrete systems), and the Lyapunov method. It can be shown that the estimation of the modulus of continuity in the theorem on continuous dependence of the characteristic roots on the matrix A elements accepts very significant root change under matrix perturbations that are so small they cannot be simulated on a computer [14]. Stability criteria based on characteristic root location are not always convenient for practical applications [18]. It is suggested to choose the value $æ(A)$, determined by all solutions of the system $\frac{dY(t)}{dt} = AY(t)$ [16]. Each such solution can be uniquely defined by the initial value $Y(0)$ of vector $Y(t)$ with $t = 0$. Here $æ(A)$ is defined as follows:

$$æ(A) = 2\|A\| \sup\left[\frac{\int_0^\infty \|Y(t)\|^2 dt}{\|Y(0)\|^2}\right], \tag{12.43}$$

where $\|A\| = \sup_Y \frac{\|AY\|}{\|Y\|}$ and $\|Y\| = \sqrt{Y_1^2 + Y_2^2 + \ldots + Y_N^2}$. If the matrix A is symmetric then $\|A\| = \max|\lambda_i|$. As $\int_0^\infty \left[\exp(-t\|A\|)\|Y(0)\|\right]^2 dt = \frac{\|Y(0)\|^2}{2\|A\|}$, therefore:

$$æ(A) = \sup_{Y(0)} \frac{\int_0^\infty \|\exp\{tA\}Y(0)\|^2 dt}{\int_0^\infty \left[\exp\{-t\|A\|\}\|Y(0)\|\right]^2 dt} \tag{12.44}$$

Using (12.44) and $\|Y(t)\| = \|\exp\{At\}Y(0)\| \geq \exp\{-\|A\|t\}\|Y(0)\|$, therefore: $\int_0^\infty \|Y(t)\|^2 dt \geq \int_0^\infty \left[\exp\{-t\|A\|\}\|Y(0)\|\right]^2 dt$, whence follows the inequality:

$$æ(A) \geq I, \tag{12.45}$$

where I is an identity matrix. If the system zero solution is not asymptotically stable; that is, the matrix A has a characteristic root $[-\lambda_0]$, such that $Re[\lambda_0] \geq 0$, then the system has a solution of the form $Y(t) = \exp\{\lambda_0 t\}Y(0) = \exp\{Re[\lambda_0]t\}\|Y(0)\| \geq \|Y(0)\|$, $\int_0^\infty \|Y(t)\|^2 dt \geq \int_0^\infty \|Y(0)\|^2 dt = \infty$, and hence: $æ(A) = 2\|A\| \sup_{Y(0)} \int_0^\infty \|Y(t)\|^2 dt = \infty$. But if zero solution of the system $\frac{dY(t)}{dt} = AY(t)$ is asymptotically stable; that is, matrix A is a Hurwitz matrix, then $æ(A)$ is finite. Here $æ(A)$ is calculated from the formula:

$$æ(A) = 2\|A\|\|H\|, \tag{12.46}$$

where H is a solution of the Lyapunov equation:

$$HA + A^*H = I \tag{12.47}$$

where I is an identity matrix. In the case of the one-valued solvability of (12.47), matrix H is strictly positive definite [16]. It is proposed in [14, 16] to compute H

in the shape of improper integral as a computational procedure (12.47):

$$H = \int_0^\infty \exp\{tC^*\}\exp\{tC\}dt, \qquad C = \frac{A}{\|A\|} \tag{12.48}$$

and C^* in the general case is the complex conjugate matrix to C. Integral (12.48) is divergent if æ(A) =æ$(C) = \infty$. Vector function $Y(t)$ must be measurable on the interval $[0, T)$ and its integration result on this interval must be finite to provide automatic control, hence, the function $Y(t)$ is defined on Hilbert space $L_2[0, T]$, in this case (12.48) is a scalar product in Hilbert space and we can write:

$$H_k = \int_0^{t_k} \exp\{tC^*\}\exp\{tC\}dt, \quad \text{where} \quad t_k = (k-1)h \tag{12.49}$$

The stability criterion takes the following form:

$$\lim_{k\to\infty} \|H_k - H_{k-1}\| = 0, \tag{12.50}$$

and in case of instability:

$$\lim_{k\to\infty} \|H_k - H_{k-1}\| = \infty, \tag{12.51}$$

With regard to the decomposition of the observation interval $[0, T]$ on N pairwise disjoint half-intervals (12.38), solution (12.41) can be represented as a finite function:

$$Y(t) = \sum_{k=1}^{N} \exp\{A(t-(k-1)h\}\chi_k(t), \tag{12.52}$$

where $\chi_k(t)$ is the characteristic function of kth quantum, which has the form:

$$\chi_k(t) = \begin{cases} Y_0((k-1)h), & \forall t \in [(k-1)h, kh) \\ 0, & \forall t \not\subset [(k-1)h, kh), \end{cases} \tag{12.53}$$

where its right bound is an upper limit of integration in (12.48). Hence, the representation of differential equation system solution in the form of (12.52) makes possible to find the system state vector at any time t given direction-finding characteristic distortion under conditions of interference. Introducing the notations:

$$\begin{aligned} dH_k = \|H_k - H_{k-1}\| = \Bigg\| \int_{t_{k-1}}^{t_k} \exp\{tC_k^*\}\exp\{tC_k\}dt - \\ - \int_{t_{k-2}}^{t_{k-1}} \exp\{tC_{k-1}^*\}\exp\{tC_{k-1}\}dt \Bigg\| \end{aligned} \tag{12.54}$$

where:

$$\begin{cases} C_k, C_k^* = const\ \forall t \in [(k-1)h, kh), \\ C_{k-1}, C_{k-1}^* = const\ \forall t \in [(k-2)h, (k-1)h), \\ t_k = kh,\ t_{k-1} = (k-1)h,\ t_{k-2} = (k-2)h \end{cases} \tag{12.55}$$

Consequently, the following expression can be written for points on the h-length interval boundaries:

$$\frac{dH_k}{dt} = \left\| \exp\{hC_k^*\} \exp\{hC_k\} - \exp\{hC_{k-1}^*\} \exp\{hC_{k-1}\} \right\| \tag{12.56}$$

Thus the criterion for the automatic control system stability estimation in real time $\forall k \in [1, N]$ can be written as:

$$\frac{dH_k}{dt} = 0 + \epsilon, \tag{12.57}$$

where ϵ is a sufficiently small right neighborhood of zero such as $\epsilon \to 0$ and in this case the condition of instability can be written in the form of:

$$\frac{dH_k}{dt} = \hat{M} \to \infty, \tag{12.58}$$

where $\hat{M}$ is a sufficiently large number limited from above by a computer's word size length.

12.6 Proposals to Improve the Interference Immunity of a Monopulse Direction Finder to the Effects of Polarization Interference

The monopulse radar as an angle direction-finder or monopulse direction-finder is presented in Figure 12.13 in the form of variant A, which contains a monopulse antenna 1 (for example, a paraboloid of revolution with two-mode feed in the mouth of which a polarization filter 19 is mounted). The working polarization for antenna 1 is a vertical one. The outputs of antenna 1 are connected to the sum-and-difference device in the form of stripline ring 2, the sum output of which is connected to mixer 3 and the difference output to mixer 4. Mixers 3 and 4 are also connected to heterodyne 5, which is also connected to mixer 13. The signal input of mixer 13 is connected to the output of additional antenna 12, having the horizontal working polarization (orthogonal relative to the working polarization of monopulse antenna 1), which is mounted on the edge of antenna 1. The output of mixer 13 is connected to the input of additional intermediate-frequency amplifier 14. The outputs of mixers 3 and 4 are connected, respectively, to the inputs of intermediate-frequency amplifiers 6 and 7, the outputs of which are connected to the appropriate inputs of phase detector 9, the output of which through error-signal amplifier 10 is in turn connected to drive mechanism 11 of antenna 1 with polarization filter 19 that is located under radome 20 and that has, for example, an ogival form. Intermediate-frequency amplifiers 6 and 14 are connected through

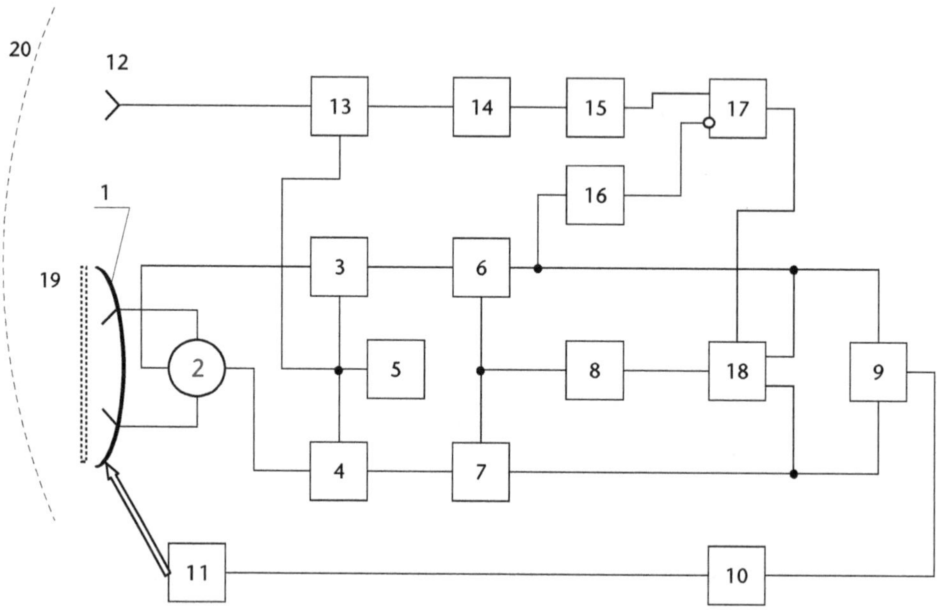

Figure 12.13 Sum-and-difference monopulse direction finder with components that realize the synthesized decision: Variant 1.

detectors 15 and 16 to the appropriate inputs of comparator 17 (compare facility), the output of which is connected to the control input of switching device 18. The outputs of intermediate-frequency amplifiers 6 and 7 are also connected to the appropriate inputs of comparator 17 the output of which is connected through automatic gain control system 8 with intermediate-frequency amplifiers 6 and 7.

The realization of units 1–16, and 19 is described in [2, Chapters 2, 3, 7].

Polarization interference was exercised on a monopulse radar with a linearly polarized receiver antenna as shown in Figures 12.3 and 12.4.

The signal polarization plane was rotated in the receiving basis of the angle pelengator monopulse antenna 1, with certain angular speed Ω_c whereby loss of automatic tracking of signal source occurred with probability close to 1. A detector of polarization interference impact was created that represented an additional channel to the receive signal on cross polarization whose output, along with the output of the sum channel, were connected to a comparator via detectors, and a logical unit was read from this comparator output on detection of polarization interference. This is simply a polarization interference detector with single-row ADC.

Figure 12.13 shows a block diagram of a sum-difference monopulse direction finder in which a synthesized solution is implemented; that is, a method of automatic angular tracking of a target using a monopulse radar in an interference condition distorting its direction-finding characteristic, which has been claimed in [24].

The numbers in Figure 12.13 correspond to the following:

1: Monopulse antenna

2: Stripline ring

3: Mixer of the sum channel

4: Mixer of the difference channel

5: Heterodyne

6, 7: Intermediate-frequency amplifiers of the sum and difference channels

8: Automatic gain control system of the sum channel

9: Phase detector

10: Error-signal amplifier

11: Monopulse antenna mechanism

12: Horn antenna of the additional (secondary) channel, which is a waveguide aperture

13: Mixer of the additional channel

14: Intermediate-frequency amplifier of the additional channel

15, 16: Amplitude detectors of the additional and sum channels

17: Comparing device (comparator)

18: Switching device (switch)

19: Polarization filter (only for prototype)

20: Radome (blister, nose cone)

24: $F_{\Sigma}^{w}(\varphi)$ is an assumption pattern of the monopulse antenna 1 on the working polarization for the sum channel

25: $F_{\Sigma}^{c}(\varphi)$ is an assumption pattern of the monopulse antenna 1 on the cross polarization for the sum channel

26: $F_{add}^{w}(\varphi)$ is an assumption pattern of the additional antenna 12 on the working polarization for additional channel

27: $F_{add}^{c}(\varphi)$ is an assumption pattern of the additional antenna 12 on the cross polarization for additional channel

28, 29: $F_{add}^{c}(\varphi)$ and $F_{\Sigma}^{w}(\varphi)$ are assumption patterns of the monopulse antenna-additional antenna system at the outputs of devices 16 and 15, respectively, during operation by the target signal on the working polarization of the monopulse antenna 1

30, 31: $F_{add}^{w}(\varphi)$ and $F_{\Sigma}^{c}(\varphi)$ are assumption patterns of the monopulse antenna-additional antenna system at the outputs of devices 16 and 15, respectively, during operation by the target signal on the cross polarization of the monopulse antenna 1

32, 33: $U_{\Sigma}(\varphi, \alpha, t)$ and $U_{add}(\varphi, \alpha, t)$ are calculated functions of the signals at the outputs of devices 16 and 15, respectively, during rotation of the target signal polarization plane with a certain constant angular velocity Ω in the receiving basis of the monopulse antenna 1 ($\dot{\alpha} = \Omega = const$)

34: $U_{com}(\varphi, \alpha, t)$ is a signal at the output of comparator 17

35: $U_{err}(\varphi, \alpha, t)$ is an error signal at the output of phase detector 9 for prior art device

36: $U_{contr}(\varphi, \alpha, t)$ is a control signal at the input of error-signal amplifier 10 during used of the synthesized decision

37: $\alpha(t)$ is rated dependence of the inclination angle of the target signal polarization plane relative to the vertical line in the receiving basis of the monopulse antenna 1

38: $\alpha(t)$ is an experimental dependence of the inclination angle of the target signal polarization plane relative to the vertical line in the receiving basis of the monopulse antenna 1

39: $U_{\Sigma}(\varphi, \alpha, t)$ is an experimental dependence of the sum channel voltage amplitude at the output of device 16

40: $U_{add}(\varphi, \alpha, t)$ is an experimental dependence of the additional channel voltage amplitude at the output of device 15

41: $U_{com}(\varphi, \alpha, t)$ is the voltage at the output of comparator 17

42: $U_{err}(\varphi, \alpha, t)$ is an experimental time dependence of the prior art error signal at the input of error-signal amplifier 10

43: $U_{contr}(\varphi, \alpha, t)$ is an experimental time dependence of the control signal at the input of error-signal amplifier 10 for the synthesized decision

44: A device for the inverse transformation of the error signal, which is a calculation device, consisting of 21, analog-to-digital converter; 22, arithmetic unit; and 22, digital-to-analog converter, as shown in Figure 12.17

The realization of units 1–16 and 19 is described in [2, chapters 2, 3, 7].

The realization of devices 8, 15, 16, 17, and 18 is shown in Figure 12.14. Signal detection of the additional and sum channels in devices 15 and 16 is carried out through diodes D1 and D2, respectively. Comparator 17 is assembled on microcircuit K140UD2A (CA3047T) with bipolar feed voltage $U_{feed} = \pm 12.6 \pm 0.5$V. Radio electronic relay 10 is used as switching device 18 with operating voltage in the range of [9V–12V], operating current 50 mA, and operating time 11 ms.

It is necessary to mention that in order to decrease operating time, any type of electronic switch instead of the relay can be used, such as transistors, thyristors, dynistors, or microcircuits.

A device realizing the operation in the mode of the proposed decision operates as follows.

Let a monopulse radar track the target, the signal polarization of which changes in time from the agreed polarization up to the orthogonal one in accordance with line 38 shown in Figure 12.15, where α is the inclination angle of the target signal polarization vector relative to the vertical line the ordinate of the diagram, and time is laid along the abscissa axis. The real changes of the signal polarization can be caused by the polarization interference jamming or by the fluctuations of the signal reflected from the target. This signal after passing through radome 20 and polarization filter 19 is received by monopulse antenna 1 having the vertical working polarization. Polarization filter 19 can be in the form of a set of thin conductors located in the monopulse antenna 1 mouth and oriented orthogonally to its working polarization, which provides the reception of vertical polarization signals without attenuation and the reception of orthogonally polarized signals with

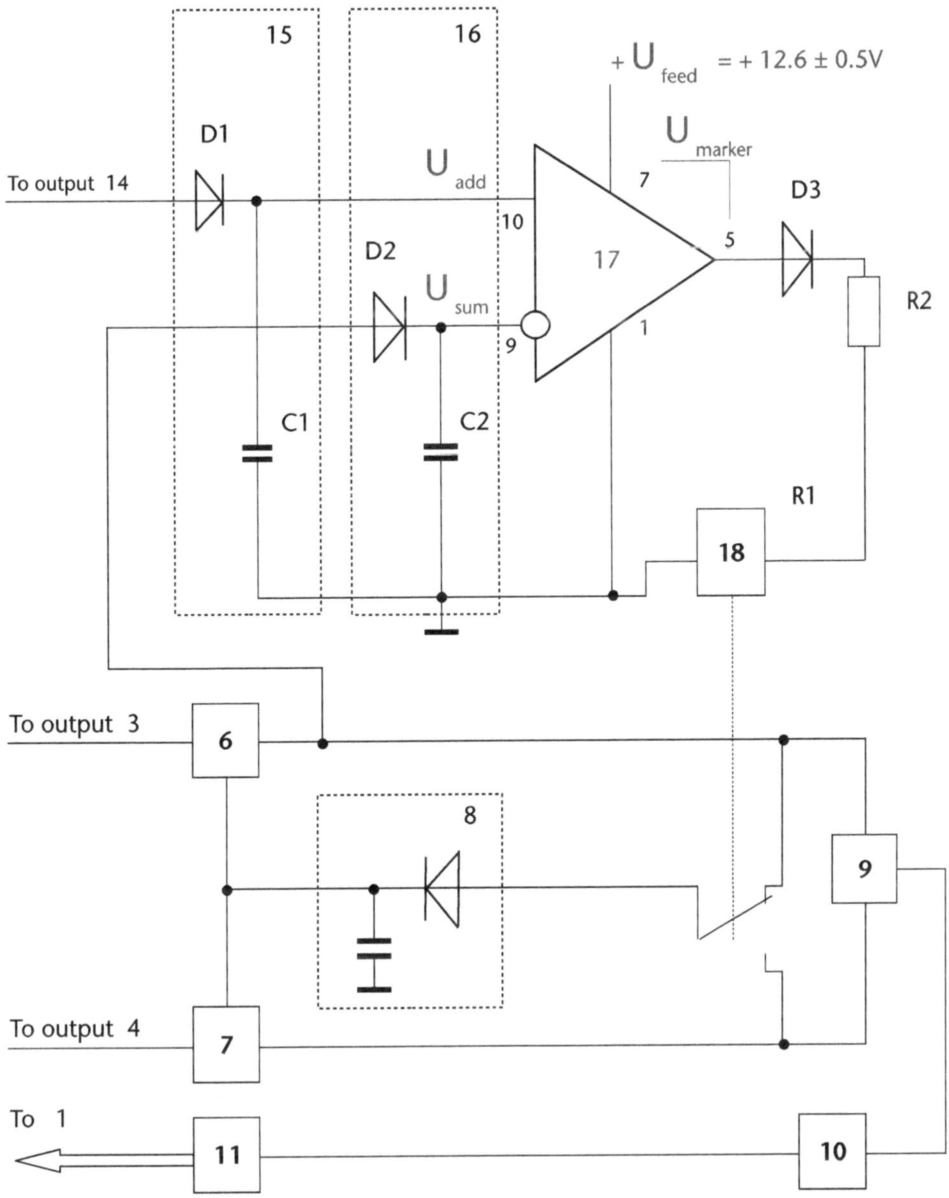

Figure 12.14 Realization of devices 8, 15, 16, 17, and 18 according to the first variant of the proposed decision.

certain attenuation. The signals from the outputs of monopulse antenna 1 come to the inputs of stripline ring 2, providing at its outputs the shaping of microwave signals of the sum-and-difference channels, the signals of which come to mixers 3 and 4, respectively, where they are transformed with the help of heterodyne 5 into the signals of intermediate frequency, which then are amplified in intermediate-frequency amplifiers 6 and 7 up to the required value and come to the inputs of phase detector 9. The difference signal amplitude $|\dot{U}_{\Delta}(\varphi, \alpha, t)|$ determines the value of the angular error signal at the output of phase detector 9, the phase difference at the input of phase detector 9 between the signals of the sum channel $|\dot{U}_{\Sigma}(\varphi, \alpha, t)|$

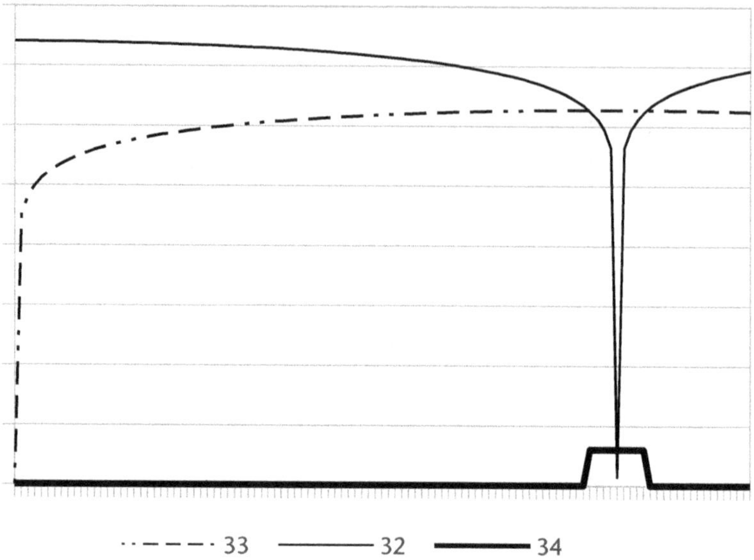

Figure 12.15 33 is $U_{add}(\varphi, \alpha, t)$, 32 is $U_{\Sigma}(\varphi, \alpha, t)$, and 34 is $U_{com}(\varphi, \alpha, t)$.

and the difference channel $|\dot{U}_{\Delta}(\varphi, \alpha, t)|$ determines the sign of the angular error signal $U_{err}(\varphi, \alpha, t)$ at the output of phase detector 9, where φ is the angular error (displacement angle between a true direction on target and radar boresight), α is the inclination angle of the target signal polarization vector relative to the working polarization vector of the monopulse antenna, and t is a time. Automatic gain control system 8 excludes the dependence of the angular error signal amplitude at the output of phase detector 9 on the level of the received signals by the connection of the input of automatic gain control system 8 through normally closed contacts of switching device 18 (Figure 12.13) to the output of intermediate-frequency amplifier 6 of the sum channel. In this case the signal at the output of automatic gain control system 8 makes a simultaneous adjustment of the amplification coefficients of intermediate-frequency amplifiers 6 and 7, providing the signal normalization of the difference channel with the help of the sum channel. Thus the error signal amplitude is the value of the quotient from the division of the measured amplitude value of the difference channel on the measured value of the sum channel, the error signal sign is the phase difference between the sum-and-difference signals, and in the general case for error signal $U_{err}(\varphi, \alpha, t)$ on the output of phase detector 9 we have

$$U_{err}(\varphi, \alpha, t) = Re\left[\frac{\dot{U}_{\Delta}(\varphi, \alpha, t)}{\dot{U}_{\Sigma}(\varphi, \alpha, t)}\right] = \frac{1}{2} \times K_{phd} \times \frac{U_{\Delta}(\varphi, \alpha, t)}{U_{\Sigma}(\varphi, \alpha, t)} \times \cos[\Psi_{\Sigma}(t) - \Psi_{\Delta}(t)], \tag{12.59}$$

where $\dot{U}_{\Sigma}(\varphi, \alpha, t)$ and $\dot{U}_{\Delta}(\varphi, \alpha, t)$ are sum-and-difference signals on intermediate-frequency, K_{phd} is the amplifier coefficient of phase detector, and $\Psi_{\Sigma}(t) - \Psi_{\Delta}(t)$ is the phase difference between sum and difference signals (see [2, pp. 13, 58]).

At the same time the reception of the signal component on the horizontal polarization by the additional channel of the monopulse radar is performed with the help of additional antenna 12, mixer 13, and intermediate-frequency amplifier 14. The outputs of intermediate-frequency amplifiers 6 and 14 are connected through amplitude detectors 15 and 16 to the appropriate inputs of comparator 17. On the output of comparator 17 formed a signal which is a result of comparing amplitude sum signal of monopulse antenna 1, which has received on the working polarization and signal from additional antenna 12, which in turn has polarization orthogonal or is close to orthogonal relative to the working polarization of said monopulse antenna. The signal on the comparator output also is a control signal for switch 18, which connects the input of the automatic gain control system to the output of intermediate-frequency amplifier 6 for prototype mode or connects it to the output of intermediate-frequency amplifier 7 for the proposed method. In any case, monopulse antenna 1 is oriented in space by drive mechanism 11, which receives control signal $U_{contr}(\varphi, \alpha, t)$ through error-signal amplifier 10.

Time dependences of the voltages on the outputs of the sum channel $U_{\Sigma}(\varphi, \alpha, t)$ and additional channel $U_{add}(\varphi, \alpha, t)$ are shown in Figure 12.15 with curves 32 and 33 in dB, respectively. Voltage of the automatic gain control system (sum channel) is shown in Figure 12.16 by curve 39 and the signal of the additional channel by curve 40.

The voltage $U_{com}(\varphi, \alpha, t)$ at the output of the comparator 17 (Figure 12.15, line 34 and Figure 12.16, line 41), will be equal to $+U_{feed}$, when $U_{add}(\varphi, \alpha, t) > U_{\Sigma}(\varphi, \alpha, t)$ and will be equal to $-U_{feed}$ when $U_{add}(\varphi, \alpha, t) \leq U_{\Sigma}(\varphi, \alpha, t)$:

$$U_{com}(\varphi, \alpha, t) = \begin{cases} +U_{feed}, & when\ U_{add}(\varphi, \alpha, t) > U_{\Sigma}(\varphi, \alpha, t) \\ -U_{feed}, & when\ U_{add}(\varphi, \alpha, t) \leq U_{\Sigma}(\varphi, \alpha, t) \end{cases} \quad (12.60)$$

If the leg 1 of microcircuit K140UD2A is grounded (Figure 12.14), the necessity in diode D3 disappears. The voltage at the output of comparator 17 is shown in Figure 12.16 by line 34 and is written in the following form:

$$U_{com}(\varphi, \alpha, t) = \begin{cases} +U_{feed}, & when\ U_{add}(\varphi, \alpha, t) > U_{\Sigma}(\varphi, \alpha, t) \\ 0, & when\ U_{add}(\varphi, \alpha, t) \leq U_{\Sigma}(\varphi, \alpha, t) \end{cases} \quad (12.61)$$

Voltage of the automatic gain control system, curve 32, and voltage of the additional channel, curve 33, are shown in Figure 12.15, and $U_{com}(\varphi, \alpha, t)$ in volts as well. Time is shown on the abscissa axis.

Voltage $U_{com}(\varphi, \alpha, t)$ comes to switching device 18 as a control signal (line 34 in Figure 12.15 or line 41 in Figure 12.16).

Figure 12.16 shows the operation of the monopulse radar in the prototype mode, and in the mode of the proposed method.

12.6.1 Operation of Prototype Mode: 1A

Conditions:

- Power supply to device 17 is switched off (microcircuit K140UD2A is disconnected).
- Relay R1 contacts are normally closed. Operation order is shown in Figure 12.16.

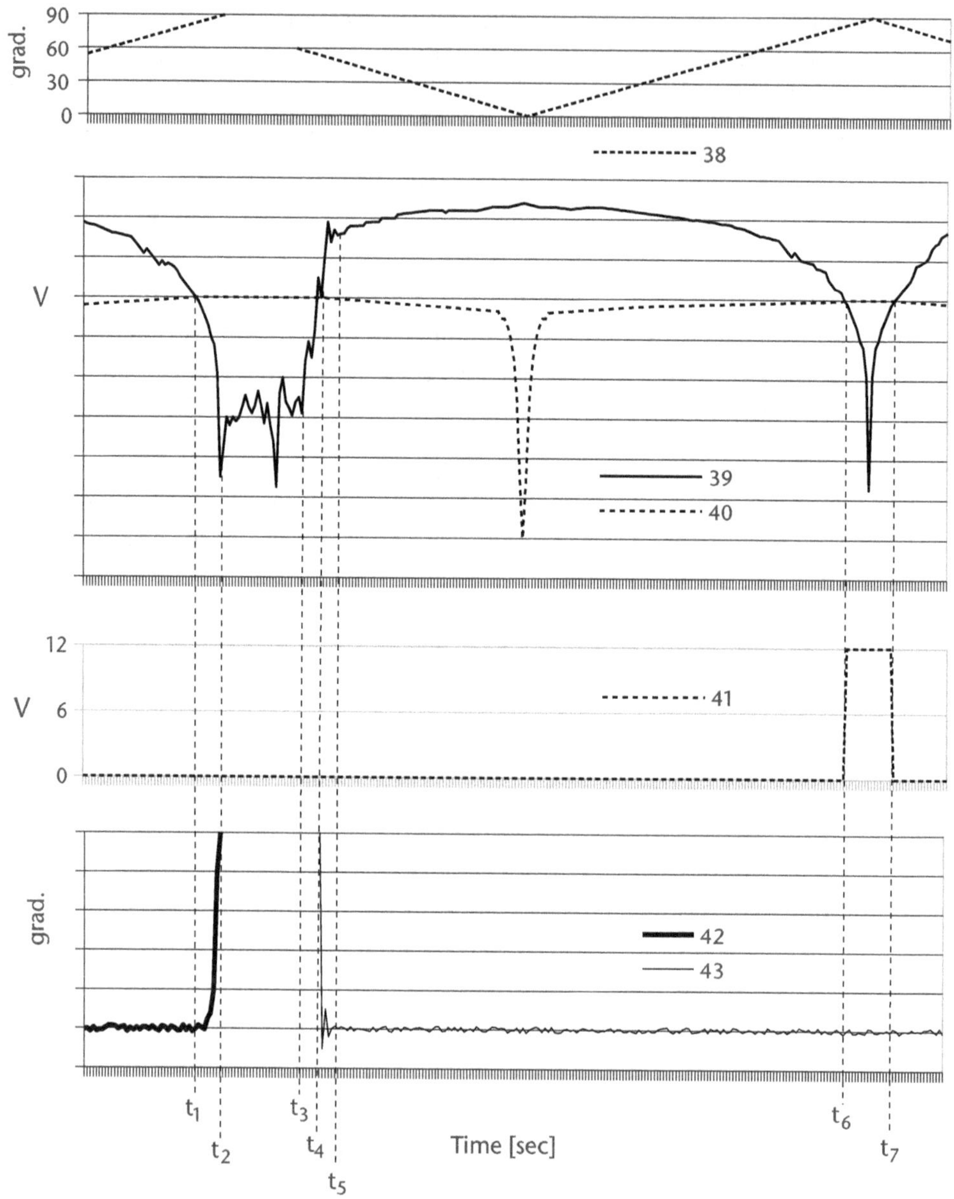

Figure 12.16 Experimental time diagrams that illustrate the direction-finder principle of operation for synthesized decision.

Up to time point t_1, the following condition is fulfilled:

$$U_\Sigma(\varphi, \alpha, t) > U_{add}(\varphi, \alpha, t) \tag{12.62}$$

therefore, $U_{com}(\varphi, \alpha, t) = 0$ and voltage at the control input of switching device 18 is zero (line 41 in Figure 12.16). In this case monopulse radar works in the prototype mode in the design mode of automatic target tracking [2, pp. 69–71]. The input of automatic gain control system 8 is connected through normally closed contacts of relay R1 (switch 18, Figure 12.14) to the output of intermediate-frequency amplifier

6 of the sum channel whereby the signal normalization of the difference channel is carried out with the help of the sum one. The error signal from the output of phase detector 9 through error-signal amplifier 10 comes to drive mechanism 11 of the monopulse antenna, which turns the antenna in such a way that its radar boresight coincides with the direction on the target and the error signal value is maintained close to zero. As the inclination angle of the target signal polarization plane of the input signal reaches the orthogonal position, the voltage amplitude of automatic gain control system 8 decreases (Figure 12.16, curve 39) and after a certain value starts the avalanche-like increase of the error signal (Figure 12.16, curve 42).

In time interval $t_1 < t < t_2$ the target signal polarization vector passes through the position close to the orthogonal position that is relative to the working polarization of antenna 1 (see Figure 12.16, curve 38). Monopulse antenna is oriented in the direction of the target using the control signal $U_{contr}(\varphi, \alpha, t)$, the amplitude and the sign of control signal are equal to the amplitude and the sign of the error signal, and we have:

$$U_{contr}(\varphi, \alpha, t) = U_{err}(\varphi, \alpha, t) \tag{12.63}$$

In this case at the output of phase detector 9 abruptly increases the angle tracking error, which leads to the loss of automatic angle tracking on the target. The sum-and-difference channels change places and the normalization condition is violated (see [2, Sections 7.3, 8.5]). The automatic tracking loss occurs because during the impact of the signal on the orthogonal polarization on the monopulse radar, the voltage of the sum channel reaches values close to zero in a certain small ε-neighborhood of the radar boresight, and, being in the denominator, turns the error signal into infinity.

12.6.2 Operation in Mode of the Proposed Decision: 1B

Conditions:

- Power supply to device 17 is switched on (microcircuit K140UD2A is switched on);
- Contacts of switching device (relay R1) are normally closed.

The operation procedure is shown in Figures 12.15 and 12.16.

During application of the proposed method the monopulse radar operates in the prototype mode up to time point t_6:

- The following condition is met: $U_\Sigma(\varphi, \alpha, t) > U_{add}(\varphi, \alpha, t)$;
- At the output of device 17 $U_{com}(\varphi, \alpha, t) = 0$ and the voltage at the control input of switch 18 is zero;
- The input of automatic gain control system 8 is connected through normally closed contacts of relay R1 (switch 18, Figure 12.14) to the output of intermediate-frequency amplifier 6 of the sum channel whereby the signal normalization of the difference channel is carried out with the help of the sum one.

Monopulse antenna 1 is oriented in the direction of the target using the control signal $U_{contr}(\varphi, \alpha, t)$, the amplitude and sign of control signal are equal to the amplitude and sign of the error signal $U_{err}(\varphi, \alpha, t)$, and we have:

$$U_{contr}(\varphi, \alpha, t) = U_{err}(\varphi, \alpha, t) \tag{12.64}$$

At time interval $t_6 < t < t_7$:

- $U_{add}(\varphi, \alpha, t) > U_{\Sigma}(\varphi, \alpha, t)$;
- At the output of device 17 is feed voltage, $U_{com}(\varphi, \alpha, t) = +12.6$V, and switch 18 is actuated: it disconnects the input of automatic gain control system 8 from the output of intermediate-frequency amplifier 6 of the sum channel and connects the input of automatic gain control system 8 to the output of intermediate-frequency amplifier 7 whereby the signal normalization of the sum channel is carried out with the help of the difference channel and the decision derived in [25] is realized.

Thus, during the time interval when the additional signal is more than the sum signal, signal on the comparator output appears, switch 18 disconnects input of automatic gain control system 8 from the output of intermediate-frequency amplifier 6 of the sum channel, and connects it to the output of intermediate-frequency amplifier 7 of the difference channel. In this case sum-and-difference signals are normalized by the difference signal and monopulse antenna 1 is oriented in the direction of the target using the control signal $U_{contr}(\varphi, \alpha, t)$, the amplitude of the control signal is the value of the quotient from the division of the measured amplitude value of the sum channel on the measured amplitude value of the difference channel, the sign of control signal $U_{contr}(\varphi, \alpha, t)$ is the phase difference between the sum-and-difference signals, and in this case we have:

$$\begin{aligned} U_{contr}(\varphi, \alpha, t) &= Re\left[\frac{\dot{U}_{\Sigma}(\varphi, \alpha, t)}{\dot{U}_{\Delta}(\varphi, \alpha, t)}\right] = \frac{1}{2} \times K_{phd} \times \frac{U_{\Sigma}(\varphi, \alpha, t)}{U_{\Delta}(\varphi, \alpha, t)} \times \cos[\Psi_{\Sigma}(t) - \Psi_{\Delta}(t)] \\ &= \left[U_{err}(\varphi, \alpha, t)\right]^{-1} \end{aligned} \tag{12.65}$$

In time interval $t_6 < t < t_7$ the loss of automatic angle tracking on a target doesn't occur because at the time of the signal influence on cross polarization in time interval $t_6 < t < t_7$ due to application of devices 12–18, drive mechanism 11 carries out orientation of antenna 1 on the target according to the location characteristic close to the location characteristic of the working polarization (Figure 12.21). In this case the voltage of the difference channel, which can reach sufficiently big values in a certain small ε-neighborhood of the radar boresight, appears in the denominator, and the values of the sum channel close to zero move to the numerator.

When the polarization plane passes the signal of the orthogonal position, the voltage of the difference channel decreases due to the change of the directivity diagram, the amplification coefficient increases correspondingly (desensitization decreases) of the sum-and-difference channels, respectively. During this process the amplitudes of the sum and additional channels are compared over time.

After passing point t_7, where $t \in [t_7, \infty)$:

- The following condition is met: $U_{\Sigma}(\varphi, \alpha, t) > U_{add}(\varphi, \alpha, t)$;
- At the output of device 17 voltage is zero: $U_{com}(\varphi, \alpha, t) = 0$;
- Switching device 18 is actuated: it disconnects the input of automatic gain control system 8 from the output of intermediate-frequency amplifier 7 of the difference channel and returns the connection of the input of automatic gain control system 8 to the output of intermediate-frequency amplifier 6 of the sum channel whereby the standard normalization of the difference channel signal is carried out with the help of the sum channel.

Monopulse antenna 1 is oriented in the direction of the target using the control signal $U_{contr}(\varphi, \alpha, t)$, the amplitude and sign of the control signal are equal to the amplitude and sign of the error signal $U_{err}(\varphi, \alpha, t)$, and we have:

$$U_{contr}(\varphi, \alpha, t) = U_{err}(\varphi, \alpha, t) \tag{12.66}$$

The circuit consisting of devices 12–17 can be characterized as a single-bit detector of the interference on the cross polarization, and device 18 connecting by the signal of the interference polarization detector the input of automatic gain control system 8 to the output of intermediate-frequency amplifier 6 of the sum channel, or to the output of intermediate-frequency amplifier 7 of the difference channel as a protector of the monopulse radar from the impact of cross-polarization signals and interferences.

Variant 2:

The monopulse radar Figure 12.17 includes monopulse antenna 1 (for example, a paraboloid of revolution with two-mode feed in the mouth of which polarization filter 19 is mounted).

The working polarization for antenna 1 is a vertical one. Polarization interference was exercised on a monopulse radar with linearly polarized receiver antenna as shown in Figures 12.3 and 12.4. The signal polarization plane was rotated in the receiving basis of angle pelengator monopulse antenna, with certain angular speed Ω_c whereby loss of automatic tracking of the signal source occurred with probability close to 1.

The outputs of antenna 1 are connected to the sum-and-difference device in the form of stripline ring 2, the sum output of which is connected to mixer 3 and the difference output to mixer 4. Mixers 3 and 4 are also connected to heterodyne 5, which is also connected to mixer 13. The signal input of mixer 13 is connected to the output of additional antenna 12 having the horizontal working polarization (orthogonal) relative to the working polarization of monopulse antenna 1, which is mounted on the edge of antenna 1, for example. The output of mixer 13 is connected to the input of additional intermediate-frequency amplifier 14. The outputs of mixers 3 and 4 are connected, respectively, to the inputs of intermediate-frequency amplifiers 6 and 7. The output of intermediate-frequency amplifier 6 is connected to the input of automatic gain control system 8, the output of which is connected to intermediate-frequency amplifiers 6 and 7. The outputs of intermediate-frequency amplifiers 6

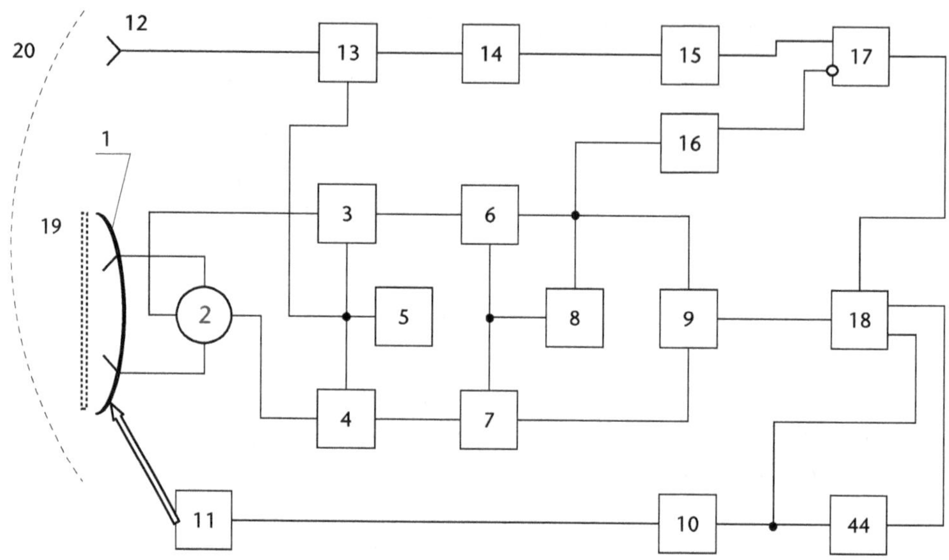

Figure 12.17 The second variant of the flow diagram of the sum-and-difference monopulse radio direction finder with the components that realize the synthesized decision.

and 7 are connected to phase detector 9, and the outputs of intermediate-frequency amplifiers 6 and 14 are connected through amplitude detectors 15 and 16 to the corresponding inputs of comparator 17, the output of which is connected to the control input of switching device 18. The output of phase detector 9 is connected to the signal input of switching device 18, one output of which is connected to drive mechanism 11 of antenna 1 through error-signal amplifier 10, the other output of the switching device through device for the inverse transformation of the error signal -44, which is a calculation device, consisting of analog-to-digital converter 21, arithmetic unit 22, and digital-to-analog converter 23, as shown in Figure 12.17, and through error-signal amplifier 10 is also connected to drive mechanism 11 of antenna 1, located under radome 20 and having, for example, an ogival form.

The realization of units 1–16 and 19 is described in [2, Chapters 2, 3, 7].

The realization of devices 15, 16, 17, 18, 21, 22, and 23 is shown in Figure 12.18. Devices 15, 16, 17, and 18 are described above. As device 21 an eight-digit analog-to-digital converter on microcircuit K1107PV4A (TDC 1025J) with the range of input voltage −2.5V to +2.5V was used, programmable read-only memory KR556RT5 was used as arithmetic unit 22, as eight-digit digital-to-analog converter (device 23)-microcircuit 1118 PA1 (MS 10318).

A device realizing operation in mode of synthesized decision.

Let the monopulse radar track the target, the signal polarization of which changes in time from the agreed polarization up to the orthogonal one in accordance with Figure 12.3 and with line 37 shown in Figure 12.19, where α is the inclination angle of the target signal polarization vector relative to the vertical line, which is the ordinate of the diagram, and time is laid along the abscissa axis.

The real changes of the signal polarization can be caused by the polarization interference jamming or by the fluctuations of the signal reflected from the target. This signal after passing through radome 20 and polarization filter 19 is received by monopulse antenna 1 having the vertical working polarization. The polarization

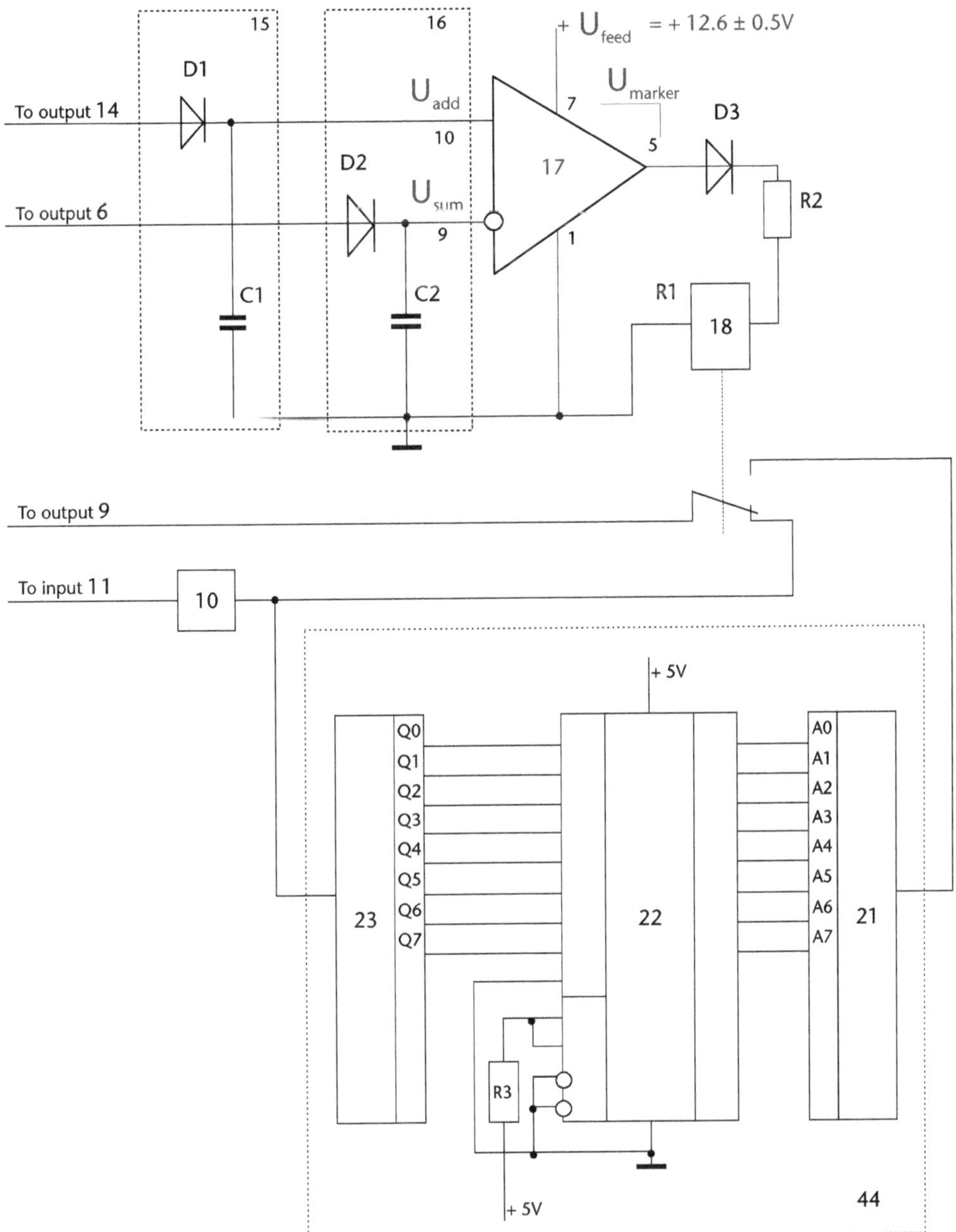

Figure 12.18 Realization of devices 15, 16, 17, 18, 21, 22, and 23 according to the second variant of the proposed decision.

filter can be in the form of a set of thin conductors located in the monopulse antenna 1 mouth and oriented orthogonally to its working polarization, which provides the reception of vertical polarization signals without attenuation and the reception of orthogonally polarized signals with certain attenuation (see Figure 12.4). The signals from the outputs of monopulse antenna 1 come to the inputs of stripline ring 2, providing at its outputs the shaping of microwave signals of the sum-and-difference

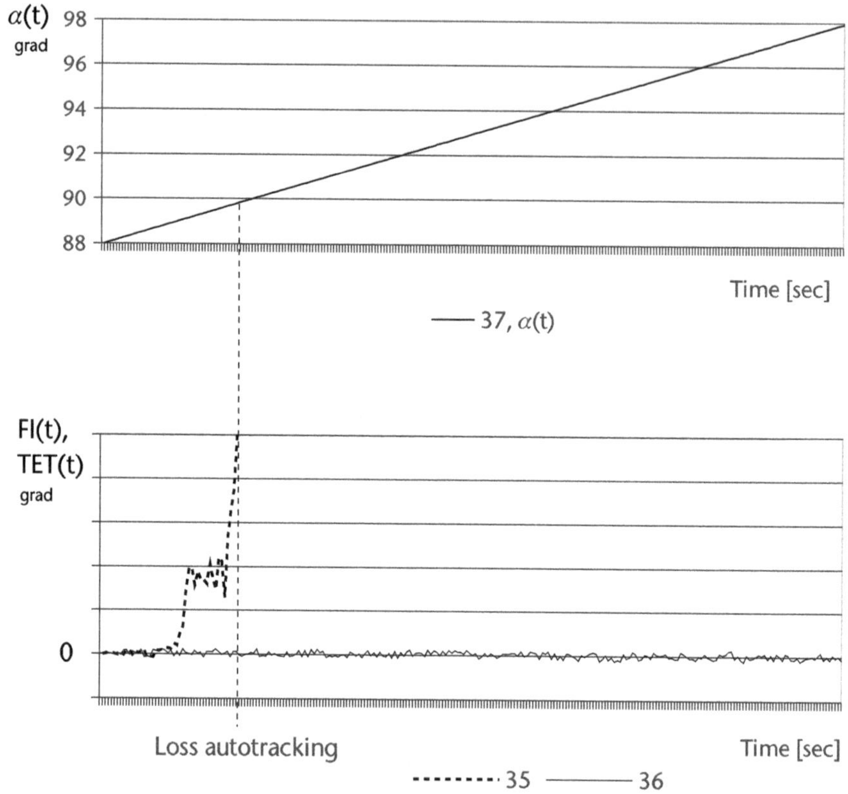

Figure 12.19 Time diagrams of the calculated functions of the error signal for the prototype and the suggested method during the ramp of the inclination angle of the signal polarization plane in the receiving basis of the monopulse antenna with a certain constant angular velocity.

channels the signals, of which come to mixers 3 and 4, respectively where they are transformed with the help of heterodyne 5 into the signals of intermediate frequency, which then are amplified in intermediate-frequency amplifiers 6 and 7 up to the required value and come to the inputs of phase detector 9. The difference signal amplitude determines the value of the angular error signal at the output of phase detector 9, and the phase difference at the input of phase detector 9 between the signals of the sum-and-difference channels determines the sign of the angular error signal at the output of phase detector 9. Automatic gain control system 8 excludes the dependence of the angular error signal amplitude at the output of phase detector 9 on the level of the received signals by the connection of the input of automatic gain control system 8 to the output of intermediate-frequency amplifier 6 of the sum channel; in this case the signal at the output of automatic gain control system 8 makes a simultaneous adjustment of the amplification coefficients of intermediate-frequency amplifiers 6 and 7, providing the signal normalization of the difference channel with the help of the sum one.

Simultaneously, the reception of the signal component on the horizontal polarization by the additional channel with the help of additional antenna 12, mixer 13, and intermediate-frequency amplifier 14. The outputs of intermediate-frequency amplifiers 6 and 14 are connected through amplitude detectors 15 and 16 to the

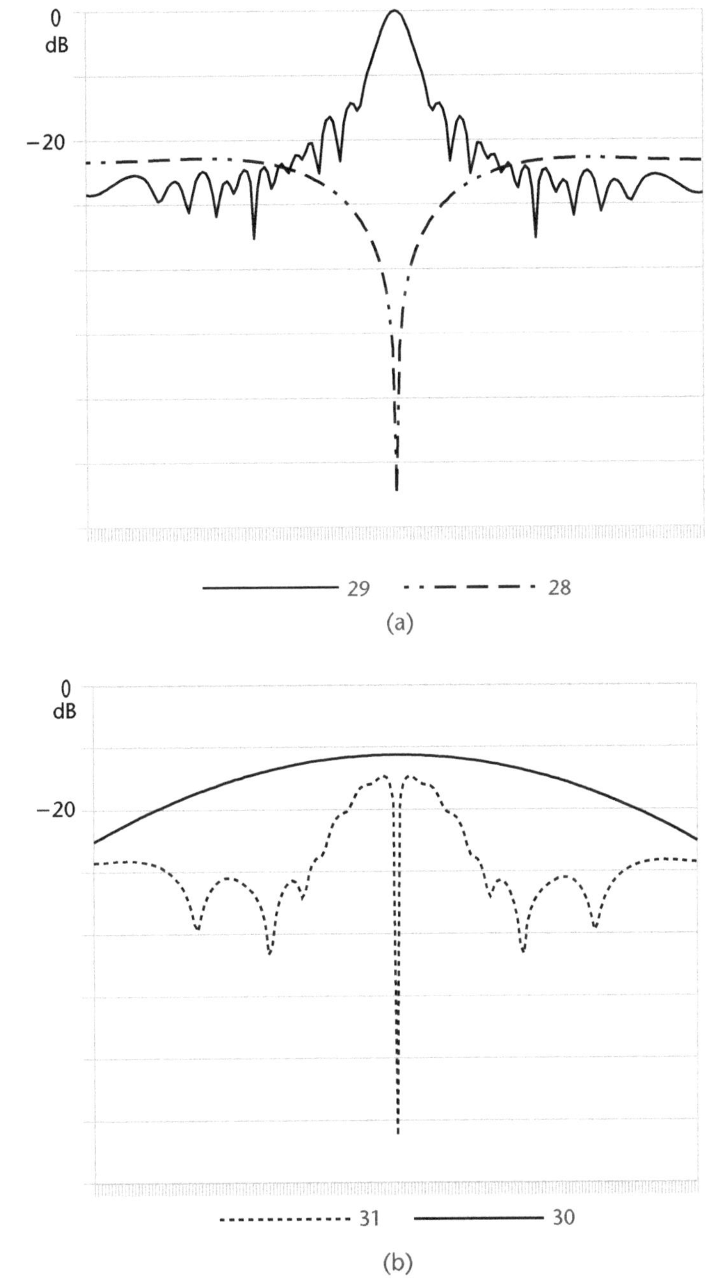

Figure 12.20 Direction-finding characteristics of the system monopulse antenna additional antenna on the working and cross polarization. (a) 29-$F_{\Sigma}^{w}(\varphi)$, 28-$F_{add}^{c}(\varphi)$ are assumption patterns of the monopulse antenna additional antenna system on the working polarization of the monopulse antenna 1. (b) 31-$F_{\Sigma}^{c}(\varphi)$, 30-$F_{add}^{w}(\varphi)$ are assumption patterns of the monopulse antenna additional antenna system on the cross polarization of the monopulse antenna 1.

appropriate inputs of comparator 17. On the output of comparator 17 a signal is formed, which is a result of comparing the signal from monopulse antenna 1 to the working polarization and signal from additional antenna 12, which have working polarization orthogonal or close to orthogonal relative to the working polarization of monopulse antenna 1. A better case of working polarization of additional antenna 12 corresponds to the cross polarization of monopulse antenna 1. The output comparator signal also is a control signal for switch 18 in which output signal from phase detector 9 switched, in prototype mode, or in the mode of synthesized decision. In any case, monopulse antenna 1 is oriented in space by drive mechanism 11, which receives control signal $U_{contr}(\varphi, \alpha, t)$ through the error signal amplifier from one of switch outputs. Time dependencies shown in Figures 12.15, 12.16, and 12.19 are the same. Expressions are also true for U_{com}, the voltage at the output of comparator 17.

12.6.3 Operation of Prototype Mode: 2A

Conditions:

- Power supply to device 17 is switched off (microcircuit K140UD2A shown in Figure 12.17 is switched off);
- Contacts of switching device 18 (relay R1 shown in Figure 12.17) are normally closed.

The operation procedure is shown in Figure 12.16: Up to time point t_1, the following condition is fulfilled:

$$U_{\Sigma}(\varphi, \alpha, t) > U_{add}(\varphi, \alpha, t), \tag{12.67}$$

therefore $U_{com}(\varphi, \alpha, t) = 0$ and a voltage at the control input of switch 18 is zero, respectively (line 41 in Figure 12.16), and the monopulse radar works in the prototype mode in the design mode of automatic target tracking ([2, pp. 69–71]). The error signal from the output of phase detector 9 through the normally closed contacts of switching device 18 comes directly to the input of error-signal amplifier 10 and then to the drive mechanism 11 of the monopulse antenna, which turns antenna 1 in such a way that its radar boresight coincides with the direction on target and the error signal value is maintained close to zero. The monopulse antenna is oriented in the direction of the target using the control signal $U_{contr}(\varphi, \alpha, t)$, the amplitude and sign of control signal are equal to the amplitude and sign of the error signal $U_{err}(\varphi, \alpha, t)$, and we have:

$$U_{contr}(\varphi, \alpha, t) = U_{err}(\varphi, \alpha, t) \tag{12.68}$$

As the inclination angle of the target signal polarization plane of the input signal reaches the orthogonal position, the voltage amplitude of the automatic gain control system decreases and after a certain value starts the avalanche-like increase of the error signal.

In time interval $t_1 < t < t_2$ the target signal polarization vector passes through the position close to the orthogonal position that is relative to the working polarization of antenna 1 (see Figure 12.19, curve 35). In this case at the output of phase detector 9 the angle tracking error abruptly increases, which leads to the loss of automatic angle tracking on the target (see [2, Sections 7.3, 8.5]).

12.6.4 Operation in Mode of the Proposed Decision: 2B

Conditions:

- Power supply to device 17 is switched on (microcircuit L140UD2A is switched on);
- Contacts of switching device 18 (relay R1 shown in Figure 12.17) are normally closed.

The operation procedure is shown in Figure 12.16: Up to time point t_6, the following condition is fulfilled:

$$U_\Sigma(\varphi, \alpha, t) > U_{add}(\varphi, \alpha, t), \tag{12.69}$$

therefore $U_{com}(\varphi, \alpha, t) = 0$ and voltage at the control input of switching device 18 is zero (line 41 in Figure 12.16).·In this case, the monopulse radar works in the prototype mode. The monopulse antenna is oriented in the direction of the target using the control signal $U_{contr}(\varphi, \alpha, t)$, the amplitude and sign of control signal are equal to the amplitude and sign of the error signal, and we have:

$$U_{contr}(\varphi, \alpha, t) = U_{err}(\varphi, \alpha, t), \tag{12.70}$$

At time interval $t_6 < t < t_7$: $U_{add}(\varphi, \alpha, t) > U_\Sigma(\varphi, \alpha, t)$; at the output of device 17 is the feed voltage, and the signal of the comparator output is $U_{com}(\varphi, \alpha, t) = +12.6\text{V}$ disconnects the output of phase detector 9 from the input of error signal amplifier 10 and connects error signal $U_{err}(\varphi, \alpha, t)$, which exist on the output of phase detector 9 to the input of analog-to-digital converter 21 (Figure 12.17). A digital signal from the output of device 21 comes to arithmetic device 22, is calculated, and then goes to digital-to-analog converter 23. The analog signal from the output of device 23 goes to the input of error signal amplifier 10. Thus, during the time interval when the additional signal amplitude is more than the sum signal amplitude, then monopulse antenna 1 is oriented in the direction of the target using the control signal $U_{contr}(\varphi, \alpha, t)$, the amplitude of control signal is the value of the quotient from the division of the measured amplitude value of the sum channel on the measured amplitude value of the difference channel, and the sign of control signal is the phase difference between the sum-and-difference signals.

When the synthesized decision is used, the loss of automatic angle tracking on the target doesn't occur because at the time of the signal influence on cross polarization in time interval $t_6 < t < t_7$ due to application of devices 12–23, drive mechanism 11 carries out orientation of antenna 1 on the target according to the location characteristic close to the location characteristic on the working polarization (see Figure 12.21).

This is achieved by the use of the control function $U_{contr}(\varphi, \alpha, t)$, calculated with the help of arithmetic unit 22, realized on the programmable read-only memory, which carries out inverse transformation of the function of error signal $U_{err}(\varphi, \alpha, t)$, by a table existing in arithmetic unit 22, having the following form:

$$U_{contr}(\varphi,\alpha,t) = Re\left[\frac{\dot{U}_{\Delta}(\varphi,\alpha,t)}{\dot{U}_{\Sigma}(\varphi,\alpha,t)}\right]^{-1} = \left[U_{err}(\varphi,\alpha,t)\right]^{-1} \tag{12.71}$$

As seen from Figure 12.16 (curve 43), the control signal value $U_{contr}(\varphi,\alpha,t)$ at the input of error-signal amplifier 10 in time interval $t_6 < t < t_7$ doesn't exceed the value of some ε-neighborhood around zero.

After passing point t_7, where $t \in [t_7,\infty)$, when the target signal polarization vector finishes to pass through a hazardous position (Figure 12.16, curve 38):

- The following condition is met: $U_{\Sigma}(\varphi,\alpha,t) > U_{add}(\varphi,\alpha,t)$;
- At the output of device 17 voltage is zero: $U_{com}(\varphi,\alpha,t) = 0$ (curve 41 in Figure 12.16);
- In this case switching device 18 disconnects phase detector 9 from the circuit of devices 21–23 and connects it directly to error-signal amplifier 10 and to drive mechanism 11 of antenna 1, the monopulse radar returns to operation in the design mode of automatic tracking, in which the error signal from the output of phase detector 9 is used to operate antenna 1 tracking the target.

In this case we have:

$$U_{contr}(\varphi,\alpha,t) = U_{err}(\varphi,\alpha,t) \tag{12.72}$$

Application of the synthesized decision will allow to:

- Reduce the location error caused by the depolarization of the signals reflected from the target to a minimum;
- Exclude the loss of angular tracking of the target from the effects of polarization interferencies;
- Increase the accuracy of angular tracking of the source of polarization interference in 8 to 10 times.

The synthesized decision does not require using a polarization filter.

The operation procedure is shown in Figure 12.14:

$[t_3, t_4]$ is the target look-up;

t_4 is the target lock-on;

t_5 is the target tracking.

Up to time point t_6 for the signals at the outputs of the sum and additional channels (curves 39 and 40 in Figure 12.14, respectively) the inequality $U_{\Sigma}(\varphi,\alpha,t) > U_{add}(\varphi,\alpha,t)$ is true and a control signal at the input of switching device 18 is absent (curve 41 is equal to 0). The monopulse direction finder works in the design mode of automatic target tracking (prototype mode) [2].

The signal influence on cross polarization in time interval $t_6 < t < t_7$ is maximally, $U_{\Sigma}(\varphi,\alpha,t) < U_{add}(\varphi,\alpha,t)$ and at the input of switching device 18 present a control signal, (curve 41 is equal to 1, see Figure 12.14) at the input of switching device 18. The switching device 18 connects error signal $U_{err}(\varphi, \alpha, t)$, which exists on the output of phase detector 9 to the input of analog-to-digital converter 21 (Figure 12.17). The digital signal from the output of device 21 goes to arithmetic device 22, is calculated, and goes to digital-to-analog converter 23 and then to error-signal amplifier 10.

Drive mechanism 11 carries out the orientation of antenna 1 on the target according to the direction-finding characteristic close to the direction-finding characteristic on the working polarization (see Figure 12.21). It is achieved by the use of the control function $U_{contr}(t)$ calculated with the error signal function $U_{err}(\varphi, \alpha, t)$ that is present in (12.71). At time point t_7, when the target signal polarization vector finishes passing through a hazardous position (Figure 12.14, line 38), $U_\Sigma(\varphi, \alpha, t) > U_{add}(\varphi, \alpha, t)$, the control voltage at the input of switching device 18 turns into zero (curve 41 is equal to 0) and switching device 18 disconnects phase detector 9 from device 44 (analog-to-digital converter 21, arithmetic device 22, and digital-to-analog converter 23) and connects it directly to error-signal amplifier 10 and to drive mechanism 11 of antenna 1, and the direction finder returns to operation in the design mode of automatic tracking.

A detailed description of the monopulse direction finder on the synthesized decision is given in [24].

It should be mentioned that a positive effect is greater when the monopulse antenna is mounted under the radome.

An additional significant advantage of the synthesized decision is the fact that its hardware implementation is based on cheap parabolic antennas and it doesn't require a great volume of additional equipment.

When this synthesized decision is used, there is no need to install expensive flat antenna arrays as a monopulse antenna 1 on an aircraft (including an unmanned aerial vehicle), which are used to solve problems associated with the loss of automatic angular tracking of the target, caused by the influence of cross-polarization signals.

The synthesized decision uses a single-bit cross-polarization interference detector (12–17), and a device of the inverse transformation of the error signal $U_{contr}(t)$ (44) consisting of analog-to-digital converter 21, arithmetic unit 22, and digital-to-analog converter 23, the signal of which is used in a monopulse antenna drive. The circuit of devices 18, 21, 22, and 23 serve as a protector of the monopulse direction finder from the impact of cross-polarization signals and interferences. The device of the inverse transformation can receive the error-signal from any type of monopulse direction finder, and therefore, this decision does not depend on the type of monopulse direction finder.

The solution (12.65 and 12.71) provides a good coincidence with the direction-finding characteristic of the monopulse direction finder on the agreed polarization on the section approximately $(0,4 \ \ldots \ 0,6)$ of its half-width taken as a unit (Figure 12.21) and polarization interference allows tracking of the source with minimal errors (Figure 12.14, curve 43).

A sum-and-difference monopulse direction-finding system in which (12.57) and (12.58) are implemented is shown in Figure 12.22.

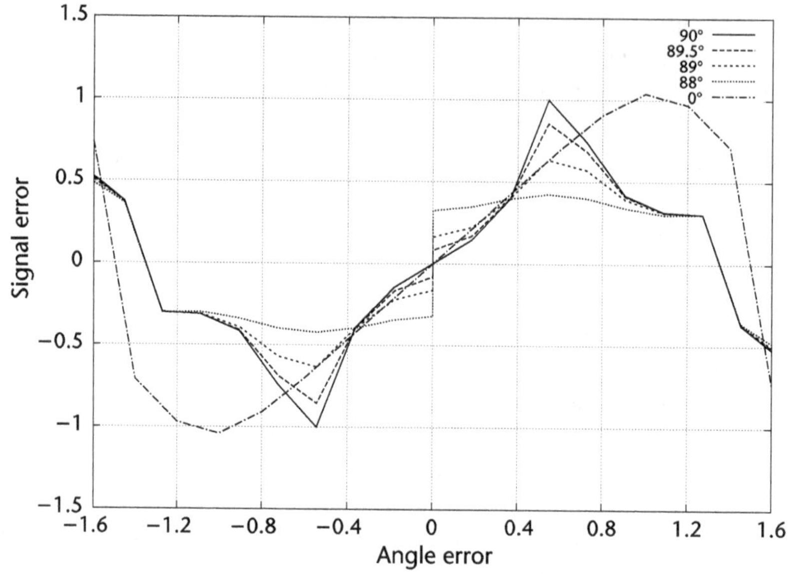

Figure 12.21 Standardized back direction-finding characteristic.

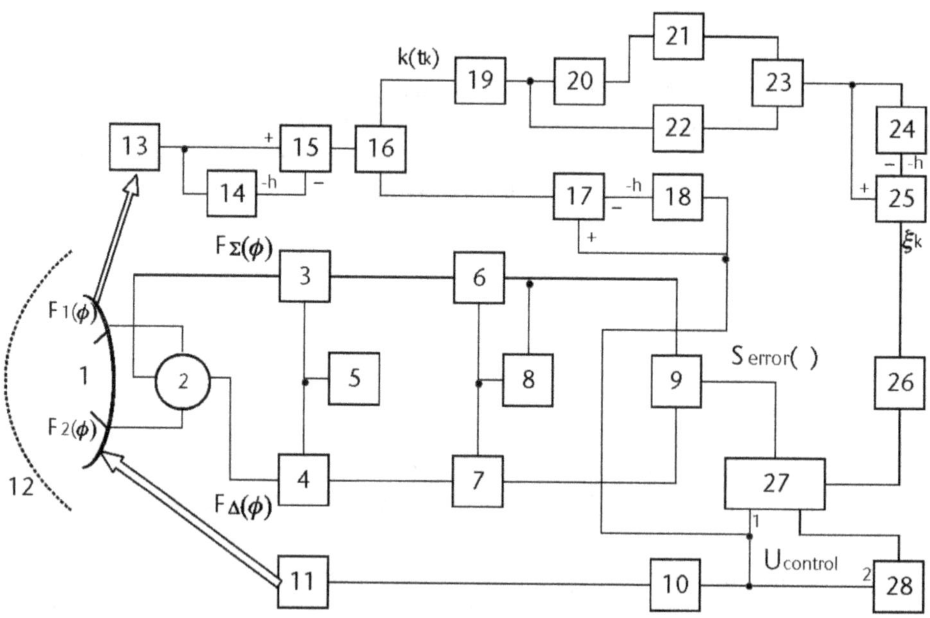

Figure 12.22 Sum-and-difference monopulse direction finding system that has implemented (14.57) and (14.58).

The following elements are presented in Figure 12.22:

1: Monopulse antenna;

2: Stripline ring;

3: Mixer of the sum channel;

4: Mixer of the difference channel;

5: Heterodyne;

6, 7: Intermediate-frequency amplifiers of the sum-and-difference channel;

8: Automatic gain control system of the sum channel;

9: Phase detector;

10: Error-signal amplifier;

11: Monopulse antenna mechanism;

12: Radome (blister, nose cone);

13: Sensor angular position monopulse antenna;

14, 18, 24: Line delay signal for one quantum of time $t_{k-1} = (k-1)h = -h$;

15, 17, 25: Subtraction device;

16: Division device;

19: Device that implements the operation:

$$C_k = \frac{A}{\left[\sum_{i,j=1}^{4} |a_{ij}|^2\right]^{\frac{1}{2}}} \tag{12.73}$$

20: Device that implements the operation:

$$C_k^* = C^T \tag{12.74}$$

21, 22: Devices that implements the following operations:

$$\left.\begin{matrix} \exp\{hC_k^*\} \\ \exp\{hC_k\} \end{matrix}\right\} \tag{12.75}$$

23: Multiplication device that implements the operation:

$$\exp\{hC_k^*\} \times \exp\{hC_k\} \tag{12.76}$$

25: Subtraction device that implements the operation:

$$\xi_k = \left\| \exp\{hC_k^*\} \exp\{hC_k\} - \exp\{hC_{k-1}^*\} \exp\{hC_{k-1}\} \right\| \tag{12.77}$$

26: Trigger that implements the operation:

$$\begin{cases} \text{if} \quad \xi_k = 0 + \epsilon & \Rightarrow \quad U_{switch} = 0 \\ \text{if} \quad \xi_k = \hat{M} & \Rightarrow \quad U_{switch} = U_{feed} = 12.6 \pm 0.5V \end{cases} \tag{12.78}$$

27: Switching device, that implements following operation:

$$\begin{cases} \text{if} \quad U_{switch} = 0 & \Rightarrow \quad U_{control\ 1} = S_{err}(\varphi, t_k) \\ \text{if} \quad U_{switch} = U_{feed} = 12.6 \pm 0.5V & \Rightarrow \quad U_{control\ 2} = \frac{1}{S_{err}(\varphi, t_k)} \end{cases} \tag{12.79}$$

28: A computing device that performs the inverse transformation of the error signal function $S_{err}(\varphi, t_k)$:

$$U_{control\ 2} = \frac{1}{S_{err}(\varphi, t_k)} \tag{12.80}$$

For the structural diagram of the angular auto-tracking system (Figure 12.22), in each k-time unit (12.10), the system of differential equations (12.14) is solved, where the matrix **A** is presented in the form (12.15). Thus, if there is a computer with sufficient speed onboard missile, it is possible to do without an additional receiving channel. The operation of the onboard direction finder in this case is based on the use of model (12.15), (12.57), and (12.58), and data on the angular position of the monopulse antenna, coming from sensor 13 of its angular position.

Under these conditions, the implementation of algorithm (12.57)–(12.79) is possible.

The circuit consisting of devices 13–25 can be characterized as a single-bit detector of the interference on the cross polarization.

12.7 The Interference Immunity Indicator

The interference immunity indicator was first proposed in [25]. Development of mass-scale deployment of measures and aids to suppress the enemy air defense in air combat operations highlights the demand for creating jammer-resistant radar systems. From this point of view, the weakest among radar systems with semiactive and active homing channels seems to be the airborne radar of an anti-aircraft guided missile. Under conditions of interference impacts, a vital issue for the airborne radar is to provide for stable automatic target tracking by angular coordinates. It is reasonable to characterize the quality of automatic tracking by the following criteria:

$$Q = \int_0^T e^2(t)dt \tag{12.81}$$

where [0,T] is the analysis interval, and $e^2(t)$ is the squared error of angle tracking. The anti-jamming capability issue boils down to the necessity to provide for the best in the sense of precision (12.81), which is target angle tracking under active jamming impacts that adversely affect radar accuracy characteristics. Two decision-making strategies seem to be relevant here: trade-off and active strategies.

Trade-off strategies can be characterized by the fact, that prior to jamming impacts specified by the jammer class $\{H\}$, the radar itself should possess certain qualities that unequivocally provide for reduction in efficiency of class $\{H\}$ jammers on their deployment. For example, when the level of the receiving antenna sidelobes is less than −48dB, cover interferences will be ineffective. Reduction of sidelobe levels for airborne radar with missile cross dimensions specified, however, would inevitably result in expansion of the radar main beam, and consequently, in its worse resolution of angle coordinates. This will degrade input signal reception and reduce the accuracy of target automatic angle tracking. To operate under conditions of blinking interference, the radar, in contrast, would require high angle resolution [1].

As a result, characteristics of airborne radar should be selected taking into account a compromise when there is a slight decrease in Q quality at target tracking under conditions of each jamming impact. It is suggested here that the active strategy should be regarded as adaptive procedure (set of adaptive procedures) that make it possible to maintain Q quality criterion at a level not less than the specified one, when affected by q number of jammer classes H_i, $\forall i = \overline{1, q}$. In these conditions, adaptation can be of any shape: parametric, algorithmic, structural, so on. It is important here that for impact of jamming signal H_i class, it would be possible to ensure the minimum criterion of quality:

$$Q^* = \arg \inf_{Q} F(H, Q), \tag{12.82}$$

where F is some operator. It is quite clear that the condition (12.82) can be complied with if the case we can correctly select the most efficient countermeasure Z_i for the detected jamming signal H_i. Once we know the types of interference that can affect the radar and actions to reduce efficiency of such an impact, we would be in position to create decision procedures that allow each class of jamming signals H_i to find correspondence to some set of $X = \{x\}$ features that characterize it and compensate Z_i impact. Thus:

$$\begin{cases} H_i = F_i(x_1, x_2, \ldots, x_n) \\ Z_i = L(H_i), \end{cases} \tag{12.83}$$

where L is some linear operator. Here, H_i jamming signals are determined apriori on the set of $X = \{x\}$ features. Let us assume that the class of jamming signals $H = \{H_i\}$ and dictionary of $X = \{x\}$ features are optimal as implied in (12.82) (i.e., the way to divide the feature space is selected so that the optimal value of quality criterion is achievable). The task then would be to find out such a rule that would provide for maximum efficiency in recognizing a jamming signal at specified criteria and restrictions [2]. In the general case, $X = \{x\}$ features can have deterministic, probabilistic, logical, and structural character [3]. Relations (12.83) will take this or that shape depending on the type of features used. As far as determined features go, classes can be specified either by values of object features assigned a priory to each class or by coordinate values of some points in the feature space that are accepted as class templates. It is quite understandable that probabilistic features would have class descriptions in the form of probability-density functions $f_i(x_1, x_2, \ldots, x_n)$ of feature values $x_1, x_2, \ldots, x_n$, provided the object belongs to class H_i. When logical features of $x_i = \{1, 0\}$, $\forall i = \overline{1, q}$ type are used, then the class descriptions will be represented by Boolean relations between features and classes. Therefore it is possible to construct some decision rules that would enable us to recognize the interference in progress from a denumerable quantity of possible ones, depending on types of features that compose the feature space $X = \{x\}$. Selecting the number and types of features is the objective of developing the feature space that can be solved based on the study of radar operation under conditions of a specified set of jamming features $H = \{H\}$. Apart from this, the synthesis of decision rules for recognition is possible only after the issue of feature space discretization is resolved to divide it into classes. Following specifications to develop the radar, the code of interference classes $\{H\}$ will be reasonably regarded as denumerable set of output $q : H_i \in \{H\}, \forall i = \overline{1, q}$. It is clear that the feature space $X \subset R^n$ will be developed based on actual capacities of a definite radar to obtain information. Resolving the issue of discretization here is understood as such way to divide the feature space X that represents it as the sum of its pairwise disjoint subspaces $\bigcap_{i=1}^{q} X_i = \emptyset$, where $\emptyset$ is an empty set, while closures $[X_1], [X_2], \ldots, [X_q]$ do not contain isolated points. In this regard, it is an essential requirement that features be consistent and noncontradictory whereby the convex feature space is developed and its structure closed at R^n. Thus, at $n = q$ each i-n class of H_i jamming can be related to some i-n detector X_i with output signal $x_i = \{1, 0\}\ \forall i = \overline{1, q}$. Let us assume that at some j-n moment, the radar is affected by H_i jamming signal, while there is a vector of observations $X_j = (x_{j1}, x_{j2}, \ldots, x_{jn})^T \in R^n$ developed at the exit of interference locator bar.

Recognition of H_i jamming is supposed to be implemented by the following decision rule:

$$\forall j, H_i : \sum_{k=1}^{n} (x_{jk} - h_{ik}) = 0, \tag{12.84}$$

where $H_i = (h_{i1}, h_{i2}, \ldots, h_{in})$ is the vector of the jamming image specified apriori.

Applying this approach to the airborne monopulse direction finder would significantly enhance its potential in the progress of interference conditions.

To implement proper comparison of angle-tracking accuracy for an airborne locator under no-interference conditions and under conditions of j-class jamming, it is necessary to provide for equal analysis intervals $[0, T]$ in each experiment. Compliance with this requirement and application of procedure (12.81) with subsequent logarithmization gives a visual presentation on the jamming immunity of its operations in the cases under consideration. The interference immunity indicator Q_j was calculated with the following formula:

$$Q_j = \ln\left(\frac{q_c}{q_j}\right) \tag{12.85}$$

After change integral (12.81) determined on integral length $[0, T]$ by N finite sums at condition of the constant step $\Delta t = const$, formula (12.81) will be presented in the form:

$$\hat{Q}_j = \sum_{i=1}^{N} e_{ji}^2 \Delta t, \qquad \forall j = 0, M \tag{12.86}$$

For the analysis, the relative interference immunity indicator will be used correctly. The master parameter for the interference immunity indicator will sensibly use the value of the automatic angle tracking quality of the target without polarization interference; that is, the signal source must have a working polarization, in this case. If we put the value of the quality of auto-tracking of the target on angular coordinates (when the signal source has working polarization) into the numerator, and the value of the quality of the angular auto-tracking of the study event in the denominator, then after contraction by Δt we can write:

$$q_j = \frac{\sum_{i=1}^{N} e_{0i}^2}{\sum_{i=1}^{N} e_{ji}^2}, \qquad \forall j = 0, M \tag{12.87}$$

where $j = 0$ - zero-index for automatic tracking without jamming. However, calculation of the interference immunity indicator by (12.87) will not give a sense about the border of its applicability. Consequently, it will be necessary to compare the results of the formula (12.87) with certain border value C. Taking into consideration that target destruction by fragment combat charge with probability $P \geq A = const$ will be possible if missile miss h will be no more than the radius of fragment distribution of the warhead R, which makes this probability of the destruction possible, we can write:

$$h \leq R \tag{12.88}$$

Thereby if it is determined apriori missile velocity V_m and missile time constant t_s as a control system, then for a highly possible error of angle tracking in degree it is possible to write:

$$e_c = \frac{180}{\pi} \arctan\left[\frac{R}{V_m t_s}\right] \tag{12.89}$$

After substitution (12.89) in (12.87) and after reduction Δt in numerator and denominator, we can write:

$$q_c = \frac{\sum_{i=1}^{N} e_{0i}^2}{\sum_{i=1}^{N} e_c^2} \tag{12.90}$$

Substituting (12.87) and (12.90) into expression (12.85) and the subsequent logarithmization gives the formula for the anti-jamming indicator for the j-variant of the analysis, which is written as:

$$Q_j = \ln \sum_{i=1}^{N} e_{ji}^2 - \ln \sum_{i=1}^{N} e_c^2 \tag{12.91}$$

The border value C for this will be in the form of:

$$C = N \sum_{i=1}^{N} e_c^2 \tag{12.92}$$

Expression (12.91) and is used as a criterion to assess the quality of interference immunity, where e_c^2 is assumed as equal to half-width $|\Delta|$ of the angle-tracking

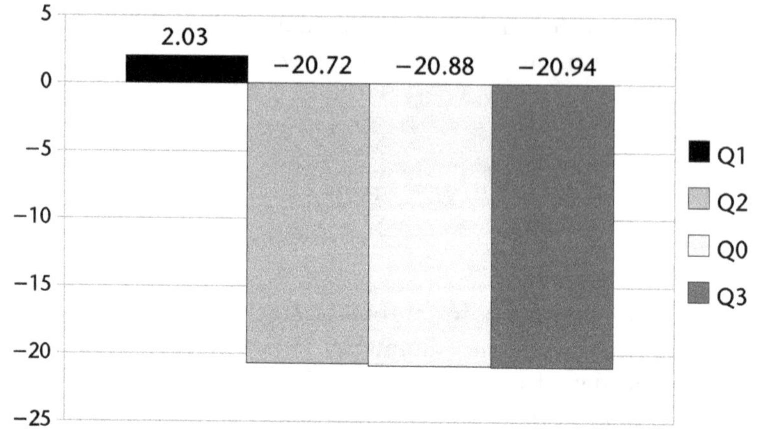

Figure 12.23 The integral interference immunity index Q.

system direction-finding characteristic on operating polarization, N is the number of time counts, e_{ji} is the tracking errors at $j = 0$ for the interference-free tracking case, where $j = 1$ for polarizing interference when there are no interference protection measures, $j = 2$ for a tracking system with a homing loop that is disconnected for the duration of the polarization interference impact, and $j = 3$ for an inversely transform, where the control signal is formed in accordance with (12.65), (12.71), and (12.80).

Values of integral performance index Q are given in Figure 12.23, where it can be seen that the indicators of interference immunity on application of the algorithm in which a homing circuit is disconnected and the best results will obtained under static conditions, including on target tracking in routine operations mode. This is explained by the null function of the tracking error on the integration in (12.81) and by the absence of random pulse disturbances induced by signals, depolarized space, and antenna protection (radome in the case of airborne radars).

In any case, when the angular tracking system is exposed to polarization interference, the interference immunity indicator with the inverse function algorithm has better characteristics compared to an unprotected radar and with other protection methods.

12.8 Conclusion

Assessment of angle tracking for an interference source that distorts the direction-finding characteristic of the angle tracker has been discussed. The assessment method consists of breaking the observation interval down to N mutually disjoint half-intervals, on each of which the differential equation system of the first order is solved. The direction-finding characteristic is approximated on each breakdown quantum by a line whose sign and inclination angle are determined by an interference signal. Polarization interference was used as a direction-finding characteristic distorting interference simulated by a linearly polarized signal. This signal polarization plane was rotated in the orthogonal basis of the angle tracker receiving antenna with certain angular speed. The resulting direction-finding characteristic presents the weighted sum of the direction-finding characteristic on the agreed and cross-polarization of the radar receiver antenna whose weight factors are sin- and cos-functions of the polarization plane inclination angle to the ortho-plane of the receiver antenna operational polarization.

A comparative analysis with deployment of experimental data showed that the synthesized decision provides satisfactory results, attesting the suitability of an advanced approach for accuracy assessment of target angle tracking in monopulse radars.

The problem of target auto-tracking stability by a monopulse radar was examined when the radar is exposed to polarization interference. From the mathematics point-of-view, this problem was reduced to the classical Lyapunov problem on the stability of the solution of a system of differential equations, by which can be used to describe any automatic control system. It is shown that the influence of polarization interference leads to negative values of the gain of the control loop of the angular position of the antenna in space. This entails a shift in the range of eigenvalues of the factor matrix of the system of differential equations to the right half-plane, which leads to instability of the automatic tracking system and, in general, to the loss target tracking on angular coordinates.

It was shown that in angle tracker operation on the polarization interference source, it is impossible to develop optimal control function after Bellman outside the system.

The solution synthesis was made in the state space, which is the optimal one in the sense that in auto-tracking a polarization interference, it was shown that the obtained algorithm exists and is unique by applying the inverse function of error signal after detection of a signal on orthogonal polarization. The obtained direction-finding characteristic corresponds well enough with the direction-finding characteristic on operational polarization up to the value of around $0,4\Delta$ relative to the boresight and provides for stable tracking of a polarization interference source.

References

[1] Leonov, A. I., and K. I. Fomichev, *Monopulse Radar*, Norwood, MA: Artech House, 1986.

[2] Leonov, A. I., and K. I. Fomichev, *Monopulse Radar*, Moscow: Sovetskoe Radio, 1970.

[3] Leonov A. I., and K. I. Fomichev, *Monopulse Radar*, Moscow: Radio i svyaz, 1984.

[4] Van Brunt, L. B., *Applied ECM*, Vol. 1, Part 4, New York: EW Engineering, 1978.

[5] Museum of the State Central Design Bureau, "Almaz-Antey," Homing Head 5G24 of the 5V21 Anti-Aircraft Missile (5G22 and 5G23 Mentioned in the Literature), https://saidpvo.livejournal.com/134389.html.

[6] "IEEE Standard Definitions of Terms for Antennas," *Transactions on Antennas and Propagation*, Vol. AP-17, May 1969, pp. 262–269.

[7] Ludwig, A., "The Definition of Cross Polarization," *IEEE Transactions on Antennas and Propagation*, Vol. 21, No. 1, 1973, pp. 116–119.

[8] Sherman, S. M., and D. K. Barton, *Monopulse Principles and Techniques*, Second Edition, Norwood, MA: Artech House, 2011.

[9] Artemyev, V. M., *Reference Manual for Study Methods of Electronic Tracking Systems*, Minsk: Vysh. Shkola, 1984.

[10] Levin, B. P., *Theoretic Foundations of Statistic Radiotechnics*, Third Amended Edition, Moscow: Radio i Svyaz, 1984.

[11] Kazakov, I. E., *Statistic Methods of Control System Design*, Moscow: Machinostroyeniye, 1969.

[12] Stearns, S. D., *Digital Signal Analysis*, Fourth Edition, Rochelle Park, New York: Hayden Book Company, 1979.

[13] Pugachev, V. S., and I. N. Sinitsin, *Stochastic Differential Systems Analysis and Filtering*, Chichester, United Kingdom: John Wiley & Sons, 1987.

[14] Godunov, S. K., *Guaranteed Accuracy in Numerical Linear Algebra*, Springer, 1993.

[15] Ostrem, K., and B. Wittenmark, *Computer Controlled Systems*, Translation from English, Moscow: MIR, 1987.

[16] Godunov, S. K., *Lyapunov Quadratic Function. Computational Difficulties of Hurwitz Problem and Ways of Their Overcoming*, Novosibirks, Russia: Novosibirks University Press, 1986.

[17] Marchuk, G. I., *Numerical Methods and Applications*, London: CRC Press, 1994.

[18] Wilkinson, J. H., *The Algebraic Eigenvalue Problem*, Oxford, United Kingdom: Clarendon Press, 1969.

[19] Corless, M. J., and A. Frazho, *Linear Systems and Control: An Operator Perspective*, CRC Press, 2003.

[20] Arato, M., *Stochastic Differential Systems*, Berlin: Springer-Verlag, 1981.

[21] Gilbert, J., and L. Gilbert, *Linear Algebra and Matrix Theory*, Second Edition, Thomson Brooks/Cole, 2004.

[22] Rosanov, Y. A., *Markov Random Fields*, Springer-Verlag, 1983.

[23] Godunov, S. K., *Ordinary Differential Equations with Constant Coefficient*, Translation of Mathematical Monographs, American Mathematical Society, 1987, Alibris ID: 9762292910/.

[24] Markin, E., U.S. Patent 8,896,483, *Method of Automatic Target Angle Tracking by Monopulse Radar Under Conditions of Interference Distorting Location Characteristic*, 2014, https://www.uspto.gov.

[25] Markin, E., "On Interference Immunity of Angle Tracking Systems Under Conditions of Interference Distorting Location Characteristic," *Proceedings of the 2009 IEEE Radar Conference*, Pasadena, CA, 2009, pp. 1-6.

CHAPTER 13

Solutions Synthesis to Increase the Noise Immunity of an Active Homing Head

13.1 Interference Immunity of a Missile Homing Head from Interference Along the Sidelobes of its Directional Pattern

In 1986, the USSR developed a model of jamming, which, most likely, could be used by a potential adversary to organize electronic suppression of air defense systems covering the most important state facilities, military facilities, and infrastructure of the defending country.

The jamming model is shown in Figure 13.1. According to this model, in order to suppress the anti-aircraft missile systems group, in the breakthrough area, in the corner sector $\pm 23°$, $\pm 17°$, relative to this group, the enemy must create a permanent broadband interference that will affect the main lobe and the first sidelobes of the radar's firing complexes of the defending side. This is achieved through the creation of three loitering zones for jammers, located at a distance of 300–350 km from the suppressed anti-aircraft missile system. In each loitering zone there are two aircraft-jammers flying at an altitude of 10 km in a box with a width of 60 km along the front and a depth of 30 km. When loitering zones are located at a distance of 350 km from the suppressed anti-aircraft missile system, the width of the loitering zones can be increased to 90 km while maintaining the distance between the loitering zones at 20 km. With such an arrangement of loitering zones, jammers find themselves outside the affected area of the S-300 PM, S-300 PMU-1, S-300 PMU-2, and S-200D anti-aircraft missile systems. This type of interference model was used in testing the S-200D anti-aircraft missile system at the Sary-Shagan test site.

At the beginning of the tests, it was necessary to determine the fundamental possibility of implementing such a scheme for constructing aircraft-jammers in space. For this purpose, a number of flights were carried out on the IL-76 aircraft. In the course of these experiments, the capabilities of aircraft-jammers were determined to maintain the specified dimensions, height, and range of the location of the loitering zones from the anti-aircraft missile system. The target illumination radar was located at test site #35, and the loitering zones were located in the area of the city of Zhezkazgan (47.78306N, 67.76667E). Since it was necessary to provide a range of 380 km from the target illumination radar, the coordinates 47.095303N, 67.179701W characterized the position of the center of the middle loitering zone, which is located on the large side of the rectangle and which has a

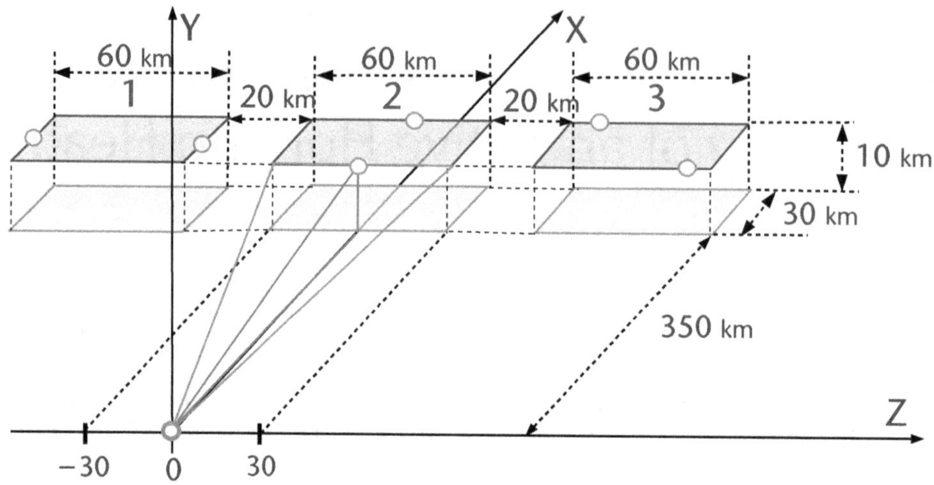

Figure 13.1 A model of cover interference from three loitering zones, located outside the defeat zone of air defense systems.

minimum distance of 354.14 km to the target illumination radar of the anti-aircraft missile system C-200D. The flights on the IL-76 aircraft were carried out from the Kambala airfield. The task was to determine the accuracy of the aircraft following the rectangle of the loitering zone. To do this, at the #35 test site, the aircraft was tracked by means of a target illumination radar and a P-14 all-round radar and target designation. There was a test officer in the cockpit who wrote down the coordinates of all turning points in a notebook. It turned out that the IL-76 plane flew along the route with great accuracy. When I again came to the Kambala airfield for the next tests to take part in the flights, I asked the ship commander how he managed to maintain the route with such accuracy. He replied that it was all about the gnome. This gnome turned out to be the onboard computer complex of the IL-76 aircraft. The navigator showed how he entered the coordinates of all turning points into the onboard computer. During the tests, the onboard navigation system has proven itself very well. It is especially necessary to emphasize that at that time the USSR did not have its own global positioning system such as GPS or GLONASS. The onboard computer of the IL-76 aircraft did all the calculations based on the azimuth angles from the corresponding gyroscopes and ground speed data generated by the air pressure receiver.

As part of the field tests of the S-200D system, numerical modeling and combat launches of missiles were carried out, which experimentally confirmed the destruction of jammers located at an altitude of 10 km and at the turn of 287 km from the target illumination radar of the S-200D missile system with a probability of at least 0.84. When aircraft-jammers were located at long ranges, the probability of their defeat dropped sharply and at a range of 370 km was equal to zero.

It was concluded that there was an extreme danger of broadband jammers deploying cover outside the area of destruction of air defense fire weapons.

During the tests of the S-200D system, a significant decrease in the combat capabilities of the anti-aircraft missile system was experimentally proven up to its complete inoperability.

Thus, cover interference aircraft-jammers, located in loitering zones outside the zone of destruction of the anti-aircraft missile system, were recognized as especially dangerous targets that should be destroyed in the first place.

This inoperability led to the issuance of a task and the subsequent development of the S-400 air defense system intended mainly for fighting with jammers located at ranges of 300–400 km from the suppressed radar. Active interference received along the sidelobes of the directional pattern of the seeker antenna worsens the signal-to-noise ratio and thereby sharply reduces the target detection range and, accordingly, the range of its stable auto-tracking. Consequently, for the homing head of an anti-aircraft guided missile, the task of reducing the effect of interference along the sidelobes is very important. In ground-based radars, this problem is usually solved with the help of the auto-canceller [1], which consists of the main channel and at least one compensation or additional channel. The main elements of an autocompensator are, as a rule, an adder, a controlled amplifier and a correlator. Noise interference signals are received simultaneously by the main and auxiliary antennas and processed in the main and compensation channels, respectively.

The first input of the adder receives the signal U^{Σ}_{main} of the total channel of the monopulse antenna of the seeker. The signal U_{add}, received by the antenna of the additional channel through a controlled amplifier having a complex gain, is fed to the second input of the adder. Due to the fact that the interference signal is received from one source simultaneously by two antennas, the voltages in U^{Σ}_{main} and U_{add} at the outputs of these antennas will differ in amplitude and phase difference caused by the difference in the signal path $\Delta\varphi$ between the phase center of the jammer's antenna and the phase centers of the monopulse antenna and the antenna of the additional channel. Therefore, the signals in the main and in the additional reception channel are correlated. The task of the controlled amplifier is to change the amplitude and phase of the interference in the additional channel so that this signal arrives at the input of the adder with an amplitude equal to the amplitude of the sum channel of the seeker, and its phase would be opposite to the phase of the signal of the sum channel. In this case, in the direction of the jammer in the resulting directional pattern of the two antennas, a zero gain for signal reception is formed and thus spatial rejection of the jamming signal is carried out.

The change in the transfer coefficient of the controlled amplifier of the additional channel occurs under the influence of the control voltage U_{contr}, which is generated by the correlator. Signals from the output of the adder and from the antenna of the additional channel arrive at the input of the correlator. The correlator multiplies these voltages and averages them over time, as a result, the amplitude and polarity (phase) of the output voltage of the correlator U_{contr} is determined by the magnitude of the residual noise at the output of the adder of the autocanceller, and the phase is determined by the phase difference of the noise signals in the main and in the additional channels receiving the homing head. Due to the negative sign of the feedback, the control voltage U_{contr} sets the minimum value of the residual noise at the output of the adder. Note that the composition of the noise signals of the sum and additional channels, in addition to the voltage of noise signals of interference, includes independent intrinsic noises of the receiving channels. In the process of multiplying and averaging voltages by the correlator, the control voltage

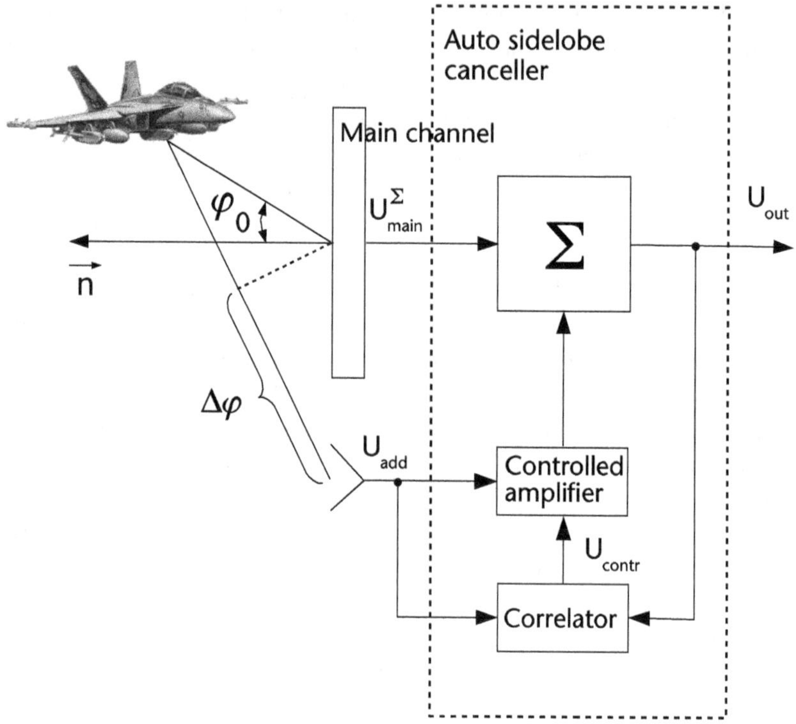

Figure 13.2 Interference auto-compensator.

U_{contr} is generated only from their common (correlated) part. That is why the feedback is called correlation. In the direction of the main lobe of the main antenna, the noise interference in the main channel significantly exceeds the interference in the additional channel, and therefore, at the output of the autocompensator, they act without attenuation. On the slopes of the radiation pattern, where the interference voltage in both channels is comparable in amplitude, the autocompensator attenuates the interference in the main channel and thereby narrows the interference reception sector along the main lobe of the directional pattern of the monopulse antenna.

It is quite clear that changing the amplitude and phase of a signal in a controlled amplifier is technically easier to implement at an intermediate frequency, which is why an autocompensator operating at an intermediate frequency is often called an autocompensator operating on the basis of a heterodyne.

The block diagram of a single-channel autocompensator with heterodyne voltage control is shown in Figure 13.3 [2]. This diagram has the following designations: 1: monopulse antenna in which only the output of the summed channel is used, 2: antenna of additional channel, 3: heterodyne, 4: mixer, 5: amplifier of the intermediate frequency of the additional channel, 6: controlled mixer, 7: integrator, 8: mixer-multiplier, 9: adder, and 10: feedback amplifier.

In this circuit, the controlled mixer plays essentially the same role as the controlled amplifier in the circuit in Figure 13.2. The parameters of a controlled mixer are determined by a change in the heterodyne voltage at the output of a narrowband filter, which is a high-quality integrator.

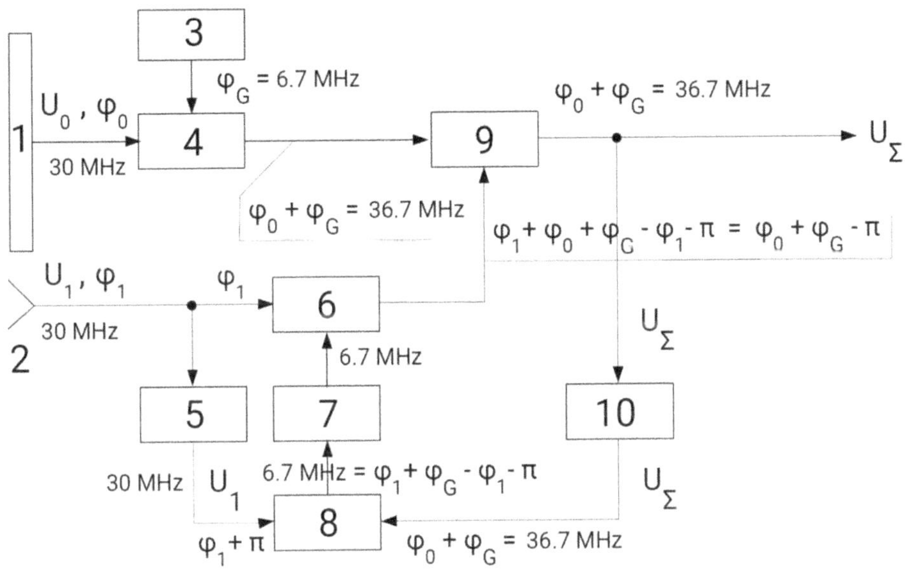

Figure 13.3 Interference autocompensator at an intermediate frequency of 30 MHz.

Mutual compensation of correlated interference of the main and additional channels in the adder is achieved by the fact that the interference of the additional channel at the output of the controlled mixer turns out to be equal in amplitude and opposite in phase to the interference of the main channel at the input of the adder. An auxiliary intermediate frequency amplifier is introduced in the additional channel path, which provides compensation for the signal delay of the main channel in the feedback circuit (adder-feedback amplifier), since the signals of the main and additional channels must arrive at the correlator inputs simultaneously. In addition, an odd number of stages in the auxiliary amplifier of the intermediate frequency provides a phase shift of the interference in the additional channel by 180°, thereby determining the nature of the correlation feedback which becomes negative and minimizes the interference signal at the output of the adder. Frequency and phase relationships between voltages are shown in Figure 13.3.

In the mixer-multiplier 8, the difference frequency f_Δ is allocated, which is accompanied by the subtraction of the phases of the multiplied voltages U_Σ and U_1. Therefore, the phase of the heterodyne voltage at the output of the narrowband filter:

$$\varphi_0 + \varphi_G - (\varphi_1 + \pi) \tag{13.1}$$

where φ_0 is the phase of interference at the input of the main channel, φ_G is the phase of the voltage of the heterodyne of the main channel, and φ_G is the phase of interference at the input of the additional channel. In a controlled mixer, on the contrary, the total frequency is extracted, which is accompanied by the addition of the phases of the input voltages of the mixer. Thus, the phase of the interference at the input of the controlled mixer of the additional channel can be written as:

$$\varphi_1 + \varphi_0 + \varphi_G - \varphi_1 - \pi = \varphi_0 + \varphi_G - \pi \tag{13.2}$$

that is, the noise at the output of the additional channel is coherent with the noise at the output of the main channel, regardless of the value of φ_1. To equalize the amplitudes of the correlated interference at the inputs of the adder 9, a negative correlation feedback circuit is used in the additional channel: feedback amplifier 10, mixer-multiplier 8, and integrator 7. A control voltage is formed at the output of the correlator 8 of this circuit, which is proportional to the mutual correlation function of the interference voltages at the inputs of the mixer-multiplier of the additional channel 8. The amplitude of the voltage at the output of the controlled mixer 6 depends on the magnitude of this voltage.

The heterodyne voltage is generated at the output of the narrowband filter (integrator) 7 only in the presence of cross correlation of the interference signals of the main and additional channels. For uncorrelated signals (internal noise, etc.) their correlation function is equal to zero, which is equivalent to automatic disconnection of the additional channel from the adder input. The integrator 7, made in the form of a narrowband quartz filter, provides the selection of the heterodyne frequency component from the frequency spectrum of the correlator 8. At the same time, being a high-quality circuit, it has a long settling time of oscillations, and therefore, the control heterodyne voltage increases in amplitude and shifts in phase gradually over $\approx 20\mu s$. This time determines the inertia of the autocompensator. The finite time of oscillation build-up in a narrowband filter ultimately determines the inertia of the autocompensator. The inertia of the autocompensator, on the one hand, should ensure the passage of useful short-term echo signals of the main channel through the adder (i.e., exclude their suppression due to the buildup of the narrowband filter). On the other hand, the buildup should not be significant, so that there are no remnants of uncompensated interference at the beginning of their action and so that the automatic compensator circuit responds in a timely manner to changes in the amplitude and phase of the interference of the main and additional channels during the rotation of the antenna and movement of the jammer aircraft. As the spatial position of the seeker antenna changes, the amplitude and phase shift of the interference signals at the inputs of the main and additional channels change in accordance with the antenna patterns. However, due to the presence of correlation feedback, the system self-regulates so that U_Σ always remains minimal.

With an increase in the amplitude of the interference in the main channel U_0 and a decrease in its phase shift relative to the interference signal in the additional channel U_1, the uncompensated remainder at the output of the adder will increase from the minimum initial value and receive a changed phase. Along the correlation feedback circuit, it will change the amplitude and phase of the control heterodyne voltage, which will lead to a change in the amplitude and phase of the interference signal of the additional channel at the input of the adder 9. This process will continue until the amplitude and phase of the additional channel interference signal change by the same amount as in the main channel, and the resulting voltage of the adder 9 returns to the minimum initial U_Σ.

Thus, when the autocompensator operates in the resulting antenna system directional pattern in the direction of the jammer φ_0, a minimum reception is created, and when the antenna rotates, the direction of the reception minimum is retained for all positions, except for the direction to the jammer along the main lobe of the total radiation pattern of the monopulse antenna homing heads. The effect of the autocompensator in the latter case is manifested in the compression of the effective suppression sector, since on the cuts of the main lobe, where the amplitude of the interference along the main channel is comparable to the interference signals of the additional channel, compensation or attenuation of the interference is possible. Thus, in the automatic noise compensator circuit, an automatic adjustment to a minimum of the noise signal at the output of the adder 9 is carried out.

Since the transmission coefficient of the controlled mixer is regulated by the correlation feedback loop according to the output effect; that is, according to the residual noise, the transmission coefficient of the circuit will differ slightly from the optimal one. As a result, at the output of the autocompensator there will always be some error in the uncompensated residual of the disturbance. It can be reduced by increasing the gain of the correlation feedback loop. However, an excessive increase will inevitably lead to self-excitation of the autocompensator, which is unacceptable and leads to the need to establish restrictions in the process of adjusting the autocompensator. The most effective suppression of active noise interference by an autocompensator is provided when the number of additional channels is equal to the number of simultaneously operating groups of jammers. This automatic jamming compensator can be used in the active homing head of an anti-aircraft missile. The implementation can be carried out by using a standard antenna and a radio fuse receiving path as an additional receiving channel. In this case, reliable suppression of interference of one group of jammers acting simultaneously from one angular direction φ_0 can be ensured, as shown in Figure 13.2.

13.2 Operation Assessment of the Air Defense Missile System for High-Speed and Low-Altitude Targets Without Interference

13.2.1 Methods for Overcoming the Defeat Zone of an Anti-Aircraft Missile System

As shown above, one of the reasons for the development of new anti-aircraft missile systems is the low efficiency of the existing air defense fire weapons against high-speed, low-altitude targets. The low efficiency of their work was due to the long finding of low-altitude targets in the area beyond their radio horizon, and consequently, the small detection range of such targets, which reduced the time of combat work on them to supercritical parameters. As a result, such targets could not be fired at all. This circumstance significantly increased the potential enemy's capabilities to overcome the air defense system and led to the nonfulfillment combat mission by air defense means.

Using the mathematics given in Section 3.2, Figure 13.4 shows the missile trajectory when firing at an aircraft-jammer moving on a collision course at an altitude of 10,000m at a speed of 200 m/s. The abscissa shows the distance from the start point of active homing (0 km, 26 km), located at an altitude of 26 km, to the point where the missile meets the target (50 km, 10 km). The ordinate is the altitude from the launch position of the missile to the start point of active homing (0 km, 26 km). Defeat probability of an aircraft-jammer with one missile under these conditions is ensured at a level not lower than 0.84.

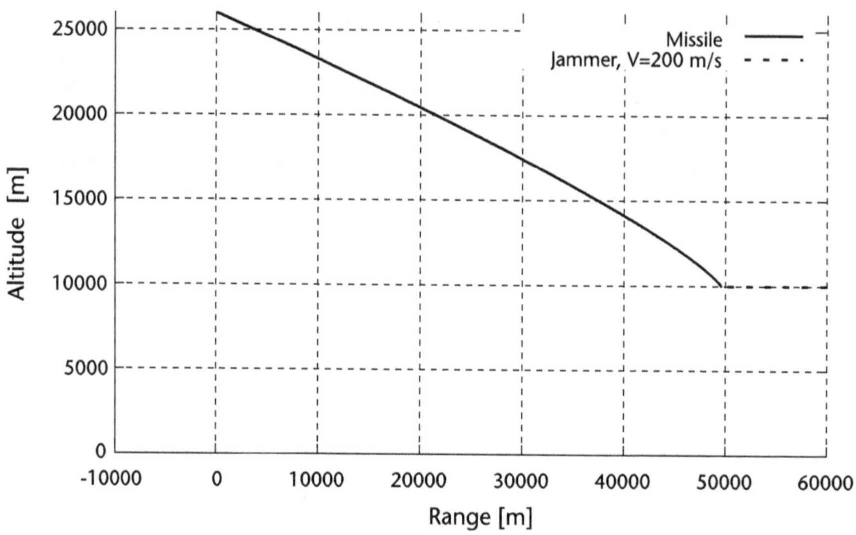

Figure 13.4 Kinematic trajectory of a missile for firing at an aircraft-jammer.

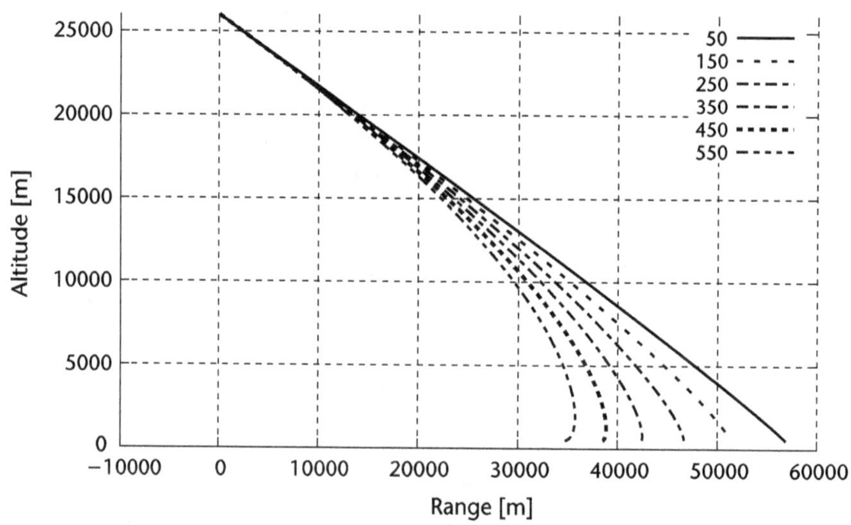

Figure 13.5 Kinematic missile trajectories for target speeds 50–550 m/s.

Figure 13.5 shows missile flight paths when firing at low-altitude targets without interference, and flying on a collision course at an altitude of 500m with speeds in the range of 50–550 m/s. The probability of their being defeated by one missile is also not lower than 0.84.

Figure 13.6 shows the trajectories of the missile when firing beyond the radio horizon at high-speed targets with a flight speed in the range of 650–1,000 m/s. It can be seen that with the speed increase of a target flying without jamming on a head-on course, the trajectory of the missile turns 180° in relation to the multifunctional radar of the anti-aircraft missile system. The missile velocity vector becomes collinear with the target velocity vector, as a result of which the firing is carried out in pursuit and the system's ability to intercept the target decreases. With such a turn of the missile's flight path, its flight characteristics sharply decrease due to

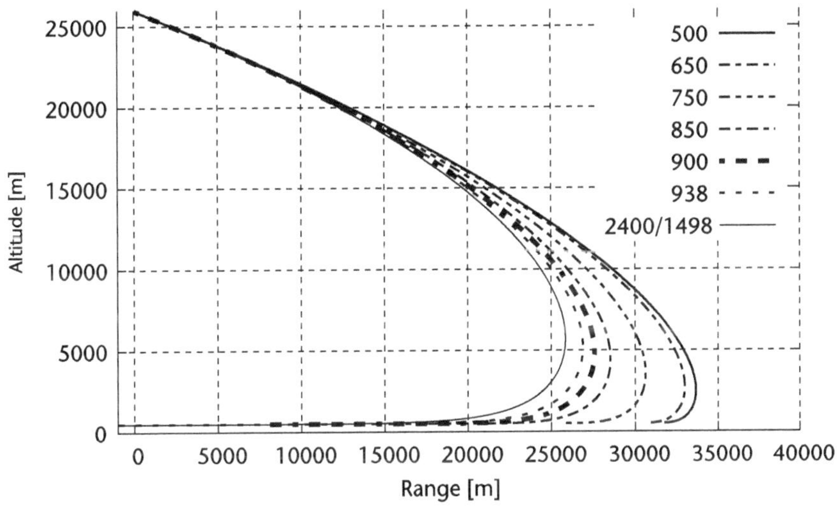

Figure 13.6 Kinematic missile trajectories for target speeds 650–2400 m/s.

significant losses in aerodynamic control (drag coefficient $C_{xb\alpha\delta} \neq 0$) and a significant increase in the medium drag force F_{air}, providing deceleration of the missile movement in units and tens of g. From Figures 13.6 and 13.7, it can be seen that at a missile speed of 1000 m/s and a target speed exceeding 938 m/s, the probability of hitting such a target by means of an anti-aircraft missile system is zero. These are ideal conditions. In real conditions, this speed will be lower by another $\approx$10% and will be 844.2 m/s or 2.49M at sound speed at an altitude of 500m up to sea level at $t^\circ = 15^\circ$C and 760 mm Hg. (101325 Pa): $\vartheta_{500\text{m}} = 338.4$ m/s [4]. This means that when flying at low altitudes on a collision course at a speed greater than 844.2 m/s, it becomes possible to overcome the affected area of a modern anti-aircraft missile system with minimal risk. Thus, when the aircraft enters the defeat area of the anti-aircraft missile system and when an anti-aircraft guided missile is detected by it, it is necessary to move to a low altitude $\approx$ 600m and overcome the defeat area of the anti-aircraft missile system in the terrain enveloping mode at a speed of $V_t \geq 2.49$M.

In addition, there is a dependence of the combat characteristics of the anti-aircraft missile system on the target's flight altitude. These dependencies are shown in Figure 13.8. The abscissa shows the time from the start of homing to the moment the warhead was detonated. The ordinate is the systematic angular error caused by the homing method at the given missile and target speeds. From Figure 13.8 it can be seen that with increasing target flight altitude, the angular error increases and starting from an altitude of 15,000m there is an unambiguous disruption of the angular tracking, which in turn leads to the failure of the combat mission. Considering the flight level of targets at a height of 16,000m as a way of overcoming that in the affected area of an anti-aircraft missile system, it is possible to plot the dependence of the angular error of the homing method on the range for different target speeds for a given speed of which was defined as 1,000 m/s. The indicated dependencies are shown in Figure 13.9.

Dependencies for velocities in the range of 250–850 m/s are considered. It can be seen that in the target velocity range of 342–650 m/s (1M–2M), angular errors

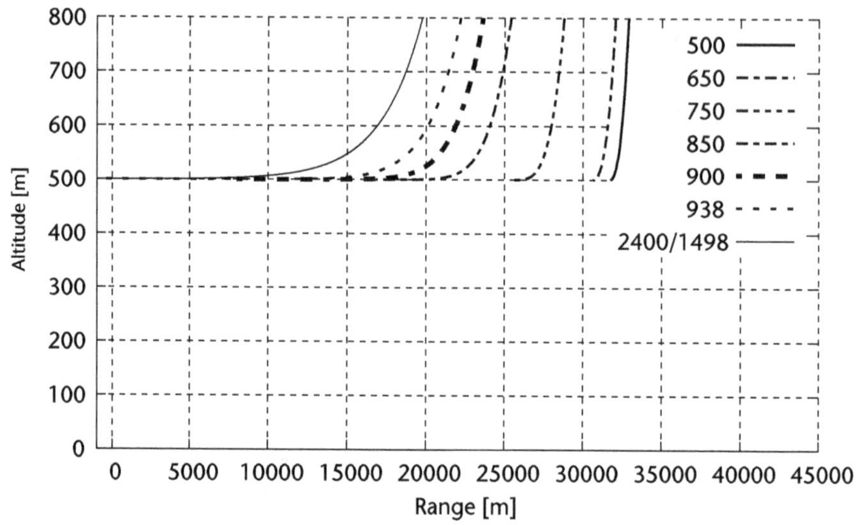

Figure 13.7 Missile kinematic trajectories for target speeds of 650–2400 m/s.

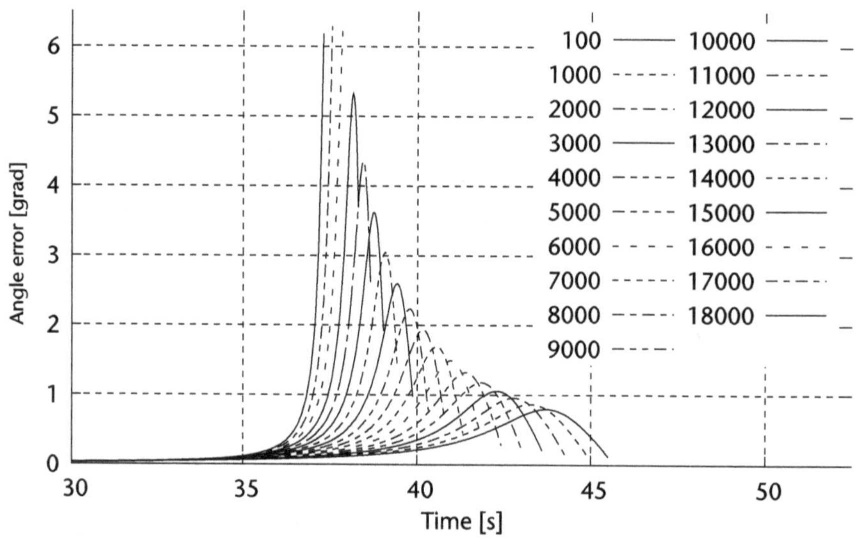

Figure 13.8 Errors of missile tracking of targets located at different flight altitudes.

lead to loss of automatic tracking with a probability close to one. This occurs as a result of high angular velocities arising directly in the vicinity of the meeting point within a radius of about 150m. Even if we detonate the warhead of the missile at the time it loses its auto-tracking of the target, the radio fuse will not be able to calculate the effective angle of inclination of the cloud of dispersion of the fragments of the warhead, and the vector of the cloud of dispersion of the fragments of the warhead, oriented perpendicular to the centerline of the missile body, will not be able to ensure the probability of hitting the target on a level not lower than 0.7 with a miss of 100m or more.

In addition, it should be especially emphasized that all firing at targets located abroad 200 km at a distance from the multifunctional radar, today can be carried out only in a two-stage mode.

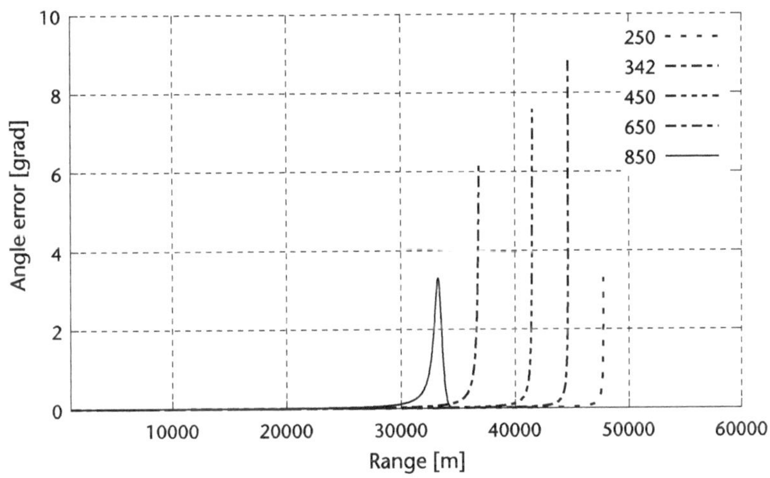

Figure 13.9 Missile tracking errors for targets flying at an altitude of ≈ 16000m in the velocity range of 250–850 m/s.

1. The missile flies to the calculated point of the start of homing on the ballistic trajectory.
2. After arriving at the calculated point:

   - The missile carries out an additional search for the target with its own homing head according to the information transmitted to the missile from the multifunctional radar, since at an altitude of 26 km the missile is in its visibility zone even at a range of 350 km, and carries out subsequent homing at the target in active mode, or in passive mode on the signal of the jammer;
   - The missile guides to the target on commands from the multifunctional radar when the missile and the target are in the visibility zone of this multifunctional radar.

The second stage of long-range firing is shown in Figure 13.10. Here, after additional search and target acquisition on auto-tracking, homing is carried out in an active mode to targets flying on a head-on course at an altitude of 16,000m with speeds of 650 m/s and 850 m/s, respectively.

Conclusions:

1. The missile ballistic trajectory, when firing at the starting point of active homing (D_t, 26 km), makes it vulnerable to being defeated by enemy air defenses. The interception of the missile before or at the moment it reaches the homing point at an altitude of about 26 km is available to any modern air defense system.
2. The single-stage scheme with a solid-propellant missile engine has reached its peak of development in the S-300 PMU-2 Favorit system, which provides the characteristics stated in the tactical and technical specifications for the system of the 48N6 family of missiles.
3. Effective firing for the radio horizon with a single-stage missile is possible at low-speed, low-flying targets, when the missile, falling from a great height,

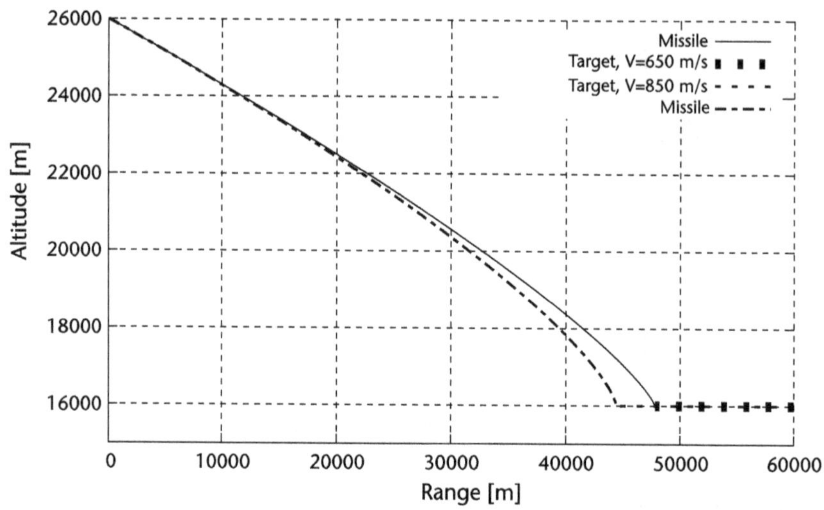

Figure 13.10 Homing trajectories of missiles to targets flying at an altitude of 16,000m.

manages to gain sufficient speed and has the ability to efficiently maneuver in dense layers of the atmosphere.

4. Altitudes greater than 16,000m do not provide the required quality of aerodynamic control of a single-stage missile in the passive phase of its flight. This fact can be considered as a way to overcome the defeat area of an anti-aircraft missile system. In this regard, it seems reasonable to develop a missile, based on a two-stage missile, made according to the carrying cone scheme, allowing overload $\geq$30g and having a flight speed $\geq$7.85M. It can be shown that this two-stage missile should have a curb weight of $\approx$4,700 kg and a length of $\approx$10m.

13.3 Solution Synthesis to Increase the Interference Immunity of an Active Homing Head During its Operation at Low-Altitude Targets and by a Group of Jammers

The work of an active seeker for low-altitude targets and for a group of jammers is complicated, as shown above, by the limitations on the spatial selection of signal sources and lack of differences in radial velocity, which makes it impossible to carry out Doppler target selection.

13.3.1 Options Analysis for Increasing the Accuracy and Formulation of the Synthesis Problem

To improve the accuracy characteristics, it is necessary to increase the resolution of the active homing head. There are several options for this:

1. Transition to a higher carrier frequency range;
2. Application of a quantum measuring device for angular coordinates;
3. Using a phased array antenna (PAA) with the ability to adaptation of its spatial radiation pattern;
4. Application of a signal with an optimal (in a given sense) uncertainty function.

The transition to carrier wavelengths less than 3 cm leads to a violation of the all-weather requirements of the onboard radar ($f_0 > 10$ GHz) and the need for its operation only in the transparency windows of the frequency range, which significantly reduces the noise immunity of an active seeker. The need to operate in the active and semiactive modes imposes additional requirements on the operating frequency range of the fairing and at $\Delta f_0 \geq 5$ GHz it is impossible to ensure the stability of its amplitude-frequency characteristic (AFC) at the level of −3 dB in the entire range. The placement of antenna active seeker on the missile body necessitates the use of nonprotruding traveling-wave antennas. In this case, achieving the same gain on four antennas as for the antenna array located under the radome is possible only by increasing the electrical length of the antenna. However, in this case, working frequency diapason drops sharply and the antenna acquires resonant properties. This, in turn, reduces the noise immunity of the active seeker in the transparency window and limits the capabilities of the active homing head for spatial scanning by the seeker antenna beam. In addition, due to the influence of the missile body, the radiation pattern of each of the four antennas is pressed, which makes it impossible to form the total radiation pattern without a dip in the direction of the equisignal direction. Application of the method of minimum radiation pattern for autoplay has low accuracy characteristics and is unacceptable in this case. Thus, the transition to a higher range of the carrier vibration is impossible due to its technical unfeasibility.

The use of a quantum meter, such as a carbon dioxide (CO_2) laser, removes many of the problems with target selection, but also faces some difficulties. Due to the fact that the area of application of anti-aircraft missiles lies in the troposphere, the use of a laser has a fairing problem. When placing both a laser gun and a measuring matrix under the fairing, it is necessary to ensure the optimal solution of the boundary problem for a given wavelength. Optimality is understood in the sense of ensuring the maximum refractive index at the minimum possible dispersion in a given solid angle. Considering the fairing as a body of rotation of the generatrix around the axis of symmetry, the fulfillment of the optimality requirements is possible only when the laser and the matrix are placed on the axis of symmetry of the fairing. However, as a result of the dynamic heating of the fairing in dense layers of the atmosphere up to 2,000°, it becomes necessary for it to be made of quartz glass completely or interspersed with the latter in the fairing nose. The first solution is unsuitable for cost, while the second option, at a slightly lower cost, also creates additional difficulties when integrating radar and quantum meters. The use of a composite radome will lead to an increase in amplitude and phase errors in the direction finding of the radar meter. In addition, placing a quantum meter in the center of the antenna array of the active seeker destroys the amplitude-phase distribution of the aperture, which entails significant changes in the active seeker antenna pattern. The placement of the laser in front of the radome faces the problem of its thermal insulation and the problem of the destruction of the field in the near zone of the active homing head antenna. Placing a quantum meter on the side surface of the missile body results in at least doubling the number of quantum meters. Thus, this option is also technically difficult to implement.

Earlier, it was assumed that the antenna design of an active seeker is an antenna array that does not have the ability to phase and weigh the signals of each of

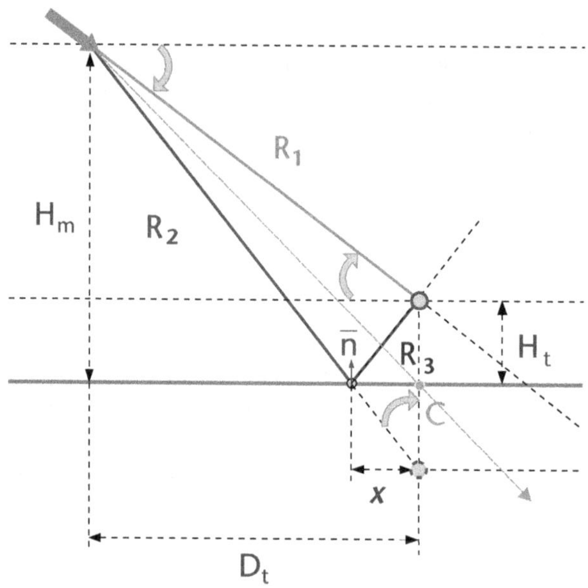

Figure 13.11 Illustration of resolution estimation.

its elements, and therefore, the adaptation of the spatial antenna pattern is also impossible for implementation in this homing head.

To select signals coming from different angular directions in conditions when the sources of these signals are in an unresolved volume in the directional pattern of the active homing head antenna seems to be expedient to determine the possible signs of their difference. As shown above, Doppler signal selection is difficult, spectral distinction are practically not used, and tracking along the angular coordinates is carried out along the harmonic with the maximum amplitude. Due to the fact that a long-range anti-aircraft missile usually approaches the target from above, its active homing head is a nadir radar. In this case, the ground (water) surface, which is also called the underlying surface, has a great influence on the reflected signal received from the target. Some results can be achieved with the polarization selection of the received signals; however, with a rather complex hardware and algorithmic implementation, it is difficult to achieve an improvement in the signal-to-noise ratio above 5 dB [3]. This is caused by the effect of depolarization, and in the general case of dispersion, of signals reflected from the Earth's (water's) surface. In this regard, it makes sense to take advantage of this phenomenon, as well as the time differences in the arrival of the direct signal reflected from the target and the first-order specular signal reflected from the underlying surface. This requires a good range resolution Δr, which is estimated as follows (Figure 13.11):

$$\Delta r = R_2 + R_3 - R_1 \tag{13.3}$$

where $R_1 = D_t/\cos\vartheta$; $R_3 = H_t/\cos\vartheta$; $R_2 = \sqrt{H_m^2 + (D_t - x)^2}$; $x = H_t \tan(\vartheta)$, and $H_m = \sqrt{R_1^2 - D_t^2} + H_t$.

Calculations using the formula (13.3) for different altitudes of the target and at a constant value of the angle of approach of the missile to the target ϑ are summarized

Table 13.1 Dependence of Missile Resolution on the Target Flight Attitude

$H_t\,[m]$	20	25	30	35	40	55	50	55	60	65
$\Delta r\,[m]$	26.19	32.74	39.30	45.85	52.40	58.96	65.52	72.07	78.63	85.19

in Table 3.1. Obviously, when aiming a missile at the energy center of radiation C located in the middle between the real target and its mirror reflection, and taking into account the fact that the radius of dispersion of the warhead fragments is R_0, at which a given target is hit with a probability of at least 0.7 is equal to 40m, the minimum target height is 40m in this case, and the range resolution Δr should be no worse than 52.40m.

Let $\Delta r = 50$m.

Thus, the problem arises of synthesizing a signal that would provide the required range resolution $\Delta r = 50$m without degrading the active seeker resolution on speed.

13.3.2 Solution Synthesis

The synthesis will be carried out in the signal space $S = \{x, P\}$, where P is a set of certain properties that are valid $\forall x \in S$.

Assuming $P_1 : \int\limits_{-\infty}^{+\infty} x^2(t)dt \le M < \infty$, it is expedient to characterize the space S by the set of signals $\{x(t)\}$ with bounded energy, that is, S is defined on the set of functions with square integrable.

For $P_2 : \quad d(x, y) = \left[\sum\limits_{i=1}^{n} |\alpha_i - \beta_i|^2\right]^{1/2}$, on $S = \{x(t)\}$ the Euclidean matrix is defined, the linear span of $\{\vec{X}_i\}$, $\forall i = \overline{1, n}$ and the class of scalars, hence the linear metric space $L^2(T)$ is defined.

For $P_3 : \quad \|\vec{x}\| = \left[\int\limits_{-\infty}^{+\infty} x^2(t)dt\right]^{1/2} \le M < \infty$; that is, the vector norm exists and it is bounded, and the space $L^2(T)$ is complete.

For $P_4 : \quad (\vec{x}, \vec{y}) = \int\limits_{0}^{T} x(t)x^*(t)dt, \quad \forall \vec{x}, \vec{y} \in L^2(T)$; that is, there is a dot product and the Schwarz inequality holds:

$$|(\vec{x}, \vec{y})|^2 \le (\vec{x}, \vec{x})(\vec{y}, \vec{y}) \tag{13.4}$$

Thus, the synthesis of the signal will be carried out in the Hilbert space; that is, in a complete linear normalized space with a given scalar product.

Due to the fact that a computer is used for the missile on board, the most preferable form of signal representation is discrete. Signal $x(t)$ in n-dimensional linear space $M_n \subset L^2(T)$ can be represented as a linear combination $\{\varphi_i(t)\}$ the only way:

$$x(t) = \sum_{i=1}^{n} \alpha_i \varphi_i(t) \tag{13.5}$$

where $\{\varphi_i(t)\}$ is a set of linearly independent basis functions of the space M_n, such that with $\forall t \in T$ the following condition is satisfied:

$\sum_{i=1}^{n} \alpha_i \varphi_i(t) = 0$ almost everywhere, if and only if $\alpha_i = 0 \;\; \forall i$. Moreover, $\{\varphi_i(t)\}$ defines the space M_n uniquely.

However, exact discrete representations of signals can be quite complex and cumbersome, and a suitable decomposition in time-shifted functions is $\varphi_i(t) = \varphi(t - iT), \;\; \forall i = 0, 1, 2, \ldots$ and is not complete in $L^2(T)$ for any original function $\varphi(t)$. Completeness of representation can be obtained using basic functions, of which the set of definitions is dense. Replacing a discrete index i in (13.5) with a continual variable $s \in S$ leads to a basis function of two variables and signal presentation in integral form:

$$x(t) = \int_S u(s)\varphi(t, s)ds \qquad \forall t \in T, \tag{13.6}$$

where $u(s)$ is the continuous representation of $x(t)$, and $\varphi(t, s)$ is the basic kernel of the integral transformation. If there is a conjugate basic kernel $\theta(s, t)$, the integral representation is valid:

$$u(s) = \int_T x(t)\theta(s, t)dt \qquad \forall s \in S \tag{13.7}$$

Substitution of (13.7) into (13.6) gives a necessary and sufficient condition for the existence of conjugate kernels:

$$x(t) = \int_S\int_T x(\tau)\theta(s, \tau)\varphi(t, s)d\tau ds = \int_T I(t, \tau)x(\tau)d\tau = I(t, \tau), \tag{13.8}$$

where:

$$I(t, \tau) = \int_S \varphi(t, s)\theta(s, \tau)d\tau. \tag{13.9}$$

Using the generalized Dirac function, the condition (13.9) will be rewritten as:

$$\int_S \varphi(t, s)\theta(s, \tau)ds - \delta(t - \tau), \tag{13.10}$$

and the reverse substitution gives:

$$\int_T \theta(s, t)\varphi(t, \vartheta)dt - \delta(s - \vartheta) \tag{13.11}$$

Basic functions are used in the Fourier transform:

$$\varphi(t,s) = \exp[j2\pi st],$$
$$\theta(t,s) = \exp[-j2\pi st], \quad \forall t \in [-\infty, +\infty], \quad \forall S \in [-\infty, +\infty],$$

which satisfy the conditions (13.9) and (13.10); that is, they are self-adjoint kernels. In this case, the transformation (13.7) of functions $x(t) \in L^2(T)$ always generates functions of $s : u(s) \in L^2(s)$ and the value of the scalar product does not change during transformations (Parseval's equality):

$$(X, Y) = (x, y) \tag{13.12}$$

Here X, Y are Fourier transforms of functions x, y.

The characteristic of the resolving power of radar signals is the radar uncertainty function defined on the plane time-frequency, which indicates those points of the domain of definition in which the dot product does not vanish.

The duality of the problem lies in the fact that the time and frequency domains are uniquely connected by a pair of Fourier transforms, in which the kernels cannot simultaneously belong to $L^2(T)$ or $L^2(f)$ as functions of t or f, respectively.

So for example, if one of the kernels is a function that is smooth and integrable with respect to t, then the other kernel is necessarily singular, or both kernels are nonintegrable as in the Fourier transform:

$$X(f) = \int_{-\infty}^{+\infty} x(t) \exp^{-j2\pi ft} dt \tag{13.13}$$

Strictly speaking, this mapping is not one-to-one. That is, for different functions of time, there can be the same Fourier transform. Mutual one-to-one conversions will be performed for set of functions: $S_x = \{x_1(t), x_2(t), \dots\}, \forall\, x_i(t) \in S_x \;:\; \|x_i(t) - x_0(t)\| < \epsilon$; that is, an equivalence relation is established almost everywhere.

$$x(t) = \int_{-\infty}^{+\infty} X(f) \exp^{j2\pi ft} df \tag{13.14}$$

Thus, by the Wiener-Khinchin theorem, the uniqueness of the Fourier transforms (13.13), (13.14) is satisfied between the autocorrelation function $R(\tau)$ of the signal $x(t)$ and its power spectrum $G(f)$. Therefore, for the uncertainty function in the time domain, we can write:

$$\dot{\rho}(f,\tau) = \exp^{-j2\pi f_0 \tau} \exp^{-j2\pi f\tau} \int_{-\infty}^{+\infty} \dot{\gamma}(t)\dot{\gamma}^*(t+\tau) \exp^{-j2\pi ft} dt, \tag{13.15}$$

where $\dot{\gamma}(t) = u(t) + j\vartheta(t)$ is the complex envelope of the signal $x(t)$, which is represented in terms of inphase and quadrature components:

$$x(t) = u(t)\cos 2\pi f_0 t - \vartheta(t)\sin 2\pi f_0 t \tag{13.16}$$

$$w(t) = \sqrt{u^2(t) + \vartheta^2(t)} \tag{13.17}$$

$w(t)$ is the modulus of the complex envelope of the $x(t)$ signal independent of the carrier frequency f_0.

Thus, the task of signal synthesis has been reduced to the task of synthesizing its envelope as a function of time, such that, with a good speed resolution, the required range resolution is achieved.

As it was shown above, the achievement of the resolving power of 50m in range will be sufficient for solving the combat mission of the active homing head. Due to the property P_1 of the signal space S_x, we can write

$$M\{w, x\} = \int_0^T (w(t) + x(t))^2 dt = const \tag{13.18}$$

The integral (13.18) characterizes the signal energy weighted by the function $w(t)$ over the time interval T. Normalizing the total signal energy $x(t)$ to unity:

$$M\{x, x\} = 1, \tag{13.19}$$

it can be shown that the limiting concentration of energy in time and frequency is simultaneously achieved using a Gaussian pulse as a function of time. This is the solution of the Schrödinger wave equation for a harmonic oscillator with a parabolic potential well $H(f) = |(f^2)|$ for the boundary conditions $x(\pm\infty) = 0$. Here $H(f)$ is the weight function in the frequency domain. In this case this requirement is desirable but not required. Given (13.19), can be represented (13.18) in the following view:

$$M\{w, x\} = \int_0^T (w(t))^2 dt, \tag{13.20}$$

and the problem is reduced to minimizing the functional (13.20), since it is necessary to ensure a minimum energy loss of the emitted signal during reception due to its processing. The minimization constraints naturally follow from the statement of the problem:

$$\int_0^F |H(f) - H_0(f)|^2 df \leq \epsilon, \tag{13.21}$$

where: $H_0(f)$ is the amplitude-frequency characteristic of the filter in active homing head implemented in the equipment; $H(f)$ is the amplitude-frequency response of the wide filter of the active homing head after envelope synthesis.

In terms of the uncertainty function, for a periodic signal with a repetition period $T_r \in L^2(T)$, the expression (13.21) is rewritten as

$$\int_0^{T_r} |\dot{P}(f, \tau) - \dot{P}_0(f, \tau)|^2 d\tau \leq \delta, \tag{13.22}$$

and the problem is reduced to minimization in the time domain of the functional:

$$M\{w(t)\} = \alpha \int_0^{T_r} |w(t)|^2 dt + \int_0^{T_k} |\dot{\rho}(f,\tau) - \dot{\rho}_0(f,\tau)|^2 d\tau, \tag{13.23}$$

here, α is the Lagrange multiplier, T_r is the signal repetition period, $T_k = \sup\{\tau\}$, and $f = const$.

Note that in this setting, no conditions are imposed on the smoothness and integrability of the function $w(t)$.

Representing $\dot{\rho}(f,\tau)$ by a series expansion in some system basis functions $\{\varphi_i(\tau)\}$ in the form (13.5):

$$\dot{\rho}(f,\tau) = \sum_{i=1}^{N} w_i \varphi_i(\tau) \tag{13.24}$$

and passing to the continuous variable t, the functional (13.23) is rewritten as

$$\begin{aligned} M\{w(t)\} = \alpha \int_0^{T_r} w(t)w^*(t)dt + \int_0^{T_k}\left[\int_0^{T_r} w(t)\varphi(t,\tau)dt - \dot{\rho}_0(t,\tau)\right]\times \\ \times\left[\int_0^{T_r} w^*(t)\varphi^*(t,\tau)dt - \dot{\rho}_0(f,\tau)\right]d\tau \end{aligned} \tag{13.25}$$

The extremum of the functional (13.25) is attained when the variation of the functional $\delta M\{w(t)\} = 0$ vanishes with a nonzero variation of the function $w(t)$; that is:

$$\delta M\{w(t)\} = 0, \qquad \forall w(t) \neq 0. \tag{13.26}$$

as a result, we can write:

$$\begin{aligned} \delta M\{w(t)\} &= \alpha \int_0^{T_r} (w(t) + \delta w(t))(w^*(t) + \delta w^*(t))dt + \\ &+ \int_0^{T_k}\left[\int_0^{T_r} (w(t) + \delta w(t))\varphi(t,\tau)dt - \dot{\rho}_0(t,\tau)\right]\times \\ &\times\left[\int_0^{T_r} (w^*(t) + \delta^* w(t))\varphi^*(t,\tau)dt - \dot{\rho}_0(t,\tau)\right]d\tau - \alpha\int_0^{T_r} w(t)w^*(t)dt + \\ &+ \int_0^{T_k}\left[\int_0^{T_r} w(t)\varphi(t,\tau)dt - \dot{\rho}_0(f,\tau)\right]\times\left[\int_0^{T_r} w^*(t)\varphi^*(t,\tau)dt - \dot{\rho}_0^*(f,\tau)\right]d\tau \end{aligned} \tag{13.27}$$

As a result of reductions and after rearrangement, the main part of the variation of the functional $\delta M\{w(t)\}$ takes the form:

$$\delta M\{w(t)\} = \int_0^{T_r} \delta w^*(t) \times \left[\alpha w(t) + \int_0^{T_r} w(t)dt \int_0^{T_k} \varphi(t,\tau)\varphi^*(t,\tau)d\tau - \right.$$
$$\left. - \int_0^{T_k} \dot{\rho}(t,\tau)\varphi^*(t,\tau)d\tau \right] dt + \int_0^{T_r} \delta w(t) \times \left[\alpha w^*(t) + \right. \tag{13.28}$$
$$\left. + \int_0^{T_r} w^*(t)dt \int_0^{T_k} \varphi(t,\tau)\varphi^*(t,\tau)d\tau - \int_0^{T_k} \dot{\rho}_0(f,\tau)\varphi(t,\tau)d\tau \right] dt$$

From the need to fulfill the condition (13.26) it follows:

$$\alpha w(t) + \int_0^{T_r} w(t)dt \int_0^{T_k} \varphi(t,\tau)\varphi^*(t,\tau)d\tau - \int_0^{T_k} \dot{\rho}_0(f,\tau)\varphi^*(t,\tau)d\tau = 0 \tag{13.29}$$

After changing variables in (13.29), an inhomogeneous Fredholm integral equation of the second kind is obtained:

$$\alpha w(t) + \int_{-1/2}^{+1/2} w(\xi)d\xi \mathcal{K}(\xi,t) = F(t), \tag{13.30}$$

where $\mathcal{K}(\xi,t) = \int_0^{T_k} \varphi(\xi,\tau)\varphi^*(t,\tau)d\tau$ is the kernel of the integral equation; $F(t) = \int_0^{T_k} \dot{\rho}_0(f,\tau)\varphi^*(t,\tau)d\tau$ is the right side of the integral equation.

Using the discrete point matching method, after multiplying the left and right sides (13.30) by a sequence of Dirac δ-functions $\delta(t - p\Delta t)$, $\forall\, p = 1, 2, 3, \ldots, N$, it becomes possible to switch to discrete representation:

$$\alpha w_n + \sum_{m=1}^{N} w_m K_{mn} = F_n, \qquad \forall n, m = 1, 2, 3, \ldots, N \tag{13.31}$$

where:

$$K_{mn} = \int_0^{T_k} \varphi_m(t,\tau)\varphi_n(t,\tau)d\tau \tag{13.32}$$

$$F_n = \int_0^{T_k} \dot{\rho}_0(f,\tau)\varphi_n^*(t,\tau)d\tau \tag{13.33}$$

In matrix notation, the system (13.31) is represented as:

$$\alpha[w] + [w][K] = [F] \tag{13.34}$$

The system of linear equations (13.34) can be solved by the iteration method, and in several cycles of the iterative process (if it converges), some approximate value of the Lagrange coefficient can be obtained, the substitution of which in (13.31) will allow us to find the function $w(t)$ that satisfies the given constraints. In order to avoid possible difficulties in the solution, it is proposed to move $[F]$ to the left side and solve the homogeneous equation. The extended matrix obtained in this case is proposed to be completed with one more row so that a matrix of dimension $(N + 1) \times (N + 1)$ is obtained. Based on the global optimality of the Gaussian momentum, it is reasonable to require the fulfillment of the boundary conditions:

$$\begin{cases} w(0) &= 0, \\ w(T_r) &= 0. \end{cases} \tag{13.35}$$

Thus, the signal envelope should be equal zero at the ends of the $[0, T_r]$ time interval. In accordance with the expansion (13.31), (13.34) at the boundary points the equality must be true:

$$w_1 + w_2 + w_3 + \cdots + w_N = 0 \tag{13.36}$$

This condition can be met using an interpolation polynomial. Application of the Lagrange polynomial gives:

$$w(t) - \sum_{i=1}^{N} w_i P_{i-1}\left(\frac{2t}{T_r}\right) \tag{13.37}$$

As a result, the matrix homogeneous equation is represented as:

$$\begin{bmatrix} w_1 \\ w_2 \\ \vdots \\ w_N \\ B \end{bmatrix} \times \begin{bmatrix} (K_{11}+\alpha) & K_{12} & \ldots & K_{1N} & F_1 \\ K_{21} & (K_{22}+\alpha) & \ldots & K_{2N} & F_2 \\ \vdots & \vdots & \ldots & \vdots & \vdots \\ K_{N1} & K_{N2} & \ldots & (K_{NN}+\alpha) & F_N \\ P_0 & P_1 & \ldots & P_{N-1} & \alpha \end{bmatrix} = \begin{bmatrix} 0 \\ 0 \\ \vdots \\ 0 \\ 0 \end{bmatrix} \tag{13.38}$$

Making the substitution:

$$\lambda = -\alpha \tag{13.39}$$

for the system of equations (13.38) we can write:

$$[w] \times [K - \alpha I] = [0] \tag{13.40}$$

Thus, the Lagrange coefficients α are determined automatically in the process of solving the complete eigenvalue problem.

Given the requirements for $\dot{\rho}(f, \tau)$, for each eigenvalue $\lambda = -\alpha$, there exists at least one vector solution $[w]$ of the equation (13.40) which is defined as follows:

$$[K] \times [w_n] = \alpha_n [w_n] \tag{13.41}$$

The vector $[w_n]$ is the nth eigenvector of the matrix $[K]$. Moreover, each nth vector $[w_n]$ is optimal in the sense (13.22) when replacing inequality with equality. Consequently, the optimal signal is realized by a variety of envelopes.

Based on the requirements of physical realizability (properties P_1 and P_3 signal space $S = \{x, P\}$ and equality (13.19)), only eigenvectors corresponding to negative eigenvalues are taken into account.

Representing the basis functions in (13.24) in the form of uncertainty functions of a certain elementary time interval, the weighted set of which will constitute the synthesized signal, we can write:

$$\varphi(t, \tau) = \dot{\gamma}(t)\dot{\gamma}^*(t+\tau)\exp^{-j2\pi f(t+\tau)}, \tag{13.42}$$

where $\dot{\gamma}(t)$ is the complex envelope of elementary time interval.

In accordance with (13.42), the kernel of the integral equation (13.30) $\mathcal{K}(\xi, t)$ and its right-hand side $F(t)$ take the form:

$$\mathcal{K}(\xi, t) = \exp^{-j2\pi f(\xi - t)} \dot{\gamma}(\xi)\dot{\gamma}^*(t) \int_0^{T_k} \dot{\gamma}(\tau + t)\dot{\gamma}^*(\tau + \xi) d\tau \tag{13.43}$$

$$\begin{aligned} F(t) &= \int_0^{T_k} \dot{\rho}_0(f, \tau)\dot{\gamma}^*(t)\dot{\gamma}(t+\tau)\exp^{j2\pi f(t+\tau)} d\tau = \\ &= \dot{\gamma}^*(t)\exp^{j2\pi f t} \int_0^{T_k} \dot{\rho}_0(f, \tau)\dot{\gamma}^*(t+\tau)\exp^{j2\pi f\tau} d\tau \end{aligned} \tag{13.44}$$

It is quite clear that the approximation $[w_n]$ will be the best in the sense of the vanishing of the Euclidean norm:

$$\Big\| [w_n] \times [K] - [F] \Big\| = 0, \tag{13.45}$$

if $[K] = I$. This is possible in case of uncorrelatedness $\varphi(\xi, \tau)$ and $\varphi(t, \tau)$, $\forall\, \xi \neq t$; therefore, the time-shifted signal values must be orthogonal:

$$\int_0^{T_r} \dot{\gamma}(t)\dot{\gamma}(t+\tau)dt = \delta(t), \quad \forall t : \tau = 0, \tag{13.46}$$

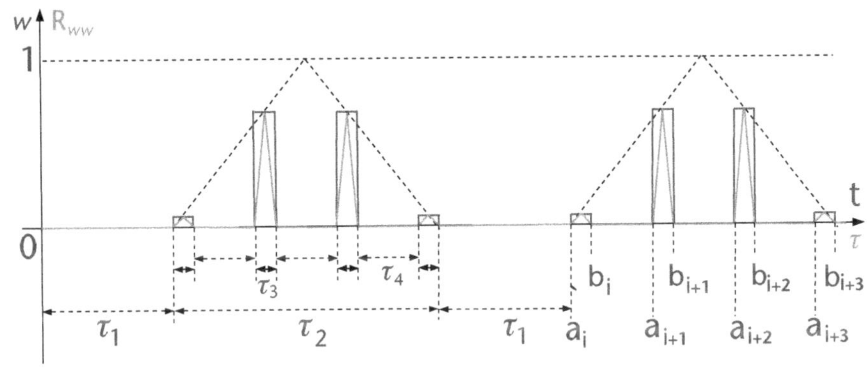

Figure 13.12 □ is the envelope $w(t)$ of the synthesized signal and its △ - correlation function $R_{ww}(\tau)$.

and this, according to the Fourier theorem, corresponds to the limit relation:

$$\lim_{W\to\infty}\int_{-W}^{+W}\exp^{j2\pi ft} = \lim_{W\to\infty}\frac{\sin(2\pi Wt)}{\pi t} = \delta(t), \tag{13.47}$$

where $2W$ is the signal spectrum width.

Thus, the structure of a signal that is optimal in the sense of (13.22) is a burst of pulses with intrapulse modulation with a wideband signal.

13.3.3 Solution for the Emitted Signal, Taking into Account the Limitations due to the Hardware Implementation of the Anti-Aircraft Missile System

Solution (13.38) for the radiated signal is subject to the additional constraints associated with the use of a single antenna for transmit and receive. Based on the interval of unambiguous determination of the range, for example, of ≈60 km and taking into account the fact that with a radius of dispersion of warhead fragments $R_0 = 50$m the probability of destroying a target with one missile $P_{int} \approx 0.95$, it is reasonable present the requirements for the range resolution $\Delta_r = 48$m. Therefore, in the case of an active radiolocation, the correlation time interval τ_0 or the duration of one discrete τ_0 will be:

$$\tau_0 = \frac{2\Delta_r}{3\cdot 10^8} = 0.32\ \mu s \tag{13.48}$$

Therefore, for the interval of unambiguous determination of the range 60 km, the time interval when the active seeker works to receive signals reflected from the target can be represented as:

$$\tau_4 = \frac{2\cdot 60000}{3\cdot 10^8} = 400\ \mu s \tag{13.49}$$

Therefore, the durations of the time intervals of the synthesized signal can be represented in the form:

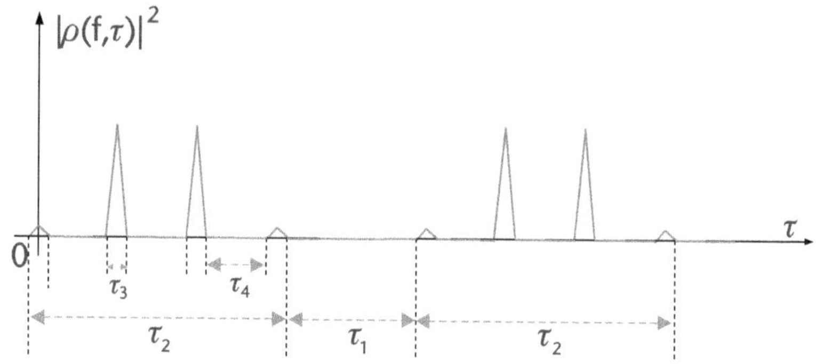

Figure 13.13 The uncertainty function $|\rho(f, \tau)|^2$.

$\tau_1 = 0.73$ ms;

$\tau_2 = 1.456$ ms;

$\tau_3 = 64\ \mu$s;

$\tau_4 = 400\ \mu$s.

The uncertainty function $|\rho(f, \tau)|^2$ can be represented, respectively, by a graph of the form, representing the signal envelope $w(t)$ in the form

$$w(t) = \sum_{i=1}^{M} a_i \chi_i(t), \tag{13.50}$$

where

$$\chi_i(t) = \begin{cases} 1, & \forall\, t \in [a_i, b_i] \\ 0, & \forall\, t \in [b_i, a_{i+1}] \end{cases} \quad \text{– the characteristic function of the } i\text{th radiation region} \tag{13.51}$$

and a_i is a weighting factor; therefore, the emitted signal can be written as follows:

$$x(t) = \varphi(t) \sum_{i=1}^{M} a_i \chi_i(t) = \sum_{i=1}^{M} \varphi_i(t) a_i \chi_i(t), \tag{13.52}$$

where

$$\int_0^{T_r} \varphi_i(t)\varphi_j(t)dt = \delta_{ij} = \begin{cases} 1, & i = j \\ 0, & i \neq j, \end{cases} \qquad \forall\, i, j = \overline{1, M} \tag{13.53}$$

Expressions (13.52), (13.53) determine the optimal radiated signal $x(t)$, and accordingly, the optimal algorithm for its processing. The signal $x(t)$ is slices from the pseudo-random sequence $\varphi(t)$ on the orthogonality interval $T_r = \tau_2$. Substitution of (13.52) into (13.13) gives an optimal signal processing algorithm in the frequency domain:

$$Y(f_k) = |H(f)| \sum_{n=0}^{N-1} \sum_{i=1}^{M} \varphi_i(n\Delta t + \tau) a_i \chi_i n(\Delta t + \tau) \exp^{[-j\frac{2\pi k(n\Delta t+\tau)}{N}]}, \quad (13.54)$$

where τ is the delay time of the reflected signal relative to the emitted one, which characterizes the range to the target.

The definition of τ can be carried out using correlation relations:

$$\tau = \max R_{xy}(\tau) = \max \int_{a_i}^{b_i} x(t)y(t+\tau)d\tau, \quad (13.55)$$

here $x(t)$ is the emitted signal (reference signal), and $y(t+\tau)$ is the received signal.

Thus, in the expression (13.55) $y(t+\tau)$ is used and the filtering in the frequency domain will be performed after the time $T_r = \tau_2$. This is enough to independently measure the range M times. However, applying correlation processing runs into the problem of synchronizing $x(t)$ and $y(t+\tau)$, which is dual to (13.55). This difficulty can be avoided by organizing P parallel reception channels, the number of which increases sharply when trying to increase the measurement accuracy. In this situation, filtering of the signal in the time domain is proposed based on the use of an adaptive transversal filter, which is filter-based on a shift register or rake delay line with a delay time between taps equal to the sampling interval. The composition of such a filter usually includes N delay elements for the time τ_0, N multipliers by weight coefficients and an adder [5–7]. This solution makes it possible to fulfill the requirement of physical realizability, since the transversal filter is casual and has a finite impulse response, which automatically removes the issue of its stability. In addition, the computation time of the covariance function is minimized due to its hardware implementation and time synchronization is not needed, as in the correlator.

From the variety of noise-like signals: frequency-modulated, phase-shift keyed, multifrequency, discrete frequency, discrete composite frequency signals, the class of phase-shift keyed signals is selected [6–10]. In this case, the duration of one sample of the phase-shift keyed signal is equal to the correlation time interval $\tau_0 = 0.32\ \mu s$. Therefore, for the interval of unambiguous determination of the range 60 km, the base of the noise-like signal is:

$$B = \frac{\tau_3}{\tau_0} = \frac{64\ \mu s}{0.32\ \mu s} = 200 \quad (13.56)$$

Using a binary phase-shift keyed signal with two values of the initial phases: $\{0, \pi\}$, the complex envelope of the signal in this case is a sequence of video pulses of different polarity:

$$\varphi(t) = \sum_{n-1}^{N} b_n U_0[t - (n-1)\tau_0], \quad (13.57)$$

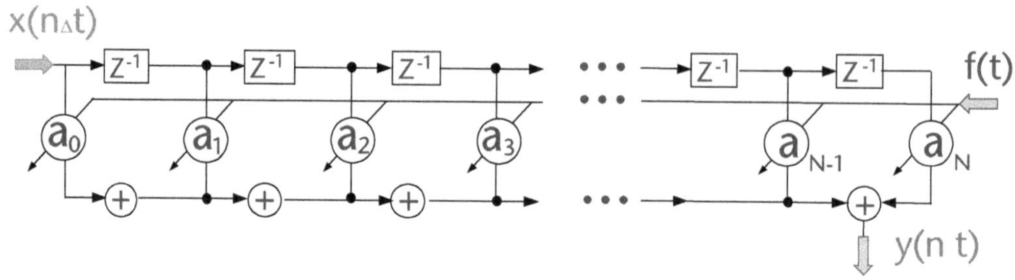

Figure 13.14 Adaptive transversal filter.

where $b_n = \{+1, -1\}$ is the amplitude of the nth pulse; $U_0(t) = \begin{cases} 1, & \forall\, t \in [n\tau_0 - (n-1)\tau_0] \\ 0, & \forall\, t \notin [n\tau_0 - (n-1)\tau_0] \end{cases}$ - piecewise constant basis function.

The correlation properties of the signal are determined by the code sequence $\{b_n\}$. A segment system of signals is defined according to (13.52).

In this sense, there are two possible implementation options:

1. Segments of the pseudo-random sequence $x_i(t)$ are transmitted into the space maximum length, which is formed continuously during the time interval $T_r = \tau_2$ and is strobed by the time function $w(t)$.
2. During the action of each gating pulse $w(t)$, one sequence $x_i(t)$ of length B is transmitted into space from the total number L of the alphabet of sequences $\{x_i(t)\},\ \forall\, i = \overline{1, L}$.

13.3.4 Proposals for the Hardware Implementation of a Synthesized Solution in a Real Anti-Aircraft Missile System

In order to solve the problem of increasing the noise immunity of the onboard radar when it is operating in active mode for low-altitude targets and for a group of jammers, the second implementation option and the Walsh function system were selected, which was chosen as the basic system of functions.

Walsh functions are a family of functions that form an orthogonal system, taking values only $+1$ and -1 on the entire domain of the function. Walsh functions are more often defined as discrete sequences of 2^n elements. Thus, Walsh functions allow us to consider sequences of 2^n elements. The group of $\{2^n\}$ Walsh functions forms the Hadamard matrix.

The alphabet of sequences $\{x_i(t)\},\ \forall\, i = \overline{1, L}$ is formed by the derivative of the system of Walsh functions by a system of sequences, the volume and composition of which must be additionally determined by a special protocol.

It can be shown that the system of Walsh functions is orthogonal, normalized, and complete:

$$\int_{-0.5}^{+0.5} wal(i,\xi)wal(j,\xi)d\xi = \begin{cases} 1, & \forall\, i = j, \\ 0, & \forall\, i \neq j. \end{cases} \tag{13.58}$$

The signal at the output of the adaptive transversal filter [5–7] (Figure 13.14) is presented in the form

$$y(t) = \sum_{k=0}^{N} x(k)h(t-k), \tag{13.59}$$

where $h(t-k)$ is the impulse response of the filter; $f(t)$ this signal sets the value of the scaling factor $\{a_i\}$ adaptive transversal filter; It is necessary to define the impulse response of the filter when processing the Walsh functions; $[x_k] = [x_{k1}, x_{k2}, \ldots, x_{kN}]^T$ is a sequence of counts of the input signal; $[y] = [y_1, y_2, \ldots, y_N]^T$ is a sequence of signal counts at the output of the adaptive transversal filter; $[a] = [a_1, a_2, \ldots, a_N]^T$ is a vector of unknown weight coefficients.

It is known that the Wiener solution for $[a]$ is written as [5]

$$[a] = R^{-1}[y], \tag{13.60}$$

where $R^{-1} = [x_k] \times [x_k]^T$.

In the case under consideration, $R^{-1} = I$, and therefore, the values of the filter weight coefficients are equal to the corresponding counts of the Walsh functions. An adaptive transversal filter is implemented on a tapped delay line, which includes operational amplifiers with gains equal to unity. At the output of each operational amplifier there is a controllable switch that allows connection to the direct or inverse output of the amplifier. Due to the fact that the system of Walsh functions is obtained from the characteristic group, a binary group, which is a topological group obtained from a sequence of binary representations of real numbers, the control bus $f(t)$ of the adaptive transversal filter through the decoder can be directly connected to the address bus and data bus of the onboard computer and organize software control [10].

Integration of the transversal filter output signal from 0 to $T = B\tau_0$ gives the value of the correlation integral:

$$Z = \int_0^{B\tau_0} wal(i,t)wal(j,t)dt = \begin{cases} B, \\ 0. \end{cases} \tag{13.61}$$

where $Z = B$ when there is matched filtering; $Z = 0$ otherwise.

Below are the signals for the eight-block Walsh functions from the output of the filter, which has the following impulse response:

$$h - [\,-1, -1, -1, -1, +1, +1, +1, +1\,] \tag{13.62}$$

In the detection mode, in the interval of unambiguous determination of the range 60 km, the correlation integral over the entire burst of pulses $T_p = 1.456$ ms is represented as

$$Z = \sum_{i=1}^{4} \int_{a_i}^{b_i} x(t)h(t-\tau)d\tau \tag{13.63}$$

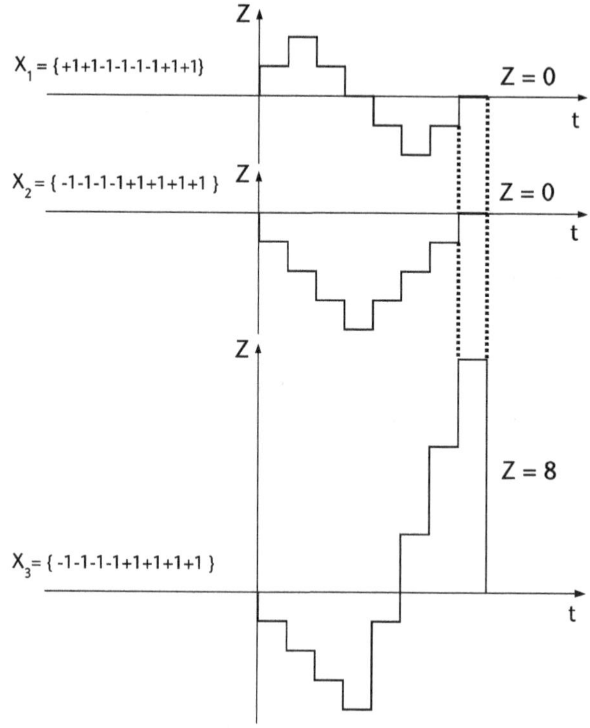

Figure 13.15 Signal processing illustration.

During the burst $T_p = 1.456$ ms, the potential capabilities of the active homing head to detect signals are $\hat{Z}^* = 10\log(4 \times 200) = 29.030$ dB.

The correlation integral $\hat{Z}^*$ in this case may look like this (Figure 13.16).

When tracking a target at a range of 30 km and less, it is advisable to switch to another radiation mode, which is shown in Figure 13.17. Here $\tau_{21} = 10.24\ \mu s$. At this time interval, a package is emitted, consisting of 32 samples with duration $\tau_0 = 0.32\ \mu s$ each and $\tau_{22} = 198\ \mu s$ is the interval of analysis of the reflected signal at a distance $R_{30} = \tau_2 \cdot 3 \cdot 10^8/2 = 29.7$ km. The signal emission period in this case is estimated as $T = \tau_1 + \tau_2 = 10.24\ \mu s + 198\ \mu s = 208.24\ \mu s$. In this radiation mode, the potential capabilities of the active seeker can be estimated by the expression: $\hat{Z}^* = 10\log(32) = 15.05$ dB. Determination of the range to the target (s) is carried out by the range determination system, which measures the time interval from the end of the next blanking pulse of the receiver to the moment of exceeding the signal from the output of the transversal filter $\hat{Z}_{lim} = 10\log(\frac{N_d}{2})$, which is determined by the trailing edge of the pulse formed by the transversal filter. The principle of determining the range to the target D_t is shown in Figure 13.18.

Here $N_{CT} = 49$ is the number of pulses with a duration of $\tau_0 = 0.32 \cdot 10^{-6}$s, counted by the counter CT, $\Delta r = \tau_0 \times C/2 = 0.32 \cdot 10^{-6}\,[s] \times 3 \cdot 10^8/2\,[\text{m/s}] = 48$m and therefore the range to the target is calculated by the formula:

$$D_t = N_{CN} \times \Delta r = 49 \times 48\text{m} = 2352\text{m} \tag{13.64}$$

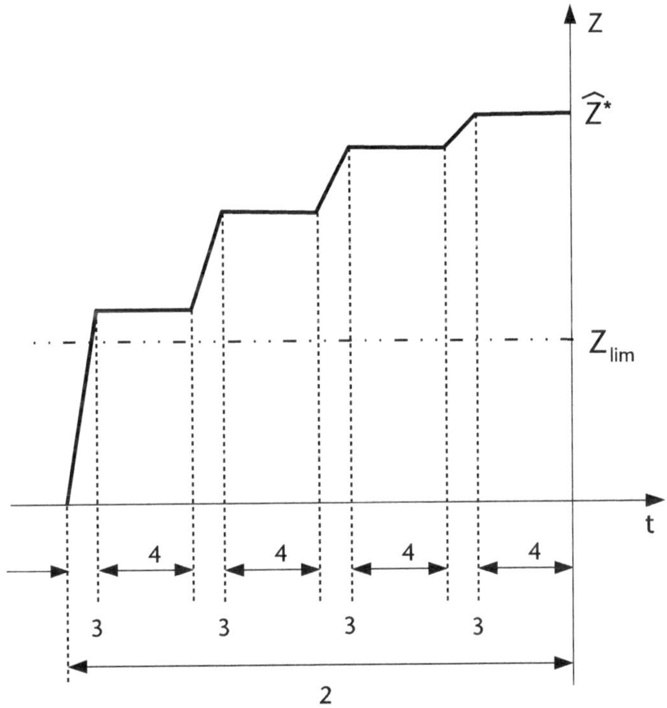

Figure 13.16 Correlation integral. $\hat{Z}^* = 29.03$ dB.

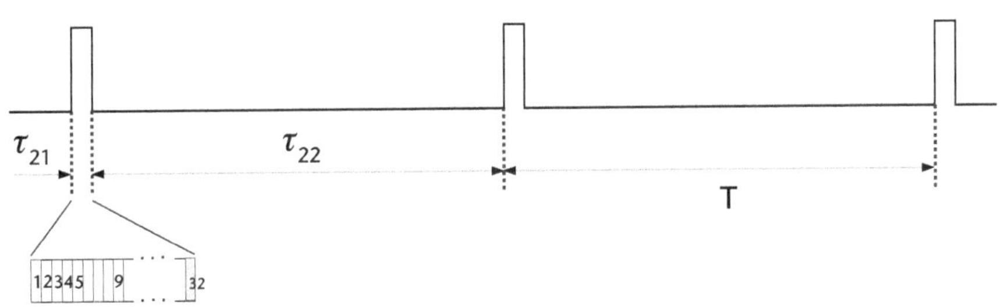

Figure 13.17 The structure of the signal emitted by the active homing head when $R_t \leq 30000$m.

N_d is the number of discretes in the transmitting package. For $N_d = 32$, the threshold value of the output of the transverse filter $\hat{Z}_{lim}$ will look like:

$$\hat{Z}_{lim} = 10 \log\left(\frac{N_d}{2}\right) = 10 \log(16) = 12.0412 \text{ dB} \quad (13.65)$$

In the case of dispersed propagation of the reflected signal (group target, reflections from the underlying surface), the range tracking strobe should capture the closest mark of the cross-correlation function.

If the coefficient of kurtosis of the cross-correlation function is greater than one, then it is necessary to make a "dense group" decision and the transition must be made to the periodic radiation of differentiated functions of the alphabet of signals.

Measuring the range to the target is carried out using the pulses of the clock generator of the onboard computer complex with a frequency of $f_G \geq 1/\tau_0 =$ 3.125 MHz.

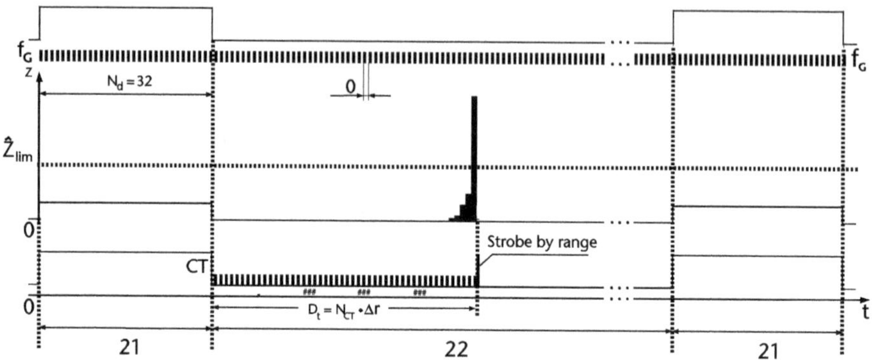

Figure 13.18 The principle of determining the range to the target by means of an active seeker.

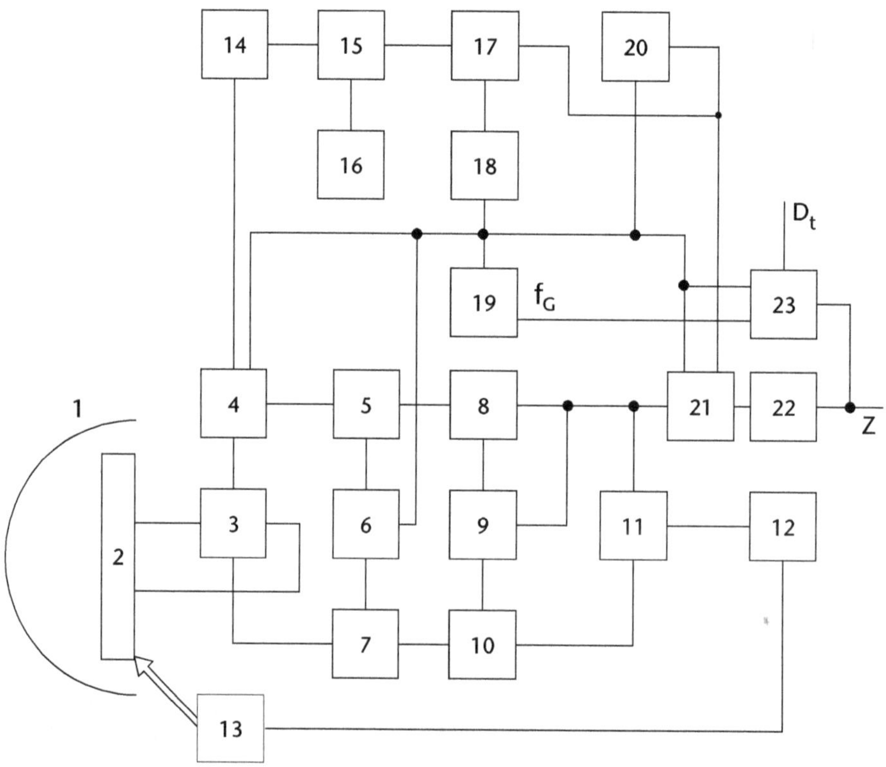

Figure 13.19 Block diagram of an active monopulse sum-and-difference homing head.

As a result, the block diagram of the active sum-difference monopulse homing head can be represented as in Figure 13.19.

The designations in Figure 13.19 are:

1: Radome;

2: Monopulse antenna;

3: Sum-and-difference device (magic-T junction or hybrid ring /rat-race/junction);

4: Receive-transmit switch;

5: Mixer of the sum channel;

6: Heterodyne;
7: Mixer of the difference channel;
8, 10: Intermediate-frequency amplifiers of the sum-and-difference channels;
9: Automatic gain control system of the sum channel;
11: Phase detector (phase-sensitive/dot-product/detector);
12: Error-signal amplifier;
13: Monopulse antenna drive (mechanism);
14: Power amplifier;
15: Modulator;
16: Carrier frequency generator;
17: Phase modulator;
18: Generator of rectangular pulses;
19: Synchronizer;
20: Generator of Walsh functions (code sequences 1 and 0);
21: Matched filter;
22: Threshold device.

Note that the realization of units 1–14, 21, and 22 is described in [3], and 15–22 in [11].

Consider the work of an active seeker [12].

The seeker active mode operation, when a microwave signal is emitted into space through monopulse antenna 2, begins after the arrival of a permitting pulse from synchronizer 19 and the same pulse is prohibitive for the operation of the seeker's receivers (τ_{21} in Figure 13.17). For the duration of this pulse, the transmit-receive switch is switched to the microwave signal transmission position. The sync pulse of the beginning of the active mode starts the operation of the code sequence generator 20 that has RAM, and that forms the Walsh function of a given order in the form of a sequence 1 and 0, which is then fed to phase modulator 17, the second input of which receives rectangular pulses generated by generator rectangular pulses 18. The phase of these pulses changes by 180° when going from 1 to 0 and back in accordance with the digital code given by the Walsh function. Further, the Walsh function in the form of a phase-modulated signal is fed to the first input of modulator 15, which modulates the carrier signal coming from the output of carrier frequency generator 16 to the second input of modulator 15. As a result, at the output of the modulator 15, a high-frequency signal of the carrier wave is obtained, modulated in phase by the digital code of the Walsh function, specified by the generator of Walsh functions 20. At the same time, at the control input of the transmit-receive switch present an enable pulse received from synchronizer 19, and therefore the transmit-receive switch is located in the microwave signal transmission position. In this case, a microwave signal is emitted into space, modulated by the Walsh function. At the same time, the Walsh function is fed from Walsh function generator 20 to matched filter 21, in which the impulse response corresponding to the digital sequence of the received Walsh function is formed. In this case, the reference function of the filter (impulse response) changes every period $T = \tau_3 + \tau_4$ or $T = \tau_{21} + \tau_{22}$ in accordance with the Walsh functions specified by Walsh function generator 20. Depending on the current radiation mode, during the action of the receiver blanking pulse on the front of synchronizer pulse 19 τ_3 or τ_{21}, the impulse response formed in matched filter 21 is valid until the next front of pulse τ_3 or τ_{21} (Figures 13.17, and 13.12 respectively). When a pulse of the active mode end appears at the output of synchronizer 19, depending on the radiation mode of the homing head, the blanking pulse of τ_3 or τ_{21} receiver ends and the transmit-receive switch switches to the receiving microwave signal position. Depending on the operating mode of radiation, during the pulse of synchronizer 19 τ_4 or τ_{22}, which allows the reception of microwave signals, the impulse response is used, formed in matched filter 21 and recorded earlier in the RAM of the active homing head throughout the entire time of sync pulse τ_4 or τ_{22}, allowing the reception of microwave signals (Figures 13.12, 13.13, and 13.17). The signals from the outputs of monopulse antenna 2 go to the inputs of magic-T junction 3, providing at its outputs the shaping of microwave signals of the sum-and-difference channels. The microwave signal of the sum channel through transmit-receive switch 4 enters the input of mixer 5 and the microwave signal of the difference channel goes to mixer 7, respectively, where they are transformed with the help of heterodyne 6 into the signals of intermediate frequency, which then are amplified in intermediate-frequency amplifiers 8 and 10 up to the required value and come to the inputs of phase detector 11. The difference signal amplitude determines the value of the angular error signal at the output of phase detector 11, and the phase difference at the input of phase detector 11 between the signals of the sum-and-difference channels determines the sign of the angular error signal $S(\varphi, t)$ at the output 11 where φ is the angular error (displacement angle between a true direction on target and radar boresight of the monopulse homing head). Automatic gain control system 9 excludes the dependence of the angular error signal amplitude at the output of phase detector 11 on the level of the received signals by the connection of the input of automatic gain control system 9 to the output of intermediate-frequency amplifier 8 of the sum channel, in this case the signal at the output of automatic gain control system 9 makes a simultaneous adjustment of the amplification coefficients of intermediate-frequency amplifiers 8 and 10 providing the signal normalization of the difference channel with the help of the sum one. The angular coordinate error signal can be represented as

$$\begin{aligned} S(\varepsilon, t) &= \mu_\varepsilon \cdot \varepsilon(t) \\ S(\varphi, t) &= \mu_\varphi \cdot \varphi(t), \end{aligned} \tag{13.66}$$

where μ_ε and μ_φ the slope of the direction finding characteristic of the active homing head in mutually orthogonal direction finding planes $\{\varepsilon\}$ and $\{\varphi\}$ with realized errors of angular automatic tracking. The error signal $S(\varepsilon, t)$ or $S(\varphi, t)$ is fed through error signal amplifier 12 to the drive of monopulse antenna 1 (mechanism), which moves monopulse antenna 1 in space to make the angular misalignment zero $\varepsilon(t) = 0$ and $\varphi(t) = 0$. The signal from the output of the intermediate frequency amplifier of sum channel 8 is fed to the input of matched filter 21 (Figure 13.14), where the operation of calculating the correlation integral takes place (Figure 13.15), during which it is determined which signal was received. If the signal emitted by the active seeker during the previous τ_3 or τ_{21} pulse was received, then correlation integral $\hat{Z} > 0$ and in the case of $\hat{Z} > \hat{Z}_{lim}$ (Figure 13.16), the system will be able to determine the range to the target (Figure 13.18). Otherwise, the signals received by the active seeker are decorrelated. In this case, the correlation integral is $\hat{Z} = 0$. If the signal-to-noise ratio in the sum channel is exceeded by more than 10 dB, the seeker can switch to the noise interference mode and will track the interference source only in angular coordinates.

It is necessary to list the new qualities that the homing head will be able to achieve if the presented proposals are implemented:

1. When working on low-altitude targets, the seeker becomes invariant to signals reflected from the Earth's (water's) surface. This is based on the fact that the signal reflected from the target is folded in a filter, and the first-order specular signal is decorrelated with respect to the expected signal by a surface area that is significant in the reflection of radio waves.
2. When working with a source of interference such as antipode sinusoidal signals will not be processed by the homing head, since the signals of the active seeker relayed by the jammer will already arrive in the next processing period and will be decorrelated, since the reference filter function (impulse response) changes every period. If the re-emitted signal falls within the same processing period, then it is also decorrelated, since it will not be the first in time.
3. High-potential signals, unmodulated in phase in accordance with the expected sequence code, are not perceived by the seeker.
4. Aircraft in a group can be resolved in range if the distance between them is $\geq$50m.
5. The potential of an active seeker to detect reflected signals has been increased by at least 2.99 times compared to the target detection range by means of technical solutions currently existing in active homing heads.
6. Using as basis Walsh functions, ordered by Hadamard, allows filtering in the frequency domain using the $N = 2^n$ point fast Walsh-Hadamard transform using the $Nlog_2N$ addition and subtraction operations. This will reduce the computation time of the amplitude spectrum $X(f)$ of the input sequence $x(n\Delta t)$ by about 10 times.

Thus:

1. The synthesized signal made it possible to increase the noise immunity of the active seeker due to the organization of its optimal processing in the receiver.
2. The homing head became invariant to the effects of signals reflected from the Earth's (water's) surface.
3. The gain in the signal-to-interference ratio is 19.03 dB more in relation to the standard solutions existing today in active homing heads.
4. The change from period to period of the code sequence of the emitted function limits the enemy's ability to create an aiming broadband interference for the active homing head, repeating the fine structure of the emitted signal.

5. Due to the high-range resolution, it is possible to defeat low-altitude targets flying at an altitude of $\geq$ 30m, with a probability of at least 0.84.

References

[1] Lewis, B. L., and F. F. Kretschmer, Jr., *Digital Sidelobe Canceller*, U.S. Patent No. 4,086,592, April 25, 1978, https://patft.uspto.gov/netacgi/nph-Parser?Sect1=PTO2&Sect2=HITOFF&p=1&u=%2Fnetahtml%2FPTO%2Fsearch-bool.html&r=25&f=G&l=50&co1=AND&d=PTXT&s1=4086592&OS= 4086592&RS=4086592.

[2] Investigation of the Auto-Canceller of Active Noise Interference, Methodical Instructions for Laboratory Work on the Discipline “Radar Systems” for Students Specializing in “Radio Engineering” and “Radio Engineering Systems,” Compiled by V. Melsitov, and G. F. Plugatar, Minsk: BSUIR, 2003.

[3] Leonov, A. I., and K. I. Fomichev, *Monopulse Radar*, Norwood, MA: Artech House, 1986.

[4] Technical Tables-Non-Commercial Electronic Directory, TechTab.ru, 2021, https://tehtab.ru/Guide/GuidePhysics/Sound/SoundSpeedAirHeight/.

[5] Dudgeon, D. E., and R. M. Mersereau, *Multidimensional Digital Signal Processing*, Englewood Cliffs, NJ: Prentice-Hall, 1984.

[6] Budge, M. C. Jr., and S. R. German, Basic Radar Analysis, Second Edition, Norwood, MA: Artech House, 2020.

[7] Wicks, M. C., E. L. Mokole, S. D. Blunt, R. S. Schneible, and V. J. Amuso (eds.), *Principles of Waveform Diversity and Design*, Raleigh, NC: SciTech Publishing, 2010.

[8] Richards, M. A., J. A. Scheer, and W. A. Holm (eds.), *Principles of Modern Radar*, Raleigh, NC: SciTech Publishing, 2010.

[9] Kingsley, S., and S. Quegan, *Understanding Radar Systems*, Raleigh, NC: SciTech Publishing, 1999.

[10] Leonov, S. A., and A. I. Leonov, *Handbook Of Computer Simulation In Radio Engineering, Communications, and Radar*, Norwood, MA: Artech House, 2001.

[11] Varakin, L. E., *Communication Systems with Noise-Like Signals*, Moscow: Radio and Communication, 1985.

[12] Markin, E., *Method of Automatic Target Angle Tracking by Monopulse Homing Head When It Works in Active Mode for Low-Altitude Targets and for a Group of Jammers*, U.S. Patent and Trademark Office, Application Number 63222357, July 15, 2021, https://www.uspto.gov.

CHAPTER 14

Determining Performance Limitations of Existing Air Defense Systems for Intercepting Hypersonic Vehicles

14.1 Introduction

This chapter describes a methodology for determining performance limitations of existing air defense systems for intercepting hypersonic vehicles. The methodology used results of scatter pattern calculations for hypersonic unmanned aerial vehicle X-51. A numerical 3-D model of the UAV was developed by me on the basis of public sources. The surface of the UAV is represented by a set of elementary flat areas with linear dimensions much smaller than a wavelength of an air defense system, as shown in Figure 14.1. In accordance with the mathematics stated in this chapter, original software for calculated UAV scatter patterns was designed. In these calculations, a vector of the electromagnetic field reflected by the UAV is represented as a complex value in 3-D space. The reflected field is calculated taking into account the scattering effect on the edges of the wings, on the edges of the intake, and in the intake cavity [1]. Special software packages, such as WIPL-D, were not used.

14.2 Field Scatter Pattern Calculation Mathematics

In this case, a UAV is an object or a target for a radar. In order to calculate a target scatter diagram, it is necessary to obtain an expression for electromagnetic field intensity provided by a target surface element and specified in R^3 space, and then to integrate the expression on all irradiated surfaces of the target S_0. It is known that when an electromagnetic wave falls on an infinite perfectly conducting surface, a surface current density on it is described by the following expression [2,3]:

$$\vec{J}_s = 2\left[\vec{n}\ , \vec{H}_s\ \right], \tag{14.1}$$

where $\vec{n}$ is a unit surface normal vector, and $\vec{H}_s$ is a magnetic field vector of an incident wave at S_0 surface. Expression (14.1) is an accurate one under the above-mentioned conditions and an approximate one in all other cases. However, this expression can be used when a target radius of curvature and geometrical dimensions are greater than the wavelength λ. In this case, locally, the waves coming from

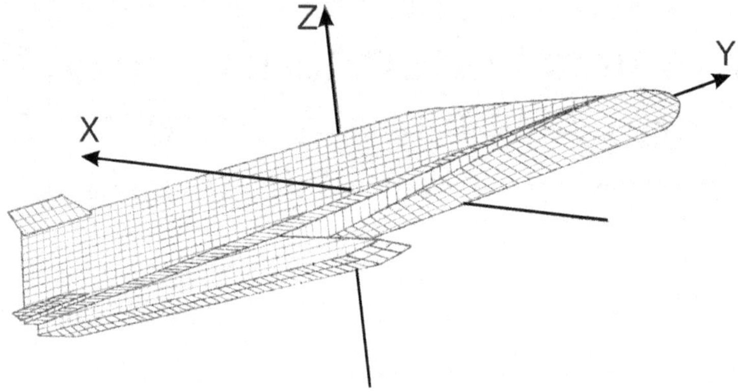

Figure 14.1 X-51 3-D-model.

the target illumination radar and reflected from the target surface S_0 can be considered as the plane ones (see Figure 14.2). Calculation of the far electromagnetic field is conducted by using Hertzian electric vector $\vec{\dot{Z}}_e$, which is determined in the point of sight P with the known volume current distribution V as follows:

$$\vec{\dot{Z}}_e(P) = \frac{1}{4\pi} \int_V \vec{\dot{J}}(q) \frac{\exp(-jkr)}{r} dV, \tag{14.2}$$

where $\vec{\dot{Z}}_e(P)$ is determined in a spherical coordinate system, and $\vec{\dot{J}}(q)$ is a current volume density vector.

$$r = R - R_q \cos \vartheta, \tag{14.3}$$

where ϑ is an angle between radius-vector $\vec{R}$ of a point of sight P and a radius-vector $\vec{R}_q$ of integration point q. Using a spherical coordination system XYZ with the origin in a current existence region, it is always possible to create a transformation T, which will enable obtaining coordinates in any other coordinate system [1]. Considering that an airborne or ground radar antenna is in the far region of the secondary scatter diagram, parallelism of vectors $\vec{R}$ and $\vec{r}$ can be assumed. Taking this into account, expression (14.2) takes on the following form:

$$\vec{\dot{Z}}_e = \frac{1}{4\pi} \times \frac{\exp(-jkR)}{R} \times \vec{\dot{F}}_e(\theta, \varphi), \tag{14.4}$$

where

$$\vec{\dot{F}}_e(\theta, \varphi) = \int_V \vec{\dot{J}}_e(q) \exp(jkR_q \cos \vartheta) dV \tag{14.5}$$

is a vector complex scatter (directional) pattern determined by electric currents. This can be written similarly for Hertzian magnetic vector $\vec{\dot{Z}}_m$:

$$\vec{\dot{Z}}_m = \frac{1}{4\pi} \times \frac{\exp(-jkR)}{R} \vec{\dot{F}}_m(\theta, \varphi), \tag{14.6}$$

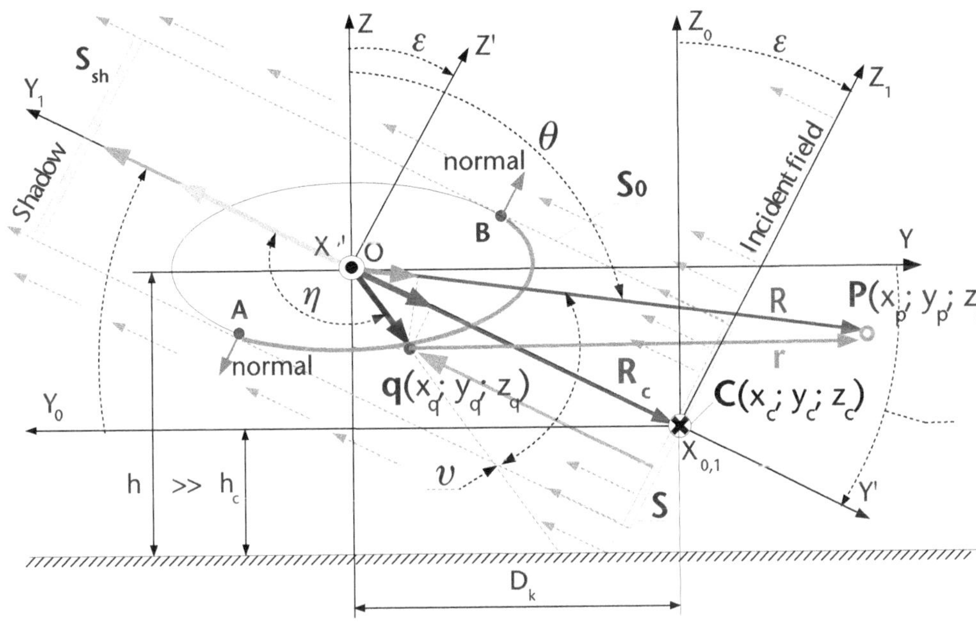

Figure 14.2 Waves coming from the target illumination radar and reflected from the target surface S_0 considered as the plane ones.

where:

$$\vec{F}_m(\theta, \varphi) = \int_V \vec{J}_m(q) \exp(jkR_q \cos \vartheta) dV \tag{14.7}$$

is a vector complex scatter (directional) diagram determined by magnetic currents, and $\vec{J}_m(q)$ is a vector of magnetic current volume density.

In this problem formulation the currents are determined on elementary flat areas S_{ij} to which the target surface is divided, and therefore, expressions (14.5), (14.7) take the following forms:

$$\vec{F}_e(\theta, \varphi) = \int_{S_0} \vec{J}_e \exp(jkR_q \cos \vartheta) ds \tag{14.8}$$

$$\vec{F}_m(\theta, \varphi) = \int_{S_0} \vec{J}_m \exp(jkR_q \cos \vartheta) ds, \tag{14.9}$$

where $\vec{J}_e$ and $\vec{J}_m$ are vectors of the electric and magnetic current densities, respectively. In a case of simultaneous existence of the electric and magnetic currents, the far field is determined by the following expressions:

$$\vec{E} = -\frac{j2\pi}{\lambda} \times \frac{\exp(-jkR)}{4\pi R} \times \left[\left(\sqrt{\frac{\mu}{\epsilon}} F_{e\theta} + F_{m\varphi}\right)\vec{i}_\theta + \right. \\ \left. + \left(\sqrt{\frac{\mu}{\epsilon}} F_{e\varphi} - F_{m\theta}\right)\vec{i}_\varphi\right] \tag{14.10}$$

$$\dot{\vec{H}} = \sqrt{\frac{\epsilon}{\mu}} \left[\vec{i}_R, \dot{\vec{E}}\right], \tag{14.11}$$

where $k = \omega\sqrt{\epsilon\mu}$ is a phase constant. $F_{e\theta}, F_{e\varphi}, F_{m\theta}, F_{m\varphi}$ are the θth and φth components of vectors $\vec{F}_e(\theta, \varphi)$ and $\vec{F}_m(\theta, \varphi)$ in the spherical coordination system and $\vec{i}_\theta$ and $\vec{i}_\varphi$ of its unit vectors. Assuming that a conductivity of each elementary flat area S_{ij} is infinite, therefore $F_m = 0$ and expression (14.10) takes the following form:

$$\dot{\vec{E}} = -\frac{j2\pi}{\lambda 4\pi} \times \frac{\exp(-jkR)}{R} \times \sqrt{\frac{\mu}{\epsilon}} \left(F_{e\theta}\vec{i}_\theta + F_{e\varphi}\vec{i}_\varphi \right) \tag{14.12}$$

Taking into account the assumptions above, components of vector $\dot{\vec{E}}$ from expression (14.12) can be determined by the following expression:

$$\begin{aligned} \dot{\vec{F}}(\theta, \varphi) &= \int_{S_0} \dot{\vec{J}}_s \exp(jkR_q \cos\vartheta) ds = \\ &= 2\int_{S_0} [\vec{n}, \dot{\vec{H}}_s] \exp(jkR_q \cos\vartheta) ds, \end{aligned} \tag{14.13}$$

where $\dot{\vec{H}}_s$ is a magnetic vector of the primary electromagnetic field created by the target illumination radar on the irradiated target part S_0. It is convenient to proceed to a rectangular coordinate system whose origin is combined with the geometric center of the target. At that the θth and φth components from expression (14.12) are calculated by formulas:

$$\begin{aligned} F_\theta &= (F_x \cos\varphi + F_y \sin\varphi) \cos\theta - F_z \sin\theta \\ F_\varphi &= -F_x \sin\varphi + F_y \cos\varphi \end{aligned} \tag{14.14}$$

Amplitude and phase directional patterns of the ground-based target illumination radar can be approximated by the following function:

$$G(\theta_R, \varphi_R) = |G(\theta_R, \varphi_R| \exp(jargG(\theta_R, \varphi_R)), \tag{14.15}$$

where θ_R is an angle between the direction of vector $|\vec{CO}|$ and vector $|\vec{Cq}|$, φ_R is an angle between projection of vector $|\vec{Cq}|$ on plane XOY and axis $\vec{OX}$. At that, it is assumed that the target illumination radar creates an electromagnetic field with linear (vertical) polarization. Using expression (14.10) the field created at the target surface can be written in the following form:

$$\begin{aligned} \dot{\vec{H}}_s &= \sqrt{\frac{\epsilon}{\mu}} [\vec{i}_\rho, \dot{\vec{E}}_s\,] \\ \dot{\vec{E}}_s &= \frac{\exp(jk\rho)}{\rho} \times \sqrt{\frac{\mu}{\epsilon}} \left\{ [\vec{i}_\rho, [\dot{\vec{F}}_{Re}\,, \vec{i}_\rho\,]\,] + [\dot{\vec{F}}_{Rm}, \vec{i}_\rho\,] \right\}, \end{aligned} \tag{14.16}$$

where $\vec{i}_\rho = \frac{\vec{\rho}}{|\vec{\rho}|}$ is unit vector $\vec{\rho}$. In accordance with expression (14.13) and assumed field polarization:

$$\begin{aligned} \dot{\vec{F}}_{Re} &= \ \ |G(\theta_R, \varphi_R)| \exp(jk[argG(\theta_R, \varphi_R)]) \times \vec{k} \\ \dot{\vec{F}}_{Rm} &= -|G(\theta_R, \varphi_R)| \exp(jk[argG(\theta_R, \varphi_R)]) \times \vec{i} \end{aligned} \tag{14.17}$$

When expression (14.17) is substituted into expression (14.16), we obtain:

$$\begin{aligned} \dot{\vec{E}}_s = \frac{1}{\rho}|G(\theta_R, \varphi_R)| \exp(jk[argG(\theta_R, \varphi_R) + k\rho])\times \\ \times\sqrt{\frac{\mu}{\epsilon}}\left\{[\vec{i}_\rho, [\vec{k}, \vec{i}_\rho]\,] - [\vec{i}, \vec{i}_\rho]\right\} \end{aligned} \tag{14.18}$$

In the rectangular coordination system vector $\vec{\rho}$ points to the center of an elementary area with coordinates $\{x_q,\ y_q,\ z_q\}$. In this case:

$$\vec{i}_\rho = \frac{(x_q - x_c)}{\rho}\vec{i} + \frac{(y_q - y_c)}{\rho}\vec{j} + \frac{(z_q - z_c)}{\rho}\vec{k} \tag{14.19}$$

$$\rho = \sqrt{(x_q - x_c)^2 + (y_q - y_c)^2 + (z_q - z_c)^2} \tag{14.20}$$

Substituting expressions (14.19) and (14.18) into expression (14.16) and after a number of vector transformations, we obtain the expression for magnetic field intensity at the elementary surface:

$$\begin{aligned} \dot{\vec{H}}_s = \frac{1}{\rho}\sqrt{\frac{\epsilon}{\mu}}|G(\theta_R, \varphi_R)| \exp(jk[argG(\theta_R, \varphi_R) + k\rho])\times \\ \times(H_x\vec{i} + H_y\vec{j} + H_z\vec{k}), \end{aligned} \tag{14.21}$$

where $\vec{H} = [\vec{\rho}, \vec{B}]$; $\vec{B} = [[\vec{k}, \vec{\rho}], \vec{\rho}] + [\vec{i}, \vec{\rho}]$; $\vec{J} = [\vec{H}, cos(\vec{n})]$. Substituting (14.21) in (14.1), we obtain the expression for surface current density:

$$\begin{aligned} \dot{\vec{J}}_s = \frac{1}{\rho}\sqrt{\frac{\epsilon}{\mu}}|G(\theta_R, \varphi_R)| \exp(j[argG(\theta_R, \varphi_R) + k\rho])\times \\ \times(J_x\vec{i} + J_y\vec{j} + J_z\vec{k}), \end{aligned} \tag{14.22}$$

where $J_x = H_y \cos(n, z) - H_z \cos(n, y)$, $J_y = H_z \cos(n, x) - H_x \cos(n, z)$, $J_z = H_x \cos(n, y) - H_y \cos(n, x)$, and $\cos(n, x)$, $\cos(n, y)$, $\cos(n, z)$ are direction cosines of normal vector $\vec{n}$:

$$\vec{n} = \cos(n, x)\vec{i} + \cos(n, y)\vec{j} + \cos(n, z)\vec{k} \tag{14.23}$$

To determine the field in the far region, it is necessary to substitute (14.22) into (14.13) and then into (14.12) for $\vec{E}$, and then we get:

$$\dot{\vec{E}} = \dot{E}_x\vec{i} + \dot{E}_y\vec{j} + \dot{E}_z\vec{k}, \tag{14.24}$$

where $\rho = R_c - R_q \cos\eta$ and for $\dot{E}_{x,y,z}$, we obtain:

$$\begin{aligned}\dot{E}_{x,y,z} = \frac{-j}{\lambda} \times \frac{\exp(-jkR)}{R} \times \frac{\exp(jkR_c)}{R_c} \times \\ \int\limits_{S_0} J_{x,y,z}\, |G(\theta_R, \varphi_R)| \times \\ \exp(jk[argG(\theta_R, \varphi_R) + R_q(\cos\vartheta - \cos\eta)])ds,\end{aligned} \tag{14.25}$$

and $[\vec{i}_{Rq}, \vec{i}_R] = \cos\vartheta$, $[\vec{i}_{Rq}, \vec{i}_\rho] = \cos\eta$, $\vec{i}_{Rq} = \frac{\vec{R}_q}{|\vec{R}_q|}$ is a unit vector directed to the point q, $\vec{i}_R = \frac{\vec{R}}{|\vec{R}|}$ is a unit vector directed to the point P. Introducing a notation:

$$k[argG(\theta_R, \varphi_R) + R_q(\cos\vartheta - \cos\eta)] = \gamma, \tag{14.26}$$

Taking into account Euler expansion expressions for projections $\dot{F}_{x,y,z}$ we obtain:

$$\begin{aligned}\dot{F}_{x,y,z} = \int\limits_{S_0} \Big\{ J_{x,y,z}|G(\theta_R, \varphi_R)| \cos\gamma + \\ + jJ_{x,y,z}|G(\theta_R, \varphi_R)| \sin\gamma \Big\} ds\end{aligned} \tag{14.27}$$

Introducing the following notations:

$$\begin{aligned}P_{x,y,z} &= J_{xyz}|G(\theta_R, \varphi_R)| \cos\gamma \\ Q_{x,y,z} &= J_{xyz}|G(\theta_R, \varphi_R)| \sin\gamma\end{aligned} \tag{14.28}$$

and replacing integral (14.27) with sums, the expressions for vector complex scatter diagram modules and phases in a direction-finder receiving antenna rectangular basis $X'Y'Z'$ will take the following forms:

$$F_{x,y,z} = \sqrt{\left[\sum_{j=1}^{M}\sum_{i=1}^{N} P_{x,y,z}\Delta S_{ij}\right]^2 + \left[\sum_{j=1}^{M}\sum_{i=1}^{N} Q_{x,y,z}\Delta S_{ij}\right]^2} \tag{14.29}$$

$$\arg F_{x,y,z} = \arctan\left[\frac{\sum\limits_{j=1}^{M}\sum\limits_{i=1}^{N} Q_{x,y,z}\Delta S_{ij}}{\sum\limits_{j=1}^{M}\sum\limits_{i=1}^{N} P_{x,y,z}\Delta S_{ij}}\right], \tag{14.30}$$

where $\Delta S_{ij} = \Delta x_i \Delta z_j$ is an area of element S_{ij} irradiated surface of the target S_0 and N, M are numbers of the total quantity of this elements across OX' and OZ'.

The field in the far zone is usually characterized by the radar cross section σ.

This is the area of the hypothetical plate perpendicular to the direction incident wave. This plate intercepts the incident field and re-emits it in the surrounding area according to the backscatter diagram $F^2(\theta, \varphi)$. As shown in [3, 4] the full cross section is

$$\sigma = \sigma^{refl} + \sigma^{sh} = 2S_{sh}, \tag{14.31}$$

where σ^{refl} is a scattering cross section of the illuminated part S_0, σ^{sh} is a scattering cross section of the shadows, and S_{sh} is a cross-sectional area of the shadow. The scattering cross section takes the following form:

$$\sigma(\theta, \varphi) = 2S_{sh}F^2(\theta, \varphi) \tag{14.32}$$

In real terms the ratio $\frac{\lambda}{\rho} \approx 0$; therefore, we can assume that the target is present in the unresolved volume of the main beam of the target illumination radar. This makes it possible to ignore its directional characteristics. Due to the fact that the power density P is a radial component of the Poynting vector [2]:

$$P = \frac{1}{2}\Re\left\{\dot{E}_\theta \dot{H}_\varphi^* - \dot{E}_\varphi \dot{H}_\theta^*\right\} \tag{14.33}$$

as $H_\varphi = E_\theta / W$ and $H_\theta = -E_\varphi / W$, then for the case $E_\theta(\theta, \varphi) = E_\varphi(\theta, \varphi)$ we get

$$P(\theta, \varphi) = \frac{|E_\theta(\theta, \varphi)|^2}{W} \tag{14.34}$$

and taking into account (14.34) to simplify the calculations by introducing $\forall \theta_R, \varphi_R :$ $|G(\theta_R, \varphi_R)| = \sqrt{P_s W}$, $arg G(\theta_R, \varphi_R) = 0$, where P_s is a transmitted power and W is a wave resistance of the medium.

Thereby, in an orthogonal basis of the receiving antenna, it is possible to find vectors of field intensity reflected from the target for any relative position of the target illumination radar, the target, and the airborne radar.

14.3 Evaluation of Interception Hypersonic Vehicles by Some Existing Air Defense Missile Systems

The combat operation can be represented by three points: t_d is a start time of target detection, t_a is a start time of target automatic tracking, and t_i is a time of target interception. Let t_a be a start time of target tracking by angular coordinates and radial velocity. Let the missile launch takes place immediately after the point t_a, then $t_m = t_i - t_a$ is a time of missile flight. Today, the battle performance of any air defense missile system is controlled by some automated command and control system (automated mode) or self-sustained assets (autonomous mode). In autonomous mode independent radars are usually used for detections and target designation. In any case the system receives target information from the outside. Today, air defense missile systems usually receive target information from the outside. As rule these are the three coordinates: azimuth ($\beta = 90^\circ - \varphi_0$), angle of elevation ($\epsilon = 90^\circ - \theta_0$) or elevation ($H_x$), and range R_0, where φ_0, θ_0, R_0 are the coordinates in a spherical coordinate system $X_0Y_0Z_0$ (see Figure 14.2). It is known that power P_r (watts) received from a target at a receiving antenna is found via the radar range equation [5]:

$$P_r = \frac{P_t G_t G_r \lambda^2 \sigma_0}{(4\pi)^3 R_0^4} \tag{14.35}$$

In this expression, P_t is the transmitted power in watts, G_t is the gain of the transmit antenna, G_r is the gain of the receive antenna, λ is the carrier wavelength in meters, and σ_0 is the mean RCS of the target in square meters [6].

In practice, the target RCS can be determined as follows:

$$\hat{\sigma}_0 = \frac{R_0^4 (4\pi)^3 P_{min} \times 10^{\frac{B_0}{10}} \times L}{P_t G_t G_r \lambda^2}, \tag{14.36}$$

where B_0 is the signal-to-noise ratio in dB at the radar receiver input that provides a stable target automatic tracking, P_{min} is the receiver sensitivity, and L is the coefficient of losses on the propagation and other losses. R_0 is the maximum range radar-target in meters for RCS $\hat{\sigma}_0$, where the following condition holds:

$$10 \log[P_r / P_{min}] = B_0 \tag{14.37}$$

Let $\hat{\sigma}_x$ be the estimation RCS for model of UAV X-51. Let $\hat{R}_{xka}$ be a maximum range "radar-target" for the kth system, where a stable automatic tracking exists for a target having a RCS $\hat{\sigma}_x$. It follows from (14.36) that for the same radar and various RCS, $\hat{\sigma}_0$ and $\hat{\sigma}_x$ ranges R_0 and R_{xka} will be different; hence, it is easy to prove that the following expression is correct:

$$\hat{R}_{xka} = \left(\frac{\hat{\sigma}_x}{\hat{\sigma}_0}\right)^{(1/4)} \times D_{0ka} \tag{14.38}$$

Let $\hat{\sigma}_0$ be the estimation of target cross section for which each kth air defense missile system has a maximum auto-tracking range R_{0ka}, such that $\forall H_x \ll R_{0ka} : R_{0ka} \approx D_{0ka}$, where D_{0ka} is the far boundary of the target engagement zone usually represented via map range. Let $\{Y\}$ be a target engagement zone. In this zone missiles can be fired from a launcher to a target. As follows from (14.38) for estimate $\hat{R}_{xka}$ it is sufficient to know any open information for $\hat{\sigma}_0$, (e.g., B-52), D_{0ka} for the kth air defense missile system, and $\hat{\sigma}_x$ for the test target. The estimation $\hat{\sigma}_x$ can be calculated for each ith map range D_{xki} by the mathematics (14.1)–(14.32) as described above. Thus, the combat capabilities of any air defense missile system can be evaluated without classified information.

Table 14.1 shows the performance characteristics of existing air defense missile systems, published in open sources [7–10]. The target engagement zone of kth

Table 14.1 Air Defense Missile Systems

k	*Air Defense Missile Systems*	*Target engagement range, km*		*Target engagement altitude, km*		*missile speed, M*	
		D_k^{min}	D_{0k}^{max}	H_k^{min}	H_k^{max}	v_k^{max}	*missile*
1	S-75M3	6	76	0.1	30	3.10	5Ya23
	(SA-2)	$\hat{R}_{m1} = 3.301 < H_x$		$\{Z_1\} = \emptyset$		cannot intercept	
2	S-200D	-	300	0.3	40	3.60	V-880M
	(SA-5c)	$\hat{R}_{m2} = 42.326 > H_x$		$\{Z_2\} \neq \emptyset$		can intercept	5V28D
		$D_{d2} = 19.914$		$h_2 = 0.085$			
3	S-300PMU	5	90	0.025	25	3.60	5V55R
	(SA-10)	$\hat{R}_{m3} = 7.983 < H_x$		$\{Z_3\} = \emptyset$		cannot intercept	
4	S-300PMU1	5	150	0.01	27	7.01	48N6,
	(SA-10D)	$\hat{R}_{m4} = 44.625 > H_x$		$\{Z_4\} \neq \emptyset$		can intercept	48N6E
		$D_{d4} = 10.921$		$h_4 = 0.052$			
5	S-300PMU2	3	200	0.01	27	7.01	48N6E2
	(SA-20B)	$\hat{R}_{m5} = 63.181 > H_x$		$\{Z_5\} \neq \emptyset$		can intercept	
		$D_{d5} = 16.601$		$h_5 = 0.043$			
6	S-300PMU2	3	250	0.01	27	8.38	48N6E3
	(SA-20B)	$\hat{R}_{m6} = 81.018 > H_x$		$\{Z_6\} \neq \emptyset$		can intercept	
		$D_{d6} = 23.405$		$h_6 = 0.078$			
7	S-400	5	400	0.005	30	3.97	40N6E
	(SA-21)	$\hat{R}_{m7} = 59.836 > H_x$		$\{Z_7\} \neq \emptyset$		can intercept	
		$D_{d7} = 26.659$		$h_7 = 0.054$			

system can be represented by a set of points: $\{Y_k\} = \{D_k^{min}, \hat{D}_{xka}\} \cap \{H_k^{min}, H_k^{max}\}$, where for UAV X-51 and any $\boldsymbol{k}$ there exists $\hat{D}_{xka}$ such that $\hat{D}_{xka}$ is the estimation of range, where a stable automatic tracking target exists on the angular coordinates and radial velocity. For UAV flight altitude H_x such that $H_x \in [35, \ldots, 60]$ ft or the same in [km]: $H_x \in [10.668, \ldots, 18.288]$ km, the UAV flight zone can be written as $\{X\} = \{D_k^{min}, D_{0k}^{max}\} \cap \{H_x^{min}, H_x^{max}\}$. In this case the wavelike flight trajectory X-51 taking into account the curvature of the Earth can be represented as shown in Figure 14.3. Hence, we obtain a necessary condition for target interception, $\{Z_k\} = \{X\} \cap \{Y_k\} \neq \emptyset$ or in another form:

$$\forall H_x < H_k^{max}, \quad \exists \hat{D}_{xka} > 0 : \hat{D}_{xka} = \sqrt{\hat{R}_{xka}^2 - H_x^2} \tag{14.39}$$

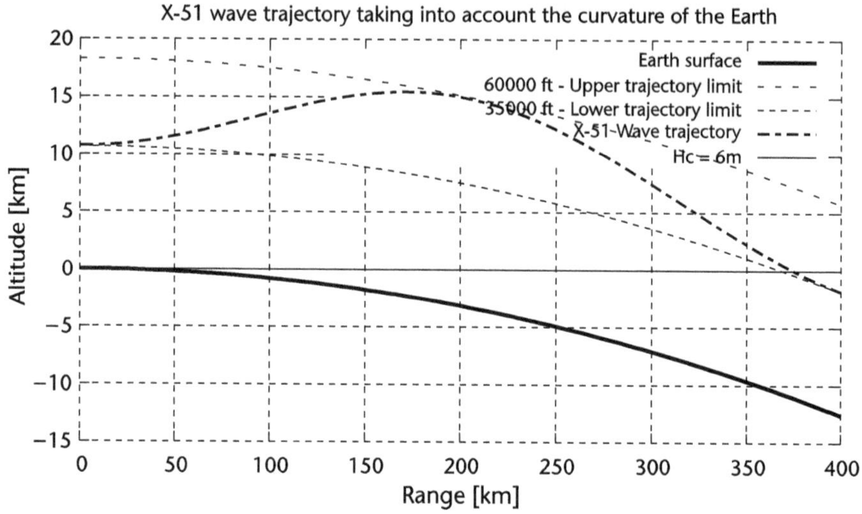

Figure 14.3 X-51 wave trajectory taking into account the curvature of the Earth.

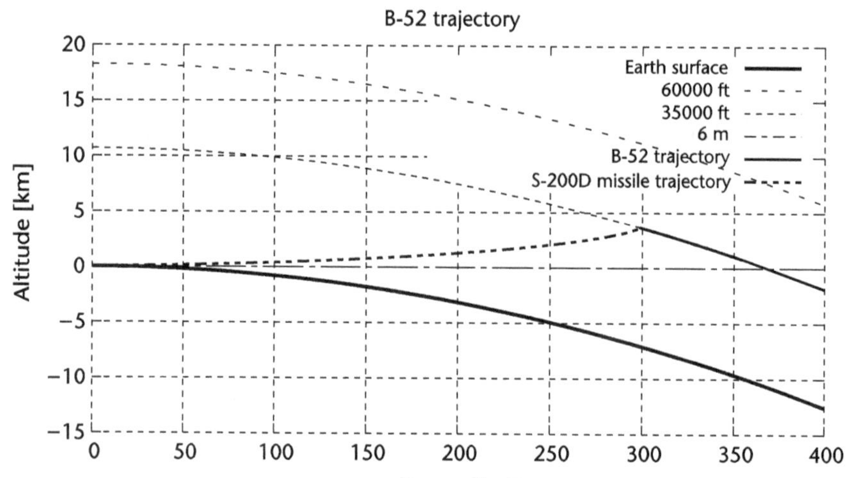

Figure 14.4 B-52 interception by system S-200D.

The Table 14.1 shows that for $H_x = H_x^{min}$ the condition (14.39) holds for air defense missile system S-200 ($k = 2$); hence, it is advisable to simulate this system. Obviously, the best characteristics of an air defense missile system will include a large value σ and flying at medium altitudes. Today, the largest value $\hat{\sigma}_0$ is on aircraft B-52 Stratofortress: $\hat{\boldsymbol{\sigma}}_0 \approx 100$ [sq.m] and cruise speed $V_0 = 723$ [km/h][11] or $V_0 = 200.83$ [m/s]. The far boundary of the target engagement zone for the system S-200 is $D_{02}^{max} = 300$ km. It is logical to assume that for missile speed the following inequality will hold: $V_{m2} \geq v_2^{max}$ (see Table 14.1); therefore, missile flight time t_{m2} to a distance of 300 km at the speed of $V_{m2} = 1078$ [m/s] $= 3.6$[M] will be $t_{m2} = t_i - t_a = 277$s. During this time, the B-52 aircraft will fly a distance 55,63 km at a speed of 723 [km/h]. Consequently, we obtain the range of a stable target

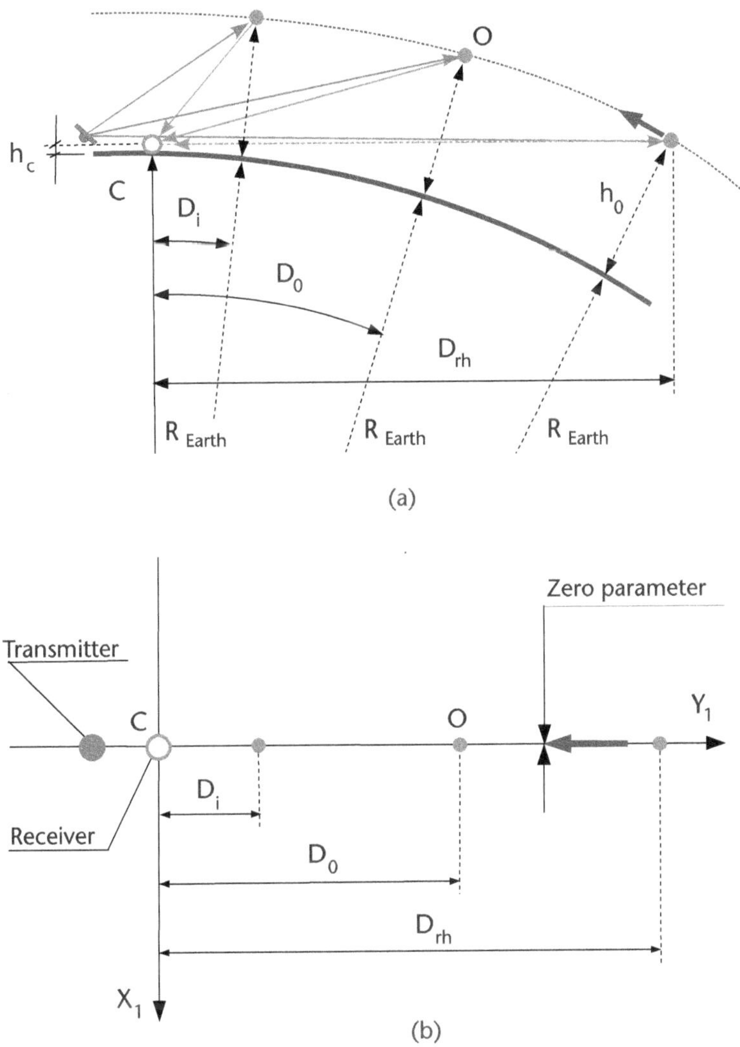

Figure 14.5 $H_0 = 10.668$ km, $D_0 = 300$ km, $D_{rh} = 377.587$ km: (a) Z_1CY_1 plane, and (b) X_1CY_1 plane.

auto-tracking for a B-52 aircraft, where level B_0 is realized:

$$\hat{D}_{0ka} = (1 + V_0/V_{mk}) \times D_{0k}^{max} \tag{14.40}$$

In this case $\hat{D}_{02a} = 355.89$ km and it is necessary that $\hat{D}_{02a} < D_{rh}$, where D_{rh} is the radio horizon range:

$$D_{rh} = \sqrt{(R_E + H_0)^2 - R_E^2} + \sqrt{(R_E + H_c)^2 - R_E^2} \tag{14.41}$$

$\forall H_0 = 10.668$ km and $H_c = 0.006$ km: $D_{rh} = 377.587$ km (Figure 14.5), where $R_E = 6371$ km is the mean radius of the Earth. A computer model of the system S-200 was created, which used a sum-and-difference monopulse radar with a mirror antenna. First, a mirror antenna was calculated for the target illumination radar.

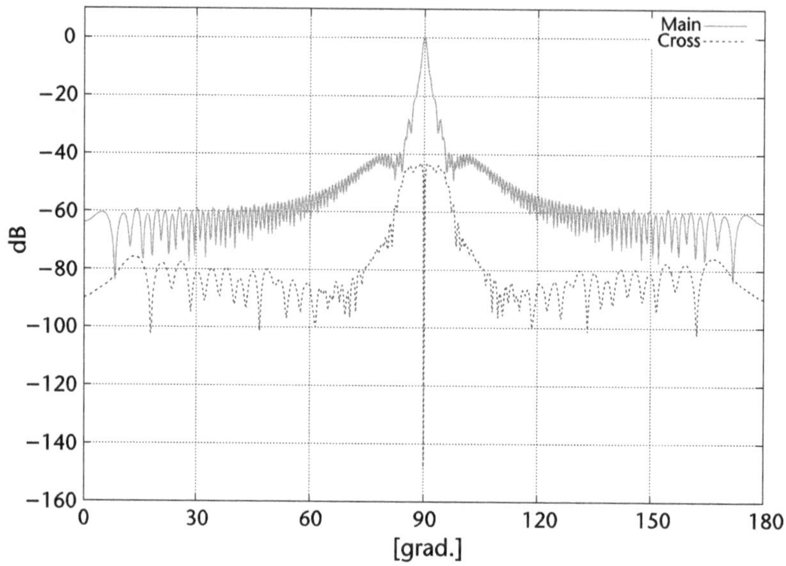

Figure 14.6 Dirrection pattern for the target illumination radar.

The antenna geometry is a paraboloid of revolution with a diameter of 75λ and focal length of 50λ. The feed is a horn antenna with vertical linear polarization (main polarization). The antenna directional pattern for main and cross polarization in the horizontal plane (azimuth) is given in Figure 14.6, where $\theta_1 = 90°, \varphi_1 = var$ are a spherical coordinates of system $X_1Y_1Z_1$ (see Figure 14.2). This antenna is designed for a centimetric wave band and has a pencil-beam directional diagram and width of the primary beam 1°. Similarly, a mirror antenna was calculated for an airborne radar based on reasonable considerations of the missile geometrical dimensions. The antenna directional pattern airborne radar for main and cross polarization was calculated for sum-and-difference channels.

A 3-D model of a hypersonic UAV was developed on the basis of public sources [12–15] and is shown in Figure 14.1. The incident field was formed on the UAV by the target illumination radar situated on the ground. The polarization is a vertical (main). The scattered field is received by the missile airborne radar. The homing missile is located on the ground, nearby the target station radar. The missile is fired from an oblique constant elevation launcher trainable in azimuth. The missile is controlled and guided by the onboard semiactive radar seeker. The flight of the UAV is at a height of 10.668 km in the direction of the target illumination radar (Figure 14.5). The value of the cross-section X-51 is calculated for each ith distance D_{x2i} by the mathematics (14.1)–(14.32). Dependence of the UAV cross section from a distance is shown in Figure 14.7. For each ith range, D_{x2i}, $i \in [1, N]$ the probability of automatic target tracking P_i is calculated based on M realizations of the random target position on the basis of the onboard radar antenna. For each ith range, D_{x2i} RCS $\hat{\sigma}_{xij}, \forall j \in [1, M]$ is calculated, and is determined by the distance D_{x2p} such that $\forall D_{x2i} < D_{x2p}$: $\exists P_i \geq P_0$, where P_0 is a certain threshold value. For the obtained value D_{x2p}, RCS is calculated as an estimate of the average for the random process on the interval $]D_{x2p}, D_{x2p} + \delta]$, where δ is a length of the interval averaging. The resulting value RCS $\bar{\sigma}_x$ is the value that

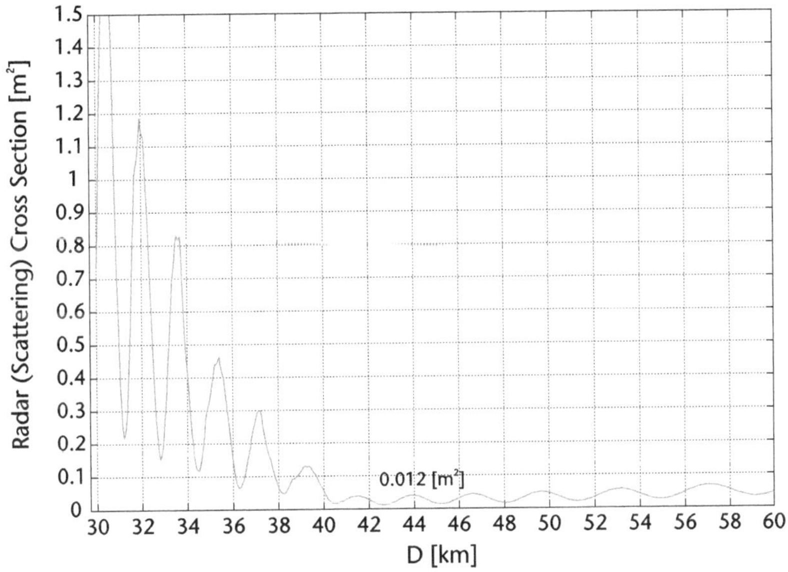

Figure 14.7 RCS X-51 without absorbing materials.

was realized in the computational experiment. Substituting the value found for $\bar{\sigma}_x$ and found early $\hat{\sigma}_0$ and $D_{02a} = 355.89$ km in (14.38), we get $\hat{R}_{x2a}$ and then $\forall H_x = 10.668$ km we get $\hat{D}_{x2a}$ as a range of stable target tracking. Substituting the value found for $\bar{\sigma}_x$, values $\hat{\sigma}_0$ and $D_{02a} = 355.89$ km in (14.38), we get $\hat{R}_{x2a}$. Then substituting $\hat{R}_{x2a}$ and $H_x = 10.668$ km in (14.39), we obtain $\hat{D}_{x2a}$ as a range of stable target tracking. If $|\hat{D}_{x2a} - \hat{D}_{x2p}| \leq \xi$, where ξ is a certain neighborhood chosen from reasonable considerations, then the estimate $\bar{\sigma}_x$ is verified and can be used for subsequent calculations. Taking into account that the function (14.38) is linear and $(\bar{\sigma}_x/\hat{\sigma}_0)^{(1/4)} = const$, hence substitution in (14.40) values D_{0k}^{max} allows to get $\hat{R}_{xka}$ and then $\hat{D}_{xka}$. For flight speed of missile-interceptor V_{mk}, flight distance $\hat{R}_{mk}$ can be estimated.

$$\hat{R}_{mk} = \hat{D}_{xka} \times \frac{V_{mk}}{V_x}, \tag{14.42}$$

where V_x is a flight speed of UAV. Hence, for target interception we obtain

$$\hat{R}_{mk} > H_x \tag{14.43}$$

Obviously, (14.43) is a sufficient condition for target interception. Hence, for estimation combat capabilities of an air defense missile system, we can use the necessary and sufficient conditions for target interception.

14.4 Calculation Results

Sustained automatic target tracking for a UAV such as the X-51 begins with a range of 39 km. It is easily show that for the detection range $D_{0d} = 355.89$ km and $\hat{D}_{x2} = 39.4$ km, the RSC average estimate for the UAV X-51 model realized in experiment is 0.0502 [sq.m]. The scatter pattern calculation is shown in figures 14.8 and 14.9 for map range $D_x = 39.4$ km and $D_x = 42.6$ km. In these figures, the model of the UAV X-51 is located in the center of the coordinate system XYZ (see Figure 14.2).

In the first case in the angular range [±8] degree failure of the signal reflected in the direction of the radar is −19 dB, and in the second, the failure signal level reflected in the direction of the radar is −24.47 dB. In accordance with formulas (14.1)–(14.32), the cross section of the UAV model at the range 42.6 km is estimated at 0.012 [sq.m]. Let the UAV surface contain radar-absorbing materials with reflection coefficient $|R| = -10$ dB. It is easy to prove that in this case the cross section X-51 is estimated at 0.0012 [sq.m]. The computational experiment for S-200 showed that the range of stable automatic tracking $\hat{D}_{x2a}$ for a target such as the UAV X-51 begins with a range of 39.4 km. As seen from Figure 14.7, it is the first extremum for the cross-section function where its value exceeded the level of 0.1 [sq.m]. As a result of the numerical experiment, the estimate of the average cross section UAV X-51 $\bar{\sigma}_x$ is 0.0502 [sq.m] without radar-absorbing materials. Substituting $\bar{\sigma}_x$ into linear function (14.38), we obtain estimates for other air defense missile systems from Table 14.1. As we can see, with the exception of S-75M3 and S-300PMU, other systems listed in Table 14.1 can intercept this UAV. The values $\hat{D}_{xka}$ do not exist $\forall k = \{1, 3\}$ and the necessary condition (14.39) does not hold. For these systems $\forall k = \{1, 3\} : \hat{R}_{xka} < H_x$, and in accordance with (14.39), values for $\hat{D}_{xka}$ do not exist. Obviously, for $\hat{D}_{x2a} = 39.4$ km, $V_x = 4.5M$, and $V_m = 3.6M$, the sufficient condition (14.43) is satisfied and air defense missile system S-200 can intercept UAV X-51. However, if we used missiles 5V28V and 5V28D, which may have a speed of 1190 m/s = 3.6M in accordance with [8], the condition (14.43) will satisfied, but only for $V_x < 13.3\ M$.

Therefore, for $V_x > 13.3\ M$, S-200 air defense missile system from Table 14.1 cannot intercept UAV X-51. Hence, the condition for overcoming the air defense zone can be written as: $H_x > \hat{R}_{mk}$. Using (14.42), we obtain

$$V_x > [\hat{D}_{xka}/H_x] \times V_{mk} \tag{14.44}$$

Figure 14.10 presents the lines for $H_x = 10.668$ km and different ranges of target tracking $\hat{D}_{xka}$. Line 1 for $\hat{D}_{xka} = 39.4$ km, and line 2 for $\hat{D}_{xka} = H_x$. Line 3 corresponds to target auto-tracking to $\hat{D}_{xka} < H_x$ and line 4 for $\hat{D}_{xka} > H_x$. Thus, reducing the auto-tracking range of a target less than the altitude of its flight reduces the requirements for its speed.

Table 14.1 shows some air defense missile systems [7]. It is easily seen that for $\hat{\sigma}_x = 0.012$ [sq.m] $\hat{D}_{x2} = 39.41$ km, $H_x^{min} = 10.668$ km, $V_x^{min} = 4.5M$, and $V_{m2} = 3.6M$, herewith $\hat{R}_{m2} = 36.72$ km, and so the condition (14.40) has been satisfied. Therefore, the systems listed in the Table 14.1 can operate against this UAV in their actual condition, except for anti-aircraft missile systems S-75M3 and S-300PMU. Figure 14.11 presents trajectories of the missiles of some anti-aircraft missile systems in the case of interception (destruction) of UAV X-51. Destruction ranges D_{dk} and misses h_k are presented in Table 14.1 for each kth anti-aircraft missile system. As shown in Table 14.1, destruction range of the UAV X-51 by anti-aircraft missile system S-200D D_{d2} is 19.914 km and the miss h_2 is 85 m. Here

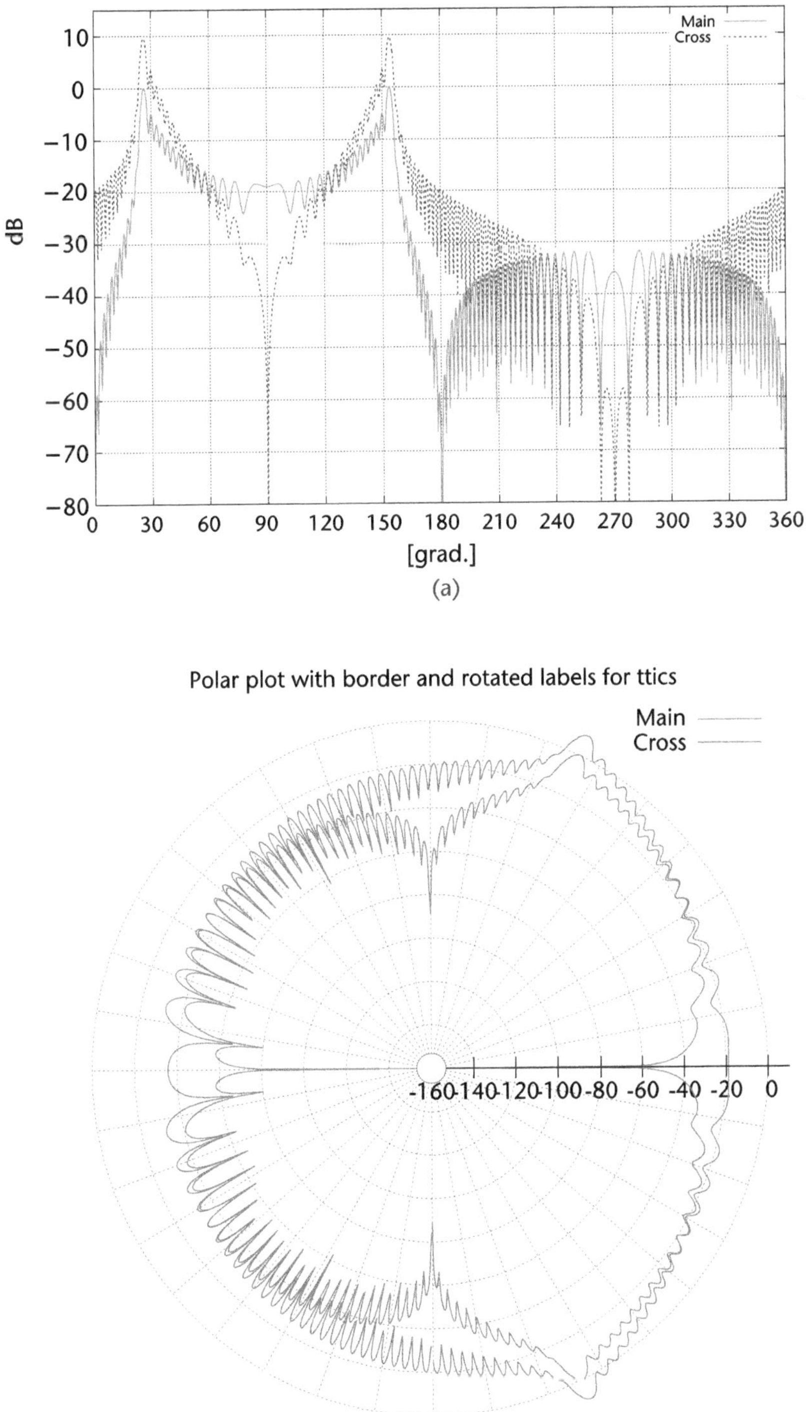

(b)

Figure 14.8 $F(\theta = 105, 15°, \varphi = var)$; $H_x = 10.668$ km; $D_x = 39.4$ km.

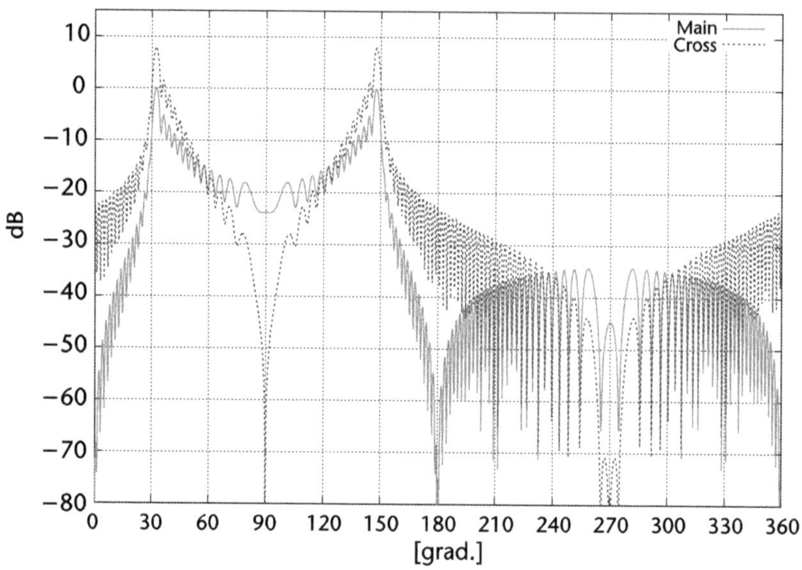

(a)

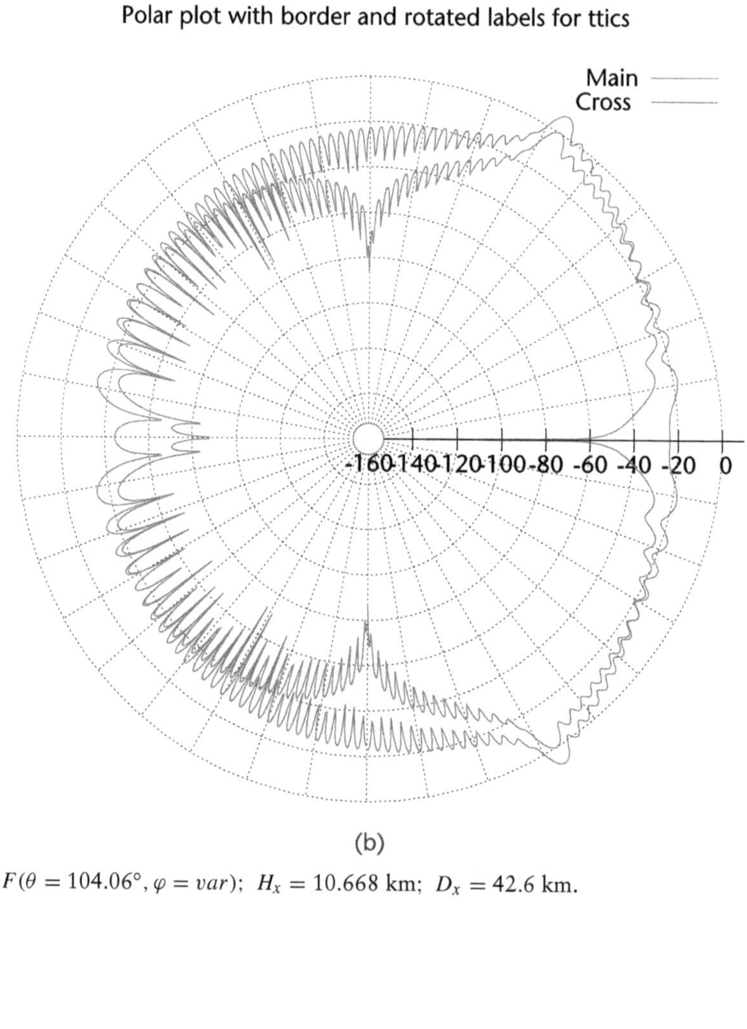

(b)

Figure 14.9 $F(\theta = 104.06°, \varphi = var)$; $H_x = 10.668$ km; $D_x = 42.6$ km.

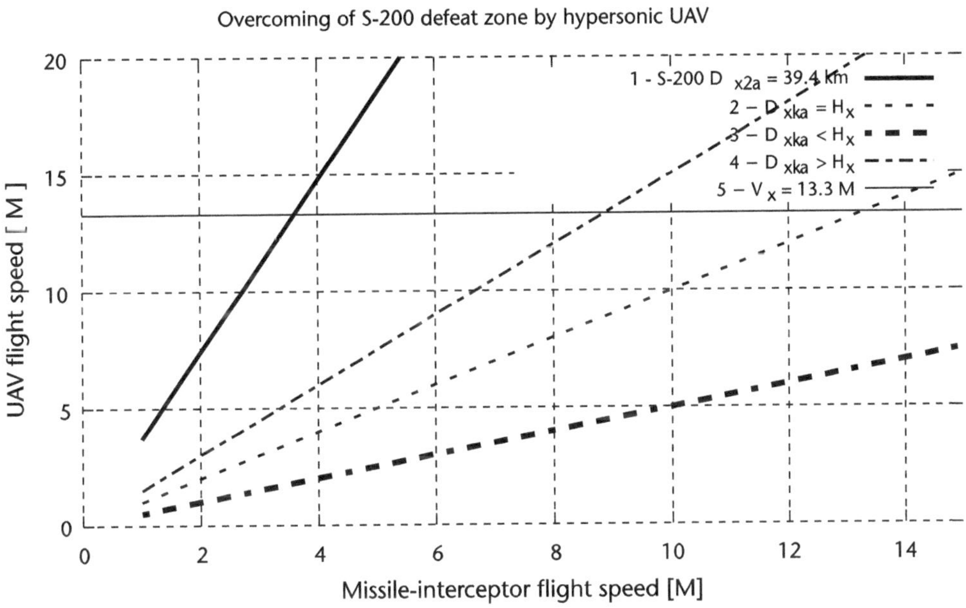

Figure 14.10 Overcoming the air defense zone.

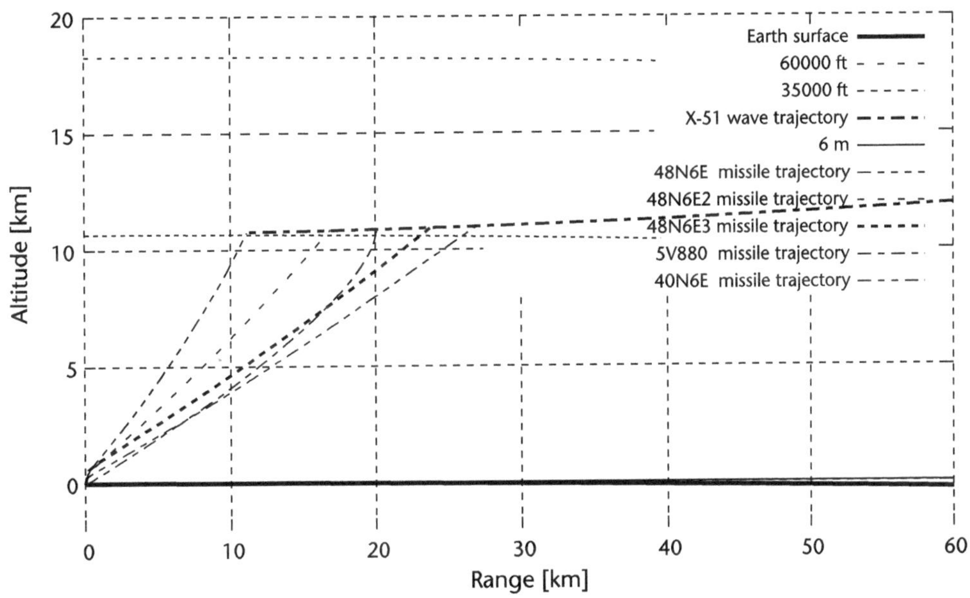

Figure 14.11 X-51 trajectory, and anti-aircraft missiles trajectories.

the miss h_k is represented by the following expression:

$$h_k = \sqrt{(X(t)_{tk} - X(t)_{mk})^2 + (Y(t)_{tk} - Y(t)_{mk})^2 + (Z(t)_{tk} - Z(t)_{mk})^2}, \quad (14.45)$$

where $X(t)_{tk}$, $Y(t)_{tk}$, and $Z(t)_{tk}$ are the instantaneous coordinates of the target in the rectangular coordinate system of the kth anti-aircraft missile system, and

$X(t)_{mk}$, $Y(t)_{mk}$, and $Z(t)_{mk}$ are the instantaneous coordinates of the missile in the rectangular coordinate system of the kth anti-aircraft missile system. It should be noted that for $k = 2$ is the anti-aircraft missile system S-200D, where the pursuit method during calculations in the guidance method. In other anti-aircraft missile systems for $k = 4, ..., 7$ when making calculations, as a guidance method the method of complete straightening of the trajectory was used.

Thus, the condition for intercepting the target, taking into account the method of pointing the missile at the target, can be written as:

$$h_k \leq h_k^{max}, \tag{14.46}$$

where h_k is the miss when shooting at a target by a selected missile with a given guidance method, and h_k^{max} is the maximum miss, at which the target is hit by one missile with a probability of at least a given. Here, a given probability of hitting a target with one missile is 0.84 and $h_k^{max} = 0.1$ km.

Thus, the necessary and sufficient condition for intercepting a given target by means of the kth air defense missile system can be represented as:

$$\begin{cases} \hat{R}_{mk} > H_x \\ h_k \leq h_k^{max}. \end{cases} \tag{14.47}$$

14.5 Conclusion

This chapter presented a methodology for calculation of a scattering pattern field reflected from complex-form objects. This can be aircraft, missiles, helicopters, and UAVs, among many others. In these calculations the vector of the field reflected by an aircraft is represented as a complex value in 3-D space. The UAV surfaces were considered as ideally conducting without any radar-absorbing materials. On the basis of public sources a numerical 3-D model of the hypersonic UAV X-51 was developed. In accordance with the mathematics stated in this chapter, software was designed and directional patterns of mirror antennas and UAV scatter patterns were calculated. The chapter described the results of scatter pattern calculations for this 3-D model and presented a comparative assessment of operation for some air defense missile systems by using a UAV X-51 model. These results can be summarized as follows:

1. The interceptor speed is a critical parameter for an air defense missile system in operating against targets with high speed high altitude and in which stealth technology is used.
2. The air defense systems discussed in the chapter can intercept an X-51, except for S-75M3 and S-300PMU.
3. Today, low altitude, high speed, and low RCS are the main trends for overcoming the SAM defeat zone.

References

[1] Markin, E., "Scatter Pattern Calculations and Determination Performance Limitations of Existing Air Defense Systems for Intercepting Hypersonic Vehicles," *2012 IEEE Radar Conference*, Atlanta, GA, 2012, DOI: 10.1109/RADAR.2012.6212189.

[2] Markov, G. T., and D. M. Sazonov, *Antennas*, Moscow: Energia, 1975.

[3] Ufimtsev, P. Y., *Theory of Edge Diffraction in Electromagnetics*, Encino, CA: Tech Science Press, 2009.

[4] Ufimtsev, P. Y., *Fundamentals of the Physical Theory of Diffraction*, Hoboken, NJ: John Wiley & Sons, 2007.

[5] Richards, M. A., J. A. Scheer, and W. A. Holm, *Principles of Modern Radar*, Raleigh, NC: SciTech Publishing, 2010.

[6] Knott, E., J. Shaeffer, and M. Tuley, *Radar Cross Section*, Raleigh, NC: SciTech Publishing, 2004.

[7] *Russia's Arms*, Catalog, Vol. V, Air Defense, Military Parade Ltd., Moscow: Russia, 1997.

[8] S-200 (Missile), https://en.wikipedia.org/wiki/S-200_(missile).

[9] S-300 Missile System, https://en.wikipedia.org/wiki/S-300_missile_system#Land-based_S-300P_(SA-10).

[10] Anti-Aircraft Missile System S-400 "Triumph," https://militaryarms.ru/boepripasy/rakety/zrk-s-400-triumf/.

[11] Radar Cross Section (RCS), http://www.globalsecurity.org/military/world/stealth-aircraft-rcs.htm.

[12] Pratt & Whitney, http://www.pw.utc.com/products/pwr/propulsion_solutions/hypersonic_scramjet_engine.asp.

[13] Waverider, http://www.boeing.com/features/2014/05/bds-aiaa-x51-05-02-14.page, http://www.boeing.com/features/2013/05/bds-x51-05-03-13.page, http://www.boeing.com/defense-space/military/waverider/index.html.

[14] Boeing X-51 Waverider, https://en.wikipedia.org/wiki/Boeing_X-51_Waverider.

[15] X-51 Wave Rider Unmanned Scramjet Aircraft, https://www.airforce-technology.com/projects/x51-wave-rider/.

Acronyms and Abbreviations

μs	Microsecond
°C	Degrees Celsius
A	Ampere
kg/m^3	Density
m^2	Square meter
m^3	Cubic meter
2-D	Two-dimensional
3-D	Tree-dimensional
5N62	Target Illumination Radar is Soviet/Russian Guidance and Illumination Radar (NATO reporting name: "Square Pair") of the SAM System S-200 (SA-5 Gammon)
A-235	Nudol—layered territorial missile defense system, USSR/Russia
A-50	Airborne early warning and control system
ADC	Analog-to-digital converter
AEA	Airborne electronic attack platform
AESA	Active electronically scanned array
AEW & C	Airborne early warning and control aircraft complex
AFC	Amplitude-frequency characteristic
AGC	Automatic gain control
AHH	Active homing head
AMS	Anti-aircraft missile system
AWACS	Airborne warning and control system
B	Billion
B-52	Boeing B-52 Stratofortress is an American long-range, subsonic, jet-powered strategic bomber
bbl	Barrel
BCM	Billion cubic meters
BRIC	Brazil, Russia, India, and China
China	People's Republic of China
cm	Centimeter
CSS	NORAD Communication System Segment
DA-ASAT	Direct-ascent anti-satellite
dB	Decibel
DOD	U.S. Department Of Defense
ECM	Electronic countermeasure
EM	Electromagnetic
EMF	Electromotive force
EMI	Electromagnetic interference

ESD	Equisignal direction
EU	European Union
EUR	Euro is the official currency of 19 of the 27 member states of the European Union
EWA	Electronic warfire airplanes
FB-111	General Dynamics FB-111 Aardvark is a supersonic strategic bomber
FCC	U.S. Federal Communications Commission
FGFA	Fifth generation fighter aircraft
G	Gain, coefficient directivity antenna
GaAs	Gallium arsenide
GDP	Gross domestic product
GHz	Gigahertz
GLONASS	Global Navigation Satellite System
GMR	Ground multifunctional radar
GPS	Global Positioning System
H-D	Height-distance
HAL	Indian company Hindustan Aeronautical Ltd.
HF	High frequency
HQ-10	Chinese analog soviet SAM system S-300PMU-1
HQ-15	Chinese analog soviet SAM system S-300PMU-2
HQ-9	Chinese analog soviet SAM system S-300PMU
Hz	Hertz
IAC	International Aviation Committee
ICBM	Intercontinental ballistic missile
IEEE	Institute of Electrical and Electronics Engineers
IF	Intermediate frequency
IMF	International Monetary Fund
INSR	Interference-to-signal ratio
J	Joule
J-11	Jianjiji-11 fighter—Chinese licensed version of the Su-27
J-15	J-15 fighter—Chinese licensed version of the Su-33
J/J	Joules per joule
JCS	Joint Chiefs of Staff
K	Kelvin
kHz	Kilohertz
km	Kilometer
km/h	Kilometer per hour
KTN	Kilotonne of trinitrotoluene, kt of TNT
kW	Kilowatt
La-350	USSR intercontinental cruise missile *Tempest*
LTAMDS	Lower tier air and missile defense sensor—360-degree missile-defense radar
m	Meter
m/s	Meter per second
Mach	Mach number
Mbbl	Thousand barrel

MDS	Minimum detectable signal
MHz	Megahertz
mi	Mile
min	Minute
MK12	American radar identification system (MARK-XII), used to determine the nationality of NATO member countries aircraft
mm	Millimeter
mph	Miles Per Hour
MRBM	Medium-range ballistic missile
Mt	Million tonnes
MW	Megawatt
NATO	North Atlantic Treaty Organization
NCS	NORAD computer system
NORAD	North American Air Defense Command
ns	Nanosecond
OTH	Over-the-horizon
P-14	Early warning radar 5N84A “Oborona-14” (NATO reporting name: “Tall King C”) is a ground-based 2-D VHF Surveillance and targeting radar, USSR/Russia
P-18	Ground-based 2-D VHF surveillance and targeting radar “Terek” (NATO reporting name “Spoon Rest D”), USSR/Russia
PAA	Phased array antenna
PCR	Pulse compression ratio
PLA	People’s Liberation Army of China
PPS	Precise positioning service
PRV-13	Radio altimeter of decimeter range, USSR/Russia
PwC	PricewaterhouseCoopers
Radar	Radio detection and ranging
RCLOS	Radar command to line of sight
RCS	Radar cross section
RF	Radio frequency
RIF	Soviet/Russian sea-based anti-aircraft missile system, US DOD designation: “SA-N-6,” (NATO reporting name: “Grumble”)
rms	Root mean square
rpm	Revolutions per minute
s	Second
S-200D	Soviet/Russian surface-to-air missile system “Dubna” (NATO reporting name: “SA-5c Gammon”)
S-300PMU	Soviet/Russian surface-to air missile system (NATO reporting name: SA-10C “Grumble”)
S-300PMU-1	Soviet/Russian surface-to-air missile system, U.S. DOD designation SA-20A (NATO reporting name SA-20 “Gargoyle”)
S-300PMU-2	Soviet/Russian surface-to-air missile system “Favorite,” U.S. DOD designation SA-20B (NATO reporting name “SA-20B”)
S-400	Soviet/Russian surface-to-air missile system “Triumph” (NATO reporting name: “SA-21 Growler”)

S-75M3	Soviet surface-to-air missile system "Volkhov" (NATO reporting name: "SA-2 Guideline")
SAM	Surface-to-air missile
SARH	Semi-active radar homing
Sary-Shagan	USSR/Russia state missile defense research site
SCC	NORAD Space Computational Center
SCR	Signal-to-clutter Ratio
SDC	Moving target selection equipment on S-75M3
SINR	Signal-to-interference ratio
SIR	Signal-to-interference ratio
SNR	Signal-to-noise ratio
SNR-75	Fire control and tracking radar (missile guidance station) of the S-75 surface-to-air missile system
SPS	Standard positioning service
SRBM	Short-range ballistic missile
Su-24M	Sukhoi Su-24M (NATO reporting name: "Fencer") is a Soviet/Russian supersonic all-weather bomber
T	Tesla—magnetic flux density unit
t	One tonne
T/R	Transmit/receive
TLC	Transport and launch container
TsIAM	Central Institute of Aviation Motors, USSR/Russia
Tu-154	Tupolev Tu-154 (NATO reporting name: "Careless") Soviet/Russian three-engine medium-range narrow-body airliner
U.S.	United States
UAC	United Aircraft Corporation
UAV	Unmanned aerial vehicle
UHF	Ultra high frequency
USA	United States of America
USD	US Dollar
USSR	Union of Soviet Socialist Republics
VHF	Very high frequency
W-s	Watt-Second
W-s/W-s	Watt-seconds per watt-seconds
W/Hz	Watt per hertz
W/MHz	Watt per megahertz
W/W	Watt per watt
W/W	Watt per watt
WGS-84	World Geodetic System
X-51	Wave rider—hypersonic unmanned aerial vehicle

About the Author

Evgeny Markin received a master's degree in radio engineering from the Minsk Higher Engineering Anti-Aircraft Missile Academy of Air Defense in 1983. In 1992, he received a master's degree in mathematics from Novosibirsk State University, and a degree in applied mathematics as a qualified mathematician. In 1982 he received the award of All-Union Competition of Young Scientists of the USSR, for his work that focused on electromagnetism. In this work, a new form of the Fredholm integral equation of the second kind was obtained, which was applied to calculate spiral antennas.

After graduating from the Minsk Higher Engineering Anti-Aircraft Missile Academy of Air Defense, he entered active military service in the USSR Air Defense Forces as an officer of an anti-aircraft missile system, where he received practical knowledge in the field of radars and guided missile systems. He gained extensive experience in the use and technical support of various radio engineering systems, including firing radars (fire control radars), surveillance radars, detection radars, radio altimeters, antennas, wire communication systems, radio relay systems, and vehicles. In an anti-aircraft missile battalion, he was in combat duty as a commander of a combat shift, and in the status of a gunner, which is an officer who makes the decision to launch a missile. During this time, 18 technical proposals were developed to improve the existing air defense system, one of which concerned the replacement of wide-beam antenna scanners in the field and in combat conditions. This proposal was subsequently included in the manual for combat work.

The Sary-Shagan state missile defense research site, located at Lake Balkhash in Kazakhstan, was the next step in his professional development. At this test site, he was an expert on anti-jamming air defense systems, where he gained extensive experience in testing air defense systems and in developing proposals for improving these systems. The developed proposals are still used in current technology. He also participated in tests at the stage of general designer, state tests, and in control tests of the first production model of a new anti-aircraft missile system S-200D.

He was the leading designer of the S-400 anti-aircraft missile system under the leadership of Alexander Lemansky, General Designer of the State Research and Development Bureau ALMAZ, Moscow, Russia. He studied electromagnetism, antennas, microwave theory, and the theory of control of automatic systems in postgraduate study at the Central Design Bureau ALMAZ, where he further developed his professional expertise.

He has been an IEEE member since 2001, as well as a member of AFCEA and EuMA.

Since 2006, he has worked as an independent expert on radar, electromagnetism, antennas and signal processing, stability analysis of automatic target tracking, including in the presence of interference, as well as in calculations of electromagnetic field scattering from objects of complex shapes.

Index

I

Q

R

S

Artech House Radar Library

Joseph R. Guerci, Series Editor

Adaptive Antennas and Phased Arrays for Radar and Communications, Alan J. Fenn

Adaptive Signal Processing for Radar, Ramon Nitzberg

Advanced Techniques for Digital Receivers, Phillip E. Pace

Advances in Direction-of-Arrival Estimation, Sathish Chandran, editor

Airborne Early Warning System Concepts, Maurice W. Long

Airborne Pulsed Doppler Radar, Second Edition, Guy V. Morris and Linda Harkness, editors

Aspects of Modern Radar, Eli Brookner

Aspects of Radar Signal Processing, Frank F. Kretschmer, Bernard Lewis and Wesley W. Shelton

Basic Radar Analysis, Mervin C. Budge, Jr. and Shawn R. German

Basic Radar Tracking, Mervin C. Budge, Jr. and Shawn R. German

Bayesian Multiple Target Tracking, Second Edition, Lawrence D. Stone, Roy L. Streit, Thomas L. Corwin, and Kristine L Bell

Beyond the Kalman Filter: Particle Filters for Tracking Applications, Branko Ristic, Sanjeev Arulampalam, and Neil Gordon

Cognitive Radar: The Knowledge-Aided Fully Adaptive Approach, Second Edition, Joseph R. Guerci

Coherent Radar Performance Estimation, James A. Scheer

Computer Simulation of Aerial Target Radar Scattering, Recognition, Detection, and Tracking, Yakov D. Shirman, editor

Control Engineering in Development Projects, Olis Rubin

Countermeasures for Aerial Drones, Garik Markarian and Andrew Staniforth

Deep Learning Applications of Short-Range Radars, Avik Santra, and Souvik Hazra

Deep Learning for Radar and Communications Automatic Target Recognition, Uttam K. Majumder, Erik P. Blasch, and David A. Garren

Design and Analysis of Modern Tracking Systems, Samuel Blackman and Robert Popoli

Detectability of Spread-Spectrum Signals, George Dillard

Detecting and Classifying Low Probability of Intercept Radar, Second Edition, Phillip E. Pace

FMCW Radar Design, M. Jankiraman

Fourier Transforms in Radar and Signal Processing, Second Edition, David Brandwood

Fundamentals of Short-Range FM Radar, Igor V. Komarov and Sergey M. Smolskiy

Handbook of Computer Simulation in Radio Engineering, Communications, and Radar, Sergey A. Leonov and Alexander I. Leonov

Handbook of Radar Measurement, David K. Barton

High-Resolution Radar, Second Edition, Donald R. Wehner

High Resolution Radar Cross-Section Imaging, Dean L. Mensa

Highly Integrated Low-Power Radars, Sergio Saponara, Maria Greco, Egidio Ragonese, Giuseppe Palmisano, and Bruno Neri

Introduction to LabVIEW™ FPGA for RF, Radar, and Electronic Warfare Applications, Terry Stratoudakis

An Introduction to Passive Radar, Hugh D. Griffiths and Christopher J. Baker

Introduction to Radar using Python and MATLAB®, Lee Andrew Harrison

Introduction to RF Equipment and System Design, Pekka Eskelinen

Linear Systems and Signals: A Primer, JC Olivier

Meter-Wave Synthetic Aperture Radar for Concealed Object Detection, Hans Hellsten

Methods and Techniques of Radar Recognition, Victor Nebabin

The Micro-Doppler Effect in Radar, Second Edition, Victor C. Chen

Microwave Radar: Imaging and Advanced Concepts, Roger J. Sullivan

Millimeter-Wave Radar Clutter, Nicholas C. Currie, Robert D. Hayes and Robert Trebits

Millimeter-Wave Radar Targets and Clutter, Gennadiy P. Kulemin

MIMO Radar: Theory and Application, Jamie Bergin and Joseph R. Guerci

Modern Radar Systems, Second Edition, Hamish Meikle

Modern Radar System Analysis, David K. Barton

Modern Radar System Analysis Software and User's Manual, Version 3.0, David K. Barton

Monopulse Principles and Techniques, Second Edition, Samuel M. Sherman and David K. Barton

MTI and Pulsed Doppler Radar with MATLAB®, Second Edition, D. Curtis Schleher

Multifunction Array Radar, Dale Billetter

Multiple-Target Tracking with Radar Applications, Samuel S. Blackman

Multitarget-Multisensor Tracking, Applications and Advances, Volume II, Yaakov Bar-Shalom and Robert F. Popoli, editors

Multitarget-Multisensor Tracking: Applications and Advances Volume III, Yaakov Bar-Shalom and William Dale Blair, editors

Non-Line-of-Sight Radar, Brian C. Watson and Joseph R. Guerci

Photonic Aspects of Modern Radar, Edward N. Toughlian and Henry Zmuda

Practical Phased Array Antenna Systems, Eli Brookner

Practical Simulation of Radar Antennas and Radomes, Herbert L. Hirsch

Precision FMCW Short-Range Radar for Industrial Applications, Boris A. Atayants, Viacheslav M. Davydochkin, Victor V. Ezerskiy, Valery S. Parshin, and Sergey M. Smolskiy

Principles of High-Resolution Radar, August W. Rihaczek

Principles of Modern Radar Missile Seekers, Evgeny Markin

Principles of Radar and Sonar Signal Processing, François Le Chevalier

Radar Cross-Section Analysis & Control, Asoke K. Bhattacharyya and D. L. Sengupta

Radar Cross Section, Second Edition, Eugene F. Knott, et al.

Radar Detection and Tracking Systems, Shaheen A. Hovanessian

Radar Engineer's Sourcebook, William C. Morchin

Radar Equations for Modern Radar, David K. Barton

Radar Evaluation Handbook, David K. Barton, et al.

Radar Meteorology, Henri Sauvageot

Radar Range Performance Analysis, Lamont V. Blake

Radar Reflectivity of Land and Sea, Third Edition, Maurice W. Long

Radar Resolution and Complex-Image Analysis, August W. Rihaczek and Stephen J. Hershkowitz

Radar RF Circuit Design, Second Edition, Nickolas Kingsley and Joseph R. Guerci

Radar Signal Processing and Adaptive Systems, Ramon Nitzberg

Radar Signals: An Introduction to Theory and Application, Charles E. Cook

Radar System Analysis, Design, and Simulation, Eyung W. Kang

Radar System Analysis and Modeling, David K. Barton

Radar System Design and Analysis, Shaheen A. Hovanessian

Radar System Performance Modeling, Second Edition, G. Richard Curry

Radar Technology, Eli Brookner

Radar Technology Encyclopedia, David K. Barton and Sergey A. Leonov, editors

Radar Vulnerability to Jamming, Stephen Lothes

Radars, Volume 1: Monopolse Radar, David K. Barton

Radars, Volume 2: The Radar Equation, David K. Barton

Radars, Volume 3: Pulse Compression, David K. Barton

Radars, Volume 4: Radar Resolution & Multipath Effects, David K. Barton

Radars, Volume 5: Radar Clutter, David K. Barton

Radars, Volume 6: Frequency Agility and Diversity, David K. Barton

Radars, Volume 7: CW and Doppler Radar, David K. Barton

Radio Wave Propagation Fundamentals, Second Edition, Artem Saakian

Range-Doppler Radar Imaging and Motion Compensation, Jae Sok Son, et al.

Robotic Navigation and Mapping with Radar, Martin Adams, John Mullane, Ebi Jose, and Ba-Ngu Vo

Russian-English and English-Russian Dictionary of Radar and Electronics, William F. Barton and Sergey A. Leonov

Secondary Surveillance Radar, Michael C. Stevens

Signal Detection and Estimation, Second Edition, Mourad Barkat

Signal Processing in Noise Waveform Radar, Krzysztof Kulpa

Signal Processing for Passive Bistatic Radar, Mateusz Malanowski

Solid-State Radar Transmitters, Edward D. Ostroff

Space-Based Radar Handbook, Leopold J. Cantafio, Richard K. Moore, and Earl E. Swartzlander Jr.

Space-Time Adaptive Processing for Radar, Second Edition, Joseph R. Guerci

Spaceborne Weather Radar, Toshiaki Kozu and Robert Meneghini

Special Design Topics in Digital Wideband Receivers, James Tsui

Statistical Signal Characterization, Herbert L. Hirsch

Surface-Based Air Defense System Analysis, Robert H. MacFadzean

Synthetic Array and Imaging Radars, Shaheen A. Hovanessian

Systems Engineering of Phased Arrays, Rick Sturdivant, Clifton Quan, and Enson Chang

Techniques of Radar Reflectivity Measurement, Nicholas C. Currie

Theory and Practice of Radar Target Identification, August W. Rihaczek and Stephen J. Hershkowitz

Time-Frequency Signal Analysis with Applications, Ljubiša Stanković, Miloš Daković, and Thayananthan Thayaparan

Time-Frequency Transforms for Radar Imaging and Signal Analysis, Victor C. Chen and Hao Ling

Transmit Receive Modules for Radar and Communication Systems, Rick Sturdivant and Mike Harris

For further information on these and other Artech House titles, including previously considered out-of-print books now available through our In-Print-Forever® (IPF®) program, contact:

Artech House
685 Canton Street
Norwood, MA 02062
Phone: 781-769-9750
Fax: 781-769-6334
e-mail: artech@artechhouse.com

Artech House
16 Sussex Street
London SW1V HRW UK
Phone: +44 (0)20 7596-8750
Fax: +44 (0)20 7630-0166
e-mail: artech-uk@artechhouse.com

Find us on the World Wide Web at: www.artechhouse.com

www.ingramcontent.com/pod-product-compliance
Lightning Source LLC
LaVergne TN
LVHW080644210926
848485LV00003B/324